Studies of Pallas in the Early Nineteenth Century

Clifford J. Cunningham

# Studies of Pallas in the Early Nineteenth Century

## Historical Studies in Asteroid Research

Second Edition

Springer

Clifford J. Cunningham
Fort Lauderdale, FL
USA

This book was originally self-published in softcover in 2004 by the author and was called The Pallas Problem. In this new edition it is substantially revised with much new material and new translations.

ISBN 978-3-319-32846-1 ISBN 978-3-319-32848-5 (eBook)
DOI 10.1007/978-3-319-32848-5

Library of Congress Control Number: 2016948619



Printed on acid-free paper

This Springer imprint is published by Springer Nature
The registered company is Springer International Publishing AG
The registered company address is: Gewerbestrasse 11, 6330 Cham, Switzerland

# Preface

In his treatise *On the Problem of the Historian*, Wilhelm von Humboldt states his views succinctly. "The problem of the historian is to tell what actually happened. The more exactly and completely he succeeds, the more perfectly he has solved that problem." In the three books of this series to date, I have tried to put Humboldt's precept into practice.

First, by publishing all of the letters and scientific papers dealing with Ceres and Pallas, the reader can examine exactly what was being thought and written by all the eminent astronomers of the time. Most of these letters have never been published before, and this is the first time they have ever appeared in English. Also, William Herschel's first two studies of Ceres and Pallas are given, along with an analysis of vicious contemporary critiques, one of which equated his work with evil.

Second, this book is deeply grounded in philosophy to establish the concepts and ideas that motivated the research and ideals of those astronomers. For example, in a study of the Paris Institute prize for a study of Pallas' orbit, I show why a mindset existed that led eminent scientists to think that all problems were solvable. Prizes offered by the academies of Paris and Berlin are surveyed to put the Pallas prize question in context. Other issues are related to such diverse philosophical discussions as Rene Descartes' *Optics;* John Locke's writing on language and ideas; Joseph Priestley's psychology of association; theories of reference to explore J.-J. Lalande's insistence on abjuring classical names for new planetary discoveries; gap theory by Adam Smith; David Hume's principle of resemblance; the concepts of anachronism, analogy and doubt; Thomas Young's analysis of hypothesis versus theory; Christian Wolff's deductive reasoning as applied to planetary distances; the theoretical and linguistic systems of Pierre Duhem; the intellectual antecedents of Laplace's concept of Intelligence in a deterministic cosmos; and the school of Scottish Common Sense Realism. The study of aesthetics by Francis Hutcheson, father of the Scottish Enlightenment, is used to relate the concepts of harmony and proportion that stretched from Pythagoras to Johannes Kepler to Capel Lofft. The relationships between Giuseppe Piazzi (discoverer of Ceres), the muse Urania, Apollo's lyre and the music of spheres are explored in part through a Latin poem by Tycho Brahe. This book shows how Pallas wrote the final movement to the symphony of celestial harmony.

The 'geometric beam' perceived at the great Library in Alexandria by the Roman traveler Ammianus Marcellinus illuminates this story of Pallas. The geometrical structure of space is shown to be fundamental in understanding how the orbit of Pallas was determined and studied. This is examined through the work done by Carl Gauss and others on the perturbations Pallas is subjected to. As the nineteenth century Prussian military theorist Carl von Clausewitz observed: "Although our intellect always longs for clarity and certainty, our nature often finds uncertainty fascinating." Nowhere is this human fascination more evident than in the work on perturbation theory. Even the glittering allure of a gold prize from the Paris Academy was not sufficient incentive to fully solve the problem in an age that had no recourse to computational tools. All the literature dealing with the gold prize is presented in this book for the first time, as well as a unique solution to the Gauss anagram, a method Gauss used to ensure the priority of his remarkable orbital discovery about Pallas. His Latin treatise on Pallas is also published here for the first time in English.

The chance that there was an unobserved planet between Mars and Jupiter led Baron von Zach to organize a search for it in 1800. Chance of a different feather, in the form of probability theory, is another major element of this book as astronomers and mathematicians sought to elucidate the orbits and origin of the asteroids.

This book opens with the long-forgotten fact that the orbital elements of a planet between Mars and Jupiter were first printed in 1681. The missing planet idea that animated astronomers from Kepler to Zach is the perfect example of what the British historian of science A. Rupert Hall (1962a: 188) meant when he said, "The history of science derives its coherence less from the temporal succession of events than from the continuity of ideas." He went on to say that "while not neglecting the work of the scientist's hands the historian has to look more deeply for the thought that guided that work and gave it a theoretical structure." This book, thoroughly revised and updated from its first edition in 2004, examines both the thought and theoretical structure created by the astronomers and mathematicians in the early nineteenth century as they grappled with the existence of not just one missing planet but two.

Fort Lauderdale, FL, USA

Clifford J. Cunningham

# Acknowledgments

As always great thanks go to my translators for providing a rich harvest of original material rendered into English: Telse Wokersien (German and French text); James Lequeux (French text); Christopher Gordon and Roger Ceragioli (Latin text); Piero Sicoli (Italian text).

Thanks to my dear friend the late A.C. Addington for his genealogical diagram in Chap. 4. David Glynn for his unique explanation of the Gauss anagram. John Ramsey and Bob Munman for their expertise in classical history, and Ari Belenkiy in mathematics. The paper by Tittel was provided by Andras Holl of Konkoly Observatory, Budapest. Valuable contributions were also made by Magda Vargha of Budapest and Paul Feldman of the National Research Council of Canada.

Inadvertently certain passages in *Discovery of the First Asteroid, Ceres,* were not properly attributed. The paragraph on pg. 66 beginning with "The various themes" is from an anonymous review in Der neue Teutsche Merkur (1802) vol. 2, no. 6. (June 1802), 135–48. The passage at the end of pg. 66, beginning with "All these" and which is in quotes, is a personal communication from Doug Stott. The phrase on pg. 66 "was quite the sensation just after the turn of the century" also should have been attributed to Stott.

Illustrations unless otherwise credited are courtesy of Wikimedia Commons. The attribution of the portrait of Maskelyne given in the *Discovery of the First Asteroid, Ceres* is not secure, but the image of Maskelyne given in this book is definitive. Various portions of this book and the other books in this series derive directly from my unpublished PhD thesis at the University of Southern Queensland. The next book in the series will deal with the discovery and early study of the asteroids Juno and Vesta.

# Contents

# About the Author

**Clifford J. Cunningham** did his Ph.D. work in the history of astronomy at the University of Southern Queensland in Australia, and he is affiliated both as an astronomer at USQ and with the National Astronomical Research Institute of Thailand. His undergraduate degrees in classical studies and physics were done at the University of Waterloo. He has written or edited 14 books on the history of astronomy, and his papers have been published in many major journals, including Annals of Science, Journal for the History of Astronomy, Culture & Cosmos, Journal of Astronomical History and Heritage, Studia Etymologica Cracoviensia, The Asian Journal of Physics and Renaissance and Reformation. He has been a history of astronomy columnist for Mercury magazine since 2001, and was appointed associate editor of the Journal of Astronomical History & Heritage in 2016. Asteroid (4276) was named Clifford in his honor by the International Astronomical Union through its Minor Planet Center at the Harvard-Smithsonian Center for Astrophysics.

# Chapter 1
# A Disturbing Inclination

## Bernoulli and the First Orbital Properties of Ceres in 1681

In 1595 Johannes Kepler (1571–1630) was the first to posit the existence of a planet between Mars and Jupiter, but it was the Swiss astronomer and mathematician Jacob Bernoulli (1655–1705; Fig. 1.1) who first offered actual numbers relating to its orbital properties. In a 16-page book about comets written in High Dutch, Bernoulli (1681: 9–10) wrote there is a primary planet revolving about the Sun, between Mars and Jupiter, whose period is about 4 years and 157 days, with a mean distance of 2.583. He expanded the treatise in Latin a year later to reach a wider audience (Bernoulli 1682; Fig. 1.2). This planet, from its smallness and great distance, he supposed not to be visible to an observer on Earth; and also, that it had several satellites belonging to it. Bernoulli couched all of this in terms of the Cartesian hypothesis, which regarded gravity as due to "a central torrent continually flung from the circumference of vortices toward the center." (Bernouilli 1986: 314)

> *And so comets are nothing other than planets of the secondary point f,* which point is without doubt occupied by another primary planet (Fig. 1.3). *But this, both because of the extreme smallness of its body and because of its immense distance [from us], perpetually escapes our gaze* (Bernoulli 1682: 33; author's emphasis).

Even though Bernoulli believed all comets performed their motions beyond the orbit of Saturn, he was not far wrong about a new planet. Bernoulli confined himself to offering just two orbital properties of the planet between Mars and Jupiter, and these are compared in Table 1.1 to the same elements proposed by Franz von Zach (1754–1832) in 1785. Zach's estimates are closer to the modern values assigned to Ceres, but both are reasonably close to the truth. The discovery of Ceres 120 years later vindicated Bernoulli's prediction, but he certainly never envisioned that a second object would also be found, with an orbital inclination and eccentricity far greater than any known planet.



C.J. Cunningham, *Studies of Pallas in the Early Nineteenth Century*,
DOI 10.1007/978-3-319-32848-5_1

**Fig. 1.1** Jacob Bernoulli

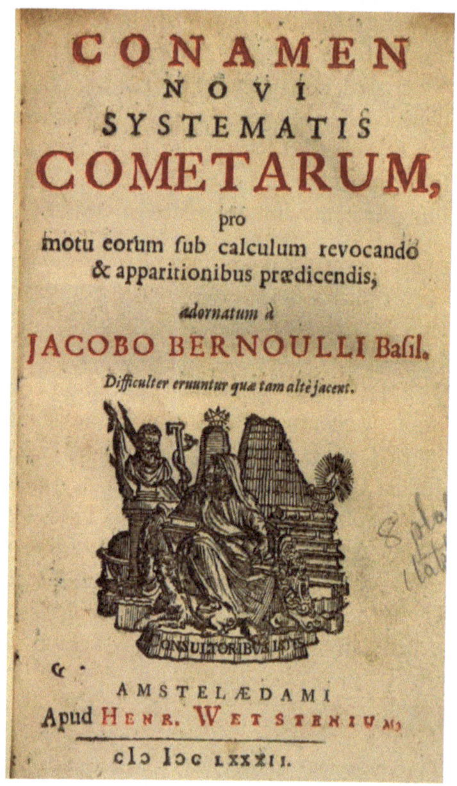

CONAMEN
NOVI
SYSTEMATIS
COMETARUM,
pro
motu eorum ſub calculum revocando
& apparitionibus prædicendis;
*adornatum à*
JACOBO BERNOULLI Baſil.
*Difficulter eruuntur quæ tam altè jacent.*
CONSULTORIBUS ISTIS
AMSTELÆDAMI
Apud HENR. WETSTENIUM,
clↄ Iↄc LXXXII.

**Fig. 1.2** The title page of Bernoulli's 1682 book on comets. Courtesy Google Books

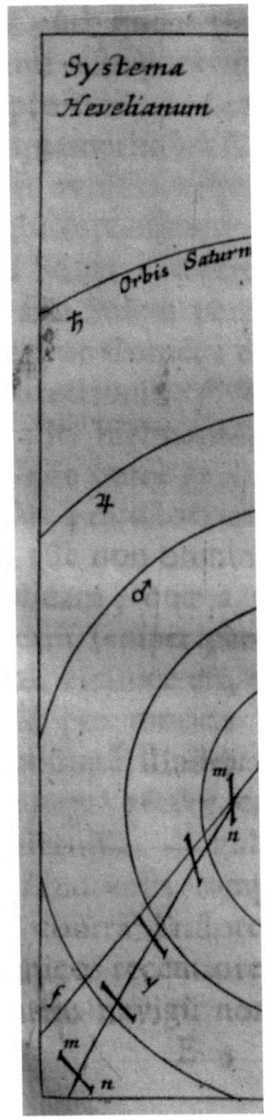


**Fig. 1.3** The letter "f" at bottom left denotes the location of the missing planet between Mars and Jupiter in this diagram by Bernoulli in 1682

**Table 1.1** Pre-Discovery orbital elements of Ceres

| Orbital Elements | Bernoulli 1681 | Zach 1785 | Modern |
|---|---|---|---|
| Distance | 2.583 | 2.82 | 2.767 |
| Period | 4 yrs. 157 days | 4 yrs. 270 days | 4 yrs. 220 days |

## The End of Heavenly Harmony: The Perturbations of Pallas

With the discovery of Ceres in 1801 and Pallas in 1802, the astronomers and mathematicians of the day were faced with two major challenges. The first, posed by Ceres, was how to predict its future position with sufficient accuracy to enable its recovery. How this challenge was met was detailed in the first two books of this series: *Discovery of the First Asteroid, Ceres*, and *Early Investigations of Ceres and the Discovery of Pallas*.

The second challenge, posed by Pallas, proved to be one that was not fully resolved until the twentieth century. Because of its unusually high eccentricity (0.231) and orbital inclination (34.84°), Pallas was subjected like no other planetary object then known to gravitational perturbations of such magnitude that they could not be ignored in a calculation of its orbit. Kepler and Isaac Newton (1643–1727) were to be tested as never before.

Thirteen years before Newton published his *Principia*, it was realized by the English natural philosopher Hooke (1635–1703; 1674: 27) that gravity acted upon all celestial bodies:

> *That all Celestial Bodies whatsoever, have an attraction or gravitating power towards their own Centres, whereby they attract not only their own parts, and keep them from flying from them, as we may observe the Earth to do, but that they do also attract all the other Celestial Bodies that are within the sphere of their activity; and consequently that not only the Sun and Moon have an influence upon the body and motion of the Earth, and the Earth upon them, but that the planets by their attractive powers, have a considerable influence upon its motion as the Earth hath upon every one of their motions also.*

Hooke is implying that the planetary orbits may not be exactly closed, a complete change in the mechanics of planetary motion accepted up to then. Thus, the parameters of attraction and velocity were "no longer reflections of pre-established heavenly harmonies to be inferred from astronomical observations." (Gall 2002: 22). The 'music of the spheres' that had begun with Pythagoras (570–495 BCE) and beguiled Kepler's thoughts on the planetary distances had finally been silenced! (See *Discovery of the First Asteroid, Ceres.* and Chap. 6 in this book for the work of Pythagoras and Kepler.) The deviations from an elliptical orbit Hooke alludes to are small in the case of the major planets—their distances from one another are large and their masses are small with respect to the Sun.

For long intervals of time, however, perturbations from the other bodies in the Solar System must be taken into account. In one last forlorn and ultimately futile hope to rescue celestial harmony, the French mathematician Urbain LeVerrier (1811–1877) in 1847 set himself the goal to "…embrace in a single work the entire planetary system, put everything in harmony if possible, otherwise, declare with certainty that there are as yet unknown causes of perturbations." (Levy 1968: 388). Perturbation theory itself was founded by Newton, as explained by Hegel (1770–1831; 1817): "It is recognized that the significant moment added by Newton to the content of Kepler's laws is the principle of perturbation. In so far as it rests upon the proposition that what is called attraction consists of an action between all the individual material parts of a body, the importance of this principle has to be adduced here." Westphal (2015) argues that Hegel ably and insightfully defends Newton's causal realism about gravitational force; a practical illustration of this was given by Newton in his discussion of the motion of a pendulum studied by the French astronomer Jean Richer (1630–1696) in 1679. Richer became the first to observe a change in gravitational force over Earth's surface (Brunschvicg 1922: 230). The relevance of causal realism is considered further in the next section.

Joseph Johann von Littrow (1781–1840; 1825: 106), director of Vienna Observatory, explained in a book on popular astronomy the value of asteroids to the study of perturbations: "For astronomers, they are of particular importance, since their large eccentricities and inclinations have made the prior, rougher calculations of the perturbations inadequate, and have led to more educated versions of these difficult calculations. The perturbations they suffer, which are mainly due to Jupiter, are large enough that the mass of Jupiter can be found with ease." The mass of Jupiter is discussed later in this chapter.

The actual application of perturbation theory to the asteroids has been widely misunderstood, even by such an eminent mathematician as Sir John Herschel (1792–1871). This is explained by the English astronomer Richard Proctor (1837–1888; Fig. 1.4), and as the finest (and most straightforward) explanation of what a gravitational perturbation is, it is worth quoting here. Proctor (1892: 301) makes mention in this passage of the grand inequality between Jupiter and Saturn, a matter considered later in this chapter.

> *Sir John Herschel wrote to me in 1870 correcting what he regarded as an error in my* Other Worlds than Ours, *where I had stated that the asteroids are more perturbed by Jupiter and Saturn than planets of greater mass would be. For, said he, a planet or a peppercorn would behave in precisely the same way under the attraction of Jupiter were Jupiter alone, and therefore must be equally perturbed by his influence when it is subordinate to that of the sun. In my reply I thought it sufficient to call his attention to the long inequality of Saturn and Jupiter, by which the smaller planet is the more affected. This presently set him on the track of the error which had misled him. So far as the relative orbits of two bodies around each other are concerned, it matters not how the masses are distributed: the sum of the masses remaining unchanged, the bodies might be equal, or one might be a grain, and the other many trillions of trillions of tons. But as regards the actual orbits of the two the distribution of the masses is all important: in the latter case one body would be appreciably affected at rest, the other would travel in a wide orbit; in the former the two bodies would travel in equal orbits, each orbit having half the dimensions of the wide orbit described by the minute body in the other case. The perturbations of small and large bodies are perfectly analogous to their relative orbital motions. When an asteroid is acted on by Jupiter, the asteroid's perturbation takes place appreciably about Jupiter as a centre – because the common centre of gravity of the immense mass of Jupiter and the very small mass of the asteroid is appreciably the same as the centre of gravity of Jupiter.*

For the case of Pallas, the application of refined perturbation methods was one of immediate, not long term, concern. As soon as the orbit of Pallas had been determined, it became apparent that perturbation effects would have to be taken into account. The planets are always observed somewhere inside a narrow belt extending 8° above and below the ecliptic. As Wilhelm Olbers (1802) wrote,

**Fig. 1.4** Richard Proctor

"Pallas strides very far north and south: for us she can reach at times 48° northern and 54° southern latitude." Zach, in a letter dated May 31, 1802, to the president of the Royal Society Joseph Banks (1743–1820), put the case clearly:

> *The new planet Pallas will call forth the utmost exertion of our analytical powers. Hitherto the two elements of a planetary orbit,* viz, *the eccentricity, and the inclination, had been considered as an infinite little quantity, and so it could be, as these two elements in all our old planets are very small; so then, the higher powers of them could be neglected without danger, as they produced no sensible term in the approximating series. But this is not the case made by Pallas, where the eccentricity of the orbit and the inclination are so very great.*

Woodhouse (1773–1827; 1804: 277) recognized early on that "In the case of the new planets, Ceres and Pallas, whose mean distances from the sun are nearly equal," a mathematical expression he developed "will be very convenient, on account of the rapid convergency of the quantities…in estimating the disturbing forces" of those two objects. Woodhouse seems to be the singular English mathematician who both had a grasp of the integration of differential expressions and saw their application to asteroid perturbation calculations. Woodhouse went on to a stellar career, holding both the Lucasian (1820) and Plumian (1822) professorships, and the first directorship of Cambridge University observatory. He began in the Cambridge and moved toward the Continental style of mathematics, which was advanced by John Herschel and others, a topic explored in Chap. 3 (Phillips 2006).

## The Temple of Natural Philosophy

Colin Maclaurin (1698–1746), professor of mathematics at the University of Edinburgh (from 1725), opens his popular 1748 account of Newton's philosophical discoveries with this over-arching theme: "To describe the phenomena of nature, to explain their causes, to trace the relations and dependencies of those causes, and to enquire into the whole constitution of the universe, is the business of natural philosophy." (Schliesser 2012: 308)

Underlying Newton's laws and his philosophy of nature was mathematics (Fig. 1.5). That mathematics would underpin natural philosophy (science) now seems indisputable, but it was not always so. In the sixteenth century, there was a concerted attack on mathematics by Aristotelian scholars, whose basic argument was that "mathematics did not possess the deductive purity of syllogistic logic and was therefore not a science because it did not reveal causal relationships." Guicciardini (2009: 4) Among these scholars were Alessandro Piccolomini (1508–1578), Pierre Gassendi (1592–1665), and the Jesuit Benedict Pereyra (1536–1610; 1562), who baldly stated "the mathematical disciplines are not proper sciences." Pererya "was a colleague of Christopher Clavius in the Collegium Romanum and for many years he ran the course in natural philosophy." (Mancosu 1996: 14)

On the matter of geometrical proofs, Newton was certainly influenced by his teacher Isaac Barrow (1630–1677; Fig. 1.6), Lucasian professor of mathematics at the University of Cambridge. Barrow was one of the English scholars, together with John Wallis (1616–1703) at Oxford University, who opposed the Aristotelians and worked to reinstate mathematics to the realm of the causal sciences. Barrow laid down the gauntlet at the outset of his lecture on geometry when he stated that the "primary and chief" way of generating magnitudes is "that perform'd by local Motion…because without Motion, nothing can be generated or produced." He then invokes the axiom of Aristotle (384–322 BCE) concerning motion: "He that is ignorant of Motion, must necessarily know nothing of Nature." (Barrow 1735: 2)

> *Barrow's defense of geometry as a model of reasoning and his idea that since geometrical magnitudes are generated by motion, a causal relationship can be captured in such mechanically based geometry must have impressed the young scholar* [Newton] (Guicciardini 2009: 4).

For Newton, the postulates of geometry "secure the relevance of mathematical generalizations to causality in Nature." (Dunlop 2012: 99) The concept of causality may seem abstruse, but it also demarcated astronomy from physics for centuries, as we can see in these quotes. First, the Danish astronomer Longomontanus (1562–1647; 1639): "Astronomy does not so much investigate the heavenly bodies

**Fig. 1.5** An allegory of astronomy, geometry and mathematics. The title page of *Astronomisches Handbuch* by Johann Leonhard Rost (1718), courtesy Google Books

themselves and their causal properties as [it investigates] their motions and definite periods; it hands over the peculiarities of the stars to physics, which treats them by means of optics." A man who studied the asteroids, Friedrich Bessel (1784–1846), agreed with him. In a lecture given in 1832 and published 16 years later, Bessel (1848) argued the business of astronomy was nothing more than the precise

**Fig. 1.6** Isaac Barrow

measurement of the positions and orbits of the celestial bodies. "Everything else that one may learn about the objects, for example their appearance and the constitution of their surfaces, is not unworthy of attention, but is not the proper concern of astronomy."

The application of geometry to the study of the orbit of Pallas is considered in the next section. It was Newton's 'mathematical way' of looking at Nature that enabled Carl Gauss (1777–1855) to tackle the challenge posed by the asteroids Ceres and Pallas. The language of mathematics is the form in which experimental results are set down, observations recorded, and the laws of nature best expressed:

> *What Newton gave the world in the body of his scientific work was like the steel frame of some great building. The mathematical laws of optics and celestial mechanics are the girders and supporting members; other men will come with the bricks, the mortar, and the cut stone to fill in the walls and lay out the partitions. The new building, the new Temple of Natural Philosophy, must be erected here, within his framework, inside the boundaries he had marked out by his mathematical laws, and not elsewhere, at some point far afield among the infinite possibilities of which Descartes had written* (Guerlac 1965: 143).

The American historian of science Henry Guerlac (1910–1985) here alludes to the concept held by the French philosopher and mathematician René Descartes (1596–1650) that God's power is infinite, so he set no limits to it, thus allowing for infinite possibilities (Conant 1991: 165). That said, for Descartes, Baruch Spinoza (1632–1677), Thomas Hobbes (1588–1679) and other seventeenth century intellectuals, their philosophical model was based on geometry, which has four

**Fig. 1.7** John Robison painting by Henry Raeburn. University of Edinburgh

impressive features: "Its results were certain, a priori, deducible from a few first principles and informative, that is, non-trivial." (Priest 2007: 33)

This book looks at some of the bricks and mortar (probability theory, least squares, the geometry of polygons and Fermat primes) and some of the master masons who built the Temple (Leibniz, Laplace, Gauss, Lagrange, Olbers). It was not, however, built without controversy. The Scottish physicist and mathematician John Robison (1739–1805; Fig. 1.7), a close acquaintance of Henry Brougham (1778–1868), who launched the most notable attack on William Herschel for his choice of the word 'asteroid' to denote Ceres and Pallas, was a very vocal opponent of Pierre-Simon Laplace (1749–1827). Robison, professor of natural philosophy at Edinburgh University from 1774 to 1805, deplored both Laplace's atheism and his account of gravity. Robison denounced Laplace's materialism in strident terms as "this stain, with which careless ignorance, or atheistical perversion, has disfigured the fair features of philosophy." (Wilson 2009; 268)

By contrast Robison was a proponent of the work of the astronomer Roger Boscovich (1711–1787), professor of mathematics at the Collegio Romano, whose work on determining the orbit of a celestial object from only a few observations preceded that of Gauss. He, according to Robison, did more than merely tinker with the Temple. He believed that while Boscovich's 1758 book *Theoria philosophiae naturalis* failed in the "very sublime attempt" to lift the "interior veil of the temple of nature," it inspired Boscovich to go beyond Descartes, Leibniz, and Newton (Heilbron 2015: 105). The work of Boscovich was consonant with the philosophical school known as Scottish Common Sense Realism (see Chap. 3), which informed the views of both Robison and Brougham, who wrote so disparagingly about both Thomas Young's wave theory of light and Herschel's noble scheme to categorize the objects of the Solar System (issues that were considered in the previous two books in this series). Robison also held strongly to the ascendancy of geometry over algebra, and it is the importance of geometry in the study of the asteroids that is examined next.

## The Orbit of Pallas and the Geometrical Structure of Space

Traditional mathematics was challenged by Gottfried Leibniz (1646–1716; 1699): "Mathematicians need to be philosophers as much as philosophers need to be mathematicians." In astronomy, the intersection of philosophy and mathematics is seen nowhere more starkly than in the work of the Flemish Jesuit Joseph (Jean) Ciermans (1602–1648; 1639): "The movements of the Planets will agree better with philosophical arguments and with the truth, if you let them run through a simple elliptical orbit." (Vanpaemel 2003: 402) This 'simple truth' became central to our understanding of the orbits of Ceres and Pallas. Not certain what these objects were, the astronomers two centuries after Ciermans tried circular and parabolic options before finally calculating the correct elliptical elements that traced them in their proper orbits and allowed ephemerides to be created for future observations—a prime topic of this book.

The importance of triangles to the work of Kepler, who was the first to realize planets follow elliptical orbits, was examined in *Discovery of the First Asteroid, Ceres*. In 1595, when Kepler noted the splendid harmony of the cosmos that suggested a missing planet between Mars and Jupiter, the models used to create planetary ephemerides were based on spherical trigonometry, which relates the angles and arc lengths of spherical triangles. Prosthaphaeresis was an algorithm used at this time for "approximate multiplication and division using formulas from trigonometry" according to Wikipedia. (Ciermans 1639; Fig. 1.8). A landmark book on trigonometry by Rheticus (1514–1574; 1551) led off with a Latin poem in hendecasyllabic meter that embodies not only that algorithm (mentioned by Copernicus in his book *De Revolutionibus Orbium Coelestium* just 8 years earlier) but many of the elements under study here, and as such it serves as a foundational text for this work. It appears here in English for the first time (tr. Roger Ceragioli):

*Those desirous of the stars, and of geometry,*
*Or whomever the more charming arts*
*(derived from first principles and their source)*
*Delight, come! Give sacrifice to the Muses,*

**Fig. 1.8** Title page of *Disciplinae mathematicae* by Joannes (Joseph) Ciermans, a 1639 book on practical mathematics (courtesy Google Books)

*And show yourselves meanwhile grateful to them.*
*For a brief theoretical treatise on triangles,*
*Here piled up by the work of Joachim*
*Rheticus, goes forth; and one briefly explicated,*
*So that far and wide it conquers the canons of*
*Triangles all, and the golden tables.*
*Although many myriads of new ones*
*Be contrived, even so, as rivulets they*
*Bubble up from this source.*
*These things are your servants promptly in practice*
*Of* [triangles] *plane as well as spherical.*
*From them seek the motions and the prosthaphaereses,*
*Or if you imagine the astronomers to embrace*
*More convoluted things in their books.*
*Nay, if you desire to understand gnomonics and the*
*Science of catoptrics, in optics the mistress of shadow*
*And light, or of the doctrines passed on in the writings of*
*Geographers, and the chords of the Muses,*
*And their instruments wondrously wrought.*
*This table, with thrice cut series, shows by*
*Easy guidance geometry above all, and as to numbers*
*Learn them at least exactly. Come! Sacrifice to the Muses,*
*And show yourselves meanwhile grateful to them.*

His opening invocation, 'desirous of the stars,' of course refers more generally to heavenly bodies including the planets, and the additional invocation of geometry harkens back to the description of the library of Alexandria by the fourth century traveler and historian Ammianus Marcellinus. Of his visits to the library from 363 to 366, he wrote

*Here everything that lies hidden is brought to light by the geometric beam. Among them music is not entirely extinct yet, nor is harmony silent, and although rare, someone revived again the study of the movements of the world and of the stars, while others are learned in the science of numbers* (Bernardi 2016: 28).

Rheticus then invokes geometry as built up by triangles, the same concept used by Kepler in his harmony of the spheres, which is alluded to by Rheticus in his line about the 'chords of the Muses.' The direct application of all this to astronomy is made clear where seeking 'the motions' can refer to the position of planets in the sky, their orbit around the Sun, and other related matters. The gnomen (shadow) is the triangular blade of a sundial, and catoptrics (light) is nothing other than the optical form of a Newtonian telescope (invented more than a century later in 1668), some of which were used in the early nineteenth century to study the asteroids.

This leads us to Newton himself (Fig. 1.9) and 'celestial mechanics,' a term coined by Laplace (1829, xxiii). It was in fact the title of his great five-volume work: *Traité de mécanique céleste* (Treatise on Celestial Mechanics). He made this bold statement in the introduction:

**Fig. 1.9** William Blake's 1795 picture of Sir Isaac Newton as the divine geometer. When Newton died in 1727, his coffin was followed by 30,000 people, headed by the king and queen. It led Voltaire to remark that the English killed their kings but gave their mathematicians a state funeral

> *Towards the end of the seventeenth century, Newton published his discovery of universal gravitation. Mathematicians have, since that epoch, succeeded in reducing to this great law of nature all the known phenomena of the system of the world, and have thus given to the theories of the heavenly bodies, and to astronomical tables, an unexpected degree of precision. My object is to present a connected view of these theories, which are now scattered in a great number of works. The whole of the results of gravitation, upon the equilibrium and motions of the fluid and solid bodies, which compose the Solar System, and the similar systems, existing in the immensity of space, constitute the object of Celestial Mechanics. Astronomy, considered in the most general manner, is a great problem of mechanics. The solution of this problem depends upon the accuracy of the observations and upon the perfection of the analysis.*

The philosopher William Gass (1985: 91) has observed that "Although philosophers have frequently and piously invoked it, rarely has geometry been gainfully employed." (Fig. 1.10) One of those rare instances it was invoked to fine effect was by Newton in the 1660s. Basic to the study of celestial mechanics is the determination of an orbit in space. Newton envisioned space as having a geometrical structure. He wrote that it is full of "of all kinds of figures, everywhere spheres,

**Fig. 1.10** Detail from the title page of Peter Appian's *Introduction to Geography* (1533). The historical relevance of triangles to the study of planetary motion is detailed in Caravella (2008)

cubes, triangles, straight lines, everywhere circular, elliptical, parabolical, and all other kinds of figures, and those of all shapes and sizes, even though they are not disclosed to sight." (Hall 1962b)

The expression of this geometrical structure in the initial study of Pallas can be understood in what Baron von Zach relates in the September 1802 issue of his journal *Monatliche Correspondenz* (hereafter referred to as *Monthly Correspondence).* The true nature of Pallas' orbit was a major challenge to Johann Karl Burckhardt (1773–1825), despite his expert knowledge of celestial mechanics and his recent experience with the orbit of Ceres:

> *When the first rumour of Olbers' discovery had reached France, the opinion there held that this new body must be a comet. And this view was stubborn and persistent which explains why they stuck longer with parabolic orbits. Dr. Burckhardt, who worked chiefly on this calculation soon found that a parabolic orbit is insufficient and that it must rather be an elliptical one, in fact twice as eccentric as Mercury's [0.206]. And since he assumed such an eccentricity the error in latitude of the second observation, which is solely responsible for the eccentricity, appeared with the same sign again after passing through zero. From*

*this Dr. Burckhardt believed he had to increase the eccentricity even more in order to serve the latitudes. And in this erroneous opinion he was even more affirmed by the parabolic hypothesis, since this matched those three basic observations* (Zach 1802b: 305).

As early as May, Maskelyne (1802) believed Burckhardt was far off the mark. "I do not think Dr. Burckhardt's notion of a periodic time of 12 years for the new planet Pallas to be probable." Wilhelm Olbers was also led astray in his early 1802 study of the orbit of Pallas, not by the parabolic orbit that bedeviled Burckhardt but by another of the geometrical structures of space enumerated by Newton, the circular. This is from one of his lectures on the asteroids, printed in Chap. 6:

*When I tried to calculate at the beginning of April the few previous observations in a circular orbit, which made a good start in the case of Ceres, I found unexpectedly that it was impossible to describe them with the Sun concentric circle. Consequently, the orbit had to be very different from a circle. After April 10 it showed that this orbit was not a parabola but an ellipse, more oval than most of the other planetary orbits.*

Crucially for this discussion, when the Paris Academy proposed its famous prize question on Pallas (detailed in Chap. 3), it invited both "*geometers and astronomers* to discuss [it]" (my emphasis). What Zach (1802c) termed "the eternal force of geometric truth" was center stage in the study of Pallas. The closely linked studies of geometry and astronomy go back to ancient times. Babylonian astronomers used trapezoids to understand the motion of the planets (Ossendrijver 2016). This ancient heritage, and the bridge between mathematical astronomy and geometrical mathematics expressed in verse at the dawn of the Scientific Revolution by Rheticus, was about to meet the greatest challenge to celestial mechanics—the determination of the perturbations of an orbit that had never been encountered before: the Pallas problem.

## The Mass of Jupiter

Why was the mass of Jupiter so important? As the English astronomer George Biddell Airy (1801–1892) wrote in 1833, "Next to the elements of the planetary orbits, the most important numerical value for the explanation and prediction of the phenomena of the Solar System, is the mass of Jupiter."

The principal disturbing effects, or perturbations, of Ceres' orbit derive from an approximate 2:5 resonance with Jupiter. That is, Ceres' revolution period of 4.6 years is nearly two-fifths that of Jupiter. The first theoretical determinations of the perturbations on Ceres were made independently as early as April 1802 by Burckhardt, Johann Wurm (1760–1833) and Friedrich Theodor Schubert (1758–1825) in St. Petersburg, Russia. A few months later Barnaba Oriani (1758–1832) at Brera Observatory in Milan published his own perturbation theory, and Gauss followed with his in September.

Gauss, in a letter to Zach on June 26, 1802, described his excitement at being able to determine the mass of Jupiter from studying Ceres and Pallas. Zach published the letter in his journal (*Monthly Correspondence* 1802: 81)

> *The very learned essay of Professor Wurm in your June issue* [Wurm 1802] *unfortunately reveals the uncertainty which is still apparent regarding the mass of Jupiter, and that we cannot determine the perturbation equations of Ceres and Pallas with any reliability as we would have wished. On the other hand, the observations of Ceres, and moreover those of Pallas, will be very beneficial in that after one or two revolutions, the mass will be easily determined, which will then affect so many other important areas of astronomy in nothing but favourable ways. In this way, in this illustrious science, which seems to have given the heavens to Man for the cultivation of his powers and to elevate him above the earthly, always holds out its hand to the discovery of another; the great discoveries never stand alone, but rather are just as important in their results as they are in themselves. Precisely these uncertainties will be a new reason for me to keep my former, previously mentioned resolve; namely, to calculate, in addition to the new ellipse for Ceres, with the most accurate consideration of the perturbations of Jupiter, yet another 'pure' ellipse, based on the same observations, but without any perturbations. I believe, it must have been very interesting to see to what extent a certain trace of Jupiter's influence will reveal itself in the movement of Ceres in the next year, for until now, not even the slightest is shown, rather, the observations of one and a half years can undoubtedly be represented very well without constraint through a 'pure' ellipse.*

Gauss was still working on the problem in 1805, but the number crunching taxed even his patience, as shown in this letter of May 10, 1805, to Olbers: "The methods according to which I had begun to compute the perturbations of Ceres I have given up again. The dead mechanical calculating which I saw ahead of me turned me away." It would be another 20 years before *The Edinburgh Philosophical Journal* (1822) printed the following important discovery by Gauss:

> *Perturbations of the New Planets.– The celebrated M. Gauss of Goettingen has determined, in an exact manner, the masses of those planets which exert a sensible influence on the four new planets. He shews clearly, that the mass of Jupiter given by Laplace is wrong by more than a tenth part. The perturbations of Pallas produced by Jupiter amount to several degrees, and afford a very certain means of determining the mass of the latter.*

The negligible effect caused by the asteroids themselves was touched upon in 1802 by the French philosopher Louis-Claude de Saint-Martin (1743–1803; Fig. 1.11):

> *Thus, before the discovery of the new planets, the ignorance we were in of their existence, did not prevent astronomers from foretelling, with tolerable exactness, the return of the comets; because those unknown planets, being either too distant or too small, could not produce any sensible perturbation on comets passing by them* (Penny 1864: 125).

Saint-Martin's book was focused on the writings of the German mystic Jakob Boehme (1575–1624). In a section about the origin of the planets, Saint-Martin tries to incorporate the asteroids into Boehme's scheme: "As to the two new planets, Ceres and Pallas, between Mars and Jupiter, they might derive, more or less, from the original causes of their two neighbors, namely, light and fire." (Penny 1864: 117)

**Fig. 1.11** Louis-Claude de Saint-Martin

Boehme identified seven fundamental powers that gave rise to the formation of the planets. He related two of these, light and fire, to the formation of Mars and Jupiter. But Saint-Martin claims "the application of his doctrine to the supposed number of seven planets was only secondary in his system," so that the discovery of more planets (Ceres and Pallas) did not overthrow his concept. "Even though the catalog of planets actually exceed the number of seven, the predominance of one or other of the seven forms of Nature would not, therefore, cease to effect in each." (Penny 1864: 109) The point made by Saint-Martin about the negligible effect of the asteroids on comets was extended by Whiting (1828: 64) to include the planets:

> *This much we are certain of: that their magnitudes or quantities of matter are very small indeed, when compared to those of the other planets: For their perturbating forces upon the motions of the other planets, must, long before their discovery, have been felt in nice* [precise] *astronomical observations, had they been sensible, but no such effects have ever been discovered. Astronomical Tables and calculations made from tables that took no notice of the disturbing forces of these planets, could never have agreed so well with actual observations, had the aggregate quantity of their disturbing forces, upon the motions of the other planets, amounted to any thing sensible. But these planets, frequently passing near each other, must have their motions greatly disturbed, by their gravitating towards the*

*larger planets; which must render the calculations of their correct places exceedingly intricate and troublesome.*

Modern research has actually been able to calculate the perturbation in the motion of Mars and comets due to the gravity of Ceres, Pallas, Juno and Vesta. While the effects are very small, they do exist (for comets: Todorovic-Juchniewicz 1983; for Mars: Mouret et al. 2009) The perturbations suffered by Pallas are explored in Chap. 4. The mass of Jupiter as it relates to Juno and Vesta will be discussed in a future book in this series.

## Laplace and the Grand Inequality

"Having shown his mastery of analysis in his first published paper, Laplace proceeded to the great outstanding problems in the application of the law of gravitation to celestial mechanics. Of these the most conspicuous was offered by the opposite inequalities of Jupiter and Saturn, which the emulous efforts of Leonhard Euler (1707–1783) and Joseph-Louis Lagrange (1736–1813) had failed to bring within the bounds of theory (Wilson 2007: 124–133). The discordance of their results incited Laplace to a searching examination of the whole subject of planetary perturbations, and his maiden effort was rewarded with a discovery that constituted the most important advance made in physical astronomy since the time of Newton. In a paper read before the Academy of Sciences on February 10, 1773, Laplace (1776) announced his celebrated conclusion of the invariability of planetary mean motions, carrying the proof as far as the cubes of the eccentricities and inclinations. It was followed up by a series of profound investigations, in which Lagrange and Laplace alternately surpassed and supplemented each other in assigning limits of variation to the several elements of the planetary orbits." (Encyclopedia Britannica 1890: 301) On November 23, 1785, Laplace read to the academy "a memoir on the secular inequalities of planets and satellites, the first of four memoirs in which he accounted for the chief remaining anomalies in the theory of the Solar System." (Wilson 1994: 1050) In this memoir he announced his discovery of hitherto unknown inequalities of Jupiter and Saturn. "The analytical tournament of these intellectual rivals closed with the communication to the academy by Laplace, in 1787, of the long-sought cause of this 'grand inequality' of Jupiter and Saturn. The solution was found in the near approach to commensurability of their mean motions (Dhombres 1992)." (Encyclopedia Britannica 1890: 301).

Observations compiled over many decades indicated Jupiter's orbit was shrinking, while that of Saturn was expanding, leading Edmond Halley (1656–1742) to introduce corrections for these movements in his 1695 tables of planetary positions: an acceleration of Jupiter of 0°57′ per 1000 years and a retardation of Saturn of 2°19′ (Steele 2012: 19). The application of Laplace's mathematical model of an idealized Solar System, along with new observations, revealed that the opposite was the case: Jupiter's orbit was expanding; Saturn's was shrinking.

In 1784, Laplace showed that because Saturn makes two revolutions around the Sun in roughly the same time that Jupiter makes five, the planets meet at nearly the same place along the ecliptic every 59 years. "For one half of the period, the orbit of Saturn is gradually enlarging or widening out, while that of Jupiter is undergoing the opposite change; and during the other 450 years, the effects are reversed. Wonderful balance or oscillation even of those most complex perturbations!" (Nichol 1851: 166); another oscillation, regarding Pallas, is at the heart of the Gauss anagram explained in Chap. 4. Laplace declared that the inequality of Jupiter and Saturn, which "seemed to form an exception to the law of universal gravitation" actually became "one of the most striking examples of its truth." (Laplace 1809: 56) Another possible exception to the law of gravitation was raised by the asteroids, a subject that will be examined in a future book in this series.

This 'Grand Inequality,' as it became known, was specifically invoked by Johann Wurm in the context of the asteroids. This is from a letter he wrote to Baron von Zach, published in the May 1802 issue of the *Monthly Correspondence*:

> *As soon as I have received Dr. Gauss's improved elements of Ceres, I will improve my perturbation calculations accordingly. Maybe Ceres does also have some kind of grande inegalité like Jupiter and Saturn, since Ceres suffers so much from Jupiter's vicinity. Only this periodic inequality correcting the mean motion cannot be as 'reciproque' as in the case of those two planets.*

It was Zach's protégé Burckhardt (1808) who worked on the Jupiter/Saturn calculations for Laplace in the late 1790s and early 1800s. After explaining the Grand Inequality in his book *Exposition du Système du Monde*, Laplace mentions the asteroids as test-beds for an extension of his research.

> The *small planets which have been discovered, are subject to very great inequalities, which will throw new light on the theory of the attractions of the heavenly bodies, and will enable us to render it perfect; but hitherto we have been unable to recognize these inequalities by means of observations* (Laplace 1830: 35).

## References

Appian, P.: 1533, *Introductio geographica* (*Introduction to Geography*). Johannes Vernerus, Ingolstadt.

Barrow, I.: 1735, *Geometrical Lectures*. Stephen Austen, London.

Bernardi, G.: 2016, *The Unforgotten Sisters*. Springer, Cham, Switzerland.

Bernoulli, J.: 1681, *Neu-erfundene Anleitung, wie man den Lauff der Comet-oder Schwantz-Sternen in gewisse grundmaessige Gesaetze einrichten und ihre Erscheinung vorhersagen koenne* (*A new Introduction, showing how the Motions of Comets may be reduced to some certain and Geometrical Rules, so that their appearance may be predicted*). Jacob Werenfelss, Basel.

Bernoulli, J.: 1682, *Conamen Novi Systematis Cometarum.* Henricum Wetstenium, Amsterdam.

Bernoulli, D.: 1986, *Quelles est la cause physique de l'inclinaision des Plans des Orbites des Planetes* (*The Mutual Inclination of the Planetary Orbits*). Die Werke von Daniel Bernoulli vol. 3, 303–326. (Original treatise written 1734). Birkhaueser, Basel.

Bessel, F.: 1848, *Populäre Vorlesungen über Wissenschaftliche Gegenstände*, 1-34. Hammerich & Lesser, Hamburg.

Brunschvicg, L.: 1922, *L'Expérience Humaine et la Causalité Physique*. Felix Alcan, Paris.

Burckhardt, J.-C.:1808, Formules Générales pour les Perturbations de Quelques Ordres Supérieurs. *Mémoires de la Class des Sciences Mathématiques et Physiques de l'Institut de France*, 36-67.

Caravella, S.M.: 2008, Triangles in the Sky: Trigonometry and Early Theories of Planetary Motion. *Convergence* 5. Online journal of the Mathematical Association of America.

Ciermans, J.: 1639, *Disciplinae Mathematicae traditae anno institutae societatis Iesu secularize*. Louvain.

Conant, J.: 1991, *The Search for Logically Alien Thought: Descartes, Kant, Frege, and the* Tractacus. *Philosophical Topics* 20(1), 115-180.

Dhombres, J.: 1992, Des Théorèmes de la Révolution, ou l'Inscription des Mathématiques dans l'Histoire. *Melanges de l'Ecole Francais de Rome* 104, 191-214.

Dunlop, K.: 2012, What Geometry Postulates: Newton and Barrow on the relationship of mathematics to nature, in *Interpreting Newton: Critical Essays* (A. Janiak & E. Schliesser, eds), pg. 79-101. Cambridge University Press, Cambridge.

Encyclopedia Britannica: 1890, Entry for Laplace. Henry G. Allen (ed.) Vol. 14, 301–304. New York.

Gall, O.: 2002, *Meanest Foundations and Nobler Superstructures: Hooke, Newton and the "Compounding of the Celestiall Motions of the Planetts"*. Kluwer, Dordrecht.

Gass, W.H.: 1985, Representation and the War for Reality, pg. 61-102. In *Habitations of the Word: Essays*. Simon & Schuster, N.Y.

Guerlac, H.: 1965, Where the Statue Stood: Divergent Loyalties to Newton in the Eighteenth Century, 317-334. In the book *Aspects of the Eighteenth Century*, E. R. Wasserman (ed.). The Johns Hopkins Press, Baltimore.

Guicciardini, N.: 2009, *Isaac Newton on Mathematical Certainty and Method*. MIT Press, Cambridge, Massachusetts.

Hall, M.B. and Hall, A.R.: 1962b, *De gravitatione et aequipondio fluidorum*. In *Unpublished Scientific Papers of Isaac Newton*, pg. 89-156. Cambridge University Press, Cambridge.

Hegel, G.: 1817, *Encyclopedia of the Philosophical Sciences in Outline* (*Encyklopädie der philosophischen Wissenschaften im Grundrisse*). A. Oswald, Heidelberg.

Heilbron, J. L.: 2015, Boscovich in Britain, in *Relocating the History of Science* (T. Arabatzis, J. Renn and A. Simoes, eds.). Springer, New York.

Hooke, R.: 1674, Attempt to Prove the Motion of the Earth, in *Lectiones Cutlerianae* (*The six Cutlerian Lectures*), 1679. John Martyn, London.

Laplace, P.S.: 1776, Recherches 1. Sur l'integration des equations differentialles. *Mémoires de mathematique et de physique* 7, 173. Paris.

Laplace, P.S.: 1809, *The System of the World* (J. Pond, tr.). Richard Phillips, London.

Laplace, P.S.: 1829, *Traité de mécanique céleste. Celestial Mechanics*, 1966: N. Bowditch (tr.). Chelsea Publishing Co., Bronx, New York.

Laplace, P.S.: 1830, *The System of the World*, vol. 2 (Henry Harte, tr.) Longman, Rees, Orme, Brown and Green, Dublin.

Leibniz, G.: 1699, Extract from a letter to Malebranche, dated 13/23 March. Quoted in *Malebranche et Leibniz* (1955), Andre Robinet (ed.). Vrin, Paris.

Levy, J.: 1968, Trois siècles de mécanique céleste à l'Observatoire de Paris. *L'Astronomie* 82, 381-393.

Littrow, J.J.: 1825, *Populäre Astronomie*. J. Heubner, Vienna.

Longomontanus: 1639, *Introductio in theatrum astronomicum*. Copenhagen.

Maclaurin, C.: 1748, An Account of Sir Isaac Newton's Discoveries. A. Millar, London.

Mancosu, P.:1996, *Philosophy of Mathematics and Mathematical Practice in the Seventeenth Century*. Oxford University Press, Oxford.

Maskelyne, N.: 1802, Letter dated 14 May to Banks. National Maritime Museum, REG09/00037; 5:166.

*Monthly Correspondence*: Letter from Gauss to Zach. pg. 81-82 (1802).
Mouret, S., Simon, J.L., Mignard, F and Hestroffer, D., 2009: The list of asteroids perturbing the Mars orbit to be seen during future space missions. *Astronomy & Astrophysics* 508, 479-489.
Nichol, J.P.: 1851, *The Planetary System: Its Order and Physical Structure*. H. Bailliere, London.
Olbers, W.: 1802. Manuscript archive number III 016, Bremen University, Bremen.
Ossendrijver, M., 2016, Ancient Babylonian Astronomers calculated Jupiter's position from the area under a time-velocity graph. *Science* 351, no. 6272, 482-484.
Penny, E.B. (ed.): 1864, *Man: His True Nature & Ministry*, translated from the French of Louis-Claude de Saint-Martin's 1802 book *Le Ministère de l'Homme-Esprit*. William Allen & Co., London.
Peterson, I.: 1993, *Newton's Clock: Chaos in the Solar System*. W. H. Freeman and Co., N.Y.
Phillips, C.: 2006, Robert Woodhouse and the Evolution of Cambridge Mathematics. *History of Science* 44, 69-93.
Priest, S.: 2007, *The British Empiricists*, 2nd ed. Routledge, London.
Proctor, R.: 1892, *Old and New Astronomy*, completed by A. C. Ranyard. Longmans, Green & Co., London.
Rheticus, G.: 1551, *Canon doctrinae triangulorum*. Wolfgang Guenther, Leipzig.
Rost, J.L.: 1718, *Astronomisches Handbuch*. P.C. Monath, Nuremberg.
Schliesser, E.: 2012, The Newtonian Refutation of Spinoza, in *Interpreting Newton: Critical Essays* (A. Janiak & E. Schliesser, eds), pg. 299-319. Cambridge University Press, Cambridge.
Steele, J.M.: 2012, *Ancient Astronomical Observations and the Study of the Moon's Motion (1691–1757)*. Springer, New York.
*The Edinburgh Philosophical Journal*, 7, 176-177 (1822).
Todorovic-Juchniewicz, B.: 1983, Perturbation in the comet's motion caused by Ceres, Pallas and Vesta. *Acta Astronomica* 33(1), 113-117.
Vanpaemel, G.H.W.: 2003, Jesuit Science in the Spanish Netherlands. In *Jesuit Science and the Republic of Letters*, pg. 389-432 (M. Feingold, ed.). Massachusetts Institute of Technology Press, Cambridge, Massachusetts.
Westphal, K.R.: 2015, Causal Realism and the Limits of Empiricism: Some Unexpected Insights from Hegel. *The Journal of the International Society for the History of Philosophy of Science* 5(2), 281-317.
Whiting, T.: 1828, *A Comprehensive System of Astronomy*. Printed for the author, London.
Wilson, C.: 1994, The dynamics of the Solar System. In *Companion Encyclopedia of the History and Philosophy of the Mathematical Sciences*, vol. 2, pg. 1044-1053 (I. Grattan-Guinness, ed.). Johns Hopkins University Press, Baltimore.
Wilson, C.: 2007, Euler and Applications of Analytical Mathematics to Astronomy. In *Leonhard Euler: Life, Work and Legacy*, pg. 121-145 (R. Bradley and E. Sandifer, eds.). Elsevier.
Wilson, D.: 2009, *Seeking Nature's Logic: Natural Philosophy in the Scottish Enlightenment*. Pennsylvania State University Press, University Park, Pennsylvania.
Woodhouse, R.:1804, On the Integration of certain differential Expressions, with which Problems in physical Astronomy are connected. *Phil. Trans*. 94, 219-278.
Wurm, J.: 1802, Versuch einer genaueren Bestimmung der Massen der Planeten, in Verbindung mit ihren Umlaufszeiten und mittlern Entfernungen. *Monthly Correspondence*, June issue, 546-570.
Zach, F.X.: 1785, Astronomische Beobachtungen und Nachrichten. *BAJ* for 1789, 156–163.
Zach, F.: 1802b. *Monthly Correspondence*, Sept. issue, pg. 305.
Zach, F.: 1802c. *Monthly Correspondence*, Nov. issue., pg. 505.

# Chapter 2
# The Great Probability Debate

*All nature is but Art, unknown to thee;.*
*All Chance, Direction, which thou canst not see.*
– Alexander Pope, 1734.

## The Legacy of Leibniz

The eighteenth century had been in a very real sense the century of mathematics. Leibniz (1677; Fig. 2.1) reduced "the essence of things to numbers" and wanted "to establish a specific calculation to account for every truth." He entered the University of Leipzig in 1661, where he was "influenced by the mathematician and *philosophe* Erhard Weigel (1625–1699), whose work was focused on reconciling Aristotle with contemporary philosophers through a mathematical method of demonstration patterned on Euclid" (Pattanayak).

Leibniz himself praised the fourteenth-century English mathematician Richard Swyneshed for introducing mathematics into philosophy (North 2013: 49). "Leibniz believed that mathematical demonstration freed philosophy from its verbal limitations and embraced the field, believing its systematic coherence could unify the sciences" (Pattanayak). But one of the greatest proponents of Scottish Common Sense Realism, Dugald Stewart (1753–1828) at Edinburgh University, weighed in heavily against the legacy of Leibniz. Writing in 1821, Stewart regarded Leibniz as a prime obstacle to truth. Stewart (1829: 262) wrote of "the disregard manifested in his writings to the simple and obvious conclusions of experience and common sense." Stewart enlists the aid of the Roman historian Tacitus (56–117) with this quote from his tract *Agricola*: "But it was soon mellowed by reason and experience, and he retained from his learning that most difficult of lessons—moderation." He was not referring here to Leibniz but rather the English philosopher John Locke (1632–1704), whose philosophy directly influenced the Common Sense school of philosophy that in turn influenced Brougham in his 1802 polemic against the word 'asteroid.' Employing a classical allusion to the battle for supremacy between the giants and the gods of Olympus, and in language reminiscent of the high baroque period, Stewart excoriates Leibniz. Compare his employment of the term 'harmony'



C.J. Cunningham, *Studies of Pallas in the Early Nineteenth Century*,
DOI 10.1007/978-3-319-32848-5_2

**Fig. 2.1** Gottfried Leibniz. His principle of pre-established harmony was inspired by the work of Kepler and his harmonic relations including the 'music of the spheres' (Bussotti 2015)

with that used by Capel Lofft in his tract about a missing planet between Mars and Jupiter, written just a year before the discovery of Ceres (See Chap. 3 for a discussion of Brougham and Lofft.).

> *How happily does this last expression* [of Tacitus] *characterize the temperate wisdom of Locke, when contrasted with that towering, but impotent ambition, which, in the Theories of Optimism and of Pre-established Harmony, seemed to realize the fabled revolt of the giants against the sovereignty of the gods!* (Stewart 1829: 262)

Leibniz would use these very theories in virtually every field he studied, including law.

> *Leibniz had learned from the law that probability is a relation between hypotheses and evidence. But he learned from the doctrine of chances that probabilities are a matter of physical propensities. Even now no philosopher has satisfactorily combined these two discoveries* (Hacking 1975: 139).

In his legal study Leibniz "assigned values of 0, 1, and ½ to conditions of laws that were impossible, necessary (absolute), or conditional" (Pattanayak). Leibniz's 1697 design of a medallion for Ernst August, duke of Brunswick (1629–1698), is the first example of the binary system used by Gauss in his anagram about Pallas (see Chap. 4). Leibniz first published its invention in 1703 (Fig. 2.2).

Mathematics became fashionable, and the European public took part. At the courts the reigning families were Leibniz's correspondents, and every prince had

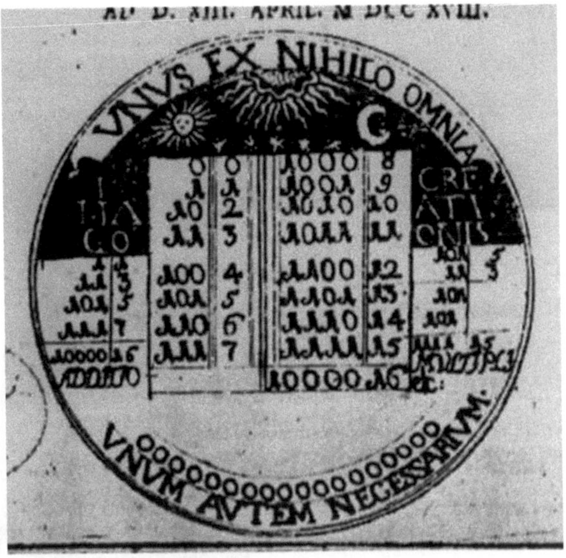


**Fig. 2.2** Design of a medallion by Leibniz, incorporating the first use of the binary system

his court mathematician. Leibniz was the archivist of the House of Brunswick; his influence with King Frederick I led to the creation of the Prussian (Berlin) Society of Science in 1700 (reconstituted as the Academy in 1746; see Chap. 3). A century later Karl Wilhelm Ferdinand, the duke of Brunswick (1735–1806), perpetuated the ducal interest in science and mathematics of his ancestor Ernst August by being the patron of Gauss. “Through mathematics, so it was believed, the confusion in the world had been reduced to order; by it the Age of Enlightenment was inaugurated” (Friedenthal 1965: 398; Greene 2004).

For Descartes (Fig. 2.3), whose influence on Leibniz cannot be overestimated, mathematics was still clothed in religious garb, as evidenced by this letter he wrote on April 15, 1630, to the French philosopher and mathematician Marin Mersenne (1588–1648).

> *I would not allow myself in my physics to touch upon metaphysical questions, and particularly this one: that mathematical truths, which you call eternal, have been established by God and depend entirely on him, as does the rest of all creation* (Aczel 2005: 139).

But his reasoned claim in the *Meditations on First Philosophy* (1641) that God insures the truth of clear and distinct ideas, including geometry, has obscured the importance of his optical theory in his foundation of a mathematical science of nature. Descartes claims to have proved that we can know physical objects insofar as they are the objects of applied geometry, which he terms ‘pure mathematics.’ In his *Dioptrics* (1637), he shows that to formulate a science of nature, we can obtain reliable information about bodies by measuring them (Maull 1978). Kepler published two works on optics, both in the service of astronomy. Descartes was intimately acquainted with these works—he even referred to Kepler as his first master in optics (Hatfield 2015).

**Fig. 2.3** René Descartes, Louvre Museum

Descartes used his knowledge of optics to solve a problem Galileo Galilei (1564–1642) could not—why stars appear no larger through a telescope than to the naked eye. He discerned that stars are an intense dagger of light that agitate the nerves in the eye to generate an image that cannot be smaller—the *minimum visible* —which is the smallest the visual system can manage (George 2006). Descartes (1637) notes "that the stars, even though they appear quite small, nevertheless appear very much larger than their extreme distance should cause them to appear." This same reasoning applies to the asteroids, a word which itself means star-like, and this should be kept in mind when reading Herschel's investigations of the physical size of Ceres and Pallas in Chap. 10. The mathematical work of both Descartes and Leibniz had roots much deeper in the past than Kepler or Galileo.

Although it presaged a mathematical way of looking at nature, it was Newton who wrote the mathematical laws that ushered in the modern scientific age.

> *Like Descartes before him, Leibniz was attracted to the work of the thirteenth- century mystic Ramon Lull* (1235–1316; Fig. 2.4). *Lull's* Great Art of combinations, *which used rotating wheels within wheels to create a large number of combinations of concepts coded by letters on these wheels, took a new and deeper meaning in Leibniz's eyes. Leibniz saw in these efforts more than just mystical play, but rather a mathematical attempt to study combinations. Leibniz developed these same concepts into a mathematical theory, and published it in a 1666 treatise titled* Dissertatio de Arte Combinatoria (Fig. 2.5). *This work developed the mathematical foundations of combinations* (Aczel 2005: 209).

**Fig. 2.4** This engraving of the planets from Athansius Kircher (1665), just 22 years before Newton published his *Principia*, is suffused with mysticism. Above is the Perpetuum Mobile of the Solar System, an enneagram from Lull's second combinatory figure that ascends into the super-celestial sphere. Although it presaged a mathematical way of looking at nature, it was Newton who wrote the mathematical laws that ushered in the modern scientific age (courtesy Google Books)

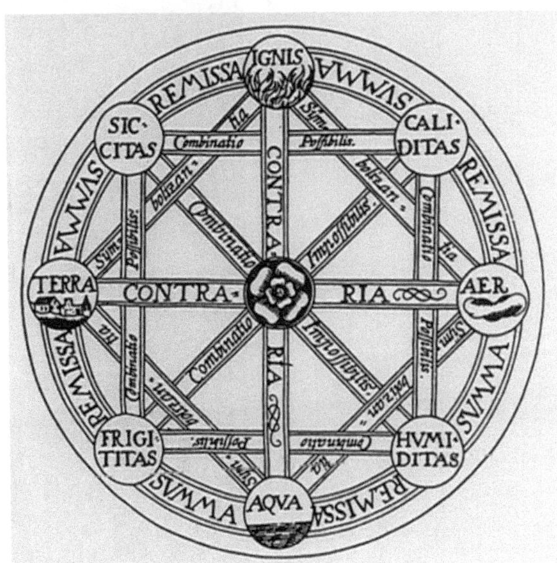


**Fig. 2.5** A diagram of the *Ars combinatorial* by Leibniz is based on four elements (earth, air, fire, water), four material states (heat, cold, dryness, moistness), and two possible rotational directions. For Leibniz, the ars combinatoria was nothing less than the "key to all the sciences." (Antognazza 2009: 63)

Lull's system represents the beginnings of formal logic, the foundation of modern science. This in turn led to work in probability theory. By the nineteenth century the role of mathematics had been transformed and largely shorn of superstition. "All the mathematical sciences are founded on relations between physical laws and laws of numbers, so that the aim of exact science is to reduce the problems of nature to the determination of quantities by operations with numbers." Maxwell (1856) When Laplace published his *Philosophical Essay on Probability* in 1814, he had firmly established the theory of probabilities as a branch of mathematics and science with none of the dissolute connotations of the doctrine of chances, as we shall see in the next chapter (Plackett 1989).

## The Most Delicate of Mathematical Theories

Throughout the first two decades of the nineteenth century there was a great clash between Wilhelm Olbers' planetary explosion hypothesis and Laplace's application of probability to physical astronomy (Fig. 2.6). How this intertwined with the

**Fig. 2.6** Pierre-Simon Laplace

mathematical work of Carl Gauss in relation to the early study of the asteroids is explored in this chapter.

In addition to his study of the Solar System—being the first to offer orbital elements for the missing planet—Jacob Bernoulli was also a crucial founding figure in the mathematics of probability. In the 1680s he showed it was possible to "extend the application of probability theory from games of chance to other fields where stable relative frequencies exist." (Hald 2004: 11) When this finally got published in 1713, "probability came before the public with a brilliant portent of all the things we know about it now: its mathematical profundity, its unbounded practical applications, its squirming duality and its constant invitation for philosophizing." (Hacking 1975: 143)

Laplace was not only the first use of the trigonometric series in celestial mechanics, but went further by linking his work on infinite series with problems of chance, a topic earlier studied by Abraham de Moivre (1667–1754). One of the great pioneers of classical probability and a disciple of both Newton and Halley, de Moivre wrote: "Another Method I have made use of, is that of Infinite Series, which in many cases will solve the Problems of Chance more naturally than Combinations." (Moivre 1756: vii) The unsuccessful application of infinite series to the orbital problem posed by Ceres was explored in *Early Investigations of Ceres and the Discovery of Pallas*. In the Leibniz-Newton dispute over the invention of calculus, Moivre sided with Newton while Bernoulli sided with Leibniz. The contemporary reach of Moivre's work in probability, including his two major books on the subject, *The Doctrine of Chances* (1718) and *Miscellanea Analytica* (1730), has been revealed by Bellhouse et al. (2009).

Laplace's interest in probability can be traced directly to Euler of the Prussian Academy of Science. His colleague Lagrange, whose 1776 work in probability theory enhanced de Moivre's efforts, was also a great admirer of Euler. Lagrange accepted Euler's ideas, believing that "algebraic operations apply to infinite series, but other mathematicians soon began to doubt the rigor of Lagrange's algebraic analysis. Among these were Abel Burja (1752–1816) in Germany and Jan Sniadecki (1756–1830)" in Poland, one of the most avid observers of the asteroids and a regular contributor to Zach's journal *Monthly Correspondence* (Synowiec 2000: 14). (Letters to Sniadecki from Zach, and his observations of Ceres, are included in this book.) Lagrange later used mathematics to study Olbers' planetary explosion hypothesis to explain the origin of the asteroids, a topic that also engaged the mind of Laplace, as shown later in this chapter, but it was all grounded in work that began in the 1770s.

> *Soon after gaining election to the* Académie des Sciences *in 1773, Laplace became seriously interested in mathematical astronomy, which was to become his dominant concern. By then two major areas of activity were evident: in celestial mechanics, the fine details of the motions of the heavenly bodies as analyzed using especially Newton's laws and allowing for perturbations; and in planetary mechanics the analysis of their shapes* (Grattan-Guinness 2005: 243).

The calculation of probabilities must have appealed to Napoleon Bonaparte (1769–1821), to whom Laplace dedicated the first edition of his work in 1812

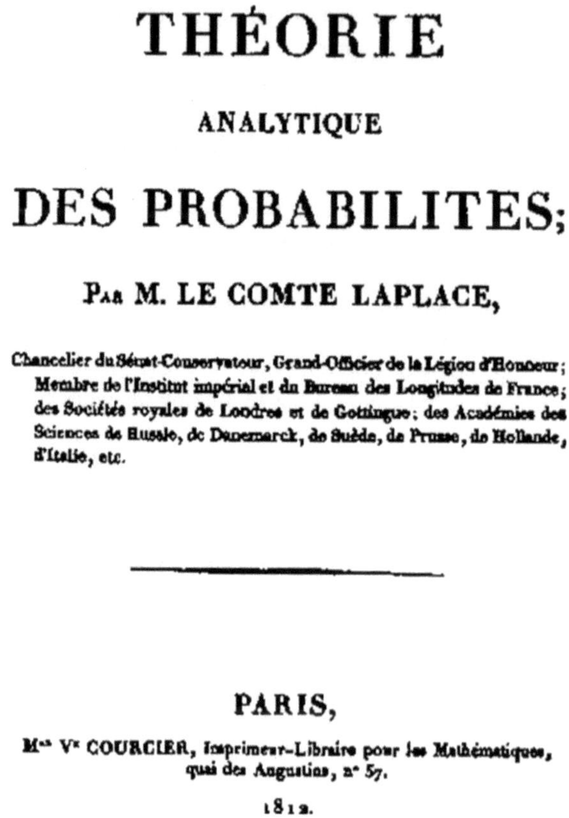
THÉORIE

ANALYTIQUE

DES PROBABILITES;

PAR M. LE COMTE LAPLACE,

Chancelier du Sénat-Conservateur, Grand-Officier de la Légion d'Honneur; Membre de l'Institut impérial et du Bureau des Longitudes de France; des Sociétés royales de Londres et de Gottingue; des Académies des Sciences de Russie, de Danemarck, de Suède, de Prusse, de Hollande, d'Italie, etc.

PARIS,

Mme Ve COURCIER, Imprimeur-Libraire pour les Mathématiques, quai des Augustins, n° 57.

1812.

**Fig. 2.7** Title page of Laplace's 1812 book on probability

(Fig. 2.7). Here is a letter dated August 12, 1812 from Witepsk in the Russian Empire, which Napoleon wrote to Laplace when he received the treatise:

> *There was a time where I would have read with interest your* Treatise on the Calculation of Probabilities. *Today I must limit myself to conveying to you the satisfaction I feel each time I see you give new works which perfect and extend the first of the Sciences, and contribute to the illustriousness of the Nation. The advancement, the perfecting of Mathematics are linked to the prosperity of the State.*

## The Historical Development of Probability

In *Prior Analytics* (70a1–10), "Aristotle defines *eikos* as that which is generally approved" (Hoffman 2008: 7). This is roughly "equivalent to the pre-Enlightenment sense of the Latin word *probabilitas,*" which was used in a sense that was true to its root verb *probo*, which means "to approve or commend." Thus anything worthy of approval could be called *probabilitas*, and this use persisted into the eighteenth century. For example, the Scottish philosopher David Hume (1711–1776):

> *...was aware of the connection between mathematical reasoning and reasoning from probabilities, but evidently unaware of the possibility of expressing probabilities mathematically. He preferred to conceive probability in terms of the 'agreeing', or concurrence, of images together to form a 'superior' idea* (Raynor 1964: 104).

The modern usage of probability arose in tandem with inductive logic. "It is connected with the degree of belief warranted by evidence, and it is connected with the tendency, displayed by some chance devices, to produce relative frequencies." (Hacking 1975: 1) Dugald Stewart made explicit what demarcated the eighteenth century from earlier times when he wrote:

> [the] *growing disposition to weigh scrupulously the* probability *of alleged* facts *against the faith due to the testimonies brought to attest them, and, even in some cases, against the apparent evidence of our own senses, enters largely and essentially into the composition of that philosophical* spirit *or temper, which so strongly distinguishes the eighteenth century from all those which preceded it* (Stewart 1829: 263).

This new divide that existed between geometrical proofs and the sort of subjects treated by moralists was highlighted by chemist and rhetorician Joseph Priestley (1733–1804; Fig. 2.8) Priestley's fame today rests chiefly in his research as a chemist, but his views raised the ire of Robison, who denounced both him and Laplace as enemies of humanity. For him, Priestley prepared the "minds of his

**Fig. 2.8** Joseph Priestley painting by Ellen Sharples. National Portrait Gallery, London

readers for Atheism by his theory of the mind." (Heilbron 2015: 104) Not surprisingly, Priestley had no use for Hume, Reid, or Common sense philosophy.

*The geometrician, when he hath laid down his proposition, proceeds, by a series of steps which terminate in a single proof, to show the agreement or coincidence of the terms of it… But in this the moralist and divine must content themselves with following them at a great and very humble distance. As the subjects they treat of are not always capable of strict* demonstration, *they are obliged to have recourse to a variety of arguments, each of which may add something to* probability *(which in its own nature admits of degrees) till the united strength of them all be sufficient to determine the assent* (Priestley 1777: 50).

Laplace explained what he meant by probability and its application to physical astronomy in 1773 (it was published in 1776). Even though it was written nearly 30 years before the discovery of Ceres and Pallas, it applies to them as well. Even though Laplace is noted for not explicitly introducing God into his scientific works, he did in fact preface his remarks about physical astronomy and probability with the concept of a universal Intelligence.

*The present state of the system of nature is evidently the result of what it was in the preceding instant, and if we imagine an Intelligence who, for a given moment, encompasses all the relations of beings in this Universe, it will also be able to determine for any instant of the past or future their respective positions, motions, and generally their disposition* (Hahn 2005: 56).

I suggest Laplace derived this in part from the work of Johann Heinrich Samuel Formey (1711–1797; Fig. 2.9), who wrote "Is it, in itself, more probable, that

**Fig. 2.9** Johann Heinrich Samuel Formey

things arranged with wonderful art, and constantly relative to evident and useful ends, are such, because they are such, than because an infinite Intelligence has presided over the arrangement of them?" (Formey 1748: 10; 1759: 279). Like Formey, Laplace capitalizes Intelligence, thus giving it the meaning 'infinite intelligence.' Formey (1759: 285) derives his interpretation of what constitutes "the logic of probabilities" directly from this. If, he writes:

> *...the enemies of Religion act like a man...who buys several tickets to a lottery in which there are 9999 white [draws] against one black one (these proportions are significantly lower than the dangers to which irreligion casts men), ...I ask that they save us from their ironic remarks and their endlessly repeated witticisms.*

How he determined the proportion of the dangers of irreligion are left unstated. So, who was Formey? None other than a professor at the French College in Berlin, perpetual secretary of the Berlin Academy from 1748, and a contributor to the French *Encyclopedie*. And with a reputed 17,000 letters flowing from his pen, he was a major figure in the Republic of Letters (Charrier-Vozel 2006: 118). Laplace takes his cue from Formey, but arrives at a very different conclusion. Instead of deriving probability and the nature of Intelligence from religion, Laplace turns to science, and finds at least a partial answer in physical astronomy. In this Laplace was directly following the views of the Marquis de Condorcet (1743–1794), who was writing in response to issues raised about "God's power to intervene in the natural laws God was credited with establishing" (Hahn, 2005: 51). The matter came to the fore in a 1756 prize question posed by none other than the Berlin Academy under the influence of Formey. Here is Condorcet (1768: 5):

> *One could conceive [of the universe] at any instant to be the consequence of the initial arrangement of matter in a particular order and left to its own devices. In such a case, an Intelligence knowing the state of all phenomena at a given instant, the laws to which matter is subjected, and their consequences at the end of any given time would have a perfect knowledge of the 'System of the World.'*

With the intellectual antecedents of the pillar of determinism in the Temple of Natural Philosophy now firmly in view, we can understand the continuation of Laplace's reasoning from his 1773 paper:

> *Physical astronomy, that subject of all our knowledge most worthy of the human intellect, offers an idea – albeit imperfect – of what such an Intelligence would be. The simplicity of the laws that move celestial bodies, and the relationship of their masses and their distances permit us to follow their motion up to a certain point with the use of calculus* [Analyze]. *To determine the state of the system of these large bodies in past or future centuries, it is enough that observations provide the mathematician with their position and speed at a given moment. Man derives this capacity from the power of the* [mathematical] *instrument he uses and the small number of parameters they include. But our ignorance of the various causes that produce these events as well as their complexities taken together with imperfections in the calculus, prevent him from making assertions about most phenomena with the same assurance. For him therefore there are things that are uncertain, and some that are more or less probable. Given the impossibility of* [total] *knowledge, man has compensated by determining their different degrees of likelihood; so that we owe to the frailty of the human mind one of the most delicate and ingenious of mathematical theories, namely the science of chance or probabilities* (quoted in Hahn 2005: 56).

He expanded on this in 1814 in the *Philosophical Essay on Probability*:

> *The regularity which astronomy shows us in the movements of the comets doubtless occurs in all phenomena. The curve described by a simple molecule of air or water vapor is regulated in a manner just as certain as the orbits of the planets; the only difference between these is that introduced by our ignorance. Probability is relative, in part to this ignorance, and in part to our knowledge* (Laplace 1814; 2007: 6).

As we saw in *Discovery of the First Asteroid, Ceres*, Laplace couched his analysis of the orbits of the planets and asteroids in terms of chance. Because the world is determined, Laplace implies, there can be no probabilities in things. "Indeed the probability isn't useful when we have at our disposal certain knowledge (p = 1), or when, on the contrary, we are in the presence of events of which the absolute impossibility is known (p = 0)." (Fortino 2002: 54) The theory of chances, then, "consists in reducing all events of the same kind to a certain number of equally possible cases, that is to say, those such that we are equally undecided about their existence." (Laplace 1814; 2007: 6) This concept of chance had a wide-ranging

**Fig. 2.10** Constantin-Francois Volney

influence (Daston 1992). We have encountered the 'doctrine of chances' from the pen of William Godwin in *Early Investigations of Ceres and the Discovery of Pallas*, but it had already been incorporated into French thought decades earlier. Witness this passage invoking harmony from Constantin-Francois de Chasseboeuf, count de Volney (1757–1820; Fig. 2.10), with his italics in Roman type:

> *When the* hidden power *which* animates *the universe, formed the globe which man inhabits, he implanted in the beings composing it,* essential properties *which became the* law *of their individual motion, the* bond *of their reciprocal relations, the cause of the harmony of the whole; he thereby established a regular order of causes and effects, of principles and consequences, which,* under an appearance of chance [sous une apparence de hasard], *governs the universe, and maintains the equilibrium of the world* (Volney 1791: 34).

## Equipossibility

The term *philosophe* will be employed several times in the pages that follow, so the question as to who a *philosophe* is must first be addressed. By the mid-eighteenth century the academies in France were in decline:

> *Internal instability left the institutions of French erudition vulnerable to external challenge, and that challenge came from the* philosophes. *To the* philosophes, *the relationship between erudition and authority was an obstacle to free inquiry and the progress of the sciences. Scholars in their own right, though excluded from the privileges of academic membership at least initially, the* philosophes *made extensive use of erudite scholarship even as they advanced a new intellectual and historical agenda exemplified by the historical essays of Voltaire and Montesquieu and ultimately by the* Encyclopédie *that appeared serially in the 1760s* (Carhart 2007: 155).

This great compendium of knowledge, the *Encyclopedie, Ou Dictionanaire Raisonne des Sciences, des Arts et des Metiers*, ran to 17 volumes and was published between 1751 and 1765. In its volume covering probability, dated 1765, we find a judicious survey of kinds of probability, but the only measurable probability is founded on the "equal possibility of several events." The author says of equipossibility that it "is to be employed when we suppose the several cases to be equally possible, and in effect it is only a supposition relative to our bounded knowledge, that we say, for example, that all the points on the die can occur equally."

As explained by Hacking (1975: 129), this concept was foreign to English writers, as they felt no need to equivocate by defining probability as possibility. "The French clearly perceived the need for both sides of probability but could not face up to it, thus taking refuge in the ambiguous concept of 'equipossibility.'" This reached its ultimate expression in Laplace, who defined probability in terms of equally possible cases. Lambert (1728–1777; 1761), who postulated the existence of a planet between Mars and Jupiter, also wrote on the matter. He maintained 'possibility' "in its old physical sense, determined by the laws of individual objects. In this early period equipossibility is not a feature of Laplace's work" either, but this changed as time went on.

Fig. 2.11 Eugene Wigner

Laplace discerned that in astronomy observations of planetary positions and motions are distributed in a regular fashion, according to the law of error. “Error was the great enemy of enlightenment, the loathsome infamy the *philosophes* had journeyed forth to slay. In that battle, mathematical science was their most fearsome weapon.” (Alder 2002: 213). Laplace concluded that astronomy “proceeded from the mathematical analysis of statistically regular empirical findings to an understanding of the constant causes that determined the phenomena of nature.” He believed that probability theory provided grounds for concluding that constant causes were at work when, as in celestial mechanics, the multiplication of observations could be shown to lead to the elimination of ‘strange effects.’ (Buck 1981: 24) How does the Laplacian concept of probability theory mesh with our modern understanding of physics and mathematics? Nobel laureate Eugene Wigner (1902–1995; Fig. 2.11) believes the insights of Laplace are fundamentally correct although not as precise as he would have wished:

> *The laws of nature are all conditional statements and they relate only to a very small part of our knowledge of the world. Thus, classical mechanics, which is the best known prototype of a physical theory, gives the second derivatives of the positional coordinates of all bodies, on the basis of the knowledge of the positions,* etc., *of these bodies. It gives no information on the existence, the present positions, or velocities of these bodies. It should be mentioned, for the sake of accuracy, that we discovered about thirty years ago that even the conditional statements cannot be entirely precise: that the conditional statements are*

*probability laws which enable us only to place intelligent bets on future properties of the inanimate world, based on the knowledge of the present state. They do not allow us to make categorical statements, not even categorical statements conditional on the present state of the world* (Wigner 1960: 5).

A related issue is unpredictability. "What is important about the systematic unpredictability of radical conceptual innovation is of course the consequent unpredictability of the future of science." (MacIntyre 1984) It is just such a radical conceptual innovation that Gauss produced to calculate the orbit of Ceres, an innovation that transformed the nascent science of celestial mechanics. Probability is also a factor in human genius itself. These words of the poet W.H. Auden (1907–1973) from 1940 are certainly applicable to Carl Gauss: "Any great man is a miracle of improbability." (Mendelson 2002: 56)

## Analogy and Probability

The concept of analogy is mentioned several times through this book. Olbers and Oriani use it (see Chap. 11), and it was invoked by Bode himself when writing of his 'law' of planetary distances. Analogy was also employed by Kepler throughout his scientific career (Gentner 1997). The use of analogy in the seventeenth century was expressed by Locke: "In things which Sense cannot discover, Analogy is the great Rule of Probability." (Schuurman 2003: 49) Kepler's use of analogy in the *Mysterium Cosmographicum* to relate the distances of the planets with geometrical entities (illustrated in Chap. 6) is something he later described as "admirable rather than happy or legitimate." (Kepler 1611) The importance of geometry to the study of the asteroids has been detailed in Chap. 1, and it was also at the heart of Kepler's agenda, where he uses the clockwork analogy (Peterson 1993) to explain the workings of the cosmos:

*I am much occupied with the investigation of the physical causes. My aim in this is to show that the celestial machine is to be likened not to a divine organism but rather to a clockwork...Moreover, I show how this physical conception is to be presented through calculation and geometry* (Kepler 1605).

Kepler here uses analogy in the Humean sense, but this raises a peril of dire consequences. Hume (Fig. 2.12) says:

*In the probability deriv'd from analogy, 'tis the resemblance only, which is affected. Without some degree of resemblance, as well as union, 'tis impossible there can be any reasoning: but as this resemblance admits of many different degrees, the reasoning becomes proportionally more or less firm and certain* (Garrett 1997: 144).

It is in the varying degrees admitted by the principle of resemblance that the problem arises. In the days of Newtonian physics and Laplacian certainty, all was well. But as celestial mechanics admitted chaotic orbits, and physics replaced certainty with uncertainty in the twentieth century, the analogies employed by Kepler and the asteroid researchers of the early nineteenth century became not only less firm but untenable.

**Fig. 2.12** David Hume painting by Allan Ramsey. National Galleries Scotland

In 1804 Olbers also raises the concept of doubt about the 'paradoxical results' found by Gauss in his study of Pallas (see Chap. 11). Here he follows the precept laid down eight years earlier by Pierre Samuel du Pont de Nemours (1739–1817) in his book *Philosophy of the Universe*. He urged one "not to definitively judge the planets and the suns. Let us doubt: it is the role of a man and a *philosophe*." (Du Pont 1796: 248)

## The Origin of the Asteroids and the Twilight of Probability

Leibniz observed that many judgments must be made in the "twilight of probability." (Daston 1988: 45) Nowhere was this more apparent than the discussion and debate between Olbers and Laplace about how the asteroids came to be. To put this in context, a consideration of contemporary opinion is in order. The following extract from Zach's *Monthly Correspondence* in July 1802 (pg. 71) sets the stage with the views of Olbers in that discovery year of Pallas.

> *In the previous issue (p. 598), we communicated Dr. Olbers' thought regarding both new planets to our readers, in which he is inclined to regard them as rubble of one single planet. This illustrious observer expresses himself in one of his later letters in the following way: 'Along with me, you are probably surprised at the curious position of the path of Pallas against that of Ceres. They are not interconnected like the links of a chain as was*

*first suspected; rather, the path of Pallas is stuck in the path of Ceres like a ring in the other. With the descending node of the path of Pallas on the path of Ceres, both paths come very close together. Whether an actual intersection really occurs or has ever occurred before the perturbations move the position of both against the other is not clear at present since we do not yet know their paths well enough.*

*From Dr. Gauss's 2nd path for Pallas and 7th for Ceres, I have found the distance of both paths in the descending node of Pallas to be = 0.07001. But if I combine the 2nd path of Pallas with Burckhardt's ellipse for Ceres, then the distance becomes only 0.06567. A small change in the elements of the path of Pallas, which is very possible, can continue to greatly reduce this distance. In this, the following question arises in me: whether Ceres and Pallas could be pieces and rubble of a formerly large planet destroyed either through natural forces working within it or through the collision of a comet? Which points to this suspicion which I offer as nothing more than a question for further investigation.*

*That Ceres and Pallas are both of very changeable luminosity – this I explain in that both planetary fragments are not round, but rather of a very irregular shape. This idea has the advantage over other hypotheses that it will soon be able to be verified. If it is true, we will find yet more rubble of this destroyed planet and this is made all the more easy since all this rubble, which describes an elliptical path around the Sun (much may have flown off in parabolas and hyperbolas)* must cross the descending node of the path of Pallas on the path of Ceres. *In general, all of these* suspected *planetary fragments have the same nodal line on the plane of the paths of Ceres and Pallas.' Again in a more recent letter, Dr. Olbers repeats this opinion and writes: 'The same orbital time of Pallas and Ceres, the positions of these paths with each other, the proximity of same in the descending node of the path of Pallas on that of Ceres; this makes it all the more probable for me that both belong together, and again and again, I return to the conjecture which I have expressed to you, that both are perhaps pieces and rubble of a former large planet.'*

From the comet collision idea given to him by the German jurist Ferdinand von Ende (1760–1816), Olbers came to believe that the asteroids were created in an explosion (Cunningham and Orchiston 2013). Laplace disagreed, and this sparked a heated debate not only about Laplacian cosmology and the origin of the Solar System, but about his application of probability theory to the whole question. The opening shot was sounded in a letter Olbers received from Burckhardt in December 1804. This is how Olbers (1804) related the matter to Gauss:

*Mr. De Laplace has communicated an objection to me* via *Burckhardt against my hypothesis on the origin of asteroids. It rests on the same orbital period as Ceres and Pallas and on the improbability that 2 fragments of a shattered planet should retain exactly the same velocity after the collision. Laplace, Burckhardt continues, explains the fact that several planets exist whose orbits have almost equal major axes very easily on the basis of his hypothesis on planet formation, namely: the planets formed as a result of the cooling of an immense atmosphere which enveloped the Sun.*

The letter then continues with a direct quote from Laplace:

*The complete volume of this atmosphere, located between Jupiter and Mars, had not been able to coalesce into a single planet for whatever reasons; it coalesced into several planets, all of which retained the same semi-major axis, namely, the distance of the Sun which the volume, had it become a solid body, would have had.*

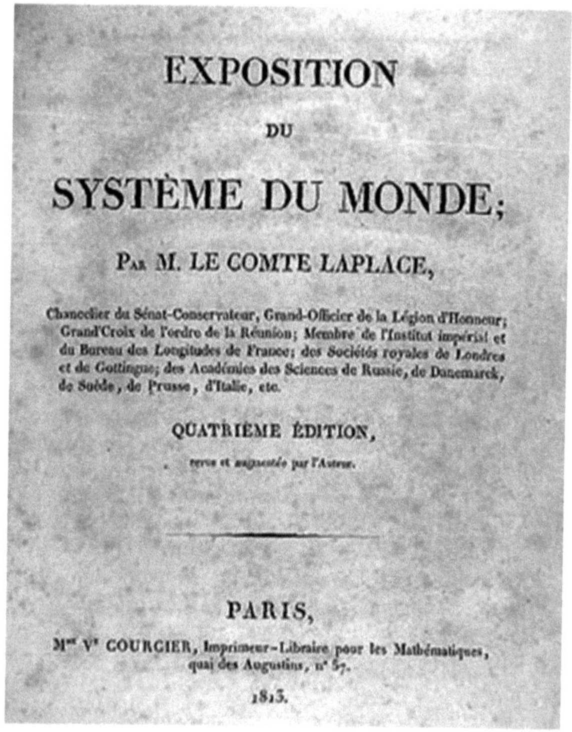
EXPOSITION
DU
SYSTÈME DU MONDE;
PAR M. LE COMTE LAPLACE,
Chancelier du Sénat-Conservateur, Grand-Officier de la Légion d'Honneur; Grand'Croix de l'ordre de la Réunion; Membre de l'Institut impérial et du Bureau des *Longitudes* de *France*; des *Sociétés royales* de *Londres* et de *Gottingue*; des *Académies* des *Sciences* de *Russie*, de *Danemarck*, de *Suède*, de *Prusse*, d'*Italie*, etc.
QUATRIÈME ÉDITION,
revue et *augmentée* par l'Auteur.
PARIS,
Mme Ve COURCIER, Imprimeur-Libraire pour les Mathématiques, quai des Augustins, n° 57.
1813.

**Fig. 2.13** Title page of Laplace's book *Exposition du Système du Monde*

Laplace went public with his objections in his 1808 book *Exposition du Système du Monde*, which was a nonmathematical version of his *Celestial mechanics* (Fig. 2.13). He applied his probability theory to the orbits of the four asteroids, estimating the odds at 4 billion to one that such a planetary arrangement is not the effect of chance. Laplace addressed the issues raised by the asteroids in his 1808 edition, and expanded on their place in the formation of the Solar System in his 1813 edition:

> *There are many discoveries still to be made in our own system. The planet Uranus, and its newly recognized satellites, give grounds to suppose that other planets, as yet unobserved, exist. It has even been conjectured that there must be one between Mars and Jupiter, in order to satisfy the double progression, which obtains (g)* [the force of gravity] *very nearly, between the intervals of planetary orbits, to that of Mercury. This conjecture has been confirmed by the recent discovery of four small planets whose respective distance from the Sun is not much different than the distance assigned by this double progression to a planet intermediate between Jupiter and Mars. Jupiter's action on these planets, increased by the magnitude of the eccentricities and inclinations of the intersecting orbits, produces considerable inequalities in their motions, which will shed new light on the theory of celestial attractions, and will give rise to perfecting it even more. The arbitrary elements of this theory, and the convergence of its approximations, depend on the precision of observations and the progress of analysis, and this should thereby acquire every day more and more accuracy* (Laplace 1808: 396-397 French; 1830: 339-340 English).

*If all the particles of a ring of vapours continued to condense without separating, they would at length constitute a solid or a liquid ring. But the regularity which this formation requires in all the parts of the ring, and in their cooling, ought to make this phenomenon very rare. Thus the Solar System presents but one example of it – namely, the rings of Saturn. In nearly every case the rings of vapor ought to be divided into several masses, which, being moved with velocities which differ little from each other, should continue to revolve at the same distance around the sun. These masses should assume a spheroidal form, with a rotary motion in the direction of their revolution, because their inferior particles have a less velocity than the superior; they have therefore constituted so many planets in a state of vapor. But if one of them was sufficiently powerful, to unite successively by its attraction, all the others around its center, the ring of vapors would thus be transformed into a single spheroidical mass, circulating about the Sun, with a motion of rotation in the same direction with that of revolution. This latter case has been the most common; however, the Solar System affords us an instance of the first case, in the four small planets which revolve between Jupiter and Mars, at least unless we suppose with Olbers, that they originally formed a single planet which a mighty explosion broke up into several portions each moving at different velocities* (Laplace 1813: 432-433 in French; [see reference for Laplace, 1808] 1830: 360, Note 7 in English).

Thus we see that Olbers disagreed with Laplace on the use of probability theory to explain the orbits of the asteroids, while Laplace disagreed with Olbers on the explosion theory to explain those very same orbits. In 1812 Olbers visited Paris, and had several personal discussions with Laplace about the origin of the asteroids. Olbers (1812a) must have been very persuasive, as witnessed in a letter to Bessel.

*Laplace gave me his recently published* Théorie analytique des probabilités. *Until now I have only been able to look randomly at the embodied treasure. To read the book thoroughly is of course work and pleasure for several months.*

*Laplace, who had always rejected my hypothesis of the origin of asteroids, appears to be more compliant. I have always emphasized the improbability that according to his idea of the development of our planetary system any planetary body might originally have an inclination of 35°. He meant against the solar equator the inclination of Pallas' orbit might be lesser. I replied that according to the current position of the nodes of Pallas' orbit exactly the contrary was occurring: Pallas' orbit still had 1¾ degree more inclination against the solar equator than against the ecliptic. The changing position of the line of nodes against each other and the small inclination against the solar equator against the ecliptic appears to me negligible. He then expressed his doubt whether Ceres' and Vesta's orbit might also intersect. I showed him that Vesta's distance from the sun at aphelion is larger than Ceres' distance at perihelion.*

Eight days later Olbers (1812b) wrote to Gauss with obvious satisfaction:

*Laplace, who, as you know, has otherwise abandoned my hypothesis of the asteroids, appears to be favorably inclined towards it. We discussed it a lot. – Some days ago, when we were returning from a meeting of the institute, he said to me on his own free will: 'I further considered your argument, that you derive from the considerable inclination of Pallas' orbit, and now I, too, have to admit that your hypothesis is most likely.' – I had told him namely that the sum of the inclinations of the seven older planets against the equator of the sun was only 35.8889° (decimal degrees), but the inclination of Pallas' orbit alone 40.4098°. If the inclination of the older planetary orbits were up to 40.4098°, according to his own formulae, the probability that the sum of the inclinations of those seven orbits is included between 0° and 35.8889° would be*

$$\frac{1}{1.2.3\ldots.7}\left(\frac{35,8889}{40,4098}\right)^7 == 0,000086474 == \frac{1}{11564}.$$

*This makes it highly likely to me that Pallas was forced by an external force off its original path.*

The Dutch astronomer Jean Henri van Swinden (1746–1823; Fig. 2.14) had earlier issued a cautionary warning on this topic. Van Swinden (1803: 28), a professor at the Athenaeum Illustre of Amsterdam, made these remarks on the dangers of developing planetary laws based on incomplete data:

*Herschel nameth both the new Planets Asteroids, that is, who are looking like the shape of Stars, or Star-like, (just like Sphere, globe, Spheroid, globular). But this giveth no clearer notion of the matter. Let us learn from this large deviation* [inclination] *of the recently discovered Planet Pallas, of a law, which was considered to be certain (i.e., that the Planets move, all, inside the Zodiac), that it is always perilous, and hardly sage, to draw up general laws from only a few observations, if one is not brought there through basic Mathematical principles.*

Philosophically, Hume argues that a successful prediction based on induction (such as the one Olbers promoted), says nothing about the truth of the theory—for him, "inductive inference is not capable of being rationally justified." (Schurz 2014: 81). The origin of the asteroids is considered further in the next book of this series.

**Fig. 2.14** Jean Henri van Swinden. Universitetsmuseum Amsterdam

## The Invention of Least Squares

In 1775 the thin volume *Opera Inedita* appeared, containing several papers of the late German astronomer Tobias Mayer (1723–1762; Fig. 2.15). Among them was a catalog of the right ascensions and declinations of 998 stars, which may be occulted by the moon and planets; the places being adapted to the beginning of the year 1756. Mayer held the chair of mathematics at Goettingen University, and he became the superintendent of the Observatory in 1754. Many of the letters in this book refer to stars in Mayer's catalogue, and when Gauss moved to Goettingen, he began to observe the asteroids in the observatory established by Mayer (Forbes 1974). "He used Mayer's 6-foot mural quadrant, a pendulum clock made by John Shelton and an achromatic refracting telescope from the John and Peter Dollond firm in England. It was equipped with micrometers to measure small angular distances from neighboring stars with great precision." (Aubin 2005: 34).

Mayer has another crucial link with Gauss—his work of 1750 is "important in the history of statistics because it contains the first successful application of *equations of condition*, a method to solve an overdetermined system of equations" (Wepster). It became known as Mayer's method. Laplace (1788) generalized Mayer's method in his paper that explains the 'grand inequality' in the motions of Jupiter and Saturn that was detailed in Chap. 1. Mayer's method was in use for about half a century until it was replaced by the method of least squares (Forbes 1980), a method Gauss

**Fig. 2.15** Tobias Mayer

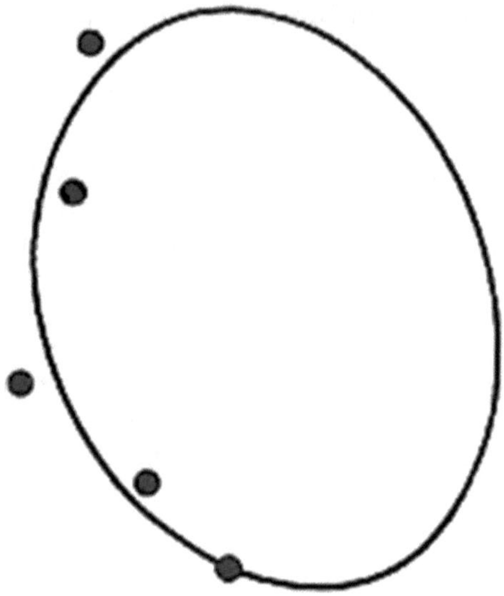

**Fig. 2.16** A best fitting curve to a given set of points

claims to have invented. Gauss himself attributes the principle of least squares to Mayer, as evidenced in this letter of January 24, 1812, to Olbers:

> *The only thing which is surprising is that this principle, which suggests itself to readily that no particular value at all can be placed on the idea alone, was not already applied 50 or 100 years earlier by others, e.g. Euler or Lambert or Halley or Tobias Mayer, although it may very easily be that the latter, for example, has applied that sort of thing without announcing it, just as every calculator necessarily invents a collection of devices and methods which he propagates by word of mouth only as occasion offers.*

The one person Gauss did not single out in this letter was Newton. But half a century before Mayer, in 1700, "Newton wrote down the first of the two so-called 'normal' equations known from the ordinary least squares method." (Belenky and Vila-Echague 2016) Newton's method forms the simplest kind of regression analysis, which was rediscovered and enhanced by Mayer.

Least squares is "a mathematical procedure for finding the best fitting curve to a given set of points by minimizing the sum of the squares of the offsets, known as "the residuals" of the points from the curve (Fig. 2.16). The sum of the *squares* of the offsets is used instead of the offset absolute values because this allows the sum to be treated as a continuous differentiable quantity. However, because squares of the offsets are used, outlying points can have a disproportionate effect on the fit, a property which may or may not be desirable depending on the problem at hand.

In practice, the *vertical* offsets from a line are almost always minimized instead of the *perpendicular* offsets (Fig. 2.17). This allows uncertainties of the data points along the *x*- and *y*-axes to be incorporated simply, and also provides a much simpler analytic form for the fitting parameters than would be obtained using a fit based on perpendicular distances. In addition, the fitting technique can be easily generalized

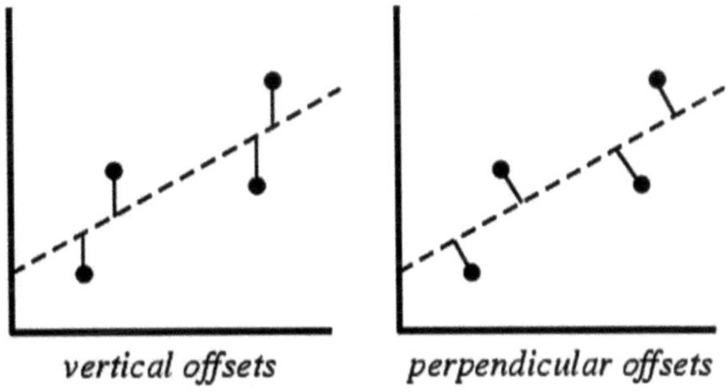


**Fig. 2.17** Vertical and perpendicular offsets from a line (Mathworld)

from a best-fit *line* to a best-fit *polynomial* when sums of vertical distances are used (which is not the case using perpendicular distances). For a reasonable number of noisy data points, the difference between vertical and perpendicular fits is quite small. The *linear* least squares fitting technique is the simplest and most commonly applied form of linear regression and provides a solution to the problem of finding the best fitting *straight* line through a set of points. In fact, if the functional relationship between the two quantities being graphed is known to within additive or multiplicative constants, it is common practice to represent the data in such a way that the resulting line *is* a straight line" (Mathworld).

The formulas for linear least squares fitting were independently derived by Gauss and Adrien-Marie Legendre (1752–1833; Fig. 2.18). The generally accepted version of events in the early nineteenth century was expressed by Bowditch (1820): "The method proposed by Professor Gauss and used by him since the year 1795 (and which was also invented by Legendre a few years afterwards) in the principle known by the name of the least squares."

This version has been the subject of heated dispute in recent decades. On the question of priority of discovery of the method, Stephen Stigler (1981) has questioned Gauss's veracity, something that has rightly been termed nothing short of libellous by Sheynin (1999). Stigler writes that while Legendre published the method in 1805, "Gauss had the temerity to claim that he had been using the method since 1795." In a letter to Laplace on January 30, 1812, Gauss wrote this about his application of least squares to the asteroids:

> *I have used the method of least squares since the year 1795 and I find in my papers, that the month of June 1798 is the time when I reconciled it with the principles of the calculus of probabilities. However my frequent applications of this method only date from the year 1802, since then I use it as you might say every day in my astronomical calculations on the new planets* (Plackett 1972).

Stigler (1999: 330) reluctantly concedes that Gauss's claim is possible:

> *Let us grant that Gauss's later accounts were substantially accurate, and that he did devise the method of least squares between 1794 and 1799, independently of Legendre or any other discoverer. There still remains the question, what importance did he attach to the discovery? Here the answer must be that while Gauss himself may have felt the method useful, he was unsuccessful in communicating its importance to others before 1805.*

**Fig. 2.18** Adrien-Marie Legendre

*He may indeed have mentioned the method to Olbers, Lindenau, or von Zach before 1805, but the total lack of applications by others, despite ample opportunity, suggests the message was not understood.* [See also Stigler (1977).]

It was Lindenau who procured the 1802 outline of Gauss's first work on the determination of planetary orbits. The text of this important document is printed in Chap. 12.

When Gauss tried to determine the orbits of the asteroids, the curve he sought was an ellipse. From the point of view of the formalism of least squares, this was not more difficult than the straight line example on the previous page; he used his method to determine the ellipse for which the sum of the squares of the deviations from certain observed points is a minimum. This is illustrated in the generic diagram Fig. 2.15, showing observed points around an ellipse.

In 1823 Gauss published a systematic and generalized presentation of his earlier theory of observational errors. In it, he develops the method of least squares with mathematical rigor as, in general, "the best way of combining observations, independent of any hypothetical law concerning the probability of error." (Kim 2009: 1128) (Hall 1970; Brand 2003). Gauss (1811) himself explained it:

*The method of determining the path to which all observations come the closest can be divided into two parts: first in the development of linear equations which give a difference between the observed and calculated positions through the functions of the sought corrections, and second in determining the most probable value of this correction through the method of the least squares, while here, at almost any time, more equations than unknessence, Gauss developown quantities occur, which therefore cannot be done strictly enough.*

There are indeed two parts in the method: (a) finding the best estimator (coefficients of the approximating curve); and (b) proving this estimator is the most efficient. So did Gauss really use least squares to compute the orbits of Ceres and Pallas as claimed by Marsden (1977)? The answer is yes.

Bessel states that in his opinion Legendre has the priority because he published the result first, but "he can testify that Gauss told him about the method some years earlier." (Krengel 2006) Also, Olbers testifies that Gauss told him about the method prior to 1805, and Wolfgang Bolyai knew of this too (Sheynin 1999). Baron von Zach (1813: 98) wrote this:

*The famous Doctor Gauss was already since 1795 in possession of this method, and used it to advantage in the determination of the elements elliptical orbits of the four new Planets* [asteroids], *as can be seen in his beautiful work* [of 1809].

Other pieces of evidence are remarks in the diary of Gauss, which was found many years after Gauss's death. He did say he used least squares when computing the orbit of Ceres, but he threw away the pages with his calculations. Furthermore, in 1799 Gauss "applied his new method of least squares to the equation of time as given in Ulugh Beg's tables." (Neugebauer 1975: 11) Gauss's least squares graduation of the differences in the equation of time by the astronomer Ulugh Beg (1394–1449) is given in *Gauss Werke* 12, pp. 64-68, and has been confirmed by Dutka (1996). (The twelve-volume collection of Gauss's complete works, known as *Gauss Werke*, was published between 1863 and 1933.) The method of least squares is still being used to refine the ephemerides of Pallas (Hestroffer et al. 1998).

## Gauss and Laplace: The Mathematical Link

There was a synergistic relationship between Gauss and Laplace. "His first proof of the method of least squares is based on inverse probability inspired by Laplace's 1774 paper." (Hald 2004: 50) In this landmark paper, Laplace

*...invented (or perfected) a calculus of statistical inference that allowed him to estimate the likelihood that a particular configuration of events would lead to a subsequent arrangement known through observation. Philosophically, its power stemmed from capturing game theory from the domain of conjecture (chance), and turning it into a method for calculating likelihood (probability)* (Hahn 1990: 379).

The full panoply of Gauss's work, presented in *Theoria Motus*, proved to be a revelation for Laplace, who then used it to bolster his own work in probability. Thus the great synthesizer of knowledge—Laplace—and the great mathematician—Gauss—came together to create the framework for understanding the orbits and perturbations of Ceres and Pallas (Plackett 1989).

> *The memoir Laplace presented in April 1810 was entitled* Memoir on Approximating Formulas That Are a Function of Large Numbers, and on Their Application to Probability. *There Laplace used the central limit theorem, but without realising that it was in any way related to the least squares method. All this fell neatly into place once Laplace had studied and absorbed Gauss's remarkable* Theoria Motus, *which appeared in 1809. At first, Laplace apparently did not fully grasp the work's relationship to his concern with finding mean values for a set of observations of comets. Gauss's treatise was studied again in December 1809, winning the Lalande medal awarded by the* Institut. *But it was only after July 1810 that Laplace became fully aware of the way it simplified his fundamental notions. In his book, not only did Gauss give a mathematically elegant explanation for finding positions of new bodies like Ceres and Pallas, but he also offered a new way of looking at the error curve – a way that became known as the Gaussian distribution. Moreover, he demonstrated how this error distribution was mathematically tied to the least square method. Seeing the light, Laplace penned a brief supplement to his 1810 memoir, in which he took full advantage of the Gaussian discovery, integrating it into his understanding of the theoretical basis for probability theory. By placing error theory on a Gaussian distribution, Laplace had integrated the principle of least squares into the now powerful tool of probability* (Hahn 2005: 112).

Gauss in turn used Laplace's work on perturbation theory, which was published in 1799 in the first volumes of his *Treatise on celestial mechanics*. But much of this work only became known in the twentieth century (Klein et al. 1919).

> *True to his principle that a work should have such a degree of completion 'that nothing more could be desired', Gauss published very little of his contributions to perturbation theory. But he communicated them to his students; for example the orbit computed for the comet named after Encke was computed by Gauss's method* [in 1818]. *Only when Gauss's posthumous papers were published in his* Collected Works *did one get a clear idea of his fundamental contributions to perturbation theory through his calculations for Ceres and Pallas. In his first calculations for the perturbations of Ceres, Gauss used his own research on the hypergeometric series and the arithmetic-geometric mean, and also made use of Laplace's work in an essential way. The calculations of the perturbations of Pallas within the bounds of accuracy of the observations consisted mostly of page after page of numerical calculations with no explanatory text. This difficult puzzle was put in order and published by the German astronomer Martin Brendel, who in his commentary points out that Gauss has almost completed an enormous problem, 'which even today an astronomer does not gladly set out upon.'*
> Hall (1970: 64)

In essence, Gauss developed "the theory of the normal distribution of the observation errors and the method of least squares to solve for orbits; the two concepts are related by a theorem, by which the probability distribution of the least square fit in the space of orbital elements is also normal in the linear approximation. He developed this theory to solve the problems of orbit determination of both Ceres and Pallas, and this method has been used by astronomers ever since" (Milani 1998: 271). How Gauss developed the fast Fourier transform and applied it to the orbital solution of the asteroids will be examined in a future book in this series.

# References

Aczel, A.: 2005, *Descartes' Secret Notebook*. Broadway Books, New York.

Alder, K.: 2002, *The Measure of All Things*. Free Press, New York.

Antognazza, M.R.: 2009, *Leibniz: An Intellectual Biography*. Cambridge University Press, Cambridge.

Aubin, D.: 2005, Astronomical Precision in the Laboratory. In *Grundsätze über die Anlage neuer Sternwarten unter Beziehung auf die Sternwarte der Universität Göttingen* (K. Beuermann, ed.), 31-36. Goettingen University Press.

Belenkiy, A. and Vila-Echague, E.: 2016, Groping Toward Linear Regression Analysis: Newton's Analysis of Hipparchus' Equinox Observations. *Observatory* 136(1250), 1-22.

Bellhouse, D.R., Renouf, E.M., Raut, R. and Bauer, M.A.: 2009, De Moivre's knowledge community: an analysis of the subscription list to the Miscellanea Analytica. *Notes and Records, The Royal Society* 63(2), 137-162.

Bernoulli, J.: 1713, *Ars Conjectandi (The Art of Conjecture)*. Thurnisius, Basel.

Bowditch, N.: 1820, *Works of the German Astronomers. The North American Review and Miscellaneous Journal* New Series 1, 267-272.

Brand, B.: 2003, Gauss's Method of Least Squares. Masters thesis, Louisiana State University.

Buck, P.: 1981, From Celestial Mechanics to Social Physics: Discontinuity in the Development of the Sciences in the Early Nineteenth Century. In *Epistemological and Social Problems of the Sciences in the Early Nineteenth Century*, pg.19-22. (H. Jahnke and M. Otte, eds.). Springer, Netherlands.

Bussotti, P.: 2015, *The Complex Itinerary of Leibniz's Planetary Theory*. Birkhaeuser, Cham.

Carhart, M.C.: 2007, The Science of Culture in Enlightenment Germany. Harvard University Press, Cambridge.

Charrier-Vozel, M.: 2006, La Bibliotheque impartiale dans les lettres de Luzac a Formey, in Critique, Critiques au 18e siecle (M. Cook and M.-E. Plagnol-Dieval, eds.), 109–126. Peter Lang, Oxford.

Condorcet, N.:1768, *Essais d'analyse ou Sur le système du monde*. Didot, Paris.

Cunningham, C. and Orchiston, W.: 2013, Olbers' Planetary Explosion Hypothesis: Genesis and Early Nineteenth Century Interpretations. *JHA* 44(2), 187-205.

Daston, L.: 1988, *Classical Probability in the Enlightenment*. Princeton Univ. Press, Princeton.

Daston, L.: 1992, The Doctrine of Chances Without Chance: Determinism, Mathematical Probability, and Quantification in the Seventeenth Century. In *The Invention of Physical Science* (M. Nye, J. Richards and R. Stuewer, eds.), 27-50. Springer, Netherlands.

Descartes, R.: 1637, *Discours de la methode pour bien conduire sa raison & chercher la varite dans les sciences: plus la dioptrique, les meteores, et la geometrie*. I. Maire, Leyden. English edition: 2001, *Discourse on Method, Optics, Geometry, and Meteorology* (Paul J. Olscamp, tr.). Hackett, Indianapolis. Quote from pg. 112.

Descartes, R.: 1641, Renati Des-Cartes Meditationes de prima philosophia. Apud Michaelem Soly, Paris.

DuPont de Nemours, P.S.: 1796, *Philosophie de l'Universe*. Chez Goujon fils, Paris.

Dutka, J.: 1996, On Gauss's priority in the discovery of the method of the least squares. *Archive for History of Exact Sciences* 49, 355-370.

Forbes, E.: 1974, The Foundations of the First Goettingen Observatory. *JHA* 5, 22-29.

Forbes, E.: 1980, *Tobias Mayer*. Vandenhoeck & Ruprecht, Goettingen.

Formey, J.H.S.: 1748, *La logique des vraisemblances*. London, no publisher given.

Formey, J.H.S.: 1759, *Philosophical Miscellanies on Various Subjects*. J. Hinton, London.

Fortino, M.: 2002, Determinism and Chance in Laplace's and Cournot's Doctrine. *Philosophical Inquiry* 24(3-4), 53-64.

Friedenthal, R.: 1965, *Goethe – His life and times*. The World Publishing Company, Cleveland.

Gauss, C.: 1811, Regarding the Elliptical Elements of Pallas. *Monthly Correspondence*, Nov. issue, 449.

Gauss, C.: 1823, "Theoria combinationis observationum erroribus minimus obnoxiae". In English: A theory for the combination of observations, which is connected with least possible error. (G.W. Stewart, tr.). Philadelphia, Society for Industrial and Applied Mathematics, 1995.

Garrett, D.: 1997, Cognition and Commitment in Hume's Philosophy. Oxford University Press, Oxford.

Gentner, D., Brem., S., Ferguson, R., Wolff, P., Markman, A. & Forbus, K.: 1997, Analogy and creativity in the works of Johannes Kepler. In *Creative Thought,* pg. 403-459. (T. Ward, S. Smith & J. Vaid, eds.). American Psychological Asssociation, Washington D.C.

George, R.: 2006, James Jurin Awakens Hume from his Dogmatic Slumber. With a Short Tract on Visual Acuity. *Hume Studies* 32(1), 141-166.

Grattan-Guinness, I. (ed.), 2005, *Landmark Writings in Western Mathematics 1640-1940*. Elsevier Science, Amsterdam.

Greene, B.: 2004, *The Fabric of the Cosmos*. Knopf, New York.

Hacking, I.: 1975, *The Emergence of Probability*. Cambridge Univ. Press, Cambridge.

Hahn, R.: 1990, The Laplacian View of Calculation. In *The Quantifying Spirit in the 18th Century*, pg. 363-380. (T. Frängsmyr, J.L. Heilbron and R. Rider, eds.). University of California Press, Berkeley.

Hahn, R.: 2005, *Pierre-Simon Laplace 1749–1827: A Determined Scientist*. Harvard Univ. Press, Cambridge.

Hald, A.: 2004, *A History of Parametric Statistical Inference from Bernoulli to Fisher, 1713 to 1935*. University of Copenhagen, Copenhagen.

Hall, T.: 1970, *Carl Friedrich Gauss: A Biography*. MIT Press, Cambridge.

Hatfield, G.: 2015, Natural Geometry in Descartes and Kepler. *Res Philosophica* 92(1), 117-148.

Heilbron, J. L.: 2015, Boscovich in Britain, in *Relocating the History of Science* (T. Arabatzis, J. Renn and A. Simoes, eds.). Springer, New York.

Hestroffer, D., Viateau, B., Rapaport, M.: 1998, Minor planets ephemerides improvement. *Astronomy and Astrophysics* 331, 1113-1118.

Hoffman, D.C.: 2008, Concerning Eikos: Social Expectation and Verisimilitude in Early Attic Rhetoric. *Rhethoric* 26(1), 1-30.

Kepler, J.: 1605, Letter to Herwart von Hohenburg, dated February 10. Quoted in Gentner (1997: 408).

Kepler, J.: 1611, *Dioptrice*. Quote is from the Preface, given in English by Holton (1988:72).

Kim, D.: 2009, *entry for Carl Gauss, in Encyclopedia of Optimization*, 2nd ed. (C.A. Floudos and P.M. Pardalos, ed), 1128-1129. Springer Science, New York.

Kircher, A.: 1615, Arithmologia. Varesij, Rome.

Klein, F., Brendel, M. and Schlesinger, L.: 1919, *Materialien fuer eine wissenschaftliche Biographie von Gauss*. Vol. 7 of Gauss Werke. B. G. Teubner, Leipzig.

Krengel, U.: 2006, Von der Bestimmung von Planetenbahnen zur modernen Statistik. *Mathematische Semesterberichte* 53(1), 1-16.

Lambert, J.: 1761, *Cosmologische Briefe über die Einrichtung Weltbaues* [Cosmological letters regarding the constitution of the planets]. Augsburg, Germany: Eberhard Klett.

Lagrange, J. L.: 1776, Mémoire sur l'utilité de la méthode de prendre le milieu entre les résultats de plusieurs observations. *Misc. Taurinensia* 5, 167-232.

Laplace, P.S.: 1774, Mémoire sur la probabilité des causes par les événements (Memoir on the probability of causes of events). *Mém. Acad. Roy. Sci. Paris* 8, 27-65.

Laplace, P.S.: 1788, Théorie de Jupiter et de Saturne. *Mém. Acad. Roy. Sci. Paris* 11, 95-239.

Laplace, P.S.: 1808, *Exposition du Systeme du Monde,* vol. 2, 3rd ed, V. Courcier, Paris; 1813, vol. 2, 4th ed., Courcier, Paris. Quotes, with slight alteration, are taken from volume 2 of an 1830 English edtion: *The System of the World*, 2 vols (H. Harte, tr.). Longman, Rees, Orme, Brown, and Green, Dublin.

Laplace, P.S.: 1812, *Théorie Analytique des Probabilités*. V. Courcier, Paris.

Laplace, P.S.: 1814, *Essaie philosophique sur les probabilities*, 2nd ed. Courcier, Paris. The first edition was 1795. This Essay became the Introduction to the *Théorie Analytique des*

*Probabilities*; in English as *A philosophical essay on probabilities* by F. Truscott & F. Emory (tr.), Cosimo, New York, 2007, pg. 6.

Leibniz, G.: 1677, Preface to the Universal Science. In *Leibniz, Selections* (Philip Wiener, ed.), 1951. Charles Scribner's Sons, New York.

Leibniz, G.: 1703, Explication de l'arithmétique binaire. *Mémoires de l'Académie Royale des Sciences, Paris.*

MacIntyre, A.: 2013, After Virtue. University of Notre Dame Press, South Bend.

Marsden, B.: 1977, Carl Friedrich Gauss, Astronomer. *Journal of the Royal Astr. Soc. of Canada* 71, 309-322.

MathWorld.: A Wolfram Web Resource. http://mathworld.wolfram.com/FibonacciNumber.html

Maull, N.: 1978, Cartesian Optics and Geometrization of Nature. *The Review of Metaphysics* 32(2), 253-274.

Maxwell, J. C.: 1856, On Faraday's Lines of Force. *Transactions of the Cambridge Philosophical Society* 10, 27-83.

Mayer, T.: 1750, *Abhandlung über die Umwältzung des Monds um seine Axe und die scheinbare Bewegung der Mondsflecken (Treatise on the circulation of the Moon on its axis and the apparent motion of the Moon spots).* Kosmographische Nachrichten und Sammlungen auf das Jahr 1748, 52-183.

Mendelson, E. (ed.): 2002, *The Complete Works of W.H. Auden. Prose*, Vol. 2. Princeton University Press, Princeton.

Milani, A.: 1998, The asteroid identification problem I: Recovery of lost asteroids. *Icarus* 137, 269-292.

Moivre, A.: 1756, *The Doctrine of Chances: or, a method of calculating the probabilities of events in play.* A. Millar, London.

Neugebauer, O.: 1975, *A History of Ancient Mathematical Astronomy.* Springer, Berlin.

North, J.: 2013, Medieval Oxford. In *Oxford Figures* (J. Fauvel, R. Flood and R. Wilson, eds.), pg. 37-50. Oxford University Press, Oxford.

Olbers, W.: 1804. Letter to Gauss, dated December 12. Goettingen archives.

Olbers, W.: 1812a. Letter to Bessel, dated July 10. Goettingen archives.

Olbers, W.: 1812b. Letter to Gauss, dated July 18. Goettingen archives.

Pattanayak, A. The Mathematical Leibniz. www.math.rutgers.edu/courses/436/Honors02/leibniz.html

Peterson, I.: 1993, *Newton's Clock: Chaos in the Solar System.* W. H. Freeman and Co., N.Y.

Plackett, R.: 1972, Studies in the History of Probability and Statistics XXIX. The discovery of the method of least squares. *Biometrika* 59, 239-251.

Plackett, R.: 1989, The Influence of Laplace and Gauss in Britain. *Bulletin of the International Statistical Institute* 53, 163-176.

Pope, A.: 1734, *An Essay on Man.* J. Wilford, London.

Priestley, J.: 1777, *A Course of Lectures on Oratory and Criticism.* J. Johnson, London.

Raynor, O.: 1964, Hume's Scepticism Regarding 'Probable Reasoning' in the Treatise. *Southern Journal of Philosophy* 2(3), 103-106.

Schurz, G.: 2014, *Philosophy of Science: A Unified Approach.* Routledge, New York.

Schuurman, P.: 2003, Willem Jacob's Gravesande's philosophical defense of Newtonian physics: on the various uses of Locke, pg. 43-58, in *The Philosophy of John Locke: New Perspectives* (P. R. Antsey, ed.). Routledge, London. Quote by Locke comes from a title in his Essay IV. xvi 12, p. 665.

Sheynin, O. B.: 1999, The discovery of the principle of least squares. *Historia Scientiarum* 8, 249-264. See also: http://www.sheynin.de/download/reviews.pdf

Stewart, D.: 1829, Dissertation, exhibiting a General View of the Progress of Metaphysical, Ethical, and Political Philosophy, since the Revival of Letters in Europe. In *The Works of Dugald Stewart*, vol. 6. Hilliards and Brown, Cambridge.

Stigler, S.: 1977, An Attack on Gauss Published by Legendre in 1820. *Historia Mathematica* 4, 31-35.

Stigler, S.: 1981, Gauss and the Invention of Least Squares. *The Annals of Statistics* 9, 465-474.

Stigler, S.: 1999, *Statistics on the Table*. Harvard University Press, Cambridge.

Synowiec, J.: 2000, Some Highlights in the Development of Algebraic Analysis. *Banach Center Publications* 53. Polish Academy of Sciences, Warsaw.

Van Swinden, J.H.: 1803, *Lessen over het Planetarium, Tellurium, en Lunarium*. P. den Hengst, Amsterdam.

Volney, C.-F.: 1791, *Les ruines*. Desenne, Paris.

Wepster, S. Mostly less-trivial links for Tobias Mayer. www.staff.science.uu.nl/~wepst101/oldpage.html

Wigner, E.: 1960, The Unreasonable Effectiveness of Mathematics in the Natural Sciences. *Communications in Pure and Applied Mathematics* 13(1), 1-14.

Zach, F.: 1813, Mémoire sur le degré du méridien mesuré en Piémont par le père Beccaria. *Mem. Acad. Imperial Sci., Literature, Beaux-Arts Turin for 1811-1812, sci. math., physics*, 20, 81-116.

# Chapter 3
# The Gold Medal

*For it is the duty of an astronomer to record celestial motions through careful observation. Then, turning to the causes of these motions, he must conceive and devise hypotheses about them, since he cannot in any way attain the true cause. But, from whatever assumptions he adopts, the motions can be correctly calculated from the principles of geometry for the past as well as the future.*

– Andreas Osiander, author of the introduction to the 1543 book *De Revolutionibus orbium coelestium* by Copernicus

## Astronomy and the Republic of Letters

The Republic of Letters grew organically as the eighteenth century progressed, and the participants themselves used the term. As the embodiment of civilisation and learning, typified by the goddess Pallas, it eventually encompassed the entire intellectual elite of Europe (Fig. 3.1). Both Scottish and English universities took full part in this informal scholarly network (Ellis 2014: 26); even though most of the people associated "with asteroid studies" were not at universities, they too were integral members of the Republic. Its most famous overseas member was Benjamin Franklin (1706–1790; Fig. 3.2) who corresponded and met with many of the people in this study and its two preceding books, including Joseph Jérôme Lalande (1732–1807), Charles Blagden (1748–1820), Joseph Banks, the Astronomer Royal Nevil Maskelyne (1732–1811), and William Herschel's long-time friend Patrick Wilson (1743–1811).

Franklin formed some of these contacts in person during his lengthy stay in Paris where the *Institut des Sciences et des Arts* was founded on August 22, 1795; it replaced the Paris Academy of Sciences founded in 1666 but disbanded during the French Revolution (George 1938). The *Institute* was meant to be, in the words of one its founders (Picrre-Claude-Francois Daunou), "the epitome of the learned world, the representative body of the Republic of Letters" (Hahn 1971: 301). Its



C.J. Cunningham, *Studies of Pallas in the Early Nineteenth Century*,
DOI 10.1007/978-3-319-32848-5_3

**Fig. 3.1** This engraving featuring Pallas Athena (*at left*) symbolizes the world of civilization and learning. The goddess of learning, whose name was chosen by Olbers for the second asteroid, holds a shield embossed with the Gorgon's head. Her sacred bird, the owl, is nearby. The Ten Commandments flanked by cherubs are in the foreground. The bearded figure with the quill is a philosopher, possibly Aristotle, but the figure is dressed more in medieval, than classical garb. And the books, too, are codices versus rolls, so not intended to recall the high classical period. The female figure likely holding a palm *may* stand symbolically for Christianity, and the banner with the little portraits on the right may recall the glory and emperors of Rome. The symbols on the sheet at the lower right (a fish and possibly the symbol for the planet Venus and a crescent for the Moon), likely represent astronomy/astrology. Frontispiece of the Dutch periodical *Boekzaal der geleerde waereld* [Book Room of the Learned World] (1812). Onder de Linden, Amsterdam. Courtesy, Google Books

**Fig. 3.2** Benjamin Franklin in London, 1767. Painting by David Martin. White House Historical Association

status may be gauged by the fact that de Moivre, when elected to the Academy, "regarded it as a conferment of nobility" (Goldgar 1995: 236). This exalted position was subjected to a series of embarrassments of astronomical proportions. Halley had predicted the return of the comet that bears his name in late 1758 or early 1759. Famed member of the Paris Academy, Charles Messier (1730–1817), searched for the comet for 18 months, only to be stunned by the announcement that an amateur astronomer, Johann Georg Palitzsch (1723–1788), was the first to see it. Worse still, Palitzsch was not a French astronomer but a German one. "The collective pride of the Paris Academy must have been deeply wounded, since the recovery of comet Halley was made by a German amateur more than three months before they (the French) knew anything about it" (Yeomans 1991: 134).

The discovery of the planet Uranus in 1781 by an English astronomer of German origin, William Herschel, stunned the Academy again. Both of the most important astronomical discoveries of the age had been made by foreigners! Even more remarkable, in light of this, is what the academics at Paris said of themselves: *Nous seuls ici (à Paris) avons de l'esprit.* [We alone have wit (here in Paris)]. (Quoted by Dr. David in 1789; ref: Seydl 1938)

This sense of French superiority was admirably expressed by the Royalist writer Antoine de Rivarol (1753–1801; Fig. 3.3), when he claimed that French was the only language in which the syntactic structure of sentences reflected the real structure of human reason, and therefore it was the only logical language in the world. According to him, German was too guttural, Italian too sweet, Spanish too redundant, and English too obscure (Lauzon 2010). "In vain do passions excite us and invite us to follow the order of the sensations: French syntax is incorruptible. Whence that admirable clarity, the eternal foundation of our language: what is not clear is not French" (Rivarol 1784: 49). His assessment was published as one of the

**Fig. 3.3** Antoine de Rivarol

two prize-winning essays in a 1783 contest held by the Berlin Academy of Sciences. It is pertinent to note the Berlin Academy published their memoirs in French; Berlin's astronomical prizes are considered later in this chapter. Rivarol made a particular point of discussing the Republic of Letters in connection with English, as reported here by *The Monthly Review* (1784: 581).

> *When he says that the 'English, accustomed to the IMMENSE credit they have obtained in public affairs, are ambitious to transport this FICTITIOUS influence into the republic of letters, and that their literature has thence contracted a character of exaggeration, which is the enemy of true taste,' we do not understand him, and we are powerfully inclined to think, that he does not understand himself.*

"Rivarol's essay was a strikingly worded but unoriginal and linguistically very poor attempt" to deal with the prize questions: How came the French language to be universal? By what title does it merit this prerogative? Is it likely to maintain it always? (Swiggers 2014) The general English sentiment of such nonsense from the French was best expressed by Dorothy Wordsworth in 1806. In a letter about a passage by the French author Blaise Pascal (1623–1662), she said "...there is always a something wanting to the fullness of my satisfaction of all elevated sentiments in the French language." If the same words had been written in English, "...there would have been more of dignity in the language of the sentences....and they would have been more impressive" (Knight 1887: 185). The German view was also somewhat jaundiced. One is reminded of the quip delivered by King

Frederick II of Prussia (1712–1786) to the French philosopher Voltaire (1694–1778), in a letter of September 5, 1743: "I admire the wisdom of France but heaven preserve me from ever imitating it!" (Carlyle 1888: 227) All this is set before the reader here to show that the Republic of Letters operated under significant strains.

Even so, few would argue with Lagrange's opinion stated in 1787: "The Academy has a great attraction for me, as it is the leading tribunal of Europe for the sciences." (Ferrone 1997: 192). During his stay in Milan, the 27-year-old General Napoleon Bonaparte (Fig. 3.4) expressed it best in an open letter of May 22, 1796 that was addressed to Barnaba Oriani (Fig. 3.5) at Milan's Brera observatory, where Oriani would later make many asteroid observations:

> *The sciences, which ennoble human intelligence, and the arts, which embellish life and transmit great events to posterity, ought to be honoured by free governments. All men of genius, all those who have attained a distinguished place in the Republic of Letters, are Frenchmen, whatever may have been their country of birth* (Antonello 2014: 7).

**Fig. 3.4** Napoleon Bonaparte, as First Consul, in 1801. Painting by Ingres, from the Chateau de la Malmaison

**Fig. 3.5** Barnaba Oriani, Brera Observatory

Oriani intended to reply to this invocation in the following words: "My General, you do credit to the distinguished savants and artists of all countries by calling them Frenchmen. Actually they belong to the Republic of Letters and hence to all of Europe, or, better, to the entire world" (Antonello 2014: 12). Fortunately for Oriani these provocative words were deleted in his letter of reply, but it shows how he and (very likely) other astronomers of the day viewed themselves. In any case Napoleon's admiration for astronomy was genuine. While still in Milan, he wrote to Lalande on December 5, 1796, that "among all the sciences, astronomy has been the most useful to reason and to commerce," adding that it is especially astronomy that both needs communications at the international level, and "the existence of the republic of letters" (ibid). The vast correspondence about the first four asteroids between the astronomers of Europe presented in this series of books certainly bears testimony to Napoleon's assessment. But how did the participants themselves actualize the Republic in the field of astronomy? For that we have the testimony of Baron von Zach (1802a).

> *A common question is: Are Ceres and Pallas really main planets of our Solar System? Since there is no King in the free Republic of science, who says* tel est notre plaisir [such is our pleasure], *the astronomers' opinions vary, which is a sign of an unfinished definition of a planet. La Lande has already awarded those two heavenly bodies the planetary diploma – in his letters and even publicly in Moniteur. Dr. Herschel on the other hand suggests in his circular letter a new species of celestial bodies named asteroids. In our view both depend on arbitrariness, agreement, or consent of all astronomers.*

Although this appears to be a very noble sentiment on his part, it is actually somewhat disingenuous. As we saw in *Early Investigations of Ceres and the Discovery of Pallas*, Zach was accused by Bode of setting himself up as a judge of others (himself in particular) in the Republic. "What tribunal allowed Mr. von Zach's personal attacks against astronomers?" Bode (1802b) asked Olbers. Zach was just one in a long line of those who was accused of abusing the unwritten rules of the Republic. The French verb *s'ériger* (to set oneself up) "was often tossed off in disgust to describe someone who took it into his head to exercise power in the Republic when he had acquired that power from no one but himself" (Goldgar 1995: 158).

## Rivalry

When Ceres was discovered by an Italian in 1801, the pride of the institute was wounded yet again. Their only hope was to be the first to calculate its orbit. Burckhardt in Paris set to work at once to calculate its orbit with the arithmetical tools developed by Laplace. But it was not to be. One can almost hear the academicians sighing Quelle—(what a pity!) "The French were miffed that a young upstart named Gauss would find the correct orbit of Ceres before they did. Burckhardt had tried to obtain the orbit by using the Laplacian method, but failed. So they probably wanted to reward him, but when they realized Gauss's approach was superior, they probably gave up on him" (Hahn, priv. comm. 2001)

The differing reactions of French and German scientists can be thought of in terms of "the political scientists' concept of 'punctuated equilibrium,' a sophisticated microversion of what British philosopher of history Arnold Toynbee (1889–1975; 1946) termed challenge and response" (Carr 2002: 45). If we posit that an underlying tendency to equilibrium is upset by events, then the momentous discoveries of Uranus and Ceres certainly represent upsets. In an analysis of the processes that maintain rivalries, Goertz et al. (2005) have found both that the most recent conflict does matter, and the longer term history of the rivalry is more important than previously recognized. In the context of the discovery of Ceres and Pallas, this can be interpreted in stark terms. Recent conflict as the race was underway in 1801 to determine the orbit of Ceres was very fresh indeed—the discovery of Ceres itself in January of that year. And as the history of embarrassments in astronomy just elucidated shows, the rivalry French astronomers had with their colleagues both in England and in Germany was decades old.

## Reason and the Axiom of Solvability

In the autumn of 1802, the year Pallas was discovered, a lecture at Elers' College in Copenhagen by the Danish Romantic philosopher Henrik Steffens (1773–1845; Fig. 3.6) raised the banner of the Enlightenment high: "Reason, the wellspring in which freedom and necessity are inseparably mingled, is that which is most central to us" (Steffens 1803: 183). In a quote attributed to the French grammarian Cesar Dumarsais (1676–1756; 1743: 175), "Reason is to the *philosophe* what grace is to the Christian. Grace causes the Christian to act, reason the philosopher."

> *Alongside the familiar sly sarcasm of this Enlightenment association of religion and rationality – domesticating and simultaneously promoting the latter in the face of traditionalist social and political resistance – what is striking about this statement* [by Dumarsais] *is its focus upon* action (Terdiman 2005: 57).

The instantiation of reasoned, rational *action* to solve problems is at the heart of the grand effort to solve the orbital problem posed by the asteroids, especially Pallas. In an effort to solve such vexing questions, the Academy began offering

**Fig. 3.6** Henrik Steffens

prizes as an incentive. But what made these *philosophers* believe that every mathematical problem that was posed could in fact be solved?

The English philosopher Joseph Glanville (1636–1680; 1668) accords François Viète (1540–1603; Fig. 3.7) a crucial place in the development of mathematics: "Viète, who by inventing the *Method* of *Extracting Roots* in the most *numerous equations*, and by converting the *Signs* used by the *Ancients* into *Letters*, brought *Algebra* to a very great *perfection*." His role in planetary theory has only recently been recognized (Swerdlow 1975). Viète (1591) traced the root idea of his art to Plato (427–347 BCE) and Theon (335–405), a teacher of mathematics and astronomy in Alexandria. Theon observed solar and lunar eclipses in 364, so he may have met Ammianus Marcellinus during his sojourn in Alexandria.

*There is a certain way of searching for the truth in mathematics that Plato is said first to have discovered. Theon called it analysis, which he defined as assuming that which is sought as if it were admitted and working through the consequences of that assumption to what is admittedly true. The ancients propounded only two types of analysis. I have added a third. The whole analytic art may be called the science of correct discovery in mathematics.*

The challenge that faced the academy two centuries later was stated succinctly by Viète (1591): "The analytic art claims for itself the greatest problem of all, which

**Fig. 3.7** François Viète

**Fig. 3.8** David Hilbert

is to leave no problem unsolved." This concept was codified as the axiom of solvability by the German mathematician David Hilbert (1862–1943; Fig. 3.8) a century after Gauss worked on the Pallas problem, but its language applies directly to the situation Gauss faced:

> *However unapproachable problems may seem to us and however helpless we stand before them, we have, nevertheless, the firm conviction that their solution must follow by a finite number of purely logical processes. Is this axiom of the solvability of every problem a peculiarity characteristic of mathematical thought alone, or is it possibly a general law inherent in the nature of the mind, that all questions which it asks must be answerable? The conviction of the solvability of every mathematical problem is a powerful incentive to the worker. We hear within us the perpetual call: There is the problem. Seek its solution. You can find it by pure reason, for in mathematics there is no* ignorabimus (Hilbert 1901:444).

For a recent study of Hilbert's account of mathematical axiomatics, see Patton (2014). Translated literally, *ignorabimus* means 'we shall never know.' Clearly the philosophes of the French Academy fully subscribed to the concept that there can be no *ignorabimus*. Even though they were themselves ineligible to enter the prize contests, they had full confidence that great minds could solve any problem they posed. Gauss (1802: 597) himself had this degree of confidence in reason, but cautioned against its unquestioned acceptance. "Nothing can be set against an exact calculation than an equally exact calculation, not empty suspicions and vague reasoning."

To fully understand the reach of mathematics into everyday thought in the eighteenth century, it is useful to consider the view of the first of the French *philosophes,* Bernard Fontenelle (1657–1757; Fig. 3.9), secretary of the *Académie*

**Fig. 3.9** Bernard Fontenelle. Chartres, Musée des Beaux-Arts

*des Sciences* from 1699–1741. Implicit in this exposition by Fontenelle is the idea that geometry can be applied to any problem. The application of geometry to nature was a central problem for Descartes as a supreme geometer, as was Gauss after him, and it was with the application of geometry that Gauss tackled the Pallas problem.

> *The geometrical spirit is not so tied to geometry that it cannot be detached from it and transported to other branches of knowledge. A work of morals or politics or criticism, perhaps even of eloquence, would be better (other things being equal) if it were done in the style of a geometer. The order, clarity, precision and exactitude which have been apparent in good books for some time might well have their source in this geometric spirit. Sometimes one great man gives the tone to a whole century; Descartes, to whom might legitimately be accorded the glory of having established a new art of reasoning, was an excellent geometer. Astronomy cannot be without optics: and both, as all parts of the mathematics, are grounded upon geometry* (Fontenelle 1699).

Historically, this goes back to the Greek mathematician Hero of Alexandria (10–70) and Claudius Ptolemy (90–168), both of whom criticize the discourses of philosophers and declare epistemological supremacy for "mathematics by asserting that geometrical demonstration is indisputable" (Feke 2014: 261).

In Chap. 1 we examined perturbation theory, in Chap. 2 we considered probability theory which is ultimately based on studies that generate numbers, and here we have looked at the concept of Reason. To finally achieve a synergy of the three, we turn to the great English polymath Brewster (1781–1868; 1831: 369; Fig. 3.10), who has figured several times in the previous two books in this series. Even the High Victorian style of writing he employs cannot obscure his meaning.

> *Recumbent on his easy-chair, the practical astronomer has but to look through the cleft in his revolving cupola, in order to trace the pilgrim star in its course; or by the application of magnifying power, to expand its tiny disc, and thus transfer it from among its sidereal*

**Fig. 3.10** David Brewster, a photo from the 1850s

*companions to the planetary domains. The physical astronomer, on the contrary, has no such auxiliaries: he calculates at noon, when the stars disappear under a meridian sun: he computes at midnight, when clouds and darkness shroud the heavens; and from within the cerebral dome, which has no opening heavenward, and no instrument but the Eye of Reason, he sees in the disturbing agencies of an unseen planet, upon a planet by him equally unseen, the existence of the disturbing agent, and from the nature and amount of its action, he computes its magnitude and indicates its place. If man has ever been permitted to see otherwise than by the eye, it is when the clairvoyance of reason, piercing through screens of epidermis and walls of bone, grasps amid the abstractions of number and quantity, those sublime realities which have eluded the keenest touch, and evaded the sharpest eye.*

When Brewster was writing, he surely had the asteroids in mind when describing a tiny disk that could be distinguished from a star, and thus achieve the rank of a planet. The disturbing agent is of course a planetary perturbation, and it all comes into focus with the application of human reason which was applied to win the prize offered by the *Institute* in Paris to solve the Pallas problem.

## The Berlin Academy Prize

In the eighteenth century the Paris Academy awarded 75 scientific prizes; the Berlin Academy, founded in 1700, offered 45 (Fig. 3.11). But the playing field was not necessarily level—when Jean-Baptiste le Rond d'Alembert (1717–1783) declined the offer of King Frederick II to hire him as President of the Berlin Academy, Euler thought it was a ploy to negotiate a larger salary. In retribution, Euler essentially banned d'Alembert from the prize competitions and journals of the Academy (Cummings and Parker 2015). As d'Alembert had already won the Berlin Academy's prize for 1746, it is quite likely he would have won more. The Berlin academy was also structured differently from its counterparts.

> *The Academy was peculiar in its structure, combining a literary section and a class of speculative philosophy with the classes of mathematics and experimental philosophy. In London and Paris, the Royal Society and the Academie Royale des Sciences were confined to natural philosophy and mathematics, while their counterpart in Berlin retained Leibniz's original design for universal scholarship* (Lifschitz 2012: 10).

**Fig. 3.11** The goddess of wisdom, Pallas Athena, takes center stage in this 1710 allegorical depiction of the Berlin (Prussian) Society of Science. With her right hand she grasps the arm of the muse of Astronomy, Urania, shown with her crown of stars. Her other hand holds the all-seeing eye of knowledge. A bust of King Frederick I of Prussia sits at upper right, while mathematics is shown at lower right

King Frederick took personal control of the Academy in 1759, appointing himself "head of the class of speculative philosophy" (McClellan 1985: 176). This interest in speculative philosophy became starkly apparent in the prize competition of 1763, which pitted evidence and certainty in metaphysics compared to the certitude available in geometry and mathematics. For the German philosopher Christian Wolff (1679–1754; Fig. 3.12), and most early mathematicians and philosophers of mathematics, geometry was the framework within which mathematical objects could be defined (Shabel 2003).

The prize competition "was the culmination of the intense debate over Christian Wolff's deductive method, which was to a large extent a controversy over the language of philosophy" (Lifschitz 2012: 166). It is relevant to this study because, as was explained in *Discovery of the First Asteroid, Ceres*, it was Wolff who was one of the first to apply his deductive reasoning to the distances of the planets from the Sun, which led to Bode's law.

> *In his now rather rare book,* Vernunftige Gedanken von den Absichten der naturlichen Dinge (Rational Ideas about the Intentions of natural Things), *which went through five editions between 1724 and 1752, Christian Wolff made lengthy efforts to show the orderly and purposeful arrangement of the whole and the parts of nature.* [The distances of the planets] *proved for Wolff that the Creator saw to it that the planets should not disturb one another either by their motion or by casting their shadows on one another* (Jaki 1972: 136).

Wolff (1728) wrote of the importance for one to "possess philosophical knowledge of celestial motions." The Berlin Academy prize for 1763 went to the German philosopher Moses Mendelssohn (1729–1786). "In his prize essay

**Fig. 3.12** Christian Wolff

Mendelssohn argued that while mathematics dealt either with quantities or with the qualities of extended objects, natural philosophy was concerned with intensive or nonextended qualities, such as speed and light" (Lifschitz 2012: 167). Runner-up for the prize was Immanuel Kant (1724–1804; 1764), who similarly claimed that mathematics dealt with mind-dependent, predetermined definitions, while philosophy must proceed by analysis of confused or inaccessible concepts.

> *His treatise marked his rejection of the mathematical ideal in philosophy...Kant's criticism swept away the very foundations of Wolff's influence. Wolff's mathematical method depended on the possibility of applying procedures of mathematical proof to all philosophical and theological questions. Now Kant argues that the basic structures of mathematics and philosophy were different. Wolff's system crumbled* (Frängsmyr 1990: 44).

For more on Kant and the opposition to his views by the asteroid observer Jan Sniadecki, see Chap. 3.

The penchant of the King for the more picturesque aspects of speculative philosophy lasted for decades. Earlier we looked at the 1783 prize contest about the French language. The award was actually won by Christoph Schwab (1743–1821; 1785) of the Caroline Academy in Stuttgart, whose work was of a scientific nature, but King Frederick insisted the prize be shared with Rivarol.

## The Paris Academy Prize

The Paris Academy (Fig. 3.13) began offering gold medals as prizes in 1671. The Prize of Eloquence was awarded for the first time that year, even though it had been founded by a bequest of the writer Jean-Louis Balzac in 1654. A gold medal was also offered for the Prize of Poesy, instituted in 1671 (Robertson 1910). The Paris Academy began offering prizes to stimulate research into particular questions in 1720, but it was a double-edged sword. At the time, the effort seemed quite noble:

> *We cannot refrain from noting that the great progress made in physical astronomy is indebted principally to the* Académie des Sciences, *through its prize competitions. If France did not first lay the building blocks of the system of the world, it can take pride in having contributed more than any other nation to the elaboration of the system in all its aspects, and in establishing its truth in an undeniable manner* (Bossut and Legendre 1787).

In hindsight, however, modern historians have taken a somewhat different view of the prize structure. Here is Seymour Chapin (1927–1995; 1968), professor of history at California State University:

> *The limitations of the Academy's principal prize program worked against astronomy as much as the other sciences. For one thing, that program was occasionally guilty of misdirection. Another restriction was the program's stipulation that Academy members could not share in the awards. Thus, although these awards represented a limited means of instruction and remained important as a stimulus – serving to call attention to significant unsolved questions – many persons who were the most capable of providing solutions were excluded from competition.*

**Fig. 3.13** Paris Academy. The Memoires of the Academy (1699) includes an oval portrait of King Louis XIV of France being held by the goddess Pallas Athena wearing her signature Medusa breastplate.

The first prize of direct relevance to the matters discussed in this book came in 1732 when the academy posed the following question: What is the physical cause of the inclination of the planes of the planetary orbits with the plane of the solar equator; and why is it that the inclinations of these orbits are different from each other? No submission was judged worthy, so the question was renewed in 1734 with double the prize money. This time it was awarded jointly to Daniel Bernoulli (1700–1782; Fig. 3.14) and his father Johann (1667–1748). Their essays were published the following year (Bernoulli 1735). "The Cartesian presuppositions of the Academie's question are plain," writes Barry Gower (1987: 441) of Durham University. "A vortex theory of planetary phenomena seems to imply that orbital inclinations to the solar equator should be at or near zero, and some such theory, therefore, could reasonably be expected to yield a solution." The extremely high inclination of Pallas and its consequent perturbations that prompted the Pallas prize question was a clear admission that whatever solution had been awarded the 1734 prize was inadequate to deal with the new reality.

Several prize questions in the ensuing decades dealt with perturbation theory. The topic for the contest of 1748 was chosen at a meeting of the prize commission

**Fig. 3.14** Daniel Bernoulli

in March 1746. The French astronomer Pierre Charles Le Monnier (1715–1799) presented evidence that Jupiter and Saturn were subject to observationally detectable inequalities that were attributable to the mutual attractions of the two planets. The prize, it was decided, should be offered for a "theory of Jupiter and Saturn explicating the inequalities that these planets appear to cause in each other's motions, especially about the time of their conjunction." The prize in 1748 went to Euler (1749); Daniel Bernoulli's essay received the runner-up. Since Euler had failed to give the derivation of his differential equations (this had been done earlier in a memoir submitted to the Berlin Academy in 1747), the commission proposed the same question for 1750, but no prize was awarded. The problem was posed again in 1752, and Euler won again.

"Lagrange won the prize of 1766. In his memoir on celestial mechanics, he arrived independently at the conclusion Euler set forth in his still unpublished memoir of 1752: the discovery that the orbital elements of a perturbed planet, as determined by observation, include major components deriving from perturbation" (Cunningham 2007). This matter is directly pertinent to the orbit of Pallas. In the event, Leonhard Euler and his son Albrecht won half the 1770 prize; the remaining half was reserved to be joined to the 1772 prize, when the same problem was posed for the third time. In 1772 Lagrange won half, the other half being awarded once again to Leonhard Euler for a more thorough development of his lunar theory (Wilson 2010).

In 1774 the academy posed a double question: (1) how one might be assured, in the calculation of the lunar motions, that no detectable error results from the quantities neglected; (2) whether, taking account if necessary of the actions of the planets and the nonsphericity of the Earth and the Moon, one could explain, on the basis of the theory of gravity alone, why the Moon appears to have a secular equation in its longitudinal motion. Lagrange won the 1774 prize, although he dealt only with the second question. In 1778 the Academy proposed the theory of perturbations of comets. In his winning essay, Lagrange developed the theory of cometary perturbations by the method of variation of orbital elements (Gray 2006).

The idea of awarding prizes for scientific achievements survived the French Revolution. Napoleon Bonaparte was a member of the *Institut* and was anxious to reward famous scientists. The concept was not solely a French innovation—*The Society of Arts*, founded in England in 1754, encouraged invention by awarding prizes, but this did not extend to the sciences because of the "antimathematical prejudice of English astronomers". The French admired mathematics and this led them to ultimate vindication with the discovery of the planet Neptune in 1846, based on the mathematical analysis of LeVerrier (Barber 1961: 599).

## The Pallas Prize

The perturbations experienced by Pallas created just the sort of problem the axiom of solvability was meant to test. The *Institute* tried repeatedly to achieve a solution to the Pallas problem by the offering of prizes. The following account relates the official pronouncements from the *Institute*, interspersed with excerpts from private letters that sometimes breathlessly reported on the latest developments.

In 1803 the members of the astronomy section of the *Institute* consisted of six people: J.-J. Lalande, Michel Lefrancais-Lalande (1766–1839; nephew of J.-J.), Pierre Méchain (1744–1805), Charles Messier, J.-D. Cassini (1748–1825, known as Cassini IV), and Alexis Bouvard (1767–1843). It is they who instigated the prize question about Pallas.

*Ars sine scientia nihil est.* (Practice is worthless without theory.) This dictum was first enunciated in Renaissance Italy, but it was still motivating the French *Institute* in the nineteenth century. As the extract shows (Fig. 3.15), the prize commission of the Academy proposed at a meeting of June 18, 1804, a "theory of the perturbations of the planet Pallas" should be the subject of the Science and Mathematics competition. At the lower right, one can also see that Giuseppe Piazzi, discoverer of Ceres, was nominated to fill a vacancy as associate member of the *Institute* (Bertrand 1869). He lost to the chemist Martin Klaproth, whose role in asteroid history was considered in *Discovery of the First Asteroid, Ceres*.

The prize was announced publicly in the Moniteur of June 30, 1804 (issue no. 281, pg. 1276) and officially published in the journal *Connaissance des tems* of 1804 (pg. 453–456):

> *Prize of Astronomy.*
>
> The Institut National *proposes this subject: To give the theory of the perturbations of the planet Pallas, discovered by Mr. Olbers.*
>
> *The geometers have treated the theory of perturbations to an extent and with sufficient exactness for all hitherto known planets, and for those to be discovered, as long as they will be part of the same zodiac and they will not possess less eccentricity. Mercury was until to date the most eccentric of all the planets and at the same time the one with the greatest*

## SÉANCE DU LUNDI 29 PRAIRIAL AN 12.

**39**

A laquelle ont assisté M.M. **Cels, Parmentier, Bossut, Pinel, Jussieu, Tenon, Rochon, Cuvier, Desmarest, Sage, Biot, Fourcroy, Coulomb, Olivier, Lamarck, Lagrange, Berthoud, Buache, Haüy, Delambre, Lelièvre, Legendre, Ventenat, Monge, Des Essartz, Portal, Guyton, Laplace, Bouvard, Fleurieu, Lalande, La Billardière, Berthollet, Messier, La Lande eveu, Tessier, Lacroix, Ramond, Charles, Desfontaines, Prony, Carnot, Deyeux, Lacepède.**

Le Président ouvre la Séance à trois heures du soir.

Le procès verbal de la Séance précédente est lu et adopté.

La Classe reçoit les ouvrages suivans:
*Mémoire sur la force active* par M. **Louis Ponchet**, douze exemplaires;
*Mémoires de l'Académie de Berlin pour 1801*;
*Practical observations on vaccination or inoculation for the Cow-pox* by **John Redman Cox**, Philadelphia. M. Hallé voudra bien en rendre un compte verbal.
*Bibliothèque italienne*, N° XII;
*Annales de chimie*, N° 149, 30 Floréal an XII;
*Journal des mines*, Germinal an XII, N° 91.

M. **Teyssier**, capitaine retiré, envoye un Mémoire sur la quadrature du cercle.

M. d'Aval envoye un second Mémoire de M. **Grundman** sur l'électricité. Renvoyé à M. M. Charles et Haüy.

M. Cels rend compte de la distribution qui a été faite des objets envoyés par M. Le comte **Moussin-Pouschkin**. Il remet sur le bureau deux notes relatives à cet envoy.

M. Berthollet rend un compte verbal de l'ouvrage intitulé *Recherches chimiques sur la végétation*, par M. **Théodore Saussure**.

La Commission, chargée de présenter un sujet de prix pour les Sciences Mathématiques, propose la théorie des perturbations de la planète Pallas, découverte par M. **Olbers**. Ce sujet est adopté par la Classe.

M. **Sage** lit un Mémoire sur les *calopodes*.

M. M. **P.** et **J. Franque** envoyent un projet de mécanique. M. Prony est prié d'en rendre compte.

M. **Coulomb** lit un Mémoire sur le magnétisme.

M. Legendre, chargé de rendre compte de deux Mémoires envoyés par M. **Arnoul**, sous-préfet de Hasselt, l'un concernant la trisection de l'angle et l'autre un graphomètre nouveau, remet ces deux Mémoires sur le bureau, en déclarant qu'il n'y a pas lieu à faire un Rapport en forme. Le premier de ces Mémoires avoit été retiré par l'auteur, dans une des Séances précédentes.

La Classe procède au scrutin pour la nomination à la place d'Associé étranger, vacante par la mort de M. de Priestley. Les candidats étoient: MM. Klaproth, chimiste à Berlin; Piazzi, astronome à Palerme; Scarpa, anatomiste à Pavie; Wahl, botaniste à Copenhague; Mascagni, anatomiste à Sienne; Watt, mécanicien à Birmingham; Werner, minéralogiste à Freyberg; Dalrijmple, hydrographe à Londres; Jacquin, botaniste à Vienne; Smith, botaniste à Londres; Humboldt, physicien à Berlin.
M. **Klaproth** ayant réuni la majorité absolue, il est

**Fig. 3.15** The proceedings of June 18, 1804, include the offering of a prize for Pallas

*inclination; but its little mass and position at the limit of the planetary system make it unlikely to cause noticeable alterations in the motion of the other planets. Uranus, discovered 20 years ago by Mr. Herschel, is placed at the other end of the system. With little mass and an average eccentricity, it also has the smallest inclination of all known; so that all formulae used for Jupiter and Saturn are more than sufficient for this modern planet. Ceres, discovered four years ago by Mr. Piazzi, having, with quite a considerable eccentricity, an inclination of 10° 38', must be subject to strong and numerous irregularities: it seems however, that all astronomers who worked on determining those, are content with the formulae known, whose development does not give products of three dimensions of the inclination and the eccentricities. Those of five dimensions were used in the* Mecanique celeste *for one particular case, according to the formula of Mr. Burckhardt. The same astronomer has presented since then to the* Institut National *the complete and general development of the third, fourth and fifth orders; but this degree of precision would certainly not suffice for the planet Pallas, whose eccentricity is even greater than that of*

*Mercury, and the inclination of 34° 37', that means, five times greater than that of any other known planet. It is even difficult to conjecture which will be the powers and which will be the dimensions of the products permitted to neglect, and the calculations could be of such a length, and the formulae of such a complexity that they will scare the astronomers working on a similar task. This consideration determined the class of mathematical and physical sciences of the* Institut National de France *to propose this subject for the prize which it has to award at its public meeting on the first Monday of Messidor year 14. Accordingly, it invites geometers and astronomers to discuss in detail all irregularities of that theory, and not to omit one not totally known to be negligible; and how these irregularities could vary so noticeably when the elliptical elements are not exactly enough known, it is indispensable that the contenders do not confine themselves to give numerical coefficients of the equations, they must also give analytical coefficients, so that one can there insert more exact values of the mean distance, of the eccentricity, of the aphelion and the inclination, when these elements will be known better. From these analytical coefficients results another advantage, that is that the planets Ceres and Pallas, being at so little different distances of the sun, that it is even today very difficult to say which of the two is the nearest or farthest, the given formula for Pallas could, without much change, serve also for Ceres, and also for any other planet to be discovered later, and of which one thus had a more complete and reliable theory. The class hopes that the question appears interesting enough to astronomers to invest a proportionate effort to the difficulty of the subject. The prize will be a gold medal of one kilogram. The works handed in for the competition must be written in French or Latin, and will not be accepted later than the first germinal year 14. This date is obligatory.*

The announcement elicited several exchanges between Olbers and Gauss in 1804:

Olbers to Gauss July 6, 1804

*Since the* National Institute Paris *has offered a prize on the theory of the perturbations of Pallas, I do hope, my dearest friend, that you will apply for this prize.*

Gauss to Olbers July 24, 1804

*I only know from hearsay that the* National Institute Paris *has put a prize on the theory of the perturbations of Pallas. If the program is given* in extenso *in the* Moniteur *or anywhere else, please give me the number. I almost believe such an extensive work too premature. If, on the other hand, quite good meridian observations will be made this year and some of the last year become known, I may possibly be inclined to tackle such a task.*

Olbers to Gauss July 27, 1804

*The programme of the* Institut National *regarding the prize for the perturbations of Pallas, which I think is easily understood, can be found in the* Moniteur *no. 281, Messidor an 12 (= June 30, 1804). If it is not available to you, I will copy it out for you.*

Gauss to Olbers Sept. 7, 1804

*I've recently read the programme in the Moniteur concerning the prize. I feel I wouldn't have the patience for such a method to handle the perturbations of Pallas as will be required here; namely, with the help of analytical formulae into which one may just simply substitute the elements of another planet, e.g. Ceres, in order to find its perturbations. But I'll certainly attend to Pallas' perturbations in my own way, although I'll probably first see to Ceres' perturbations. For even with Ceres it seems not much is to be gained from what*

*I've taken along from the perturbations. It seems somewhat off to me that the* National Institute *itself admits that the analytical formulae, which could solve the problem once and for all, might be so complicated that they would frighten off even the most serious astronomers.*

## The Prize Is Doubled

After reiterating its initial question, and emphasizing the potential importance of the recent discovery of Juno (in 1804), the *Institute* doubled down on the prize 2 years later. This English translation of the renewed award for the Pallas question is given in *The Monthly Magazine* (1806), but no Englishman was thereby persuaded to attempt a solution, an issue considered later in this chapter.

*This consideration* [the complexity of the formula] *induced the class, two years ago, to propose this subject for the prize to be adjudged in August 1806; but the time appearing too short, and the number of planets encreased* [sic] *by the discovery of Juno by M.* [Karl] *Harding* (1765–1834), *whose eccentricity seems to be still greater than that of Pallas, and whose inclination of 13° exceeds that of all the other planets, excepting Pallas; the class thought fit to renew the subject with certain modifications, and a double premium. Accordingly, it invites geometricians and astronomers to a complete discussion of all points of this theory, so as not to omit any inequality that can possibly be of any consequence; and as these inequalities cannot be accurately determined unless the elliptical elements are perfectly well known, it is indispensably necessary that the competitors should not confine themselves to a statement of the numerical co-efficients of the equations: it is the analytical formulae which it is of the most importance to know, that the more exact amounts of the mean distance of the eccentricity, or the perihelion, and of the inclination, may be successively substituted in their stead; in proportion as these elements shall be perfected. The competitors may even dispense with giving any numerical amount, provided the analytical expressions be presented in a manner sufficiently detailed to enable an intelligent calculator to follow their development, and to reduce them into tables. From these general formulae will result another advantage; namely, that the planets Ceres, Pallas, and Juno, being at distances from the sun so nearly alike, that it cannot yet be decided with any certainty which of the three is the farthest or the nearest, the formula given for Pallas may likewise serve for the two others, as well as for any other planet that may hereafter be discovered, and whose eccentricity and inclination may be contained within the same limits. The memoirs intended for the competition must be written in French or Latin, and will not be received after the 1st of October, 1808. The prize will be adjudged on the first Monday in January 1809.*

Over the next few years, Zach, Gauss, and Olbers exchanged views on the prize:

Zach to Gauss Feb. 5, 1806

*Are you entering the contest for the Paris Prize? If you enter, who would stand a chance against you? It is an easy matter to predict the outcome. There are bound to be other contestants, one of which must be Oriani. Schubert is still recovering from his Chinese journey* [he was a member of a failed Russian expedition to China in 1805], *and the* National Institute *is not in a position to put up a 'contender with a chance' against you.*

Gauss to Olbers Aug. 6, 1810

*If the deadline of the Parisian prize question were 4 months later, I would probably apply; but there is so little time and for 10 months I have been lacking energy and mind to start any complex work.*

Gauss to Olbers Oct. 24, 1810

*If I knew that the institute would prorogue the prize question I would not be reluctant and choose Pallas otherwise I have to prefer Ceres which will be however more satisfying since soon the opposition is observable.*

Gauss to Olbers Nov. 26, 1810

*I was curious whether these six oppositions would be combined much more easily if the perturbations by Jupiter were taken into account. I admit, that such an unpleasant work holds little interest to me and that even the prize of 6,000 francs could hardly have made it more tempting as long as all observations were represented by one ellipsis.*

Once again the *Institute* was disappointed—it received no entries by 1808. Undeterred, it extended the deadline yet again to 1811 (Mercure de France 1809: 34). Two entries were received by then (it being quite likely Burckhardt was one of them), but the *Institute* was not impressed, as reported in *The Annals of Philosophy* (1816):

*Two memoirs only have been received, the authors of which have not conformed sufficiently to the intentions expressed in the annunciation of the prize. Both (especially one) have left too many analytical developments to be executed by the mathematicians who should wish to put themselves in a situation to understand and judge of the solution of the problem which they have given. They have neglected too much to come down to the level of the calculator, who should wish to form tables of Pallas, or of any other planet. The supplements sent at different times are very far from removing all the difficulties.*

*The Class perceiving by these supplements, and by the notes transmitted by the anonymous authors, that they had not time to enter into all the developments necessary; and considering further, that the same cause might prevent other mathematicians, possessed of the requisite knowledge and abilities to treat so difficult a subject, from coming forward as candidates, has thought proper to prolong the time for deciding the prize for another year. The prize will be voted in the meeting of January 1817, to the paper which shall fully satisfy the conditions above stated. The prize shall be double, that is to say, a gold medal of the value of 6000 francs (250 lb). The Essay sent to the* Institute *must be written in French or Latin, and will not be received after the 1st of October, 1816.*

The announcement in 1811 elicited one final flurry of letters that ended in 1816. The first is from Gauss to Heinrich Christian Schumacher (1780–1850), Director of Manheim Observatory from 1813 to 1815. In 1811 he was adjunct professor of astronomy in Copenhagen.

Gauss to Schumacher Jan. 6, 1811

*In case the Parisian institute poses Pallas' perturbations for the fourth time as a prize question I am not averse to reflecting on that because the matter has become interesting to me.*

Olbers to Gauss January 26, 1811

*With the greatest pleasure I have seen in the* Moniteur *that the prize for the perturbation of Pallas at 6000 franc has been prorogated until October 1, 1816, in the same measure as it will be given to the first paper which is submitted during this period and meets the intentions and requirements of the competition.*

Schumacher to Gauss June 30, 1811

*Only now I had the chance to read the* Moniteur *of June 9, 1811 and I rush to inform you – in case you do not know yet – about the details of the prizes. Regarding Pallas' perturbations two treatises were submitted – one has been neglected while the other shows a detailed understanding of analysis but was nevertheless insufficient. Thus the deadline has been extended by 5 years until June 1816 and the institute will crown the first treatise submitted within this time that fulfils the requirements. The prize is still the same, namely 6,000 francs.*

Gauss to Olbers January 8, 1816

*Should the Pallas prize be awarded or withdrawn now I would publish my theory piece by piece in the* Comm. *or in any other special paper.*

Olbers to Gauss March 7, 1816

*I am in a hurry to give you the details of the Parisian prize question just in case you have not heard yet. The deadline for the prize regarding the planets' perturbations, namely Pallas, has again been extended until 1817. But the papers have to be submitted before October 1, 1816. The question was: 'the theory of planets whose eccentricity and inclination are too great for us to be able to calculate exactly their perturbations according to the already known methods.' The class does not require a numerical application but analytical formulae only, but in such a way that someone good at arithmetic is able to apply them either to the planet Pallas or to any other of the new planets or any planet yet to be discovered. Only two treatises were handed in but whose authors had not sufficiently considered the pronounced intention of the prize question.*

*They failed to condescend to the point of view of a calculator, wishing to compile tables for Pallas or any other planet. Since the class could see from the appendices and the submitted notes of anonymous senders that they did not have enough time to take into account all the necessary developments and they became aware that at the same time there may be other mathematicians too, possessing the capability and skills to deal with this difficult matter, were deterred from trying to win the prize, the award has been postponed till January 1817. – The prize has been doubled, a gold medal and 6,000 Francs.*

In the event, there were no entries, so the prize lapsed. The subject of Fermat's Last Theorem, formulated in 1637 by Pierre de Fermat (1607–1665) was also a favorite topic for the *Institute*, and one that dragged on even longer than the Pallas prize question. In 1823 Olbers urged Gauss to compete for this prize as well, but Gauss replied that he had little interest in it. A gold medal and 3000 francs was finally awarded in 1857 to Ernst Kummer (1810–1893), but the first successful proof of the theorem was not published until 1995.

To put the currency in context, a British pound was—in the late eighteenth century—worth about 23 francs. So the Pallas prize was worth about 250 £, or roughly 15,000 £ (US $21,000) in 2016 currency. A head footman of the time might earn 10 £ a year, so the prize offered by the Academy would pay the salary of such a person for 25 years! The average income of a teacher or farmer was around 40 £ a year (Williamson 1982). Another way of comparing French and British currencies is the amusing fact that a dozen French lessons would cost an Englishman one pound.

## The Prize and the Perturbation Calculations: A Timeline

*March 1802:* Pallas discovered.
*June 1804:* The *Institute* offers a prize for a theory on the perturbations of Pallas.
*1806:* The closing date for the initial offering of the prize.
*October 1808:* The closing date for the second offering of the prize.
*December 1810:* Gauss makes a numerical integration of the orbit of Pallas from 1803–1811.
*1811:* Two mathematicians submit entries to the *Institute* in hopes of winning the prize.
*January 1811:* The *Institute* extends the prize date by 5 years to June 1816.
*August 1811:* Gauss begins work on a theory of the general perturbations of Pallas.
*April 1812:* Gauss begins the second iteration for the solution and publishes his anagram.
1812: Damoiseau writes a paper on the variations of the elliptical elements of Pallas and Ceres
*July 1813:* The second iteration is completed.
*1814:* Gauss's assistant Nicolai begins work on the perturbations of Pallas by Saturn.
*March 1816:* The *Institute* extends the prize deadline a final time to October 1, 1816.
*January 1817:* As no entries were received by the *Institute*, the prize lapses.
*1817:* General tables for the motion of Pallas are completed by Gauss.
*1843:* Gauss writes for the last time about his perturbation work on Pallas.

The 1812 entry here introduces a dark horse in the saga of orbital work on Ceres and Pallas. It is by the Frenchman Marie Charles Théodore de Damoiseau de Monfort (1768–1846), Captain-Lieutenant of the Royal Navy Brigade. Damoiseau was elected a member of the Lisbon Academy of Sciences in 1806 during his tenure as assistant director of Lisbon Observatory, and 6 years later he published a French-language paper about Ceres and Pallas in the Academy's *Memoirs,* an otherwise entirely Portuguese-language publication (Saraiva 2008). The introductory section of his paper is given in Chap. 11.

## Lalande and the Asteroids

Lalande was clearly captivated by the asteroids, ever since he attended the astronomical congress organized by Zach in 1798, where the existence of an unknown planet between Mars and Jupiter was surely a topic of discussion (Chapin 1983). Even though no gold medal was awarded for asteroid work, the *Institut de France* did award a prize established by Lalande for the progress of astronomy, and all the asteroid discoverers were recipients. Its origin and a complete list of winners can be found in Maindron (1881: 66–69).

*The order of the government of the 13 floreal year 10, which authorises the* Institut National *to accept a donation of 10,000 Fr offered by Mr. Lalande, stating, art. II, that 'Accordant to the intentions of the donor, the annual profit of the capital will be used by the* Institut *to award each year a gold medal of such a weight the revenue will allow, or the value of that medal, to that person who, in France or elsewhere, excluding the members of the* Institut, *will make the most interesting observation, or the most useful memoir for the progress of astronomy'* (Connaissance des tems, 1804, pg. 455).

The awarding of the first two prizes, for the years 1803 and 1804, was noted in a manuscript by Charles Messier (undated). For his own observations of Ceres, see Messier (1801/1802).

*The 1st was awarded on the 1st Messidor of the year 11 (June 20, 1803) to M. Olbers for having discov. the Planet Pallas in the year 10. The second prize to M. Piazzi, for having discov., on January 1, 1801, the Planet Ceres.*

The citation for Piazzi's award in 1804 was for the development of his star catalog; it noted in passing that the discovery of Ceres Ferdinandea occurred while he was engaged in working on the catalog. Karl Harding was awarded the prize in 1805 for the discovery of Juno, and Olbers won again in 1807 for Vesta's discovery. Gauss won the Lalande prize in 1810. Here is how Olbers' first award was announced in *The Moniteur* (1803):

*PRIX D'ASTRONOMIE. The government's rule, article II, of the date 13 floréal year 10, which authorizes the* Institut National *to accept the gift of 10,000 Francs offered by the citizen Lalande says that according to the donator's intentions that the annual product of this capital should be employed by the* Institut National *to give each year a gold medal of the weight allowed by the revenue or the value of that medal to the person who, in France or elsewhere, except the members of the* Institut, *has made the most interesting observation or the most useful memorandum for astronomy's progress. On the commissioners' report who have been appointed for this cause have awarded the prize to Guillaume Olbers, medical doctor, member of the* Académie impériale des Naturalistes *and correspondent of the* Royal Society of Goettingen *for his discovery in the course of the year 10 of the planet which has been given the name Pallas Olbersiana by all astronomers. Before this discovery Mr. Olbers was already known to astronomers by his dissertation on the easiest and most comfortable method to calculate a comet's orbit and for the exact calculations and observations he delivered for Ceres' or Piazzi's planet's rediscovery, which he actually achieved and announced first.*

The Institut went further in 1810, by electing Olbers a member. France was not alone in offering a gold medal for asteroid work. In November 1824, the Royal Astronomical Society in London proposed the following prize question, but the medal was never awarded:

*The gold medal for approved formulae, for determining the true place of either of the four newly discovered planets, Ceres, Juno, Vesta and Pallas; within such limits as the Council may think sufficiently correct for the present state of astronomy; such formulae in each case to be accompanied with comparisons of the observed places at various periods. All answers must be received before the 1st of February, 1827* (The London Journal of Arts and Sciences, 1824).

The Lalande medal for 1814 was awarded to the young Pierre Daussy (1792–1860) for his work on the perturbations and elliptical elements of Vesta; see the next book in this series for the details.

The fact England did *not* bestow medals for asteroid discoveries was lamented in an article in *The Critic* (1857) when describing the history of the Royal Society's Copley Medal. After listing many winners, we find this passage:

> *Yet we still miss the discoverers of Ceres, Pallas, Juno and Vesta...we must confess that we should have been better pleased at having to point out the impartial justice with which every great 'contribution to science' had been recognised, than to catalogue these grave and unaccountable sins of omission.*

## The Lack of Theoretical Work in England

Newton's invention of calculus in England was one of the greatest mathematical achievements of all time, and certainly the foremost by any English mathematician. But the independent development of calculus on the Continent by Leibniz, while beneficial for the long-term development of calculus, left English mathematicians behind because of its very different formulation. Even though the state of affairs in mathematics in England around 1800 is the prime concern here, a century earlier its health was scarcely any better (Newton alone excepted). The lack of theoretical versus practical achievements was commented upon by the English astronomer Henry Crozier Plummer (1875–1946):

> *If the eighteenth century appears a dull and unenterprising period in English science, there are reasons for it. Newton established no school, and his country took no significant part in the triumphs of Euler, Lagrange and Laplace. The Royal Society passed into the hands mainly of dilettanti, and the English universities equally failed to see the opportunity for scientific studies. But the record is far from being one of unrelieved gloom. It is only necessary to remember that astronomy at least had two giants in James Bradley* [1693–1762, Astronomer Royal] *and William Herschel* (Plummer 1943: 92).

Plummer goes on to give the names of such instrument makers as James Short (1710–1768), John Bird (1709–1770), John Dollond (1706–1761), and Jesse Ramsden (1735–1800), who made the great circle Piazzi used to discover Ceres. So, "Admittedly England lagged behind in the field of mathematics, but compensated for this grievous error by the quality of the artists, such as those named" (Ibid.)

For a contemporary look, here is the view from 1802 by the French mathematician Edme Marie Joseph Lemoine d'Essoies (1751–1816), casting his eye back to the previous century. It includes a reference to Jacob (or James) Bernoulli, whose role in disseminating Leibniz's differential calculus has been examined by Wahl (2014):

> *The methods which Newton discovered, and which enabled him to investigate all the great questions in mechanics and astronomy, were for some time a hidden treasure, of which he*

*was the sole proprietor. And it is singular, that the English geometricians knew nothing of the new calculi, except what they collected in the* Acta Eruditorum *of Leipzig. James* [Jacob] *Bernoulli (a Swiss) was the first geometrician whose eyes were opened, and who began to second efforts of Leibniz* (Lemoine 1802).

Were English astronomer/mathematicians too complacent to work on the orbital problems posed by the discoveries of Ceres and Pallas? To posit the question so boldly invites the insidious French trap, the *question mal posée*—that is, to distort the problem by oversimplifying it. The answer lies not just in mathematics, but in philosophy:

*The brilliant achievements of British mathematicians, astronomers and physicists under the influence of Isaac Newton were followed by a long period of comparative inactivity. This was largely due to the fact that, during a considerable part of the eighteenth century, members of the British school were, more or less, out of touch with their continental contemporaries. A free exchange of views is essential to vigour and, the more varied the outlook and training of those concerned, the more fruitful is the intercourse. The effect of this isolation, moreover, was intensified by the manner in which English writers strove in their demonstrations to follow Newtonian forms. If Newton, in his* Principia, *confined himself to geometrical proofs, it was because their validity was unimpeachable; and, since his results were novel, he did not wish the discussion as to their truth to turn on the methods used to demonstrate them. But his followers, long after the principles of the calculus had been accepted, continued to employ geometrical proofs, whenever it was possible, even when these did not offer the simplest and most direct way of arriving at the result. In short, we may say that, in the course of English mathematical science, the last seventy years of the eighteenth century form a sort of isolated backwater* (Ward and Waller 1907–1916: 37–38).

To some extent, the malaise just alluded to can be traced to the philosophical works of George Berkeley (1685–1753; Fig. 3.16), who had a great influence on British thought during the enlightenment. His views were also well known in Germany, as evidenced by the reference to Berkeley by Karl Seyffer in his 1801 treatise about Ceres (see *Discovery of the First Asteroid, Ceres*). Berkeley (1734: 93) pointedly asks toward the end of *The Analyst*:

*Whether mathematicians, who are so delicate in religious points, are strictly scrupulous in their own science? Whether they do not submit to authority, take things upon trust, and believe points inconceivable? Whether they have not their mysteries, and what is more, their repugnancies and contradictions?*

One consequence of the controversy begun by Berkeley was a dramatic increase in the publication of English textbooks on calculus, especially between 1736 and 1758 (Guicciardini 1989: 58). During this period 12 textbooks on fluxions were published. Berkeley's attacks on Newtonian mathematics and the parallel developments on the Continent that advanced mathematics in ways that differed from the English (i.e., Newtonian) method worked in tandem to create the 'isolated backwater.' Elizabeth Garber of the State University of New York has argued that the concept of such an 'isolated backwater' is unfounded, but she does draw a distinction between British and Continental mathematics:

**Fig. 3.16** George Berkeley. National Portrait Gallery, London

> *Mathematicians* [in England] *were neither isolated nor cramped. However, during the eighteenth and early nineteenth centuries, mathematics served very different cultural purposes in Britain than those on the continent. As the foundation of a 'liberal education' at Cambridge, mathematics served a social and cultural function unknown in continental Europe. The type and level of mathematics taught was directed towards developing standards of logical consistency rather than technical competence* (Garber 1998: 188).

In France, the intellectual *milieu* of the mathematicians was dominated by Descartes, for whom clarity and distinctness were warrants of truth (Graham and Kantor 2009). It was not so in England, so when the discoveries of Ceres and Pallas demanded a novel mathematical approach, there was no one in England capable of rising to the challenge. More generally, it is also crucial to recognize the inherent attributes of the Germans, the French, and the English. This triad, which was famously recognized by the philosopher Georg Wilhelm Friedrich Hegel (1770–1831), was examined in great detail in *The British Quarterly Review*. The paper may have been written by the *Review's* editor, Robert Vaughan (1795–1868). Casting the argument in an astronomical perspective, the author notes that England is the most

> ... *comfortable* [country in the world], *so that if we wished to give some starveling inhabitant of Saturn or Georgium Sidus a fair idea of what is to be done in the way of*

*making things 'snug' in the planet Terra, we should direct him to an English hotel* (The British Quarterly Review 1851: 350).

In remarks that can surely be associated with William Herschel, who was born in Germany, the author notes that the English were colonized by peoples from Germany, and that "… German thought, through various channels, does exercise a decided influence on the opinions and principles of at least the three other leading nations of the world—France, England, and America" (ibid: 356). In tracing the flow of this thought from Germany, he says "… the ideas which wound a slow and tortuous course through long sentences of German introversion, which were clipped and cut down and suited to the powers that be, in English, will come out clear, terse, and sparkling in French" (ibid, 340). This concept may be kept in mind as we consider some of the philosophical underpinnings that influenced the development of mathematics in these countries. But first we take a detour to Scotland.

## Scottish Common Sense Realism: Brougham, Stewart, and Reid

It is important to understand the fluid nature of disciplines at the time. The case of Dugald Stewart (Fig. 3.17) is a case in point. His father was a Professor of Mathematics at the University of Edinburgh, a post Dugald Stewart assumed on his

**Fig. 3.17** Dugald Stewart painting by Sir Henry Raeburn. National Galleries Scotland

father's death. Then in 1785 he swapped places with Adam Ferguson (1723–1816)—Stewart took the Chair of Moral Philosophy, and Ferguson took Mathematics. In his new role Stewart studied the work of Locke within the framework of the Newtonian methodological heritage (Mortera 2010). Despite the reputation of Locke and Stewart's effort to support his work, not all Scottish philosophers were of the same mind. Thomas Reid (1710–1796), who was also a Professor of Moral Philosophy (at the University of Glasgow until 1781) argued strongly against the Theory of Ideas advocated by Locke and Descartes.

The links between Stewart, Henry Brougham, and Thomas Thomson have been noted in *Early Investigations of Ceres and the Discovery of Pallas*. Stewart, Reid, and Brougham were proponents of what became known as Scottish Common Sense Realism (Olson 1975), which figured largely in the German 'Republic of Letters' (Kuehn 1987: 5). It is this philosophy that was promoted in Poland by the asteroid observer Jan Sniadecki (1821). He was a staunch defender of eighteenth-century heritage (against Immanuel Kant) and "remained firmly in the empiricist tradition, seeing all knowledge as gained through the sensations" (Bujarksi 1972: 7). The basic tenet of this philosophy has a proud antecedent. "A modern physicist on examining Galileo's works is surprised to find how little experiment had to do with the establishment of the foundations of mechanics. His principal appeal is to common sense" (Pierce 1891: 163).

It was Daniel Bernoulli, the nephew of Jacob, who applied probability theory to moral, political, and economic subjects. Another scion of the Bernoulli family, Nicholas (1709), applied mathematical probability to legal questions. Such catholic use of probability theory was all the rage in the eighteenth century, a mindset best codified by the French philosopher Claude Buffier (1661–1737): "The most natural and general use of probability is to supply the place of truth… Probability, however, must make various impressions on us, according to the subject and occasion" (Buffier 1724; 1780 English edition: 121).

Buffier anticipated both Reid's common sense approach, and the application of probability to human experience promulgated by Daniel Bernoulli (Marcil-Lacoste 1982). Buffier (1724) defined common sense as that disposition which nature has placed in all or most men so that they can form a common and uniform judgment with respect to objects (such as Ceres and Pallas) different from the internal sentiment of their own perception. Buffier claims that these truths of common sense possess the highest probability of being the correct ones. Common sense certainly dictated that Ceres and Pallas were, as Herschel discerned, *not* planets, and it is this perception which eventually became the uniform judgment of astronomers.

The Scottish contribution to the problems of British mathematics cannot be overestimated. Nearly all renowned British mathematicians of this time were both Scottish and ardent Newtonians. Scottish mathematicians are seen to have presided over the era in which British mathematics fell behind because of their collective adherence to geometry (Withers and Wood 2002: 150). However, as Wickman (2009: 121) has perceptively pointed out, "Geometry in eighteenth-century Scotland was a cultural rather than simply a mathematical battleground, and the mistake in perceiving it as retrograde is a function of seeing it too narrowly."

## Contemporary Views on the State of Astronomy in Great Britain

In *Early Investigations of Ceres and the Discovery of Pallas* a paper by Brewster (1802a: 287) was quoted regarding the recent discovery of Pallas. In the final section of the paper dated April 23, 1802, he laments the state of astronomy in England and Scotland:

> *We cannot conclude this paper without remarking the great progress of astronomical science, which has lately taken place. The discovery of three planets within the space of about twenty years, affords some reason for supposing that we are but little acquainted with the system to which we belong; and that some future period will disclose to our view, a plan more extensive and magnificent than the imagination of man has, as yet, conceived. But while we thus rejoice in the rapid progress of the noblest of sciences, we are concerned to state, that this progress is chiefly owing to the assiduity and genius of continental philosophers. It is a disgrace to our own country, that practical astronomy is so much neglected by men in a private capacity, and that so little has been done in Britain to forward this branch of physical science. We except, however, from this remark, the labours of Dr. Herschel, as being the result of regal munificence, and not of public spirit. But to Scotland it is most particularly applicable. There is scarcely in this country an observatory of any importance, and some of those which are honoured with the name, are left in the most deplorable and ruinous condition. Who has not lamented, and lamented, too, with indignation, the ruinous state of the observatory in the metropolis of the kingdom? – fitted scarcely for the amusement of ladies and children, and far less for the improvement of Astronomical Science. Let it not be supposed, however, that this neglect of practical Astronomy arises from the indolence and inability of our countrymen. It is owing, without doubt, to the want of a Court to countenance and reward her deserving sons, and to that defect of literary talent and public spirit, which is so strikingly visible in the nobility of our land.*

Brewster alludes to the observatory situation in Edinburgh, a sore point that goes back to 1785 when, due to the influence of Henry Dundas, the Regius chair in astronomy was created. Robert Blair (1748–1828) was appointed, but he would not teach because he had no observatory and had none because the town council of Edinburgh refused to fund one since a private group was preparing to build its own observatory in the capital on Calton Hill. Blair kept the astronomy chair for 43 years without doing any astronomy (Emerson 2008:334) When Brewster was writing, the Calton Hill observatory was being leased to opticians, actual astronomy not having been done there since 1788. Astronomy in England was more robust, but mathematics was not.

The state of mathematics in England may be gauged by a contemporary assessment of a two-volume book published in 1814, which was directed at the British student. *An Easy Introduction to Mathematics* was written by Charles Butler (1750–1832; 1814) and critically reviewed in *The Monthly Review* (1816: 316). One passage is of particular relevance to the analysis in this section:

> *The chapter on Numbers seems to have been principally drawn from an old edition of Bonnycastle's Arithmetic* [John Bonnycastle, 1751–1821]: *but all the more important numerical theorems which we owe to Fermat, Euler, Waring* [Edward Waring, 1736–

> 1798], *Lagrange, &c. are totally omitted; which certainly could never have happened if the author had been aware of the distinct treatises on this subject by Gauss and Legendre.*

So the book, which was supposed to offer students a modern introduction to mathematics, was actually oblivious of the Continental advances. More generally, this synopsis from 1807 of what a student should expect from an education at Cambridge is telling:

> *The purpose for which a young man is taught natural philosophy at the University of Cambridge, is at once to invigorate the intellect by an exercise specially adapted to strengthen and subtilize its operations, and to furnish that kind of information respecting the qualities and affections of objects of daily occurrence, as shall conduce to the pleasure, advantage, and respectability of his future life* (*The Critical Review* 1807: 50).

Thus an advanced education at England's foremost university is described in terms that might be most appropriate for a 'finishing school' to prepare men for the pleasures of a refined life. The use of *subtilize*, which means 'to elevate in character', clearly has precious little to do with training in mathematical or astronomical research.

A writer (possibly Brougham) in *The Edinburgh Review* (1819: 393) elucidated the reasons for the sad state of English mathematics in the review of an astronomy book by Robert Woodhouse (1773–1827). In a lengthy digression from the review itself, the reviewer attacked the Royal Society: "It is indeed much to be lamented that this Institution, intended for the advancement of science, should hold out so little encouragement to Mathematical learning." He also fingered as a main culprit

> *...the very extensive dissemination of* general knowledge, *which is so much the case over the whole of this kingdom. Literature and the Arts give abundant occupation to the mind of a man of liberal curiosity, and leave less inducement to attach himself to abstract studies ... Now-a-days, a man must be conversant in chemistry, mineralogy, entomology, modern languages, history, politics, and fifty hard-worded studies beside – so that, in fact, unless he chuse to devote himself almost exclusively to Mathematics, he has little chance of aspiring to discovery, or even to eminence, in that pursuit.*

In 1821, Stewart linked Leibniz with what he called "... the continental philosophy ...", suggesting that a habit of "... deference to the authority of Leibniz ...", combined with a short-sighted mistrust of honest John Locke (Fig. 3.18), might be responsible for the "... striking contrast between the characteristical features of the continental philosophy... and those of contemporary systems which have succeeded each other in our own island" (Rée 2009: 25).

That men of science on the Continent might have looked askance at Locke, the most influential Enlightenment philosopher England produced, is an intriguing insight in the thinking of the time. His 1690 book, *An Essay Concerning Human Understanding* became, in the words of the American historian Becker (1932), "the psychological gospel of the eighteenth century." Locke was often mentioned, by the English, in the same breath as Newton, as is well attested by some lines penned by David Garrick (1717–1779; 1755):

> *In the deep mines of science though Frenchmen may toil,*
>
> *Can their strength be compar'd to Locke, Newton, and Boyle?*

**Fig. 3.18** John Locke painting by Sir Godfrey Kneller. State Hermitage Museum, St. Petersburg

The approaches of Leibniz and Newton were certainly shaped by their differing intellectual environments. The concept that different regions would produce genii of different natures had long been recognized in England (Barclay 1633: 26):

> *That every age almost, hath a particular Genius different from the rest; that there is a proper Spirit to every Region, which doth in a manner shape the studies, and manners of the inhabitants, according to itself. That it is worth the labor, to find out those Spirits.*

The 'Region' inhabited by the English has for centuries been characterized as unintellectual. Here is the assessment of Sir Lewis Bernstein Namier (1888–1960), one of the most eminent British historians of the first half of the twentieth century, writing in a book about the late eighteenth century: "... the characteristically English attempt to appear unscientific springs from a desire to remain human. What is not valued in England is abstract knowledge as a profession" (Namier 1961: 14–15) Nothing could be considered more abstract than a knowledge of mathematics.

The year Juno was discovered, a broadside was fired at the state of science and mathematics in England by Rev. John Toplis, Fellow of Queen's College, Cambridge; a decade later he would publish an English translation of the first book of Laplace's *Mecanique Celeste.* Toplis (1805: 26–29), writing in late 1804, contrasted what should be taught at Oxford with what was actually being taught.

*We seem, as a nation, for this last half century, to be sunk into a great degree of supineness with respect to the sciences, regardless of our former fame...I cannot forbear making a few remarks upon the method of study made use of in the university of Oxford and the principle seminaries of this kingdom...Are the sciences which 'weigh the sun and his revolving stars;' and measure the velocity of light and distance of the fixed orbs;...to be utterly neglected, for barren unceasing attempts at imitating the style of Xenophon, Cicero, or Horace?*

The dearth of mathematical knowledge amongst British astronomers and instrument makers was examined by James David Forbes (1809–1868; 1856: 848), Professor of Natural Philosophy at the University of Edinburgh:

*Until very recent times, with few exceptions, our observers, however industrious, have merely used mechanically the apparatus put into their hands by intelligent and conscientious artists, who yet, never having occasion to apply their own handiwork to the purposes for which it was made, could not be expected to detect deficiencies of construction, nor to possess the mathematical knowledge required to remedy them.*

More generally, Forbes (1856: 806) uses a quote from d'Alembert, one of the greatest of all mathematicians, to place the state of affairs in England in proper context:

*Some idea may be formed of the negation of mathematical talent in Britain during the later portion of the last century, when we find d'Alembert declaring, in 1769, that if an Englishman is to be elected one of the eight foreign associates of the Academy of Sciences, he will vote for* [Philip] *Earl Stanhope* [1714–1786], *as the best mathematician there, as he believes, not having read any of his works! If the choice was to be free, he should prefer M. de Lagrange!! A more cutting, though unintentional satire on the state of Mathematics in this country could not have been written.*

Forbes (ibid.) agrees with the view put forward in Ward and Waller (1907–1916: 37–38) that the prime reason for this lack of knowledge was the use on the Continent of the

*...Leibnitzian notation of differentials which was absolutely unfamiliar in England,* [which] *deterred almost every one even from perusing the writings of Clairaut* [Alexis Clairaut, 1713–1765] *and d'Alembert*], *Lagrange and Laplace.*

For their part, the French were acidly derisive of the English in this regard. The mathematician François-Joseph Servois (1768–1847; 1814) wrote "...the Leibnizian notation for the differential calculus ought to be preserved. Leave the dotted letters to the English; let us preserve accents for the useful work of extending our alphabets."

The notation of differentials, "... which could alone break down the barrier between British and foreign mathematicians, was first introduced at Cambridge by the efforts of John Herschel and Dean Peacock [George Peacock, 1791–1858] about 1816." (Garber 1998: 190) They, along with Edward Thomas Ffrench Bromhead (1781–1855), were the core of the Analytical Society which published its first and only book in 1813 (Becher 1980). It was written by Charles Babbage (1791–1871) and William Herschel's son John (1792–1871). The state of mathematics at that point may be gauged by the reaction to this book by Bromhead, who wrote that the Memoirs "... were too profound to do us any good and not one mathematician in

1000 can understand them" (Cannell 2001: 38). To remedy the situation, Babbage, Herschel, and Peacock published an English translation of Sylvestre Lacroix's (1765–1843; 1816) calculus book, followed in 1820 by a book of worked examples in the new notation. Credit must also be given to Charles Hutton (1737–1823), Peter Barlow (1776–1862), and William Wallace (1768–1843; Professor of Mathematics at the University of Edinburgh), who introduced the French calculus to British mathematicians.

The year 1830 was the nexus of a heated debate over the decline of science in Britain; Babbage and Brewster were at the center of it (Koutalis 2014). It was in that year Babbage wrote his famous book *Reflections on the Decline of Science in Britain*. But England was not the only nation in a scientific doldrum. Baron von Zach, residing in Paris at the time, gives us a penetrating insight that takes a riff off of Babbage's controversial book.

> *The state of astronomy is quite pitiful in France, it is absolutely nil. It is in Paris that there would be need for another Babbage in order to write some* Reflections on the Decline of Science in France; *for really, sciences are at the lowest ebb. I hope it is not so in England; at least observations are made in Greenwich, in Cambridge, in Dublin, while in France, nowhere* (Zach 1830).

The view from England on the subject of mathematics was equally dire, as commented on by John Herschel in 1830:

> *In England, whole branches of Continental discovery are unstudied, and, indeed, almost unknown even by name. It is in vain to conceal the melancholy truth. We are fast dropping behind. In mathematics we have long since drawn the rein, and given over a hopeless race* (Quoted in Sime 1900: 97).

This ignorance by the English of "branches of Continental discovery" did not go unnoticed by others in that decade:

> *Our Continental neighbours the Germans and the French, by no means neglecting to investigate the works of nature, and certainly gaining ground upon us in the processes of manufacture, have thought it also worth their while to study the philosophy of history, the philosophy of the fine arts, and the history of philosophy itself, of which three great branches of knowledge we scarcely possess even the idea* (*The Quarterly Review* 1838: 501).

The anonymous reviewer in *The Quarterly Review* contrasts this absence of intellectual rigor to the English preference for the solution of more practical problems, an attitude he makes clear does not motivate the greatest scientists, French, German, or English:

> *Astronomy may occur to our readers. It is well known that an accurate knowledge of the motions of certain satellites is useful to the masters of vessels. But was this the object which animated Laplace in his profound mathematical studies, or was it this which pointed our Herschel's telescope at the Georgium Sidus or the binary stars?* (*The Quarterly Review*, 1838: 501)

Specifically in the field of asteroid studies, a lot of work was done in the first few decades of the nineteenth century on the theories of the four asteroids. This was accomplished—first and most famously—by Gauss (1809) in Germany, and later extended by Giovanni Sante Gaspero Santini (1787–1877) and Francesco Carlini (1783–1862) in Italy, Ernst Schubert (1813–1873) in Germany and later the United States, Friedrich Bernhard Gottfried Nicolai (1793–1846) in Germany, and Pierre Daussy in France.

Again, England was notably absent (Airy 1833: 156-159). In his survey of astronomy of the nineteenth century, George Biddell Airy (1801–1892; Fig. 3.19) gave a very curious summary of asteroid studies to date, taking several backhand swipes at the English involvement, or lack of it. While giving the discovery circumstances, Airy (who would become Astronomer Royal just 2 years later) completely omits any mention of William Herschel and his physical study of the asteroids, concentrating almost entirely on orbital and perturbation calculations.

Speaking of Ceres, he notes that after its recovery in 1802 "... it has since been regularly observed at most observatories (at least the continental ones)." (Airy 1833: 156) On the subject of mathematical work on the four asteroids, he further notes that "... the elements of all these orbits have been successively improved (entirely by the Germans)" (Airy 1833: 157). Thus, in two parenthetical asides and one glaring omission, he totally dismisses all English involvement, studiously

**Fig. 3.19** George Biddell Airy painting by John Collier. Royal Museums Greenwich

calling them planets so as not to introduce Herschel by way of explaining the introduction of the word 'asteroid.'

Airy's deliberate use of 'planet' instead of 'asteroid' can be characterized as an 'abusive anachronism,' one of five types of anachronism identified by Thomas M. Greene (1926–2003), Professor of English and Comparative Literature at Yale University. In his typology, Greene (1986: 221) characterizes such an anachronism as one which single-mindedly adheres to a model that clashes contemptuously with contemporary usage. Few people by 1833 could really regard the four tiny objects between Mars and Jupiter as being in the same league as planets such as Jupiter, or even little Mercury. The word 'asteroid' was already in widespread use.

## References

Airy, G.B., 1833, *Report on the Progress of Astronomy During the Present Century, section VII.* British Association for the Advancement of Science, 125-189. John Murray, London.

Antonello, E.: 2014, *Bonaparte and the astronomers of Brera Observatory.* http://Goldgararxiv.org/ftp/arxiv/papers/1405/1405.6841.pdf

Barber, B.: 1961, Resistance by Scientists to Scientific Discovery. *Science* 134, 596-602.

Barclay, J., 1633, *The Mirror of Minds.* Translated into English by Thomas May from the original Latin ed. of 1614. Printed by I.B. for Thomas Walkley, London.

Becher, H.W.: 1980, Woodhouse, Babbage, Peacock, and Modern Algebra. *Historia Mathematica* 7, 389-400.

Becker, C.L.: 1932, *The Heavenly City of the Eighteenth-Century Philosophers.* Yale University Press, New Haven.

Berkeley, G.: 1734, *The Analyst, or a discourse addressed to an infidel mathematician.* J. Tonson, London.

Bernoulli, N.: 1709, *De usu artis conjectandi in jure.* Johannes Conradi a Mechel, Basel.

Bernoulli, D.: 1735, Recherches physiques et astronomiques sur le problème…Quelle est la cause physique de l'inclination des plans des orbites des planètes. *Receuil des pieces qui ont remporté le prix de l'Académie Royale des Sciences* 3, 93-122 (French version), 123-144 (Latin version).

Bertrand, J.: 1869, *L'Académie des Sciences et les académicians de 1666 à 1793.* J. Hetzel, Paris.

Bode, J.: 1802b, Letter to Olbers, 15 May. Bremen University Archives.

Bossut, C. & Legendre, A.: 1787, Report to the Académie des Sciences on *Introduction à l'étude de l'astronomie physique* by Jacques Cousin. Quoted in R. Hahn (2005), pg. 146.

Brewster, D., 1802a, Notice concerning another NEW PLANET discovered by Dr. Olbers of Bremen, in March 1802. *The Edinburgh Magazine*, 19, New Series, April, 287.

Brewster, D.: 1831, *The Life of Sir Isaac Newton*, vol. 1. J. Murray, London.

Buffier, C.: 1724, *First Truths and the Origin of our Opinions Explained.* Translated from *Traite des Premières vérités* into English in 1780. J. Johnson, London.

Bujarski, G.T.: 1972, Polish Liberalism, 1815-1823. *The Polish Review* 17(2), 3-37.

Cannell, D.M.: 2001, *George Green: Mathematician and Physicist, 1793-1841. Second Edition.* Society for Industrial and Applied Mathematics, Philadelphia.

Carlyle, T.: 1888, *History of Friedrich II of Prussia*, vol. 5. Chapman & Hall, London.

Carr, R.: 2002, Challenge and Response. *The Spectator*, 25 May issue, 288(9068), 45.

Chapin, S.: 1968, The Academy of Sciences During the Eighteenth Century: An Astronomical Appraisal. *French Historical Studies* 5, 371-404.

Chapin, S.: 1983, What Better Way to go to Gotha: Lalande, Blanchard, & the Balloon. *Griffith Observer*, November issue, 2-19.

Cummings, S. and Parker, A.: 2015, D'Alembert, Lagrange, and Reduction of Order. *Convergence* 12. Online journal of the Mathematical Association of America.
Cunningham, C.: 2007, Euler at 300. Mercury 36(4), 9.
Damoiseau, M.C.: 1812, Memoire sur les variations séculaires des elémens elliptiques de Pallas et de Cérès. Memorias de Mathematica e Phisica da Academia R. das Sciencias de Lisboa, Vol. 3, Part 1, 15-67.
Dumarsais, C.: 1743, *Nouvelles libertés de penser.* Piget, Amsterdam.
Ellis, H.: 2014, Enlightened Networks: Anglo-German Collaboration in Classical Scholarship, in *Anglo-German Scholarly Networks in the Long Nineteenth Century* (Heather Ellis and Ulrike Kirchberger, eds.). Brill, Leiden, The Netherlands.
Emerson, R.: 2008, *Academic Patronage in the Scottish Enlightenment.* Edinburgh University Press, Edinburgh.
Euler, L.: 1749, Recherches sur les irregularités du mouvement de Jupiter et de Saturne. In *Pièce qui a remporté le prix de l'Académe Royale des Sciences*, pg. 1-123. Martin, Coignard & Gurein, Paris.
Feke, J.: 2014, Meta-mathematical rhetoric: Hero and Ptolemy against the philosophers. *Historia Mathematica* 41(3), 261-276.
Ferrone, V.: 1997, The Man of Science. In *Enlightenment Portraits* (Michel Vovelle, ed.), pg. 190-225. University of Chicago Press, Chicago.
Fontenelle, B.: 1699, Préface de utilité de mathématiques (The Utility of Mathematics). *Academie Royale des Sciences*, Paris. For English translation, see Marsak, L.:1961, Bernard de Fontenelle: The Idea of Science in the French Enlightenment. *American Philosophical Society Transactions*, new series 49 (7), 1-64.
Forbes, J.D.: 1856, Dissertation Sixth. *Exhibiting a General View of the Progress of Mathematical and Physical Science, principally from 1775 to 1850.* Adam and Charles Black, Edinburgh.
Frängsmyr, T.: 1990, The Mathematical Philosophy, in *The Quantifying Spirit in the 18$^{th}$ Century*, pg. 27-44. (T. Frängsmyr, J.L. Heilbron and R. Rider, eds.). University of California Press, Berkeley.
Garber, E.: 1998, *The Language of Physics: The Calculus and the Development of Theoretical Physics in Europe 1750-1914.* Birkhäuser Boston, New York.
Garrick, D.: 1755, On Johnson's Dictionary. *Public Advertiser*, 22 April.
Gauss, C.: 1802, *Monthly Correspondence*, June issue, 597.
Gauss, C.: 1809, *Theoria motus corporum coelestium in sectionibus conicus solem ambientium Theory of the Motion of the Heavenly Bodies Moving about the Sun in Conic Sections).* Perthe and Besser, Hamburg.
George, A.: 1938, The genesis of the Académie des Sciences. *Annals of Science* 3, 372-401.
Glanville, J.: 1668, *Plus Ultra.* Reprint in 1958. Scholars' Facsimiles Reprints, Gainesville, Florida.
Goertz, G., Jones, B. and Diehl, P.: 2005, Maintenance Processes in International Rivalries. *Journal of Conflict Resolution* 49 (5), 742-769.
Goldgar, A.: 1995, *Impolite Learning: Conduct and Community in the Republic of Letters, 1680-1750.* Yale University Press, New Haven.
Gower, B.: 1987, Planets and Probability: Daniel Bernoulli on the Inclinations of the Planetary Orbits. *Studies in History and Philosophy of Science Part A* 18(4), 441-454.
Graham, L., and Kantor, J.-M., 2009, *Naming Infinity.* Harvard University Press, Cambridge.
Gray, J.: 2006, A History of Prizes in Mathematics. In *The Millennium Prize Problems*, pg. 3-30. (J. Carlson, A. Jaffe and A. Wiles, eds.). American Mathematical Society, Providence, Rhode Island.
Greene, T.M.: 1986, History and anachronism. In *The Vulnerable Text, Essays on Renaissance Literature.* Columbia University Press, New York. Pp. 221.
Guicciardini, N.: 1989, *The Development of Newtonian Calculus in Britain, 1700-1800.* Cambridge University Press, Cambridge.
Hahn, R.: 1971, *The Anatomy of a Scientific Institution: The Paris Academy of Sciences 1666–1803.* Univ. of California Press, Berkeley.

Hilbert, D.: 1901, Mathematical Problems. *Archiv der Mathematik und Physik*, 3rd set, pg. 44-63. English translation in: Bulletin of the American Mathematical Society (1902), vol. 8, pg. 437-479.

Jaki, S.: 1972, The Original Formulation of the Titius-Bode Law. *JHA* 3, 136-138.

Kant, I.: 1764, *Untersuchung über die Deutlichkeit der Gründsätze der natürlichen Theologie und der Moral*. Haude u. Spener, Berlin.

Knight, W. A.: 1887, *Memorials of Coleorton*, vol. 1. D. Douglas, Edinburgh.

Koutalis, V.: 2014, Making discoveries for a better life vs bringing fruits to the national treasury: Davy, Babbage, Brewster and the (ongoing) struggle for the soul of science. *Almagest* 5(1), 52-109.

Kuehn, M.: 1987, *Scottish Common Sense in Germany, 1768-1800*. McGill-Queen's University Press, Montreal and Kingston.

Lacroix, S.: 1816, *Traité du Calcul Différential et du Calcul Intégral (An elementary treatise on the differential and integral calculus)*. C. Babbage, G. Peacock, and J. Herschel (tr.). J. Deighton and Sons, Cambridge.

Lauzon, M.: 2010, *Signs of Light*. Cornell University Press, Ithaca.

Lemoine, E.M.J., 1802. A sketch of the history of Pure Mathematics, translated from Traite Elementaire de Mathematiques Pures. *Monthly Magazine and British Register*, 14(2), 122.

Lifschitz, A.: 2012, *Language & Enlightenment: The Berlin Debates of the Eighteenth Century*. Oxford University Press, Oxford.

Locke, J.: 1690, *An Essay Concerning Humane Understanding*. Thomas Basset, London.

Maindron, E.: 1881, *Les foundations de prix a l'Académie des Sciences*, Les laureates 1714–1880. Paris.

Marcil-Lacoste, L.: 1982, *Claude Buffier and Thomas Reid: Two Common-Sense Philosophers*. McGill-Queen's University Press, Kingston and Montreal.

McClellan, J.: 1985, *Science Reorganized: Scientific Societies in the Eighteenth Century*. Columbia University Press, New York.

Mercure de France, *Institut de France: Prix de Mathematiques*. Vol. 35, pg. 34 (1809).

Messier, C.: 1801/1802, Astre de Piazzi et conjonction de plusieurs planets observes an mois de Nivose An. 10. Mem. Des Societes Savantes Et Litter. *de la Republ.* France 2, 114.

Mortera, E.L.: 2010, Dugald Stewart on innate ideas and the origin of knowledge. *Journal of Scottish Thought*, 4, 123-140.

Namier, L.: 1961. *England in the Age of the American Revolution. Second Edition*. Macmillan, London.

Olson, R.: 1975, *Scottish Philosophy and British Physics 1750–1880*. Princeton University Press, Princeton.

Patton, L.: 2014, Hilbert's Objectivity. *Historia Mathematica* 41(2), 188-203.

Pierce, C.S.: 1891, The Architecture of Theories. *The Monist*, 1, 161-176.

Plummer, H.C., 1943. An eighteenth century correspondence. *The Observatory*, 65, 92-94.

Rée, M.R., 2009. Dispersed and distracted. A book review of Leibniz, an intellectual biography. *London Review of Books*, 31(12), 25 June, 25.

Rivarol, A.: 1784, *Discours sur l'universalité de la langue française*. Bailly, Paris.

Robertson, D.: 1910, *A History of The French Academy 1635–1910*. G. W. Dillingham Co., NY.

Saraiva, L.M.R.: 2008, Mathematics in the Memoirs of the Lisbon Academy of Sciences in the 19th Century. *Historia Mathematica* 35(4), 302-326.

Schwab, J.C.: 1785, *Über den Ursachen der Allgemeinheit der französischen Sprache und der wahrscheinlichen Dauer ihrer Herrschaft*. Jacob Friedrich Heerbrandt, Tübingen.

Servois, F.-J.: 1814, Reflections on the various systems of exposition of the principles of the differential calculus and, in particular, on the doctrine of the infinitely small. *Annales des Mathématiques pures et appliquées* 5, 141-170. R. Bradley & S. Petrilli (tr.): 2010. *Convergence* 7. Online journal of the Mathematical Association of America.

Seydl, O.: 1938, *Briefe Franz Xaver Freiherrn von Zach, Direktors der herzoglichen Sternwarte am Gotha-Seeberg, und seines Nachfolgers Bernhards von Lindenau von 1791–1861 an*

*P. Martin Alois David, Adjunkt und Direktor der Königlich Prager Sternwarte.* [Letters by Zach & Lindenau to David]. Gessellschaft der Wissenschaften, Prague.
Shabel, L.: 2003, *Mathematics in Kant's Critical Philosophy: Reflections on Mathematical Practice*. Routledge, New York.
Sime, J.: 1900, *William Herschel and his Work.* C. Scribner's & Son, New York.
Sniadecki, J.: 1821, *The Philosophy of the Human Mind* (in Polish). Zawadski, Wilno (Vilnius).
Steffens, H.: 1803, *Indledning til philosophiske Forelaesninger*. Seidelin, Copenhagen.
Swerdlow, N.: 1975, The Planetary Theory of François Viète. *JHA* 6, 185-208.
Swiggers, P.: 2014, Ideology and the 'clarity' of French, in *Ideologies of Language*, pg. 112-xxx (John E. Joseph and Talbot J. Taylor, eds.). Routledge, Abingdon, England.
Terdiman, R.: 2005, *Body& Story: The Ethics and Practice of Theoretical Conflict*. Johns Hopkins University Press, Baltimore.
*The Annals of Philosophy*, 7, 164 (1816).
*The British Quarterly Review*, 13, 331-357 (1851).
*The Critic*, 16, 1 May issue, pg. 7 (1857).
The Critical Review, 1807, *Bryan's Lectures on Natural Philosophy*. Vol. 11, 50-57.
The Edinburgh Review, 1819, *Woodhouse's Astronomy*. Vol. 62, 375-394.
*The London Journal of Arts and Sciences*, 8, 323 (1824).
*The Moniteur*, 28, 1238 (1803).
*The Monthly Magazine, December 1st issue*, vol. 22, pg. 481, (1806)
The Monthly Review, 1816, *Butler's Introduction to Mathematics*. Vol. 80, 314-317.
Toplis, J.: 1805, On the Decline of Mathematical Studies, and the Sciences dependent upon them. *The Philosophical Magazine* 20, 25-31.
Toynbee, A.: 1946, *A Study of History*. Oxford University Press, Oxford.
Viète, F.: 1591, *In artem analyticam isagoge. Introduction to the Analytic Art* (1983): edited and translated by T. Witmer. Kent State University Press, Ohio.
*The Quarterly Review*, vol. 61, 501 (1838).
Wahl, C.: 2014, Between cosmopolitanism and nationalism. The role of expatriates in the dissemination of Leibniz's differential calculus. *Almagest* 5(2), 40-68.
Wickman, M.: 2009, Scottish Newtonianism and modern(ist) history. *Journal of Scottish Thought*, 2(1), 113-126.
Williamson, J.: 1982, The Structure of Pay in Britain 1710–1911. *Research in Economic History* 7, 1-54.
Wilson, C.: 2010, *The Hill-Brown Theory of the Moon's Motion*. Springer, New York.
Withers, C., and Wood, P. (eds.): 2002, *Science and Medicine in the Scottish Enlightenment.* Tuckwell Press, East Linton.
Wolff, C.: 1728, Praemittitur Discursus Praeliminaris de Philosophia in Genere (Preliminary Discourse on Philosophy in general); this was the introduction to *Philosophia rationalis sive Logica.* Renger, Frankfurt. English version in *Philosophy: An Innovative Introduction*, quote from pg. 300 (M. Boylan and C. Johnson, eds.), 2010. Westview Press, Boulder, Colorado.
Yeomans, D.: 1991, *Comets: A Chronological History of Observation, Science, Myth and Folklore*. Wiley, New York.
Zach, F.: 1802a, *Monthly Correspondence*, Sept. issue., pg. 310.
Zach, F.: 1830, Letter to Eliza, the wife of British Admiral William Smyth. American Philosophical Society, Philadelphia.

# Chapter 4
# The Gauss Anagram

## The Origin of the Anagram

In the eleventh to the seventeenth centuries, mathematicians who were not independently wealthy made their living by selling their service to/as reckoners (computers), accountants, bookkeepers, and astrologers. There was a dispute between Niccolo Fontana (1499–1557), known as Tartaglia ('the Stamerer') and Girolamo Cardano (1501–1576) about who first discovered the formula for roots of third-order polynomials. Tartaglia confided his solution to Cardano in 1539 in the form of a *capitolo* poem (Katscher 2006). Six years later, Cardano (1545) published the solution without permission. This discovery had some use in computing the internal rate of return (a practical accounting problem at that time), but the main reason why each wanted to claim priority was for marketing reasons: being its inventor would have enhanced their reputation and allowed them to charge higher fees or to get new clients for their livelihood, the bookkeeping business.

Thus a technique emerged which was widely used by mathematicians to claim priority. They would write out their result (or the proof); they would make an anagram, or take the first letter of each (Latin) word, etc.; and they would mass-mail it to every mathematician they knew, so all of them would receive it at about the same time. Then the author would publish it at leisure, knowing that once he decoded the anagram or published the proof whose first letters coincide with their broadcast, it will be evident that he had the proof on the date it was first mailed.

But "the use of anagrams had severe limitations. They had to be short, otherwise opponents could claim that the letters might be rearranged to give more than one meaning. More importantly, a short sentence could not include all the evidence for a claim, which might therefore just be a lucky guess." (Meadows 2007: 31) Galileo twice employed anagrams to conceal his discoveries. His observations of the phase changes of Venus were sent in the form of an anagram (intended for Kepler) to the Tuscan ambassador of Prague, Giuliano de Medici. *Haec immatura a me iam frustra leguntur o y* is unscrambled to *Cynthiae figuras aemulatur mater amorum.*



C.J. Cunningham, *Studies of Pallas in the Early Nineteenth Century*,
DOI 10.1007/978-3-319-32848-5_4

In English this means 'the mother of love' [Venus] emulates the figures of Cynthia [the Moon] (Van Helden 1989: 107).

Galileo used an anagram again to describe his observations of Saturn: SMAISMRMIMLEPOETALEUMIBUNENUGTTAURIAS. This anagram of a Latin sentence (*Altissimum planetam tergeminum observavi*) was erroneously deciphered by Kepler, who believed Galileo had discovered two moons around Mars (Gingerich 1970). It was not realized until the 1650s that Saturn has a ring; Galileo's telescope was so poor he had merely observed two circular lumps, or anses, perched on each side of the planet. The Latin loosely translates into English: I have seen the highest planet [Saturn] to be triple in form.

Christiaan Huygens (1629–1695); Fig. 4.1 was not an astronomer, yet he made two important discoveries in seventeenth-century astronomy. With a 12-foot telescope he and his brother had made, he was the first after Galileo to discover a new satellite around a planet (Andriesse 2005). In June 1655 he saw Titan, the largest moon of Saturn; he quickly sent an anagram to the mathematicians John Wallis in Oxford and Gottfried Kinner von Loewenthurn (c1610–1670) in Prague, consisting of a line from Ovid: *Admovere oculis distantia sidera nostris* (Faraway stars move towards our eyes), and the letters uuuuuuu ccc rr h b q x. Uncoding the anagram forms the sentence *Saturno luna sua circumducitur diebus sexdecim horis quator* (A moon revolves around Saturn in sixteen days and four hours). The discovery of Titan was made public in *De Saturni luna observatio nova* (1656) (Andriesse 2005: 10). In the form of an anagram, this sheet also contained his second discovery: the explanation of the strange anses of Saturn that had stupefied astronomers since Galileo. According to Huygens, as he explained fully in *Systema Saturnium* (1659), these anses are the appearance of a solid ring around the planet. His anagram read: aaaaaaacccccdeeeeeghiiiiiiilllmmnnnnnnnnnoooppqrrstttttuuuuu. He wrote this as an anagram in Latin in honor of Galileo. It reads: *Annulo cingitur, tenui, plano, nusquam cohaerente, ad eclipticam inclinato*, which translates as Saturn is girdled by a thin, flat ring, touching it nowhere, and tilted to the ecliptic (Alexander 1962: 94).

Beginning in 1673 Leibniz began corresponding with Newton's colleague, Henry Oldenburg (1618–1677, secretary of the Royal Society from 1661), on the latest work in English mathematics. In 1676 he got keenly interested in Newton's work and asked Oldenburg for further information. Newton replied to Oldenburg in two letters, requesting Latin translations be sent to Leibniz. The first letter, dated June 13, 1676, dealt with the binomial theorem, summary of series, and a hint of an important 'method.' The second, dated 24 October 1676 discussed a treatise on infinite series. At the point where an explanation of the 'method' would have followed, Newton utilized an anagram: 6accdæ13eff7i3l9n4o4qrr4s8t12vx.

Later Leibniz would be accused of cracking the anagram to steal Newton's work, but it was really not that revealing ("Given an equation involving any number of fluent quantities, to find the fluxions, and conversely"). The British mathematician Augustus De Morgan (1806–1871; 1914) pointed out that cracking it and interpreting it would have been a greater triumph than independently discovering calculus.

**Fig. 4.1** Christiaan Huygens painting by Caspar Netscher. Museum Hofwijck, Voorburg

Anagrams have been used for many covert purposes through the years. In the seventeenth century, they were often used to disguise the name of an author of a controversial book. Hiding identities was very important in an age when a book could get an author thrown in jail, or even executed. In 1620, a treatise entitled *Kanones peuriles,* condemning the Rosicrucians and his colleague the German mathematician Johann Faulhaber (1580–1635), was published under the name Kleopas Herennius. This was in fact an anagram of Kepler's name in Latin: Iohanes Keplerus.

## Concinnity

In 1803, Gauss made a ringing declaration. "Astronomy and pure mathematics are now once and for all the magnetic poles to which my soul's compass shall always point" (Dunnington 2004: 61). Nine years later, his soul's compass compelled him

**Fig. 4.2** Concinnity is harmony in the inter-arrangement of parts with respect to a whole. How sections are arranged to achieve the correct form is the key to understanding the Gauss anagram. This sculptural figure, composed of sections, can be thought of as an allegory to represent this concept. Each segment can be arranged in a variety of interlocking patterns, but only one reveals nature in its true form. (Photo by C. Cunningham in Budapest in 2004)

to create a code with a magnetic seal that researchers have tried to unlock for two centuries. The code contains Gauss's greatest discovery about the orbit of Pallas. Appendix A gives a complete translation of all efforts that attempt to unravel the mystery of the Gauss anagram (Fig. 4.2). A new interpretation, printed here for the first time, is based on Gauss's work with the 17-sided polygon.

The first entry in Gauss's Latin journal (Fig. 4.3) relates to his discovery of the 17-sided polygon (the heptadecagon). Dated 30 March 1796 it simply states "The principles upon which the division of the circle depend, and geometrical divisibility of the same into seventeen parts, and so forth." (Dunnington 2004: 469) Gauss (1796) describes the discovery in his first published article, written at the age of 19

> *Every beginner in geometry knows that it is possible to construct different regular polygons, for example triangles, pentagons, 15-gons, and those regular polygons that result from doubling the number of sides of these figures. One had already come this far in Euclid's time, and it seems that since then one has generally believed that the field for elementary geometry ended at that point, and in any case I do not know of any successful attempt to extend the boundaries beyond that line.*
>
> *Therefore it seems to me that this discovery possesses special interest, that besides these regular polygons, a number of others are geometrically constructible, for example the 17-gon. This discovery is really only a corollary of a theory with greater content, which is not complete yet, but which will be published as soon as it is complete* (Dunnington 2004: 28).

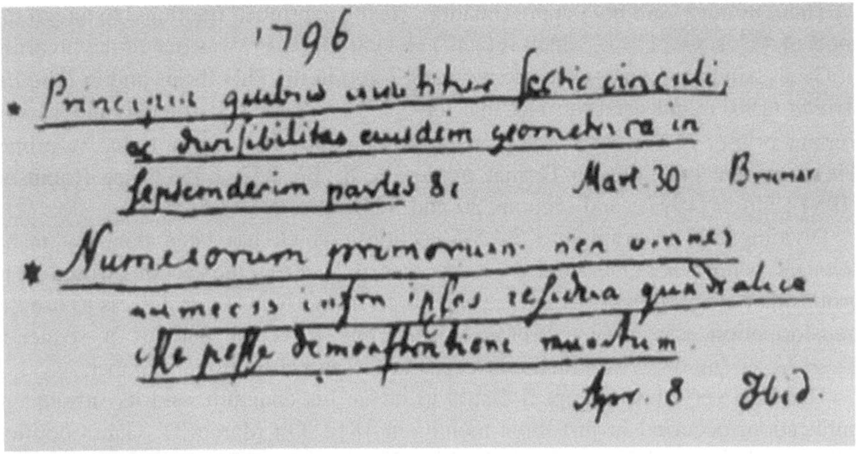

**Fig. 4.3** First three lines are a note by Gauss giving the division of a circle into 17 equal parts

Gauss considered his discovery to be of such importance that he requested a 17-sided polygon should be inscribed on his tombstone. On his monument in Brunswick is a 17-pointed star; Howaldt, the sculptor who carried out the work, maintained that everyone would confuse a regular 17-sided polygon with a circle (Archibald 1920; Hall 1970).

## The Solution of the Gauss Anagram

The anagram, in the form of a sequence of the numbers 1 and 0, contains the most important result about the movement of Pallas

**1111000100101001**

The binary 'code' has traditionally been read as

| | | | |
|---|---|---|---|
| 111 | 1000 | 10010 | 1001 |
| (7) | (8) | (18) | (9) |

However, it should be read as follows:

| | | | |
|---|---|---|---|
| 1111 | | 0001 | 00101001 |
| (**15**) | + | (**1**) | (**41**) |

The last 8 binary digits refer to the number 41. The first 4 represent 15 and the second 4 the number 1. Adding the first two numbers gives 16, which gets divided by 41. A calculation of $^{16}/_{41}$ is extremely close to $^{7}/_{18}$ (0.39024 versus 0.38888).

These numbers are both approximations (using continued fractions) to one of the roots of $4x^2 + x - 1 = 0$, which is $(\sqrt{(17)} - 1)/8$. The $\sqrt{(17)}$ was one of the favorites of Gauss, since he constructed the regular 17-gon in his PhD thesis and in *Disquis. Arithm* (1801), and used the fact that the $\sqrt{(17)}$ could be constructed, and 17 is a Fermat prime. A Fermat prime is a Fermat number $F_x = 2^{2^n} + 1$ that is prime. These are the only known Fermat primes: 3, 5, 17, 257, 65537 [see Robinson (1954); Singh (1997); and sections 20 and 37 of Dorrie (1965).]

Dividing the anagram into 3,4,5,4 sections, which has been done by many scholars, is not what Gauss did. 16 is a power of 2, so that the sections would be in groups that are all powers of 2. One way to go from $x/y = {}^7/_{18}$ to ${}^{16}/_{41}$ is to use the transformation $x/y \rightarrow (y + 2x)/(3y + 4x)$. Thus it is all part of a sequence ${}^1/_4 \rightarrow {}^3/_8 \rightarrow {}^7/_{18} \rightarrow {}^{16}/_{41}$, which converges to the above irrational number.

The discovery that prompted Gauss to create his anagram, and its subsequent publication, occurred in just three months in 1812. On March 31, Gauss notified Olbers and on April 5 he started checking his discovery. On April 25, the publication followed and on May 5 he informed Bessel about the meaning of the codification. According to his letter to Bessel, the anagram contains the theorem that the quotient of the centered movements of Jupiter and Pallas moves backwards and forwards around a fixed rational value, namely 7:18, so that it consists of an oscillation.

> *I was mainly occupied with the perturbations of Pallas by Jupiter. You probably read about that in the* Gelehrten Anzeigen *no. 67. I am pleased to tell you the result, given there in a cipher, but under the condition that it remains just between you and me for the time being. It means that the mean motions of ♃ and ⚴ are in the ratio 7:18, which, by the influence of Jupiter, is always exactly reestablished, like the rotational period of our Moon. I have already begun a second calculation of the periodic perturbations* (Gauss to Bessel: May 5, 1812).

In light of this, Gauss's opinion of theorems, expressed on March 4, 1824, to Bessel, takes on added meaning: "All of the measurements in the world do not outweigh one theorem by which the knowledge of eternal truths is really advanced" (Dunnington 2004: 120).

## Gauss and the Royal Society

Gauss gained admittance to the Royal Society of London on 12 April 1804, based on the work he did on the first two asteroids (Royal Society 1803).

> The text of the letter reads:
>
> *Frederick Gauss, MA of Brunswick in Germany, author of a learned treatise on the higher parts of arithmetic, which he has presented to this society. Who by instituting nice calculations upon the observations of the planet Ceres, made by Mr Piazzi within the small period of six weeks, enabled Baron Zach and Dr. Olbers, each separately* [sic], *to rediscover it, after it had been lost ten or eleven months; and, who by similar calculations of the*

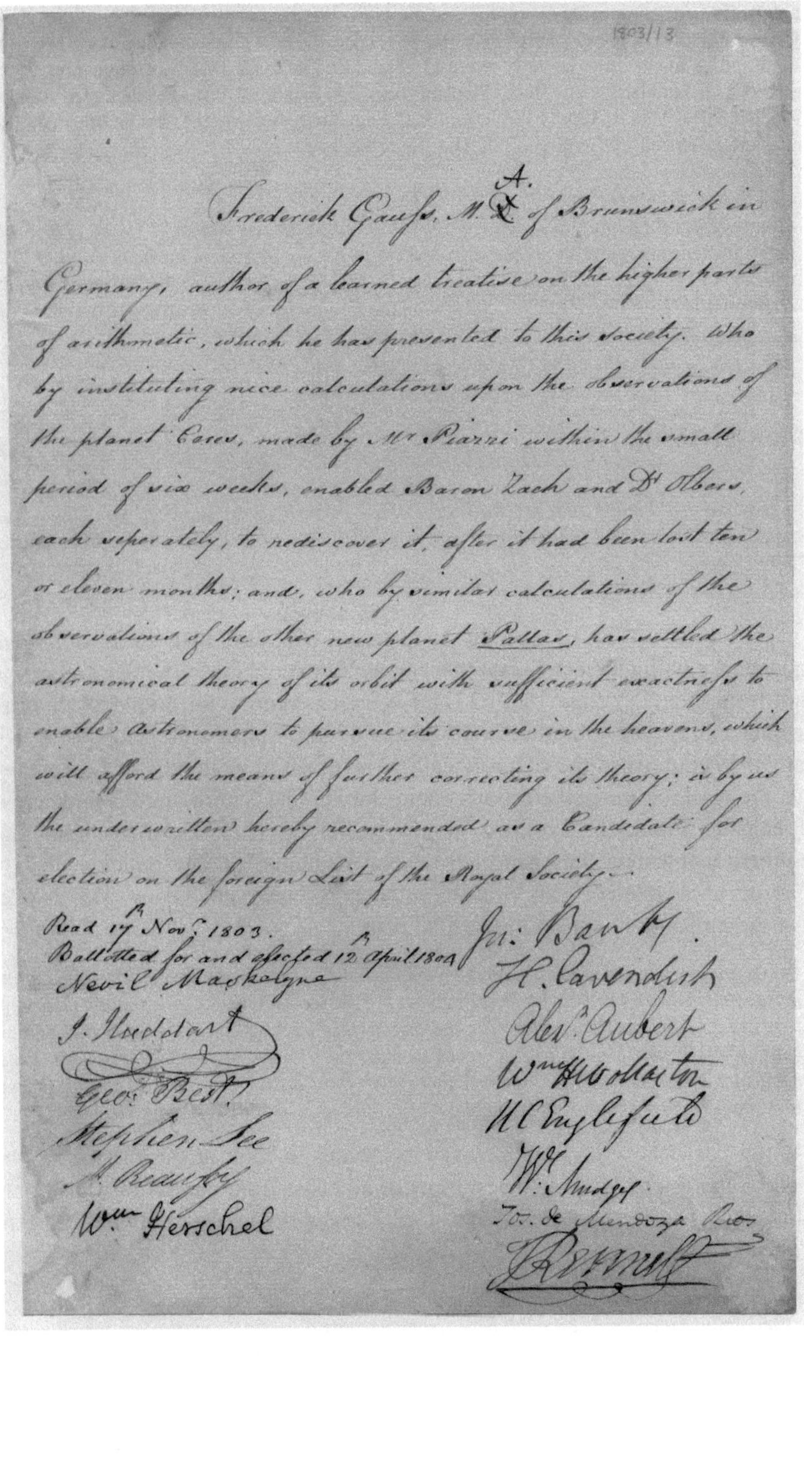

1803/13

Frederick Gauss. M.A. of Brunswick in Germany, author of a learned treatise on the higher parts of arithmetic, which he has presented to this society. Who by instituting nice calculations upon the observations of the planet Ceres, made by Mr Piazzi within the small period of six weeks, enabled Baron Zach and Dr Olbers, each seperately, to rediscover it, after it had been lost ten or eleven months; and, who by similar calculations of the observations of the other new planet Pallas, has settled the astronomical theory of its orbit with sufficient exactness to enable Astronomers to pursue its course in the heavens, which will afford the means of further correcting its theory; is by us the underwritten hereby recommended as a Candidate for election on the foreign List of the Royal Society.

Read 17th Novr. 1803.
Balloted for and elected 12th April 1804.

Nevil Maskelyne
[illegible]
Geo. Best
Stephen Lee
[illegible]
Wm Herschel

Jos: Banks
H. Cavendish
Alexr Aubert
Wm H Wollaston
H C Englefield
[illegible]
Jos. de Mendoza Rios
[illegible]

◀ **Fig. 4.4** The letter proposing Gauss for election to the Royal Society, specifically noting his work on Ceres and Pallas. It is especially interesting as it bears the signatures of many of the Englishmen who played a major role in early asteroid research. The total roster of signatures is Nevil Maskelyne; J Huddart; Geo Best; Stephen Lee; M Beaufoy; Wm Hershel; Jos Banks; H Cavendish; Alex Aubert; Wm H Wollaston; H C Englefield; Wm Mudge; Jos de Mendoza Rios; J Rennell. Reproduced by permission of The Royal Society

> *observations of the other new planet Pallas, has settled the astronomical theory of its orbit with sufficient exactness to enable astronomers to pursue its course in the heavens, which will afford the means of further correcting its theory; is by us the underwritten hereby recommended as a Candidate for election on the foreign list of the Royal Society* (Fig. 4.4).

## Gauss gives up

> *It is a painful thought that my work on Pallas' perturbations, made 20 years ago, has remained discontinued, undeveloped and unpublicised ever since and will – like many a thing – perish with me.*
>
> Carl Gauss 1834

Cicero (107–43 BCE) said in his treatise *On Fate* that "it is only by an acquaintance with efficient causes that can we foreknow the particular events which result from them" (Yonge 1853: 277). Even Gauss could not divine all the causes affecting Pallas, and thus the future course of that object was denied him. Klein et al. (1919) published the rich calculations involved in Volume 7 of *Gauss Werke*. After the work on the perturbations due to Jupiter and Saturn were completed with the help of Friedrich Bernhard Gottfried Nicolai (1793–1846), whom Gauss identified as a young man especially expert in numerical computation, the work broke up uncompleted. The virtuoso calculating energy of Gauss was unable to solve the Pallas problem up to the final results.

> *Gauss had progressed very far with the work on Pallas.... [He published] his second major astronomical work* On Secular Perturbations [1818; see Hill 1882], *a standard theory of first order approximations. However the astronomer* [Georg] *Struve showed much later* [Struve 1910] *that for Pallas one needs at least third order approximations. This gives some idea of the magnitude of the task that Gauss faced with Pallas, and of the still larger task that preparing the work for publication would have entailed* (Young 2011: 147).

Gauss has been criticized for even bothering to look at the orbital problems posed by the asteroids. The lowest blow of all came from the Scottish historian of mathematics Eric Temple Bell (1883–1960); Fig. 4.5 who accused him of grasping at 'cheap fame.'

> *A new planet had been discovered in a position which made it extraordinarily difficult of observation. To compute an orbit from the meagre data available was a task which might have exercised Laplace himself. Newton had declared that such problems are among the most difficult in mathematical astronomy. The mere arithmetic necessary to establish an orbit with accuracy sufficient to ensure that Ceres on her whirl round the sun should not be*

**Fig. 4.5** Eric Temple Bell

*lost to telescopes might well deter an electrically-driven calculating machine even today; but to the young man whose inhuman memory enabled him to dispense with a table of logarithms when he was hard pressed or too lazy to reach for one, all this endless arithmetic – logistica, not arithmetica – was the sport of an infant.*

*Why not indulge his dear vice, calculate as he had never calculated before, produce the difficult orbit to the sincere delight and wonderment of the dictators of mathematical fashion and thus make it possible, a year hence, for patient astronomers to rediscover Ceres in the place where the Newtonian law of gravitation decreed that she must be found – if the law were indeed a law of nature? Why not do all this, turn his back on the insubstantial vision of Archimedes and forget his own unsurpassed discoveries which lay waiting for development in his diary? Why not, in short, be popular? The Duke's generosity, always ungrudged, had nevertheless wounded the young man's pride in its most secret place; honor, recognition, acceptance as a "great" mathematician in the fashion of the time with its probable sequel of financial independence – all these were now within his easy reach. Gauss, the mathematical god of all time, stretched forth his hand and plucked the Dead Sea fruits of a cheap fame in his own young generation.*

*For nearly twenty years the sublime dreams whose fugitive glimpses the boyish Gauss had pictured with unrestrained joy in his diary lay cold and all but forgotten. Ceres was rediscovered, precisely where the marvellously ingenious and detailed calculations of the young Gauss had predicted she must be found. Pallas, Vesta and Juno, insignificant sister planets of the diminutive Ceres were quickly picked up by prying telescopes defying Hegel, and their orbits, too, were found to conform to the inspired calculations of Gauss [see the next book in this series regarding Hegel]. Computations which would have taken Euler three days to perform were now the simple exercises of a few laborious hours. Gauss had prescribed the method, the routine. The major part of his own time for nearly twenty years was devoted to astronomical calculations* (Bell 1937: 241).

## The Royal Patrons

Although Bell may censure Gauss for succumbing to the Duke's generosity, it was in fact the prime avenue for 'scientists' of true genius to attain a place in society that would allow them to work. Many rulers (Fig. 4.6) competed to attract reputable scientists to their courts, in part because some skills could come in handy, but largely just to show off. The competition for the best and brightest scientists between European rulers required open science as the only way to identify the truly great. The situation a century earlier, as described in this letter from Leibniz to the Duke of Hanover in 1675, had scarcely changed by the time Gauss accepted the patronage of his own Duke.

*As it is from grand princes that one may await remedies to public evils and as they are the most powerful instruments of divine benevolence, they are necessarily loved by all those who have disinterested sentiments* (Stewart 2006: 143).

The competition of different institutions (many established by the rulers of principalities) for the most distinguished scientists—those of 'disinterested sentiments'—meant that genii such as Gauss could set their own research agendas and appropriate the benefits of research (Mokyr 2005). Thus Bell ["a very careless historian of mathematics" in the view of Feferman (2006)] was very much off the mark in his criticism of Gauss, who was working within the framework of the science of his time. So why did Gauss devote nearly all his efforts to astronomy, and especially the asteroids, in the first decade of the century? "The most likely explanation, crudely put, is that there was not much of a market for the sublime truths of higher arithmetic" (Bucciarelli and Dworsky 1980: 127). In light of this grim reality, one cannot begrudge Gauss for accepting the generosity of the Duke of Brunswick. In both Scotland and England, many professors relied not on royal but on aristocratic patronage. Scotland in particular was dominated Henry Dundas, 1st Viscount Melville (1742–1811) who claimed in 1801 "Every professor in the universities of St. Andrews and Edinburgh has been appointed for more than 20 years either actually by myself or upon my recommendation" (Emerson 2008: 11). A recent study dealing with the leading chemist of the age, Joseph Priestley, explains

*Natural philosophy sustained the 'arts' upon which cultivated life depended, making it fundamental for the refinement enjoyed by the aristocracy. 'Without that knowledge, rank and fortune would be of little value,' Priestley wrote; in other words, the 'elegant enjoyment of life' depended on ingenious people like him. At the same time, natural philosophy deserved patronage because of it utility to all society. The aristocracy and gentry were in a position to support research simply because they had money; they were not caught in the*

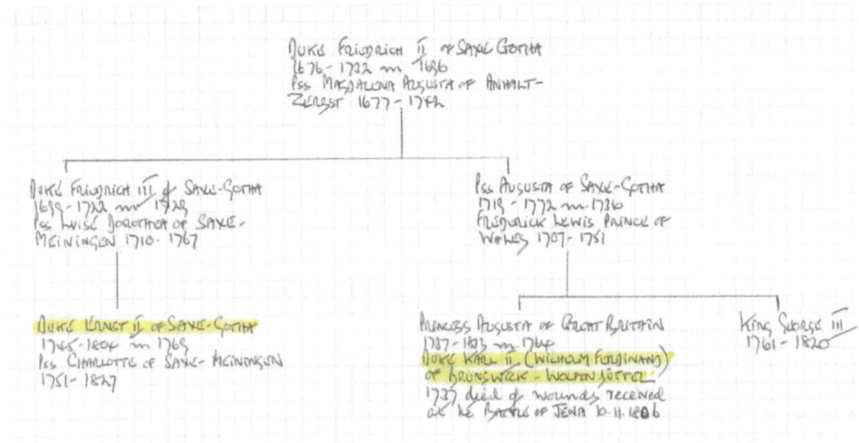


Fig. 4.6 There was a close family connection between the royal patrons of the early asteroid researchers. Duke Ernst II supported Baron Franz von Zach and financed construction of Seeberg Observatory. Duke Karl II enabled Gauss to begin his career without financial worries, and King George III of Great Britain made William Herschel his Court Astronomer

*mad economic scramble that seemed to define this age. 'Being free from most of the cares particular to individuals, they may embrace the interests of the whole species,' Priestley claimed* (Jackson 2005: 45).

Even though Gauss accepted the patronage of the Duke, he was well aware of his responsibilities and obligations. When the Duke raised his stipend in the momentous Ceres discovery year of 1801, his friend Eberhard von Zimmermann (1743–1815) gave Gauss the happy news. He is reported to have exclaimed "But I certainly haven't earned it. I haven't yet done anything for the nation" (Dunnington 2004: 60).

## References


Alexander, A.F.: 1962, *The planet Saturn: a history of observation, theory and discovery*. Faber & Faber, London.

Andriesse, C. D.: 2005, *Huygens: The Man Behind the Principle*. Cambridge Univ. Press, Cambridge.

Archibald, R.: 1920, Gauss and the Regular Polygon of Seventeen Sides. *Amer. Math. Monthly* 27, 323-326.

Bell, E. T.: 1937, *Men of Mathematics*. Simon & Schuster, New York.

Bucciarelli, L.L. and Dworsky, N.: 1980, *Sophie Germain*. D. Reidel, Dordrecht.

Cardano, G..: 1545, *Ars magna*. Petreius, Nuernberg.

Cunningham, C.: 2004, The Planet Hercules. *Mercury* 33(5), 10.

De Morgan, A.: 1914, *Essays on the Life and Work of Newton,* (P. Jourdain, ed.). Open Court Publishing Co., Chicago.

Dorrie, H.: 1965, *100 Great Problems of Elementary Mathematics.* Dover, New York.

Dunnington, G. W.: 2004, *Carl Friedrich Gauss.* The Mathematical Association of America, Washington D.C. (Reprinted from 1955).

Emerson, R.: 2008, *Academic Patronage in the Scottish Enlightenment.* Edinburgh University Press, Edinburgh.

Feferman, S.: 2006, Provenly Unprovable. *London Review of Books*, February 9 issue, pg. 30.

Gauss, C.: 1796, Intelligenzblatt der allgemeinen Litteraturzeitung. April issue. Quoted in Dunnington, 2004: 28.

Gingerich, O.: 1970, The Satellites of Mars: Prediction and Discovery. *JHA* 1 (2), 109-115.

Hall, T.: 1970, *Carl Friedrich Gauss: A Biography.* MIT Press, Cambridge.

Hill, G.W.: 1882, On Gauss's Method of Computing Secular Perturbations. *Astronomical Papers Prepared for the Use of the American Ephemeris and Nautical Almanac.* 316-361.

Huygens, C.: 1656, *De Saturni luna observatio nova.* In *Opera varia,* 1724, pg. 523-526. Janssonios Vander, Leyden.

Huygens, C.: 1659, *Systema Saturnium.* Adriani Vlacq, The Hague.

Jackson, J.: 2005, *A World on Fire.* Viking, New York.

Katscher, F.: 2006, How Tartaglia Solved the Cubic Equation. *Convergence* 3. Online journal of the Mathematical Association of America.

Klein, F., Brendel, M. and Schlesinger, L.: 1919, *Materialien fuer eine wissenschaftliche Biographie von Gauss.* Vol. 7 of Gauss Werke. B. G. Teubner, Leipzig.

Meadows, J.: 2007, Limitations on the Publishing of Scientific Research. In *The History of Information Security* (Karl de Leeuw and Jan Bergstra, eds.), 29–52. Elsevier, Amsterdam.

Mokyr, J.: 2005, The Great Synergy: the European Enlightenment as a factor in Modern Economic growth. In *Understanding the Dynamics of a Knowledge Economy* (W. Dolfsma and L. Soete, eds.). Edward Elgar, Cheltenham.

Robinson, R.: 1954, Mersenne and Fermat Numbers. *Proc. Amer. Math. Soc.* 5, 842-846.

Royal Society: 1803. Letter nominating Carl Gauss for election to the Royal Society of London. Manuscript number EC/1803/13.

Singh, S.: 1997, *Fermat's Enigma.* Walker & Co., NY.

Stewart, M.: 2006, *The Courtier and the Heretic: Leibniz, Spinoza and the fate of God in the Modern World.* W.W. Norton, New York.

Van Helden, A.: 1974, Saturn and his anses. *JHA* 5(2), 105-121.

Van Helden, A. (ed.): 1989, *Sidereus Nuncius or the Sidereal Messenger*, by Galileo. University of Chicago Press, Chicago.

Yonge, C. D.: 1853, *The Treatises of M. T. Cicero.* Henry G. Bohn, London.

Young, L.: 2011, *Mathematicians and Their Times.* North Holland Publishing Co., Amsterdam.

# Chapter 5
# Hypothetical Planets

In the long history of astronomical observation, there have been many errors, but claiming to find a new planet when there really isn't one is certainly a mistake greater than most. The cosmos shown in Fig. 5.1 were often the subject of change, real or imagined.

## Ludwig's Star

Sidus Ludovicianum, or Ludwig's Star: the planet that was not. On December 2, 1722, Johann Georg Liebknecht, professor of theology and mathematics at the University of Giesen (located in a town south of Hannover, Germany), turned his 6-foot non-achromatic telescope to Ursa Major. There he saw the star Tycho 3850-257-1, between Alcor and Mizar. Believing he had detected a motion of this object, Liebknecht (1723) came to the startling conclusion that it was a planet, even though it was in the exact same position as it had been when Benedetto Castelli observed it in 1617. Liebknecht named his new 'planet' in honor of the Landgrave Ludwig V of Hesse-Darmstadt (1596–1626). "The name survived, but the professor's reputation did not" (Ashbrook 1957). Even though Liebknecht was mistaken, the precedent for naming a new planet after the patron of its discoverer was established (Cunningham 2017).

## Vidal's Planet

Jacques Vidal (1747–1819) became director of Toulouse Observatory in southern France in 1791. He earned the admiration of Lalande, who wrote that by his extensive observations of Mercury "he has done more than all the other astronomers



C.J. Cunningham, *Studies of Pallas in the Early Nineteenth Century*,
DOI 10.1007/978-3-319-32848-5_5

**Fig. 5.1** A view of the cosmos in 1801, the year Ceres was discovered. The seven planetary orbits are shown with the orbits of both parabolic and hyperbolic comets. All is surrounded by the starry heavens and encircled by a serpent symbolizing the totality of existence, infinity and the cyclic nature of the cosmos. The *word at bottom* reads Omnipotence. Frontispiece of a new edition of *A Hundred Years Calendar* published a century earlier by a doctor in Erfurt, Christoph von Hellwig (1663–1721; 1801) Courtesy, Google Books

of the world, ancient and modern together, and in this respect Citizen Vidal must make all others despair." (Tisserand 1882: 223) On April 24, 1802, Vidal was elected a nonresident associate in the astronomy section, first class, of the National Institute. When he apparently discovered a new planet in 1805, Lalande thought it tantamount to a criminal act to call it anything other than Vidal. Here is how Olbers (1805) related the tale to Gauss.

*Today I must tell you a few words about the best French discovery, even though you might receive an account of it from our friend Harding himself. The day before yesterday Mr. Harding obtained the following letter of 9 January 1805 from Lalande:*

*A planet was discovered at Mirepoix by Mr. Vidal. He instructs Lalande to assign it a name, but he (Lalande) believes it to be a crime if he names it anything other than planet Vidal. It appeared to be 8$^{th}$ to 9$^{th}$ magnitude. Here are the meridian observations:*

| | | |
|---|---|---|
| *9 Dec* | *0 h 11 m 40 s* | *Decl. 10° 32′S* |
| *16* | *0 15 42* | *9 31* |
| *19* | *0 17 26* | *9 4* |
| *23* | *0 19 47* | *8 29* |
| *26* | *0 21 33* | *8 3* |
| *27* | *0 22 8* | *7 54* |

*Harding wanted to sketch the movement of planet Vidal on his chart and found immediately – that it was Ceres. I would have wished that the French had published this incomparable discovery; but I find nothing about it in the* Moniteur *until January 11. It brought this shame upon itself by its thoughtlessness. Isn't it annoying for over a whole month to mistake a moving star for a new planet in the region where, as everyone knew, Ceres must be, without not once investigating whether it might not be Ceres itself? – But the worst is still that the RA on December 9 is too small by precisely one minute of time, that is, 15 minutes of arc. Might not the mistaken account of the first observation have occurred intentionally? This first observation would certainly have been needed for the calculation of the orbit, and hence we never would have been able to find anything worthwhile about the orbit until this error had been discovered.*

Notice of the 'new planet' did not escape the attention of Zach (1805), as he relates in a letter to Gauss. It includes a pithy commentary by Jacques Joseph Thulis (1748–1810), who became the director of Marseilles Observatory in 1801

*On another occasion, I shall send you a notification of the appearance of a new planet, which was found by the French. During my stay in Marseilles, I heard of Vidal reporting such a discovery in a journal in a rather pompous manner. We, Thulis and I, should have liked to observe this newcomer immediately, but we were forced to conclude, that this 'new' planet was none other than Ceres. In Paris, this bit of news was not given printed publicity and in Germany, it also remained unpublished. Thulis voiced his opinion of this miraculous discovery: 'Such will ever be the boys of Gascony.'*

## Olbers' Giant Planet

In May of 1804 a very curious report was published in *The Times* of London. This normally reputable newspaper printed the extraordinary claim that Olbers had discovered the largest planet of the Solar System (Cunningham 2004). Here is the report from the May 9 edition of the newspaper

> *A letter from Germany states that Dr. Olbers has discovered a planet, which, from its immense size, he has called Hercules. It is three times the size of Jupiter, and goes round the Sun in the space of 211 years, because it is supposed to be 3,047,000,000 of miles from the Sun; it looks to the naked eye like a star of the sixth magnitude, and is now in the sign Gemini. Dr. Olbers observed, on the 8th of December last, that it moved, and, on the 6th of February, that it was a planet, attended by seven satellites, one of which is twice the size of earth. It is inclined to the plane of the ecliptic, in an angle of 30 degrees. It is in 13 degrees North amplitude; its eccentricity is .1100, and the Sun to an inhabitant of the Earth placed in it, with our powers of vision, would appear no larger than the smallest of the fixed stars.*

This remarkable report could scarcely be topped, but just three days later *The Times* managed to do it. In an editorial comment published on May 12, one might expect that the discovery of such a planet would be welcomed as a major scientific advance. *The Times* was having none of it. In a statement that perhaps reflects the attitude generally prevailing towards the discovery of Ceres and Pallas, the newspaper issued this rebuke

> *German news state the discovery of another, and an immensely large Planet. One would imagine, that when Europe is so troubled with "malignant stars," men would have quite employment enough in discovering the modes of relieving themselves from earthly distress.*

The reference to malignant stars is a swipe at Napoleon, who began his first reign six days later. The article prompted one of William Herschel's closest correspondents, the University of Glasgow astronomer Patrick Wilson, to bring it to his attention on May 19, 1804

> *By a late paragraph, in the London papers, it is mentioned that Olbers has discovered a Planet 3 times larger than Jupiter, at a period of 211 years; and with many satellites* etc. etc.*–all seems very apocryphal – I suppose there is no reality whatever in this account.*

## Huth's Moving Star

Johann Huth's observations in December 1801 of what he thought was Ceres were mentioned in *Discovery of The First Asteroid, Ceres*, where Johann Bode briefly surveyed some letters he had received from Huth (1763–1818). The subject reappears in this book in a series of letters from Bode to Olbers: December 19, 1801; and four letters in early 1802: January 12, January 19, February. 27, and April 6. The original paper appears in Huth (1802). Here is how the matter was viewed 74 years later in the pages of the journal *Nature* (1876).

*HUTH'S 'MOVING STAR' of 1801–02. At the beginning of the present century, when, although Bode and some few others had been looking forward to such a discovery, astronomers generally were startled by Piazzi's accidental detection of the small planet Ceres, we read of observations of more than one so-called 'moving star,' which, after progressing slowly for a short interval, finally disappeared. The most singular narrative refers to an object said to have been remarked by Councillor Huth, at Frankfort-on-the-Oder, on the night from December 2 to 3, 1801, particulars of which were communicated to Bode in several letters during the ensuing five weeks. If the observations are* bona fide, *there is yet a mystery attaching to the object to which they relate. Huth was one of the three independent discoverers of the periodical comet now known as Encke's, on October 20, 1805, Pons and Bouvard sharing with him an almost simultaneous discovery, and he did other astronomical work. Writing to Bode on December 5, he says: 'In the night from the 2*$^{nd}$ *to the 3*$^{rd}$ *of this month, I saw with my 2½-feet Dollond, in a triangle with* $\theta$ *and* $\delta$ *Leonis to the south-west,* a star with faint reddish light, round, and admitting of being magnified. *I could not discern any trace of it with the naked eye; it had three small stars in its neighbourhood.' He writes again on the 15*$^{th}$*, that unfavourable weather had allowed of his observing the object only on three occasions, which appear to be on the early mornings of the 3*$^{rd}$*, 13*$^{th}$ *and 14*$^{th}$*, and he concludes from his observations that it had a slow retrograde motion to the south-west. From the 13*$^{th}$ *to the 14*$^{th}$*, by eye-estimates, it had retrograded 4' of arc, and from the 3*$^{rd}$ *to the 13*$^{th}$ *at most 30'. He forwarded to Bode at this time a diagram of the neighbouring telescopic stars. On December 21 he writes again that he had only succeeded in observing his moving star on one additional night, that of December 19–20, when he found it 'near four stars apparently situate to the westward, about half a diameter of the full moon below a smaller one.' Its path appeared directed towards* $\iota$ *Leonis and towards the ecliptic. He adds: 'Of the motion of this planet-like star I can now no longer doubt, since I have observed a difference of 5/6° nearly, between its positions on the 3*$^{rd}$ *and 20*$^{th}$*. In a fourth letter, dated 1802, January 12, he informs Bode that he had seen the star on two later nights, those of the 1*$^{st}$ *and 2*$^{nd}$ *of the same month from 11 h to 14 h, with many telescopic stars in its vicinity, of which he enclosed a diagram, by eye-estimate only, with the path of the object.*

*He mentions that on January 1 the star was even smaller than one of the satellites of Jupiter, and on the following night he had difficulty in perceiving it in close proximity to a star towards which it was moving. On the 5*$^{th}$ *he could discern only now and then, to the right of the star, on the left of which it was situated on the 1*$^{st}$ *and 2*$^{nd}$ *of January, and at a very small distance from it, a glimmer, but the star's former place on the left was vacant. He concludes that the object must have been receding from the earth, and might perhaps have been more distinct and larger before December 3. On the night of January 6 there was no trace of it. He closes this final letter by saying that he would have gladly learned that some other astronomer had observed this star and confirmed its motion, and expressing his regret that Bode had not succeeded in finding it. On the latter point Bode remarks that the weather during December had been but very rarely favourable for observation, and in the few moments that the sky was clear he had occupied himself with his 'Seeker' and Dollond, partly in giving attention to the neighbourhood of Huth's star, and partly to the region in which Ceres was expected to be recovered on her second appearance. He also remarks on the imperfect manner in which the star's positions had been communicated to him, but concludes that 'without doubt it was a distant comet,' and its great distance caused it to appear without nebulosity. He supposes it on December 3 to have been in longitude 156° 20', with latitude 10° 40' north, and on January 2 in 154° 20', with latitude 8° 50'. Huth's rough diagrams are reproduced in the* Berliner Jahrbuch, *1805, but they are on a very small sale, and no two persons are likely, perhaps, to agree as to the inferences to be drawn from them* (Fig. 5.2). *We may remark, however, that the arc of great circle between Bode's extreme positions exceeds the length of the path, as described in*

**Fig. 5.2** *One of the small diagrams in Huth's* 1802 *paper, from the BAJ for 1805*

*Huth's letters. The following places result from an examination of the figures with the particular view to identify several of the telescopic stars entered in the larger diagram:*

| | | | | | |
|---|---|---|---|---|---|
| *1801* | *Dec 3* | *Longitude* | *157°.0* | *Latitude* | *... +10°.5* |
| *1801* | *Dec 14* | *Longitude* | *156.7* | *Latitude* | *... + 9.9* |
| *1802* | *Jan 1* | *Longitude* | *156.2* | *Latitude* | *... + 9.1* |

*Calculations founded upon the deductions from Huth's diagram lead to no satisfactory, indeed no probable, results. The ordinary formulae fail, but the distance of such an object could hardly have been great.*

*With regard to the* bona fides *of Huth's observations, it is worthy of remark that he wrote several letters to Bode, while according to his own showing, observations would have been very practicable, but for the unusual prevalence of clouded skies; while there is no doubt of the looseness with which he gave its positions.* [end of Nature article]

Huth theorized about the origin of the asteroids in a letter to William Herschel on September 10, 1802, which was printed in *Early Investigations of Ceres and the Discovery of Pallas*. He expanded on his views in a letter to Bode almost exactly two years later in which he alludes to Ludwig's star

*I saw Pallas and Ceres several times again and found, that both follow accurately the path on the maps of the MC drawn by Mr. Harding. I congratulate our gifted Mr. Harding from the bottom of my heart on the discovery of the third planet [Juno]. I hope that this planet is not the last one that will be found between Mars and Jupiter. I think it very probable that these little planets are as old as the others and that the planetary mass in the space between Mars and Jupiter has coagulated in many little spheres, almost all of the same dimensions, at the same time in which happened the separation of the celestial fluid and the coagulation of the other planets.*

*That they were discovered only now is no proof of a newer origin, but has its reason partly in the hitherto incomplete knowledge of smaller stars, partly in the current use of better telescopes and the present eagerness of astronomers to observe. I will not be surprised, if Ceres and Pallas will get at least 10 other coplanets – I rather call these celestial bodies this and I believe this name more appropriate than asteroids. But because of their smallness they will always be hard to discover and find again, until we have very detailed special maps, at least of the zodiac. That these small planets roam individually on almost the same path (region), seems to me to be caused by the fact that in this region or orbit is no large planet, whose dominant gravitational force would gather and keep the smaller ones around. I already expressed these thoughts in a letter of September 1802 to Mr. Herschel, which I sent from London to Slough. But I do not know what this famous and gifted astronomer thinks of it, since he neither replied to them nor to the remarks I made regarding his refusal to regard Ceres and Pallas as true planets. – As soon as the sky is clear again, I will look for the new arrival and inform you what I have found.*

*The so-called Ludwig's star, which forms together with Mizar and Alcor a scalene triangle, always appears of white light and never with clear but nebulous boundaries, even then, when Mizar and Alcor are visible as perfectly determined disks. The other day, when Prof. Wrede was visiting, he convinced himself at my Dollond of the disk-like shape of the fixed stars, like small suns without irradiation. A trained eye can, using this instrument, compare the diameter as well as the distance of close stars accurately* (Huth 1804).

The visitor Huth mentions is Ernst Friedrich Wrede (1766–1826), a Protestant theologian who was natural history teacher at the *Gymnasium* in Berlin. He later became Professor of Mathematics at the University of Königsberg in 1806. He wrote extensively on meteors and geology, including the asteroids, and in this context we will encounter him again in the next book in this series. The letter by Huth shows even educated observers at the time deluded themselves into believing they could resolve stellar disks. It is no wonder that so many ghost planets have populated the Solar System over the centuries.

It appears that 1802 was a prime year for the discovery or speculation about another planet. In 1764 Peter Roedkiaer at Copenhagen Observatory made several observations of what he thought was a satellite of Venus. In 1802, the medical doctor and amateur astronomer Julius August Koch (1752–1817) in Danzig revived interest in these observations, suggesting Roedkiaer had observed the planet Uranus. Other observations made at Copenhagen, Koch said, were quite likely due to an undiscovered planet. "I cannot deny that I strongly believe in such a planet, especially since the discovery of Pallas." (Koch 1802) Observations of Juno by Koch will be given in the next book in this series.

## Cassini's Planet of 1668

Another planet from the vault of forgotten discoveries was unearthed in an English publication in 1839, and it specifically references the asteroids as one of the ways an astronomer could achieve 'sidereal fame.'

*Ever since the discovery of the Georgium Sidus and the four Asteroids, every astronomer is eager to find a new planet, on which, like Herschel and Olbers, he may pin his ticket for sidereal fame. The star-hunters will, therefore, be glad to learn that in* Le Journal des Scavans *for 1667 and 1668 an account is given of a planet, that has often played at bo-peep with astronomers, by appearing and disappearing in the neck of the Whale. It was seen by Cassini, on Jan. 7, 1668, and continued visible till March 10, when it took French leave. It remained, however, long enough during its visit to enable Bonillau to ascertain that it revolved upon its axis, that it was composed of two unequal parts, of which the larger was dark and the smaller luminous, and that it completed its revolution in 333 days* (The Church of England Quarterly Review 1839).

## Cacciatore's Planet

Finally on offer here is a report in the inaugural issue of a New York publication, *The Plaindealer*, in 1836

> *New Planet – The notion for some time past entertained by certain astronomers, that there is another planet, similar to Ceres, Pallas, Juno, and Vesta, is in some measure confirmed by further observations recently laid down before the French Academy of Sciences. The orbit of the new planet is not even conjectured.*

The subject of this report is none other than Niccolò Cacciatore (1770–1841; Fig. 5.3), the co-discoverer of Ceres! Many people were convinced a fifth asteroid had finally been discovered, and it even featured in a revised map of the Solar System in an 1837 book (Fig. 5.4) by the Boston-based author Samuel Griswold Goodrich (1793–1860; Fig. 5.5), writing under the pseudonym Peter Parley

> *But* newer *still, than either of the four little planets of which I have just now told you, is a fifth little planet of this group, discovered only last year (1836) by the Italian astronomer Cacciatore.*

**Fig. 5.3** Niccolò Cacciatore

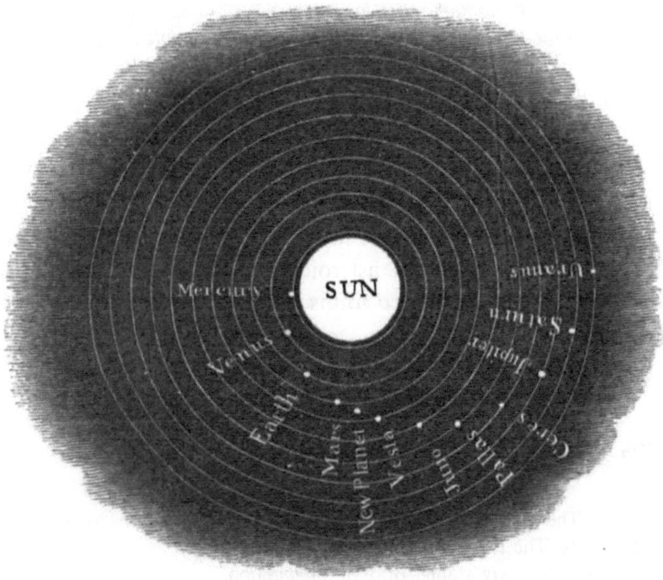


**Fig. 5.4** A map of the Solar System in 1837 includes five asteroids. From Parley (1837)

**Fig. 5.5** Samuel Griswold Goodrich

*I have not yet heard of any name that has been given to this planet; but its existence is confirmed to us by the testimony of the well-known and estimable astronomer and natural philosopher, M. Valz, of Nismes; who estimates its year, or period of revolution in its orbit, at three of our terrestrial years.*

*In my representation of the planetary system, at page 176 and page 202, I have ventured to place this New Planet of Cacciatore between Mars and Vesta* (Parley 1837: 226).

The astronomer he refers to is Benjamin Valz (1787–1867), who was born in Nimes. In 1836 he became Director of Marseilles Observatory. His name was later associated with the discovery of the asteroids 20 Massalia and 25 Phocaea, but the discoverer for both is now attributed to his Marseilles Observatory colleague Jean Chacornac (1823–1873).

## References


Ashbrook, J.: 1957, The Field of Mizar and Alcor. *Sky & Telescope* 16, April issue, 265.

Cunningham, C.: 2004, The Planet Hercules. *Mercury* 33(5), 10.

Cunningham, C.: 2017, *Ludwig's Star*. Book in preparation.

Hellwig, C. von: 1801, *Neu bearbeiteter hundertjähriger Haus-Kalender*. Johann Andreas Heinrich, Grätz.

Huth, J.: 1802, Beobachtung eines beweglichen Sterns im Löwen. *Berlin Astronomical Yearbook* for 1805, 215-217.

Huth, J.: 1804, Letter of 21 September to Bode. *Berlin Astronomical Yearbook* of 1807, 264.

Koch, J. A.: 1802, Letter dated 24 August. Quoted in Kragh, H.: 2008, T*he Moon That Wasn't: The Saga of Venus' Spurious Satellite*, pg. 112. Birkhauser, Basel.

Liebknecht, J.G.: 1723, *Colegae uberior stellae Ludovicianae noviter detectae*. Gissae, litteris Joh. Mülleri.

*Nature,* Aug. 3 issue, pg. 291-292 (1876).

Olbers, W.: 1805. Letter to Gauss, dated January 20. Goettingen archives.

Parley, P.: 1837, *Tales about the Sun, Moon and Stars*, second edition. Thomas Tegg, London.

*The Church of England Quarterly Review*, 6, 134 (1839).

*The Plaindealer*, December 3 issue. Vol. 1(1), 11 (1836).

Tisserand, F.: 1882, Notice of French Observatories Towards the End of the Last Century. *The Astronomical Register* 19 (225), 218-226.

Wilson, P.: 1804. Letter to William Herschel, dated 19 May. RAS, Herschel archives, 13.W.145.

Zach, F.: 1805. Letter to Carl Gauss, dated 25 November. Goettingen University archives.

# Chapter 6
# New Planets: The Transition from 1745 to 1804

## The Planetary Gap

In the first two years of the nineteenth century man's entire outlook on the cosmos underwent a transformation with the discovery of Ceres and Pallas. The Solar System of the eighteenth century was outdated (Fig. 6.1). As we saw at the beginning of this book, it was not entirely unexpected, as Kepler, Bernoulli, Zach, Bode and Lofft had posited the existence of a new planet. This chapter begins with two eighteenth century views of the planetary distances from an English perspective, followed by an early nineteenth century overview by Wilhelm Olbers on the reality of not one but two new planets.

The Scottish metaphysician Andrew Baxter (1686–1750) wrote a book in English in 1745 that was an expanded version of his Latin text from 1738. It consists of a dialog between two people, Matho and Philon. Here they discuss the planetary distances (italics in original):

> M. And now methink I see the Reason why *Jupiter* is placed at such a Distance from *Mars*, and *Saturn* from *Jupiter*; as also, why the next inferior Planet to *these two* ought to have been but small.
>
> P. What do you take to be the Reason of these Things?
>
> M. Those two huge Bodies [Jupiter and Saturn], I suppose, with their Train of Satellites, must by their Attraction have disturbed each other, had not the Space between them been large enough to prevent that Inconvenience: And had the next inferior Planet been equal to either of them, that must have increased the Disorder. What you said a little before seems to me extremely reasonable, *That there can be no Chasms, nor unnecessary vacant Spaces in the System*...Then it seems, placing the little Planet Mars in the Gap between our Earth and Jupiter, as also the whole Disposition of the Planets, was not merely arbitrary, and indifferent, as is generally thought, but the Result of pre-contrived Order and Convenience.
>
> P. You will perhaps see more of this as you go on. When we have weighed Things maturely, and considered the Forces of the Sun and Planets, according to the known Laws of Gravitation; we shall find that this large Interval, which we may think employed to little Purpose, is in the main the greatest Frugality of Space.



C.J. Cunningham, *Studies of Pallas in the Early Nineteenth Century*,
DOI 10.1007/978-3-319-32848-5_6

**Fig. 6.1** An eighteenth century engraving of the Solar System

This passage, with its specific invocation that gaps or intervals in the distances of the planets cannot be allowed, is quite remarkable. Baxter wrote about the large interval between the Earth and Jupiter; whether the author had read Kant (1755: 179) or not, the wording used to express his sentiments about this interval is nearly the same used by Kant about the Mars-Jupiter interval: "It would be as if one wished to complain that the space between Jupiter and Mars was unnecessarily empty." (See *Discovery of the First Asteroid, Ceres,* for the full quote by Kant, and Zach's efforts described in the following section.) The problem actually goes back to Ptolemy. In his system of nested spheres, he assumed there are no empty spaces between the spheres, but Ptolemy "himself violated it in one instance by leaving an empty space between the spheres of Venus and the Sun. This space was disturbing to many medieval astronomers who used any number of devices to get rid of it." (Goldstein 1985: 369) The one instance mentioned is in his text *Planetary Hypotheses*, written after the *Almagest*, in which "Ptolemy addressed wider physical and cosmological issues and developed a coherent scheme of sizes and distances of all the heavenly bodies." (Van Helden 1985: 21)

The philosophical concept of a gap is just as important in this respect as an actual gap between the planets. In his *History of Astronomy*, Adam Smith (1723–1790; Fig. 6.2) wrote about finding one or more objects that elicit surprise and wonder. Most of the work was written between 1740 and 1746 when Smith was a student at Balliol College, Oxford, but it was not published until half a century later. His analysis is directly applicable to the discovery of Ceres and Pallas, even though he wrote long before they were found. He talks first about a coherent chain of things, which can be likened to the order the planets, and then moves to the effect on the human imagination when that order is interrupted.

**Fig. 6.2** Adam Smith, National Galleries Scotland

> *But if this customary connection be interrupted, if one or more objects appear in an order quite different from that to which the imagination has been accustomed, and for which it is prepared, the contrary of all this happens. We are at first surprised by the unexpectedness of the new appearance, and when that momentary emotion is over, we still wonder how it came to occur in that place. The fancy is stopped and interrupted in that natural movement or career, according to which it was proceeding. Those two events seem to stand at a distance from each other; it endeavours to bring them together, but they refuse to unite; and it feels, or imagines it feels, something like a gap or interval betwixt them* (Smith 1795: 18).

For Smith, the successful outcome of surprise and wonder leads directly to a third sentiment, admiration of the natural order, relatable for the analysis in this book as the incorporation of the asteroids into the natural order of the Solar System. In a study of Smith's *History of Astronomy*, the Scottish historian of philosophy at the University of Glasgow Alexander Broadie (2012) concludes that "Having discovered a new principle of unity amidst the diversity of natural phenomena, nature presents itself to us an even more magnificent spectacle than we had previously found it to be. Where we had seen disparity we now see coherence." The idea of a more magnificent spectacle echoes the words of Herschel, who believed the discovery of a new class of objects was just as important as the discovery of another planet.

## A Wild and Romantic Idea

The late eighteenth century witnessed several suggestions about a missing planet. Johann Bode gave a chart of the Solar System in his 1778 book that showed its position (Fig. 6.3). In 1785 Franz von Zach wrote down the orbital elements of the

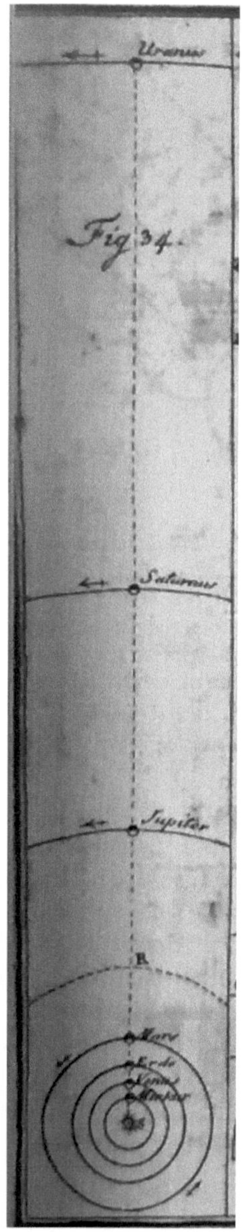


**Fig. 6.3** This diagram in Johann Bode's 1778 book *Terse Explanation of Astronomy* shows the orbital location of an unknown planet between Mars and Jupiter, denoted by the *letter R*, 23 years before the discovery of Ceres. In the following lecture delivered in 1802, Wilhelm Olbers credits Bode with suggesting a missing planet between Mars and Jupiter

supposed planet. And in the time between these events, in 1781, William Herschel actually discovered another planet beyond Saturn.

At the very end of the eighteenth century Zach and others in Germany formed a society to search for the missing planet, and in England the scientific dilettante Lofft (1751–1824; 1800) fully expected a planet would be found in the gap between Mars and Jupiter. At the outset of his exposition, Lofft employs the concepts of proportion and harmony. In the field of mathematics, he attributes this to Colin Maclaurin, but it is equally important to realize that in the field of philosophy, Francis Hutcheson (1694–1746; Fig. 6.4) expounded the belief that humans are endowed with a special sense that enables us to perceive harmony and proportion.

Hutcheson, whose work was grounded in Locke's philosophy, is regarded as the father of the Scottish Enlightenment. As we saw in *Discovery of the First Asteroid, Ceres,* Kepler believed the structure of the planetary system was derived from the Platonic solids, known as regular polyhedra (tetrahedron, cube, octahedron, dodecahedron, and icosahedron). Each of these can be constructed by triangles. Hutcheson (1753: 17–18) used these solids to illustrate his concepts of harmony and proportion, which he terms here "equal uniformity." His alternative spelling of these Platonic solids is retained in the following quote; the words he italicized are given here in regular type.

**Fig. 6.4** Francis Hutcheson painting by Allan Ramsay. Hunterian Museum and Art Gallery

> *Since it is certain that we have* Ideas *of Beauty and Harmony, let us examine what* Quality *in Objects excites these Ideas, or is the Occasion of them...First, the* Variety *increases the Beauty in equal Uniformity. The Beauty of an equilateral Triangle is less than that of the Square; which is less than that of a Pentagon; and this again is surpass'd by the Hexagon. When indeed the Number of Sides is much increas'd, the Proportion of them to the* Radius, *or* Diameter *of the Figure, or of the* Circle, *to which regular* Polygons *have an obvious Relation, is so much lost to our Observation, that the Beauty does not always increase with the Number of Sides; and the want of Parallelism in the Sides of* Heptagons, *and other Figures of odd Numbers, may also diminish their Beauty. So in* Solids, *the* Eicosiedron *surpasses the* Dodecaedron, *and this the* Octaedron, *which is still more beautiful than the* Cube*; and this again surpasses the regular* Pyramid: *The obvious Ground of this, is greater* Variety *with equal* Uniformity.

Hutcheson is writing here about aesthetics, the idea of beauty that also infused the work of Kepler, whose "outlook and method were as fully dominated by an aesthetic as by a purely theoretic interest, and the whole of his work was overlaid and confused by crude inherited superstitions which the most enlightened people of his time had already discarded." (Burtt 1932: 60) In his deployment of the ideas beauty and harmony, Lofft also embraces this concept of aesthetics in his tract about the planetary distances. Professor George Arabatzis of the University of Athens cogently links the work of Hutcheson to physics, geometry, and the philosophy of Locke.

> *The demand for a definition of the specificity of the aesthetic sense emerge* [from Hutcheson] *as a consequence of the epistemological distinction between nature given in sensation and nature conceived through mathematical formulas and verified by experimental physics in the work of Galileo and Newton. It was Galileo that first posed and, later, Newton that stressed the distinction between primary and secondary qualities which became central in Locke's philosophy. Primary qualities are geometrical properties not sensed; secondary qualities are impressions that need an observer in order to be taken into account* (Arabatzis 2004: 108).

Hutcheson assumes even greater importance when one considers that his body of work was the link between the philosophy of Locke and the Scottish school of Common Sense Realism that formed the basis of Brougham's writings on science, including the asteroids. In his *History of Astronomy*, Hutcheson's pupil Adam Smith has this to say on the elements of harmony, proportion, and Pythagoras that open Lofft's tract. After outlining a universe in which Earth finds itself at the center, Smith writes:

> ...[it is] *surrounded by the elements of Air and Ether, and covered by eight polished and cristalline (sic) Spheres, each of which was distinguished by one or more beautiful and luminous bodies, and all of which revolved round their common centre, by varied, but by equable and proportionable motions. It seems to have been the beauty of this system that gave Plato the notion of something like an harmonic proportion, to be discovered in the motions and distances of the heavenly bodies; and which suggested to the earlier Pythagoreans, the celebrated fancy of the Musick of the Spheres: a wild and romantic idea, yet such as does not ill correspond with that admiration, which so beautiful a system, recommended too by the graces of novelty, is apt to inspire* (Smith 1795: 45).

Here we see that Smith, most famous for his landmark 1776 book *The Wealth of Nations*, invokes the concept of beauty twice in his discussion of the planetary system. This aesthetic sense derives directly from his teacher Hutcheson, and influenced his admirers including Dugald Stewart. The tract by Lofft follows. It is entitled *The Athenian Letters* (italics in original appear in regular type):

> *The turn of the genius of* Pythagoras, *the circumstances, probably, of his times, his conferences with the Egyptian sages, and the objects which impress'd themselves during his travels in that country, all gave an allegorical form, partaking of the nature of* hieroglyphic, *to his mode of expression. Hence the precept* not to stir fire with a sword; – not to turn back on a journey; *– and many others.*
>
> *This has been believ'd, and particularly by a very great philosopher (Maclaurin, in his View of the Discoveries of Sir Isaac Newton, p. 34), to account for the Pythagorean language concerning the music of harmony of the spheres. A musical chord gives, says Maclaurin* (Fig. 6.5), *the same notes with one double in length, when the tension or force with which the latter is strecht is quadruple; and the gravity of a planet is quadruple of the gravity of a planet at a double distance. From this similitude of proportions, he adds, the celebrated harmony of the spheres is suppos'd to be deriv'd.*
>
> *Another way a certain harmonical proportion between the planetary distances may be consider'd.*
>
> *If these distances, according to the latest observations, reduc'd to the lowest terms, be thus taken,*

**Fig. 6.5** Colin Maclaurin

| Mercury | Venus | Earth | Mars | Jupiter | Saturn | The Herschelian |
|---|---|---|---|---|---|---|
| 4 | 7 | 10 | 15 | 52 | 95 | 180 |

*then double the distance of Venus from the SUN will be equal to the sum of the distances of the Earth and Mercury; double the distance of the Earth from the Sun will be very nearly equal to the sum of the distances of Mars and Venus; the difference being only 1–10 of the double distance of the Earth; double the distance of Jupiter from the SUN will be nearly equal to the sum of the distances of Saturn and Mars, the difference being less than 1–17 of the intermediate doubled; four times the distance of Mars from the Sun will be nearly equal to the distance of Jupiter and Saturn from the Sun added to each other; while the NEW PLANET will very nearly equal the distances of all the others added together: they exceeding it by 3 only; which is but 1–60 of the distance of the newly-discovered Planet.*

*If, as many have suspected, an intermediate planet be suppos'd between Mars and Jupiter, whose distance to that of Mars should be as 33 to 15, the proportion of the distance of each intermediate planet added to itself, being equal to the sum of the two extremes added together, would very accurately obtain for five of the planets nearest the Sun; and in that supposition the sum of the distances of the four inferior planets would be nearly equall'd by the distance of the 5$^{th}$, the sum of the distances of the 8$^{th}$ planet would exactly equal the sum of the distances of the three next to it, and the distance of the 7$^{th}$ planet, which would thus be Saturn, would almost exactly equal the distance of the three planets next below him.*

*If a planet were taken at ¼ a* greater distance than Mars from *the Sun, the double distance of Mars would equal the sum of the distance of its two extremes on either side, consider'd as a mean term, the suppos'd planet and the earth. If it be objected that the attraction of Jupiter would be too considerable to allow of our supposing a Planet thus situated, it would be small indeed to that of Jupiter and Saturn on the Herschelian Planet.*

*There seems, therefore, to be much reason to expect the discovery of such an intermediate Planet. (In Long's Astronomy, Art. 938, there seems considerable presumption that two stars set down were a Planet with ecliptic progress; and by the date it seems that the Herschelian would not then have been in that part of the ecliptic.)*

*For it seems very improbable, when even the Planets most distant from the Sun, and most liable therefore to disturbance from the attraction of neighbouring planets, (as their gravitation toward the Sun diminishes by their distance, and is as its square inversely,) never amount quite to double the distance of their nearest planet from the Sun, that Mars, so much nearer, should have the nearest planet beyond him above quadruple of his distance from the Sun.*

*If such a Planet be suppos'd to exist, and to be 1/9 less in diameter than Mars, about equal to the differences of the latest computed proportion between Jupiter and Saturn, (and it cannot well be imagin'd more than this,) its diameter in PERIGEE, at five times the nearest distance of Mars from the Earth, would be about 5", which is less than half the diameter of Mercury; it would perhaps more probably be to Mars, as the diameter of Mars to that of the Earth; and then, being not much more than half the diameter of Mars, and at five times the* perigean *distance, it could be seen from the Earth under an angle of only about 2 ½ or 3 seconds. Now, the new discover'd Planet [Uranus], which is scarcely at all discernible by the eye, and with some difficulty by a good telescope of considerable power, though its position and appearance have been for nineteen years back since its discovery nearly ascertain'd, is observ'd to have an apparent diameter of four seconds. A planet, therefore, of not more than three seconds might well be expected to elude the eye, and hitherto even the telescope; considering the total uncertainty near what part of the ecliptic we should be*

> *to look for it, and that its apparent motion, if at twice the distance of Mars, would be small. If its superior light above that of the Herschelian planet, from its comparative nearness, be objected, that light would depend on other circumstances also: if subject to such refraction as Mars appears to suffer, it would be very weak and obscure at such a distance, and, if seen under an angle of five or six seconds only, perhaps hardly discernible. Mars, when near to his APOGEE, and seen under an angle of two or three seconds, would disappear to us, if his place were not so well known. Now, even in perigee, or nearest distance to the earth, the apparent diameter of such a planet can hardly be suppos'd to exceed Mars in its apogee; and in apogee such a planet, even if its place were known, would hardly indeed be visible even to the best telescope.*
>
> *But whether such a planet exist or not, some judgement may be form'd from what* does *appear of the principle and application of the* Pythagorean *harmony of the* planetary *spheres. The* idea *indeed of* harmony *in the full extent of the term,* adaptation, accord, congruity, *beautiful and sublime* fitness *of the means to the* best *and most* perfect *end, with the charm and excellence of those means (speaking now of the most general and prevailing), when consider'd in themselves, powerfully must strike the mind of an observer of nature. In the Pythagorean philosophy this expression of* Harmony *is applied as significant of* health, beauty, wisdom, virtue, *all* excellence, *even DEITY itself. Friendship is beautifully term'd* an harmonious equality. *And thus of Harmony it is affirm'd, that* by it all things consist.
>
> Proportion *of every kind must have been interesting to a mind thus attun'd. The law of musical chords, and the relation of length or tension in their several harmonic parts, has already been notic'd as discover'd by* Pythagoras; *– to him also is ascrib'd the completing of the* octochord *of the lyre; – to him the introduction, which must mean apparently the more precise regulation, with a just theory of their mechanical principle, of* weights and measures *among the* Greeks; *– to him the adjustment and* harmonizing *of the* civil *polity, and, what is much more, the* manners *of* Greek *and* Italian *states; – to him the appeasing, by an influence which seem'd divine, of the dissentions [sic] of rival Nations.* [end of Lofft's tract]

In classical mythology, the lyre (shown in Fig. 6.6 being held by Apollo) is closely associated with astronomy, whose muse is Urania (at the right, resting her elbow on a crystal sphere studded with stars). In 1806, James Leslie even suggested the word uranoscopist as a synonym for astronomer! Like the discoverer of Ceres, Giuseppe Piazzi, Tycho Brahe (1546–1601) felt he was personally touched by Urania, as he wrote in his 1573 Latin poem *In Urania:* "Young man, you should not turn away from me. I am the goddess who carries the name Urania. I pass under this name because astronomy is under my protection. Among the muses of Apollo, the sweetest deity, no one is more beloved by magnificent Jupiter than I am." (Bjork 2017)

The reference to the octochord accorded with the belief Pythagoras "added an eighth string in order to produce the overall range of an octave…The discovery of the relationship between the planets and the seven strings of the lyre is attributed to Orpheus." (Mathieson 1999: 244, 246) The poet Campbell (1777–1844: 1800: 84), who met with Herschel in 1813 (see the next book in this series), penned a couplet that links Benjamin Franklin's capture of lightning with Herschel's discovery of Uranus that added the seventh string to the heavenly lyre.

**Fig. 6.6** Apollo, God of Light, Eloquence, Poetry and the Fine Arts with Urania, Muse of Astronomy. This painting by Charles Meynier (1768–1832) was done between 1789 and 1800 (Cleveland Museum of Art)

*With Franklin's grasp the lightning's fiery wing,*

*Or yield the lyre of heaven another string.*

This idea of the music of the universe was transmitted from ancient times through the so-called Dark Ages by the philosopher Boethius (480–524) who wrote about *musica mundana* (see Boethius 1867). Lofft of course is alluding to the harmonic proportion between the planetary distances, which was famously noted by Olbers.

**Fig. 6.7** The nexus between Renaissance art and science is shown in this painting of the mathematician Fra Luca Pacioli (1447–1517) by Jacopo de Barbari. The glass polyhedron (actually constructed by Pacioli), with its reflection, refraction and perspective, has been interpreted as "a grand symbol of humanity's quest for perfection." (Harrison 2005: 34). Leonardo da Vinci (1452–1519) sketched this polyhedron for Pacioli's 1509 book *De Divina Proportione.* On the table at right is the dodecahedron made of wood, one of the polyhedrons Kepler used in his model of the solar system that led to his suggestion there was a planet between Mars and Jupiter. Museo di Capodimonte, Naples (For more on the painting's history, see Taylor, 1980: 204)

He also links the mind of Pythagoras to the concept of proportion; the animating influence of Pythagoras can further be found in the writing of Bartholomew Prescot (see Chap. 10). That such notions are not mere alchemical philosophy was clearly elucidated by Mark Peterson, Professor of Physics and Mathematics at Mt. Holyoke College, who claims it was the mathematics of Renaissance arts—not Renaissance science—that became modern science (Fig. 6.7):

> *When Galileo finally summarized in* Two New Sciences *the results of those experiments in music completed with his father so long ago, he did it in terms of proportionalities, although not the proportions of Pythagorean harmonies ... The notion of proportion, central in all the arts, took on a new significance in Galileo's work. The existence of unsuspected proportions in nature, waiting to be discovered, became a unifying theme in his thought* (Peterson 2011: 289).

Thus Lofft (in England) and Bode (in Germany) were not just playing with numerology but working in the scientific tradition of Galileo. While modern science gives little credence to Bode's Law, their search for unsuspected proportions in nature cannot be dismissed. Furthermore, the search for beauty has become an essential tool in modern physics, as explained by the 2004 Nobel laureate in physics Wilczek (2015). The role of Renaissance thought on the harmony of the spheres will be considered further in the next book in this series.

Aside from the flights of fancy about Pythagoras expressed at the end of his text, Lofft lays out a compelling case to expect a new planetary discovery. Even he might have been surprised to know it would actually happen at the beginning of the following year, but his text can also be seen as a missed opportunity. Even a century earlier Bernoulli offered precise numbers for two orbital elements of the missing planet. The fact Lofft did not do even this robbed him of the accolades that would certainly have accrued to him if his numbers had been reasonably accurate. For Lofft's views of the asteroids after 1800, see *Discovery of the First Asteroid, Ceres.*

## Olbers' Asteroid Lectures

Here follows two lectures about the asteroids, delivered by Wilhelm Olbers to the *Museum Society of Bremen* (Bremer Gesellschaft Museum). Although he does allude to the ancient Egyptians, Olbers, as a sober man of science, steers clear of any mention of the Pythagorean influence that animated the more poetically inclined Lofft. The handwritten manuscripts of these lectures, in the Bremen University archives, have never before been published.

These texts represent unique documents in the history of asteroid research, as they are the only lectures ever given by a discoverer of one of the first four asteroids. Even the papers read to the royal societies are of a different genre. Herschel's papers to the Royal Society in London were not even read by him, and any address by Lalande to the Paris Academy typically just included the asteroids in his all-encompassing annual report on the history of astronomy.

One must try to imagine Olbers speaking before a relatively small audience in Bremen, one composed not of astronomers but a few learned men in a town far away from the main centres of learning such as Goettingen or Berlin. At the time he was speaking, its population was less than 35,000. His well-reasoned presentations in 1802 and 1803 were the careful product of a rhetorically astute mind. Here we see him elucidate the latest findings and persuasively argue one of the most startling hypotheses ever promulgated—the explosion of an ancient planet. How he approached it was to build a case of particulars (the two asteroids) and bind them in an associative way to garner the force needed for his generalization. According to Priestley, judgment happens by a similar premise and is no less rooted in the psychology of association (Hawhee and Holding 2010). In another use of analogy (see Chap. 2), Priestley (1777: 164) writes that:

> *[T]he moment we perceive that the parts of any object are analogous to one another, and find, or are informed, what that analogy is, the sight of a part, without any farther investigation, suggests the idea of the whole; and the judgment is most agreeably and successfully employed in completing the image.*

In the case of Olbers in his second lecture, he showed that a protoplanet now consists of various parts, and he informs his audience of the analogy that leads him to his planetary explosion hypothesis: "Why here instead of one larger planet, which we had been expecting and suspecting due to analogy, at least two?" asks Olbers rhetorically. While there is no first-hand account from a member of the audience, it seems quite likely they used their judgment to complete the image Olbers was painting—the image of an exploding planet.

## The First Lecture (October 25, 1802)

*I would like to inform you today about the most important part of these astronomical discoveries, a small treatise on the two new planets of our Solar System, Ceres and Pallas. The five bright major planets, visible to the naked eye, were already known in ancient times. All of these five planets are so bright and appear, at least to the naked eye, of equal size as any fixed star, often even brighter and bigger. Their motion and changing positions soon distinguished them from the fixed stars. Most likely, Venus was the first to be recognized as a planet, then soon Jupiter and Mars and later Saturn. It is strange that the calculators of Babylon's plains and the banks of Egypt's Nile knew Mercury so early* (Fig. 6.8) *as a planet since it is always so close to the Sun and almost always only visible for a short time in dusk and dawn* (see Ossendrijver 2012: 68)

*It is questionable whether the Greeks knew all 5 planets before Eudoxus' and Plato's times, more than 400 years BC, or whether Eudoxus brought this knowledge from Egypt. Deceived by the brightness most of the old astronomers considered Earth the center of the universe, and believed all those 5 planets among which they also counted the Sun and the Moon, revolved around the Earth on orbits of different size. The number 7, which is thus obtained, was a holy number: those 7 planets were handed down together with all dreams and Chimaeras of astrology which had been ruling for so long and during the Dark Ages the question whether there might be some more planets could not arise.*

*Finally, Copernicus taught us around the middle of the 16th century the true nature of the universe and his system, at first so contrary to all sensual perception, which was rejected by the Church as heretical and absurd by the Aristotelians has become little by little mathematic certainty. According to him the Sun and not the Earth is the center of the planetary system: the Earth is a planet just like the others and it moves accompanied by its moon between Venus and Mars with them on an almost circular orbit around the Sun.*

*So there were 6 main planets and Copernicus' idea, concordant with truth, soon delivered means to determine the dimensions of these orbits and the ratio of their distances from the Sun. Kepler's genius, based on Tycho's* (Fig. 6.9) *exact observations, discovered the eternal laws of their motion, and now the ratio of the diameters of these 6 planetary orbits became known with certainty.*

**Fig. 6.8** Cuneiform tablet from the 1st millennium B.C. regarding the visibility of Mercury. The Metropolitan Museum of Art, New York. Accession number: 86.11.287d

**Fig. 6.9** Tycho Brahe

*At the beginning of the 17^th^ century the telescope was invented and soon pointed at unknown celestial objects and by and by perfected. Very soon the 4 moons of Jupiter were discovered and from 1655 till 1684 the satellites of Saturn. The knowledge of our Solar System remained on this level for about 100 years: 6 known main planets and 10 minor planets.*

*Kepler's lively and fancy imagination had produced many speculations about the arrangements of the planetary system* (Fig. 6.10). *At first he believed to have found that it was possible to sketch between the 6 planetary orbits exactly the 5 only possible regular geometric bodies and this fact seemed to be reason and cause for their number of 6 and their different distances from the Sun. Then he compared these distances with musical chords, which were in beautiful harmony except for the ratio of the distance between Mars and Jupiter, where there was a dissonance. This gap between the orbits of Mars and Jupiter appeared too great to several astronomers and philosophers, who were either taken aback by its existence or tried to explain this supposed anomaly.*

*In a translation of Bonnet's* Contemplation of Nature *by professor Titius, published in 1767* [actually 1766], *our attention was drawn for the first time to the simple progression that the planets observe in their distances from the Sun. If the distance of Mercury = 4 the distance of Venus is consequently 4 and 3 = 7, of Earth 4 and 6 = 10, of Mars 4 and 12 = 16. Now*

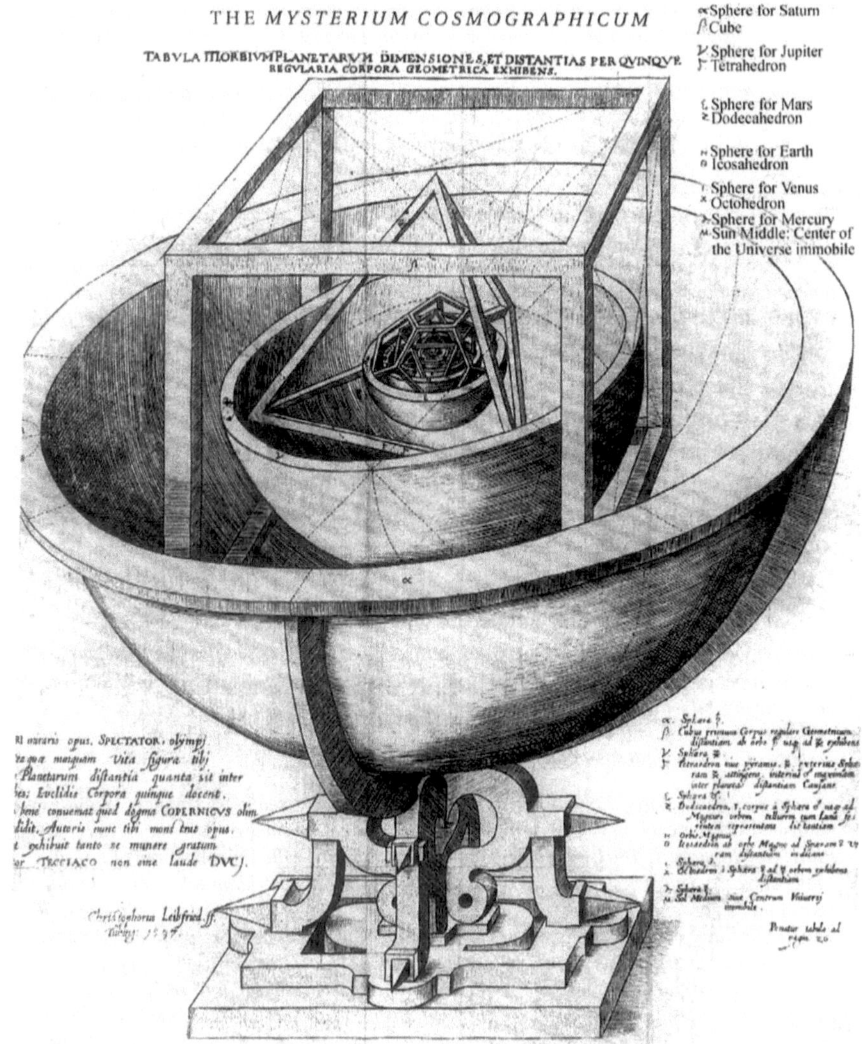


**Fig. 6.10** Kepler's diagram of the relationship between the planetary orbits and geometric bodies, from his book *Mysterium Cosmographicum*, with the English translation. From mifami.org

*there should be a planet whose distance is 4 and 24 or 28: but it was missing. Then the progression continues orderly since the distance of Jupiter is 4 and 48 or 52 and of Saturn 4 and 96 or 100.*

*Mr. Bode has been favouring this idea of the progression since 1772 and expressed repeatedly that there might be a (due to its faintness) undiscovered planet between Mars and Jupiter and also that beyond Jupiter's orbit several very distant planets might exist which had either not been discovered because of their great distances or not been distinguished from fixed stars.*

*On March 13, 1781, Herschel, who became so famous afterwards, then musician in Bath in England, discovered with his telescope in the region between Taurus and Gemini, which he examined searching for something else, a star of the 6$^{th}$ order of magnitude, which distinguished itself from the other fixed stars by a discernable diameter at a magnification of 227 times. Two days later he noticed that the new star had changed position. Now he announced his discovery to other astronomers and on March 17 Maskelyne began observing the new-found star, and on April 16 Messier.*

*At first it was considered a strange comet, and only continued observations led to the assumption it was a planet, until after approximately 6 months the general conviction was that Herschel's discovered star was really a planet which moved beyond Saturn on an almost circular orbit around the Sun like the other planets do and which is now known as Uranus. It was very peculiar that also Uranus agreed in his distance from the Sun with the law of distances that was derived from the other planets. His previously known distance must have been 4 and 192 or 196 of those above-mentioned parts from the Sun, and really, the radius of his orbit deviated only slightly. This new proof of the harmonic progression gave the assumption that there must be another planet between Mars and Jupiter more value and prestige. It was searched after diligently. Herschel examined the whole zodiac five times without finding any trace of the suspected planet. La Lande ventured also under this aspect the observation of his 50,000 stars and von Zach his new catalogue of zodiacal stars: but neither found a new planet.*

*When in September of the year 1800 several astronomers met in Lilienthal they talked among other things about this suspected planet and the means to find it and this was the trigger for founding the* Astronomical Society of Lilienthal, *whose aim was to discover this planet. It was agreed to invite 24 members to this society, to divide the zodiac into 24 parts or departments, of which each was assigned to one member for examination and inspection. Hardly had this plan become public when Burckhardt and Le Francais in Paris explained that they had already made a pact with each other to find those planets that perhaps had been undiscovered.*

*During all these efforts a typo in Wollaston's catalogue of fixed stars* [Francis Wollaston published his star catalog in 1789] *prompted the royal astronomer at Palermo in Italy, Mr. Piazzi, on January 1, 1801, to observe the 87$^{th}$ star of La Caille* [Nicolas-Louis La Caille, 1713–1762] *in Taurus. He also saw at the same time in his meridian telescope a smaller star of the 8$^{th}$ order of magnitude, preceding the other, and determined its position as well. The light of this small star was faint, of the same color as Jupiter, but it appeared to Mr. Piazzi as a regular fixed star of the 8$^{th}$ magnitude. The following day he repeated his observation and the position of the small star was different than that of the previous day. Piazzi suspected an error in his first observation but on January 3 he convinced himself that this small body was not a fixed star but was wandering among the fixed stars rather planet- or comet-like. He followed this increasingly suspicious body, which was initially retrograde, but became direct between January 10 and 11 till February 11, when he was refrained from observing partly by twilight and partly by illness. Piazzi publicised this moving star as a strange comet. But when Mr. Bode learnt about the Piazzian star, he immediately expressed his assumption that this was not a comet but the planet between Mars and Jupiter that has so long been expected.*

*Piazzi withheld his observations for quite some time but finally they became known and Mr. Bode's assumption was completely confirmed. This star, which was so different from a comet in appearance because it did not appear with any nebulosity was not a comet but a planet, it did not move on a parabola but on an almost circular orbit slightly eccentric around the Sun between Mars and Jupiter at exactly that distance where the analogy expected it. Piazzi's observations described only a very small period of time, the star had only covered an arc of 9° around the Sun. For most astronomers this arc seemed too small to derive more exact dimensions of the orbit and mainly contented themselves with assuming a concentric circle with the Sun for the orbit and calculating the radius of this circle from the observations. Only one young ingenious scientist at Brunswick, Dr. Gauss, who was already at the age of 25 among the greatest European mathematicians, had more confidence in Piazzi's observations and found a matching ellipse. Fortunate success crowned these efforts and proved that the ellipse was already very well determined. This did not only redound to Mr. Piazzi's honor because of the accuracy of his observations but to Dr. Gauss's as well because he derived from those the orbit with the utmost skilfulness. And furthermore, one has to admit that we would hardly ever have found Ceres Ferdinandea, as Piazzi named his planet, again. What if Dr. Gauss had failed to calculate? All circular hypotheses give Ceres' position too far west and probably the astronomers would not have looked for her that far east, where she really was.*

*When Piazzi's observations became known in Germany and France, Ceres was still under the rays of the Sun, from under which she would come out again approximately in September. From September on she was searched after diligently and impatiently at all observatories in Germany, France, England, and Italy, although the extremely bad and foggy weather of the autumn of 1801 put obstacles in our way. For a long time all efforts were in vain. Finally, Mr. von Zach in Gotha, observed on December 7 in the region where Ceres should be according to Mr. Gauss's ellipse, four small stars. One of these stars really was Ceres. But continuous overcast skies impeded Mr. von Zach to turn his observation into certainty before January. Here in Bremen the skies were more favorable. The first and second day of January 1802 were completely bright and clear, and on the 2$^{nd}$ I saw to my happy surprise that one small star which I noticed on the 1$^{st}$ had changed its position considerably. In the morning of January 6, I found the star had moved as much as theory demanded for Ceres. On January 6 German and foreign astronomers were informed about the rediscovery of Ceres. Hereupon Bode (on January 15), the astronomers of Paris (January 24), Maskelyne (on February 9), and Piazzi himself (on February 23) saw and started observing.*

*Now the existence of this planet between Mars and Jupiter was reality and the previous gap was filled. Now it was believed our planetary system was known completely up to Uranus, but an unexpected discovery showed that this certainty came too early.*

The next passage by Olbers invokes the concept of chance, a matter explored in Chapter 2. In this context it provides a fascinating contrast with Herschel's statement to Lalande on his discovery of the planet Uranus in 1781. He wrote on September 5, 1782, that "the discovery was not owing to chance" since the planet "must sooner or later fall into my way, and as it was that day the turn of the stars in that neighborhood to be examined I could not very well overlook it." (Schaffer 1981: 15) As for Pallas, Lalande described the discovery by Olbers as a "fortunate accident." (See Chap. 11 for the full quote.)

*When I, as usual, observed Ceres on March 28, I searched with my comet searcher the smaller stars of the northern wing of Virgo, among which it was roaming, in order to become more acquainted with her position. As if by chance my eye fell on that region where I found Ceres on January 1 again, and saw to my surprise at that position a small 7$^{th}$ magnitude star, which certainly had not been there in January. I started examining it more closely and the observations of that evening already showed the motion of the little stranger, which was confirmed the next evening. The newly discovered star was similar to Ceres in light, size and outer appearance, retrograde like Ceres and only more increasing in latitude. In the telescope it appeared slightly less bright or smaller than Ceres. Piazzi had concealed his observations for a long time, but this discovery was publicized so fast,* *that this little moving star could be observed on April 4 in Gotha, April 5 in Berlin, on April 9 in London, and on April 10 in Paris.*

*What was one to think of this unexpected, peculiar star? Its slow regular motion, its appearance, so similar to Ceres' and so very different from all comets, led to the assumption of a planet. But when I tried to calculate at the beginning of April the few previous observations in a circular orbit, which made a good start in the case of Ceres, I found unexpectedly that it was impossible to describe them with the Sun concentric circle. Consequently, the orbit had to be very different from a circle. After April 10 it showed that this orbit was not a parabola but an ellipse, more oval than most of the other planetary orbits. A more accurate calculation was postponed to a time when more observations covering a greater period of time have been obtained and thus could promise a greater reliability of the results of the calculations. But the tireless Dr. Gauss beat all other astronomers and geometers and this incomparable calculator determined already on April 17 the main dimensions of the orbit of the newly discovered celestial body, which was named Pallas. As soon as further observations were made and improved he determined the orbit that made Pallas, just like Ceres, a planet between Mars and Jupiter.*

*Most astronomers did not believe in Pallas' planetism. Bode declared Pallas was a distant comet and the great La Place did not want to accept her as planet since it seemed her motion could not be brought into agreement with his favourite hypothesis. Besides, the French astronomers tried for a long time to determine her orbit in vain. But Pallas was observed and at last her orbit was calculated which put Pallas on the same level as the other planets and finally it was calculated by Burckhardt in Paris at the end of March just as Dr. Gauss had.*

*Here Pallas was observed until July 9 since in our northern latitudes the long lasting twilight of the summer renders the small faint planet invisible in the telescope. Maskelyne still saw Pallas on July 18. Oriani, undisturbed by nocturnal twilight under more southern skies, continued his observations in Milan till August 8. But Mechain's undaunted diligence surpassed all other astronomers. He followed her until the beginning of September. Piazzi was unable to find Pallas.*

*Thus, instead of one planet that had been expected and searched for so long, we found within the short time of 15 months two of them. Now let us have a closer look at these two celestial bodies as far as we know them to date.*

*The orbits of these two heavenly bodies are situated between the orbits of Mars and Jupiter. Their mean distances from the Sun are about of equal size, i.e. 2770 of such parts, of which the mean distance of the Earth from the Sun contains 1,000. This mean distance is thus approximately as great as it should be according to the above-mentioned progression and is about 53,000,000 geographic miles. The period of revolution around the Sun is approximately the same for both: 4 years and 7 months. Both orbits are ellipses differing not much from a circle, but the Sun is not in the center of these circles just as with the other planets. Ceres' orbit is not as oval as that of Mars: The ratio of her greatest distance from the Sun to the smallest is 6:5. Whereas the orbit of Pallas is much more oval – more oblate – than the orbit of Mercury. The greatest distance from the Sun to the smallest is in the ratio of 3:2 in the case of Mercury and 5:3 in the case of Pallas. They differ most from the other planets with regard to the dimension of the inclination of their orbits against the plane of the Sun's orbit. All previously known planets had an inclination of only a few degrees and even Mercury is only 7°. But Ceres' orbit is 10½ and Pallas' 35° inclined against the ecliptic and thus both reach beyond the limits of our previous zodiac. Especially Pallas strides very far north and south: for us she can reach at times 48° northern and 54° southern latitude.*

*Both planets are very small in comparison with the others, Pallas apparently even smaller than Ceres. It is still difficult to determine their size and probably never will be possible, we can only to estimate it. Even at a magnification of 240 times of my telescope I was never able to notice a distinct enlarged disk, distinguishable from a fixed star. Two of our most reliable astronomers, Schroeter and Herschel, tried to measure her but surprisingly their results differ. Schroeter finds the diameter of Pallas more than 20 times greater than Herschel. It can be shown that Herschel must have found the diameters of these two celestial bodies much too small, but on the other side probably prediction and natural imperfection of all visual tools have increased the diameters of the small planets in Schroeter's telescope. I consider Ceres at most 1/3 of our Moon's diameter, and that of Pallas one third smaller than that of Ceres.*

*Schroeter and Herschel both believe to have noticed a dense atmosphere around these planets. It is not yet clear whether this was just an optical feature due to the telescope. Strange is their peculiar change in light that was noticed by almost every astronomer. From one evening to the other they appear with very different brightness. I believe that it was more striking and conspicuous in the case of Ceres. Very strange is the position of both orbits of these planets to each other. As mentioned above, for both the major axis, or the mean distance, is equally great and sometimes Pallas is farther away from the Sun and much closer to the Sun respectively, than Ceres can come to it. But nevertheless, the two orbits do not entwine round each other like two chain links but the more oval orbit of Pallas stands in the rounder orbit of Ceres. – At the descending node of Pallas' orbit on Ceres' orbit, i.e., at that point where Pallas' orbit becomes more southern than that of Ceres, both orbits are extremely close together and it is always possible that formerly both orbits, before they were slowly changed by the attracting forces of the other planets, did really intersect. Now both planets do not come very close there. According to my calculation, Pallas passes this node 184 days later than Ceres.*

In his presentation, Olbers laid out his analysis of the criteria for what constitutes a planet. In this anfractuous passage he criticizes Herschel but admits rather perversely that a distinction *should be made* between these newly discovered objects and the planets. This is akin to what the English historian Edward Gibbon (1737–1794) termed 'condemning the innovation but acquiescing in the sentiment':

*According to the general opinion of all astronomers both celestial bodies are considered planets. If they are named once in a while differently, this only stems from the fact that previously the definition of a planet was too strict. That they are much smaller than the rest of the planets cannot exclude them from those: Pallas might be several 1,000 times smaller in volume than Mercury but Mercury is also 27,000 times smaller than Jupiter. The extreme inclination of their orbits proves only that planets revolve on very inclined orbits around the Sun as well. But these differences of Ceres and Pallas from the older planets induced Mr. Herschel to create a new class of celestial bodies he wants to call asteroids. I do not like Mr. Herschel's reasons why he wants to make a distinction between Ceres and Pallas and the planets: But I do believe for very different reasons that the astronomers could really be persuaded to distinguish these two heavenly bodies, together with similar ones we might discover in the future, from the planetary gender as a species of its own.*

*The orbit of Ceres, which we have seen for the second year, is now so well known, we will always see her again as soon as she is distant enough from the rays of the Sun. But it will be much more difficult to find Pallas again next year since she has only been observed for 5 months. In 1803 and 1804 she will be close to her aphelion and is also far away from Earth at her opposition. Besides, Poniatowski's Taurus, the Eagle, and Cerberus, which Pallas will be passing through next year, are so studded with small telescopic stars in uncountable numbers that it appears an almost impossible task to find the small Pallas among those.*

*But we certainly will find her again, if not in 1803 and 1804, then in 1805 when she is closer to Earth and Sun than a 7th magnitude star in Eridani. Ceres and Pallas have also been given symbols or calendar signs.*

*Since ancient times Mercury has been represented by his caduceus, Venus by her mirror, Mars by his shield and spear, Jupiter by his thunderbolt and Saturn by his scythe, and now for Ceres has been chosen the sickle and for Pallas the tip of her lance. This is about what we know of these two heavenly bodies. This discovery will always be incredibly important and memorable for astronomy. We do not only know two more planetary bodies of our Solar System but Ceres and Pallas will cause a lot of speculations, questions about and insights into the origin, formation and history of our planetary system. Why here instead of one larger planet, which we had been expecting and suspecting due to analogy, at least two? Why the strange position of their orbits? Why this great inclination of their orbits against the orbit of the Earth, which distinguishes them from the other planets? It is still too early to answer these questions. But apparently, either it was impossible for the laws, which were applied by the creating Almighty to the active natural forces, due to unknown reasons, to keep the matter between Mars and Jupiter in one large sphere rather than to form several smaller ones, during the formation of our planetary system, like everywhere else; or, and this seems more likely to me, Ceres and Pallas are only parts, only debris, of a formerly greater planet, which burst either by an unexpected explosion or the collision with a comet. I am not in the position to explain the reasons for the latter hypothesis today in more detail, which are derived from the closeness of both orbits at the node of Pallas, from the great inclination of these orbits against the orbit of the Sun, from the strange change in light of these celestial bodies, which appears to me to result from their irregular, not round shape. In both cases we have to expect the discovery of several such small planet-like bodies between Mars and Jupiter.*

*And indeed, it happened due to the increased attention which was paid to small fixed stars observing these two new planets, that several of those are missing where they were found and observed some years ago. Some of these missing stars might be only periodically changing stars, but others can be considered such planets and asteroids like Ceres and Pallas. – Consequently, maybe the discovery of Ceres and Pallas opens up a new field for the greedy researcher: And even on perfecting our astronomical theories it will have a lasting effect. Ceres, but especially Pallas is so little perturbed by the great and powerful Jupiter that our hitherto existing methods are not sufficient to calculate these perturbations and we are forced to revolutionize this calcul completely in order to apply it to Pallas as well* (Olbers 1802).

## The Second Lecture (Late 1803)

*With your permission, dear gentlemen, I would like to talk again about the history of astronomy in the two lectures I will be holding this year. It is impossible for me, due to the short time I dare to take up your attention each time, to communicate in due detail everything that has happened with regard to the history of astronomy since my last lecture. It seems both new planets, with which we got acquainted at the beginning of the 19th century, still deserve our full attention, and consequently I am counting on your kind approval, if I am talking only about these two interesting celestial bodies, Ceres and Pallas, today* (Fig. 6.11).

| Names. | Mean diameters in English miles | Mean distances from the Sun in English miles. | Inclinations of orbits to the Ecliptic, 1780. | Inclination of axis to orbits. | Diurnal rotation on their axis. | Tropical revolutions. | Sidereal revolutions |
|---|---|---|---|---|---|---|---|
| | | | ° ′ ″ | ° ′ ″ | d. h. m. s. | d. h. m. s. | d. h. m. s. |
| Sun | 883,246 | — | — | 82 44 0 | 25 14 8 0 | — | — |
| Mercury | 3,224 | 37,000,000 | 7 0 0 | — | — | 87 23 14 33 | 87 23 15 44 |
| Venus | 7,687 | 68,000,000 | 3 23 35 | — | 0 23 21 0 | 224 16 41 28 | 224 16 49 11 |
| The Earth | 7,911.7 | 95,000,000 | 0 0 0 | 66 32 0 | 1 0 0 0 | 365 5 48 49 | 365 6 9 12 |
| Mars | 4,189 | 144,000,000 | 1 51 0 | 59 22 0 | 1 0 39 22 | 686 22 18 27 | 686 23 30 36 |
| Ceres | 160 | 260,000,000 | 10 37 57 | — | — | 1681 12 9 0 | — |
| Pallas | 80 | 266,000,000 | 34 50 40 | — | — | — | 1703 16 48 0 |
| Jupiter | 89,170 | 490,000,000 | 1 18 56 | 90 nearly | 0 9 55 37 | 4330 14 39 2 | 4332 14 27 11 |
| Saturn | 79,042 | 900,000,000 | 2 29 50 | 60 probably | 0 10 16 2 | 10,746 19 16 16 | 10,759 1 51 11 |
| Herschel | 35,112 | 1800,000,000 | 0 46 20 | — | — | 30,637 4 0 0 | 30,737 18 0 0 |

**Fig. 6.11** A tabular view of the Solar System, with the new discoveries Ceres and Pallas. Even though this table appeared in a book published after the discovery of Juno and Vesta, it includes only the first two asteroids. The author Dougall (1760–1822; 1810), who studied at Edinburgh University, adopts the small diameters given by Herschel, not the large ones of Schroeter

*Ceres, which was discovered by Piazzi on January 1, 1801 and observed until February 11, was, as I mentioned already the last time, found again in December 1801 in Gotha and on January 1, 1802, in Bremen and was observed by all major European observatories for 8 months until August 8 or 9, since it was invisible even in Italy's bright skies and it hid under the rays of the Sun. The amount and exactness of the observations and the time interval of now 17 months, made it possible for the astronomers to calculate Ceres' orbit more accurately and to no longer content ourselves with an elliptical hypothesis. Oriani, Wurm, Burckhardt, but foremost Dr. Gauss in Brunswick, developed the complex perturbations which this small planet must suffer from the mighty Jupiter in its motion and calculated those in tables. The astute Dr. Gauss created such convenient tables, it is possible to calculate all these perturbations with great accuracy each time.*

*Because one could be certain to find Ceres any time now, careless astronomers failed to find her again this year for a long time, when she could have been accurately observed at the meridian circle. And her orbit was considered more accurately determined than it really was, although Dr. Gauss himself explained that there was still a considerable uncertainty remaining in the elements and he asked for early observations on Ceres, which were even more important since the small planet was passing through its descending node. But the unexpected indolence of the astronomers missed this peculiar moment. I was too busy observing Pallas that I could dare to observe also Ceres regarding the small amount of time I am able to dedicate to astronomy for leisure and besides, I believed those observations, which were easier, in more capable hands. Much to my regret I learnt later that this was not the case. Ceres was seen from March 22 until 27 at Celle by Mr. Ende and the visiting Inspector Harding, but only seen, since they did not have a proper instrument at hand to observe her. Piazzi started observing Ceres only around mid March.*

*At the end of June no observations of Ceres of this year became known in Germany. Because Dr. Gauss had promised me the pleasure of a visit, I searched for Ceres on June 22 and easily found her. To my surprise I learnt later on that my observation of June 22*

*was the first made of Ceres in Germany. Dr. Gauss had not seen Ceres yet: and on June 24 this famous mathematician saw through my telescope for the first time the planet that had cost him so much effort and calculations and earned him so much fame.*

*From the end of June on, Ceres was also observed at Milan, Seeberg, Paris* etc. *The low position of the planet in Sagittarius forced the German astronomers already at the beginning of August to give up their meridian observations. For how long she was observed in other latitudes, especially Palermo, is not known to me.*

*Actually, Ceres was found even 6' farther west than according to Dr. Gauss's calculation, and 9' more east than she should be according to Dr. Burckhardt's determinations. A very slight improvement of the previous elements neutralized this difference. But the elements of this year's observations on the orbit of Ceres show a slight advance that the hitherto developed perturbation equations do not account for. So much for Ceres. Last year Pallas was observed by me only from March 28, the day of her discovery, until July 9 when her light became too faint for the bright twilight of our summer nights. She was observed until August 8 at Milan. Discouraged by her faint appearance and the twilight I failed to search for her in the darker nights of August. At Paris they were bolder, more persistent and fortunate, and Messier could recognize Pallas even on October 16. By this our hope to find her again this year was revived – I had doubted it before due to her much greater distance from Earth and Sun. An early rediscovery of Pallas was even more desirable because some great astronomers still doubted the planetism of Pallas despite all mathematical evidence just because this planet does not seem to match their hypotheses about the formation of our Solar System and they needed the rediscovery of this planet to convince themselves of its nature. The capable and agile assistant of our famous Schroeter, Inspector Harding, was diligently working on the rediscovery of Pallas. He sketched two charts with all hitherto observed small stars of the region that Pallas had to pass this year with exemplary elaborateness, which supplemented von Zach's* Monatliche Correspondenz *as copperplate prints. This celestial region is very rich in stars and this map comparatively still very poor in stars. He wanted to get acquainted with even those missing in our catalogues by continuous observations. The efforts of this astronomer of merit were crowned with success: On February 18 he had the pleasure finding Ceres. He saw her as a small 13$^{th}$ magnitude star close above n.36 of Poniatowski's Taurus, almost exactly where she should be according to Dr. Gauss's calculation.*

*But the observations beyond the meridian were difficult. The sky in Poniatowski's Taurus, the Eagle and Cerberus are so studded with stars, that I oftentimes counted together with Pallas 37 to 40 stars in the field of the telescope, among which she had to be searched by her position.*

*She appeared in February as a 13$^{th}$ magnitude star, at her greatest brightness at the end of January as an 11$^{th}$ and in October as a 14$^{th}$ magnitude star, in February and October certainly more distinct than the 4$^{th}$ satellite of Saturn. Usually it is assumed, and Herschel bases a lot of his sentences about the concept of the skies on this, that the distances of the fixed stars from us are in the same ratio as the classes among which we count them, and that for instance a star of the 7$^{th}$ order of magnitude is seven times, a star of 13$^{th}$ order of magnitude 13 times more distant from us than a star of the 1$^{st}$ order of magnitude. Pallas has refuted this faulty assumption completely. In April 1802 she appeared similar to a 7$^{th}$ magnitude star. In February 1803 she was, due to her greater distance from Earth and Sun, 10 times less luminous than in April of the previous year. So she should, if this assumption is correct, have appeared as a 22$^{nd}$ magnitude star, but she was similar to a 13$^{th}$ magnitude star. Consequently, the distances of the fixed stars increase in a far greater ratio than the orders, which is estimated by our eye, and that I can notice stars with my telescope which are 30 times more distant than Sirius. What spaces must not reach those huge powerful telescopes of Schroeter and Herschel, and to what degree must not widen the measures* *which Herschel gave for the system of our fixed stars? Also the orbit of Pallas was further improved by this year's observations and Gauss and Burckhardt are now busy calculating her far more complicated perturbations than Ceres', which her elliptical orbit suffers by the attraction of the other planets, especially Jupiter.*

*We know the orbits of the two new planets well enough for a general overview and for a determination of the positions to a few minutes in the coming years: but we are far from that accuracy the impatient astronomers wishes for because for this the intervening periods are too small. Ceres has only completed 3/5 and Pallas about 1/3 of one revolution around the Sun since the first discovery. When Uranus was discovered in 1781, it soon became clear that in 1756 Mayer and in 1690 Flamsteed* [John Flamsteed, 1646–1719, Astronomer Royal] *had already observed this planet as a fixed star and these older observations enabled the orbit of Uranus to be determined very accurately. It would be desirable to find such older observations for Ceres and Pallas as well. But to date all research in this regard has been in vain. Very early, when Ceres was recognized as planet, la Lande claimed boldly that we will find her among his 50,000 fixed stars. But this brazen, although not very unlikely claim, did not come true: and it seems almost certain that by an unfortunate and stubborn coincidence both planets hid from the telescopes of the Parisian astronomers for ten years, as close as they came to them at times. For instance on September 5, 1795, Pallas went only 40' below and on March 1, 1797, even only 14' south of Mr. la Lande's field of the telescope. Actually, Ceres must have been in it but her observation was missed in favor of other stars. And also among the stars Flamsteed observed, which are now missing, there is none that would match Ceres or Pallas. Time will tell if we find them among La Caille's or Bradley's stars* [James Bradley].

*Both planets described in the gap between Mars and Jupiter at exactly that distance from the Sun, where one had been expecting one since long according to the harmonious progression, their almost circular orbits in almost the same time of 4 years and 7 months around the Sun. I am saying almost equal time: for we do not know for certain yet, whether this time is exactly identical and it still seems that Ceres moves slightly faster than Pallas. The orbit of Ceres is not as oblate as that of Mars: the orbit of Pallas is more oblong and eccentric than any other planetary orbit. But we are certain to see Pallas at any time in the future if she is far enough from the Sun since even this year, so close to her aphelion and under unfortunate circumstances, we were able to observe her still in October. I already mentioned the last time that these orbits have unexpectedly, and against analogy of the other planets, a very great inclination against the orbit of our Earth. The inclination of all other planets is only a few degrees and even of Mercury only 7°. But Ceres is 10½°, Pallas even 35° inclined against the Earth's orbit.*

*The positioning of both orbits against each other is extremely peculiar. For both the mean distance from the Sun is equal: but due to the greater eccentricity of Pallas' orbit the distance of Pallas from the Sun is sometimes greater, sometimes smaller than the distance of Ceres. Now Pallas is further away from the Sun than Ceres: but for two years she will be much closer to the Sun. The orbits do not interlink like chain links but the more oval orbit of Pallas stands in the rounder orbit of Ceres. And because of this it will be difficult to determine in which order to put these two planets among the others. At the descending node of Pallas' orbit on Ceres' orbit, i.e. at that point where Pallas' orbit becomes more south seen from the Sun, both orbits are very close. I find that this planet seen from the Sun is now at the 8$^{th}$ degree of Taurus, with about 10 1/3° latitude south, and that the distance of both orbits is there only 73 of such parts of which the Earth from the Sun has 1,000. This is still about 1,400,000 German miles. But although both orbits come very close at that point, both planets do not. For Ceres passes 184 days, or more than 6 months, before Pallas through this node.*

*The strange change of light of these planets could be seen this year as well, especially in the case of Pallas. Mr. Harding, who kept an eye on it, saw the small planet' s light de- or increase considerably within 40 or 50 min. He believes to have recognized something regular in Pallas' change of light, which probably depends on her rotation. Harding assumes according to his observations that this rotation occurs within 23 or 24 h, just as with all other planets within Jupiter's orbit.*

*As I mentioned already the last time both planets are very small compared to the others. It is hardly possible to measure their apparent diameter: it is too small for our senses and our instruments. And the measurement attempts of two great astronomers differ a lot. But I believed in April 1802 by comparison of Ceres' and Uranus' luminosity and due to photometric reasons, to give Ceres' apparent diameter 3/5 s then. Consequently its diameter is only the 18th part of the diameter of our Earth, thus not 100 miles. I consider Pallas' diameter 1/3 less, about 50 or 70 miles. So one could form from our Earth almost 6,000 bodies of Ceres' size and more than 20,000 bodies of the size of Pallas.*

*These two planets are so small, their orbits are so close together that these facts will cause a lot of speculation about the formation and history of our planetary system in scientists and philosophers. Admittedly, I dared to suspect, keeping all these circumstances in mind, that these two might only be parts, only debris of one larger planet which burst due to some catastrophe. But do not get me wrong, I am expressing this only as a possible hypothesis, which I am stating only with due mistrust with which all should be met which is not a result of observation or calculation, and which truth remains to be seen. But I would like to give some reasons for this hypothesis with your permission.*

*The harmonious progression of the distances of the orbits of the planets from the Sun, mentioned first by professor Titius, and later so often admired and expressed by Bode, has been treated due to a minor animosity* [he first called it 'personal, against Bode,' but crossed it out] *too contemptuously. We do of course not know the causes for this law of the distances and of course the progression of the distances is not entirely a geometric journey: but experience shows that the distances of the orbits of the planets are in accordance with this law. And bear in mind this law was phrased before the discovery of Uranus and Uranus was in accordance with this law. Maybe human ingenuity will discover the causes why the distances of the planets from the Sun grow in an almost geometric progression and why something was missing in this progression. The reasons for this law, this almost geometric progression, are not entirely known. But it is a fact. According to it we were entitled to anticipate between Mars and Jupiter a planet of medium size. But we found two small bodies here instead, which form together the 1000th part of the mass which we would grant according to analogy an existing planet. The arbitrariness of the creating Almighty, keeping in mind unfathomable purposes can of course explain anything: but we think in other respects that this omnipotent wisdom applied certain general forces to the matter, and that according to the laws of nature, determined by those, everything was formed. We could not think of a general law of nature, according to which exactly the matter between Mars and Jupiter formed several smaller spheres and not large ones like everywhere else. If from this it follows that it is fairly likely that previously here one larger planet, not such little bodies, was rolling around the Sun, this hypothesis is confirmed by the great inclination of Pallas' orbit.*

Olbers next muses about the origin of the Solar System, a subject that was addressed in 1778 by Georges-Louis Leclerc, Comte de Buffon (1707–1788; Fig. 6.12). His hypothesis that the planets were created by the collision of a comet with the Sun was discounted by Laplace (1809: 362). Buffon's idea is also mentioned in Chapter 10 of this book by Erasmus Darwin (1731–1802).

**Fig. 6.12** Comte de Buffon painting by F.-H. Drouais. Musee Buffon, Montbard

*All of the seven older planets move on orbits, which are only slightly, not more than 7°, inclined against the plane of the Sun's equator. If this inclination was not restricted to this limit by an acting cause during the formation of the planetary system, a general law of nature, an inclination of any size was possible, so the odds were at least 53,000,000:1 that not all seven planetary orbits would be so close to the plane of the Sun's equator: Before the discovery of Pallas it was most likely, almost certain, that during the formation of our Solar System some natural power brought all of these planetary orbits into this position. Buffon's vivid imagination and La Place's brilliancy have tried to explain this striking phenomenon, the origin of this slight inclination: The first rather colorful than thorough and the second perhaps not satisfying enough. One might accept La Place's cosmogenesis or not; one thing is certain, that one cannot think of any mechanic cause of this phenomenon, which would not have affected Pallas as well.*

*At the moment Pallas is digressing 40° from the plane of the Sun's equator and this seems to hint at an external determination that afterwards brought this part of planetary mass out of the plane, in which all planets had their orbits and necessarily had to have. And bear also in mind that the much greater eccentricity of Pallas' orbit must confirm the assumption that at that time when all planetary orbits were formed, her orbit could not have been initially formed as it is now, but that it was shaped by a previous force. Looking at the closeness of Pallas' and Ceres' orbits, which most likely were really intersecting at one point: the same period of revolution, which hints at an immediate relation to each other, and bearing also in mind the strange change of light which is easily and naturally explained by the fact that Ceres and Pallas are not round spheres but debris of irregular shape, which revolve round one of their axes showing us now a narrower and then a wider profile, this hypothesis does not seem to be without apparent reason.*

*But is such an explosion of a planet thinkable? Is it possible? I do not believe we can deny this possibility. We see what great revolutions our ground has suffered. Exactly the force which divided our Earth's granite skeleton, moved it, could 10 or 20 times stronger blast it to pieces and thus transform it into asteroids and fire spheres. Could this force not have been greater on another planet? And is it not just provable that the destroying impact of a comet on a planet is not impossible, but more than likely to happen once in such an immense period of time of several 100 million years?*

*But let me repeat it, I still consider the idea that Pallas and Ceres might really be only parts of a planet which burst several thousand or million years ago just a hypothesis, which truth has not been proven until research and calculation examine it further. But such an examination is possible with time. If this hypothesis is correct and Ceres and Pallas are barely the thousandth part of* the *mass of their former main planet, there must be a lot of other fragments of it. A lot of parts of the burst planet might have flown away on hyperbolas and parabolas to never return again, others might forever remain invisible to us because they are too small or be the material of our fireballs and meteor showers* [Olbers wrote 'stone showers']: *but all those parts which describe elliptical orbits around the Sun due to the initial speed, need to return, due to the laws of gravity, during each revolution to that point where they initially started, that means to that point where the original planet exploded. This of course is to be understood as far as these elliptical orbits have not been by and by changed by the gravity of other planets. But all this moving does probably not exceed a certain limit, and consequently, one result of this hypothesis is that all orbits of such asteroids in a certain celestial region must come very close to each other. This celestial region is determined by the descending node of Pallas' orbit on Ceres' orbit and falls for us into the sign of the Whale, and with less hope into the opposite celestial region, the northern wing of Virgo and* [Coma] *Berenice's hair; we have to search these thoroughly for such small planetary fragments and asteroids. And really now, caused by my above reasons, several astronomers have started searching through the Whale often and thoroughly and since now there are several previously observed stars missing, which might very well have been asteroids, this fuels the hope to find new asteroids there. If there will really be found one or more small asteroids according to this theory based on that hypothesis, their orbits would have such an inclined position, they would approach Ceres' orbit in the same region where Pallas comes close to her, experience would give my conjecture a high degree of probability, which could be verified by calculation. After having developed all perturbations of the various orbits one would have to determine whether all these orbits did not intersect at one point at the same time and this point of time would give away the epoch when this formerly huge planet burst into the actual smaller asteroids. But enough of these assumptions, which you, dear gentlemen, will perhaps rightly regard as dreams.*

*Pallas and Ceres cannot be lost again – we will always find them again if one wants to see them. Ceres has already received the honors and rights of the other planets and can be found in astronomical calendars and yearbooks. That Pallas has not been equally treated is caused by the fact that no tables for her have been publicised. Next June Ceres will be visible again in the mornings, and be in opposition with the Sun on October 1, and can be observed until February 1805. She will pass through the Whale during this time. Pallas, slightly more luminous than this year, will be found again around June in the neck of Pegasus. She will be at opposition with the Sun around mid August and goes backwards from Pegasus to Aquarius. We will be able to see her until December. Hopefully, I will be able to inform you about these observations next year again* (Olbers 1803).

In reality, Ceres is an airless body like Pallas. Neither is enveloped by the clouds alluded to by the Swedish astronomer Lars Regnér (see *Discovery of the First Asteroid, Ceres*). The magnitude of Ceres is actually quite constant, exhibiting an amplitude of only 0.04 magnitude; the amplitude of Pallas is just 0.15 magnitude. Their period of rotation is also far different from that assumed by Harding: 9.1 h for Ceres and 7.8 h for Pallas. The variations noted by these astronomers were due entirely to the poor optics of their telescopes, and perhaps some wishful thinking.

## Hypothesis and Theory

The English scientist Thomas Young (1773–1829; Fig. 6.13) wrote cogently about the relationship between hypothesis and theory. The ancient Greek usage of the term "hypothesis" meant the foundations of an intellectual structure. In the light of Olbers' two lectures, dealing extensively with his hypothesis on the origin of the asteroids, it will not be out of place to consider here the contemporary view of what was meant by this term (in Chap. 11 see Zach on hypothesis in the MC, June 1802).

> *Although the invention of plausible hypothesis, independent of any connexion with experimental observations, can be of very little use in the promotion of natural knowledge; yet the discovery of simple and uniform principles, by which a great number of apparently*

**Fig. 6.13** Thomas Young

*heterogeneous phenomena are reduced to coherent and universal laws, must ever be allowed to be of considerable importance towards the improvement of the human intellect* (Young 1802:12).

In Young's second foray into hypothesis, we see explained the synthetical method Olbers used by first laying down the cause—a planetary explosion—and deducing from this that the effect would be small objects remaining in the space between Mars and Jupiter. Olbers never attempted to go from hypothesis to theory, and his effort can now be viewed like the philosopher Karl Popper's famous picture of the progress of science as conjecture followed by refutation.

*There are two general methods of communicating knowledge; the analytical, where we proceed from the examination of effects to the investigation of causes; the other synthetical, where we first lay down the causes, and deduce from them the particular effects. In the synthetical manner of explaining a new theory we necessarily begin by assuming principles, which ought, in such a case, to bear the modest name of hypothesis; and when we have compared their consequences with all the phenomena, and have shown that the agreement is perfect, we may justly change the temporary term hypothesis into theory. This mode of reasoning is sufficient to attach a value and importance to our theory, but it is not fully decisive with respect to its exclusive truth, since it has not been proved that no other hypothesis will agree with the facts* (Young 1804).

Young himself was kept apprised of the latest asteroid research by his Royal Society colleague Sir Henry Englefield (1752–1822; Fig. 6.14). Englefield was one of the first to be notified—by William Herschel—about the discovery of Ceres.

**Fig. 6.14** Sir Henry Englefield

He was much interested in the smaller bodies of the Solar System, having written two books on comets: in 1788 *Tables of the Apparent Places of the Comet in 1661, Whose Return is Expected in 1789;* and in 1799 *The Determination of the Orbits of Comets, According to the Methods of Father Boscovich, and M. De La Place, with New and Complete Tables.* Early in 1802, Englefield wrote with some excitement to Young about his observations of Ceres from Blackheath, south of Greenwich:

> *Friday*
>
> *I have seen the new planet twice, on Sunday night, and again last night. It is just visible to a common night-glass. With a power of 90 in my great telescope it was less bright than the 34 Virginis, near which it is. With a power of 200 no disc is visible, and with 300 I can scarcely say that it has a sensible diameter, more than what arises from irradiation, for small stars seen with such powers always appear dilated.*
>
> *I looked at the Georgian* [the planet Uranus] *soon after the new planet, but clouds came on, and I did not try 300 on it last night. With 200 the Georgian is, I am sure, the brighter, and it was a very much more visible object in the night-glass.*
>
> *Sunday*
>
> *Last night I again saw the new planet, and observed it with a power of 400. With this great power it seemed to have an apparent magnitude, but was extremely small, faint, and ill defined. I then turned the telescope to the Georgian (which as you know is very near), and the superiority in size and brightness was very striking. The Georgian was not well defined, but I am sure it was full four times the diameter of the new planet, and much brighter in proportion to the different size. Indeed the brightness of the Georgian is very surprising, its vast distance from the sun being considered. I really think that the diameter of the new planet cannot exceed a second; and it is of a very faint light even for that diameter. I looked then at the double star gamma Virginis, and saw the two stars distant from each other full three times their apparent diameter, a proof of the good adjustment and high power of my telescope.*

## Lalande's Planetary Nomenclature

Another important element that began in the eighteenth century and continued in the nineteenth was the issue of what to call the new planetary discoveries. When Herschel discovered a planet in 1781, he named it Georgium Sidus in honor of his patron King George III of Great Britain, but the French astronomer Lalande (1789) insisted it be named Herschel. He was just as adamant a generation later that the newest planetary discoveries (Ceres, Pallas, Juno) be named for their discoverers. This imperative came from Lalande's belief that the days of naming planetary objects after mythological figures had passed. Here is Lalande (1804: 130) explaining his position in his annual review of the history of astronomy.

> *For my part I shall never call the planets recently discovered by any other names than Herschel, Piazzi, Olbers, and Harding. I shewed, in the* Journal des Savans *for 1789, that mythological names are perfectly unmeaning. If we call the planets of the ancients by the names by which they were known to them, this is necessary for the purpose of*

*understanding them, and the names indicate the attributes of the deity whose name was given to the planet. The most brilliant planet was called Venus, because she was the most beautiful of the goddesses. Can it be said that the planet Olbers was called Pallas because that planet is the emblem of wisdom and knowledge? Why should that of Piazzi be called Ceres?– Has it any connection with the harvest?– With respect to the new planets, therefore, these names are insignificant and misplaced. They appear to me to have been derived from ancient fable only to deprive astronomers of the recompence of their labours, of the gratitude which is due to them, and of the emulation which may thence result. It is, in my opinion, a contemptible jealousy and offensive ingratitude, as I have already said in the* Journal des Debats *of the* $8^{th}$ *November, 1804.*

In this he certainly would have agreed with the Scottish philosopher Ferguson (1767: 114), who invokes the concept of common sense that was examined in Chap. 3. This is part of his discussion on Greek mythology:

*A mythology borrowed from abroad, a literature founded on references to a strange country, and fraught with foreign allusions, are much more confined in their use: they speak to the learned alone; and though intended to inform the understanding, and to mend the heart, may, by being confined to a few, have an opposite effect: they may foster conceit on the ruins of common sense.*

But was Lalande's firm dismissal of classical names just a personal predilection? I would suggest there is a deeper motive at work. Lalande's paper, dated 25 November 1788, was published in 1789, the year that witnessed the storming of the Bastille and the beginning of the French Revolution. The tumult that engulfed France induced a sort of gestalt about the use and misuse of language. Enlightenment critics found in the philosophy of Locke's discussion on the abuse of words "a guide to the origins of philosophical obscurantism or the impediments to scientific progress." (Rosenfeld 2001: 18) The *philosophes* expressed a desire to explain the workings of the world in linguistic terms—what better way to extend this to the universe at large than to make a break with the scholastic past by rejecting the names of classical deities and substituting the very prosaic names of astronomers? A key problem with this is that while astronomers would certainly recognize the names Piazzi and Olbers, many people of the day would not, and a century or two later very few people indeed would know those names. The classical Gods, on the other hand, are so entrenched in the Western psyche that their names are essentially eternal. A principle about proper names that achieved widespread acceptance was elucidated by the Oxford Professor of Metaphysical Philosophy Peter Frederick Strawson (1919–2006; 1959: 20).

*...it is no good using a name for a particular* [object] *unless one knows who or what is referred to by the use of the name. A name is worthless without a backing of descriptions which can be produced on demand to explain the application.*

While the 'backing of descriptions' may not be the sole method of connecting a name to an object, "some such principle governs the referential function of proper names." (Donnellan 1970: 335) In the era under discussion here, classical education was widespread, so the names Ceres and Pallas would be immediately understood based on their descriptions from ancient mythology, a cachet that could never be attached to the names Piazzi and Olbers, or, in the case considered in Chapter 5, Vidal.

## Pallas: Chemistry in 1803

"The goddess Athena made a wooden image of Pallas, her playmate who was accidentally killed by Athena. This image was flung out of heaven and landed before the tent of Ilus, founder of Troy. It was placed in a temple known as the Palladium, and became the guardian of the safety of Troy." Avery (1972: 403)

The handbill depicted here (Fig. 6.15) was published in April 1803 by William Hyde Wollaston (1766–1828; Fig. 6.16), England's leading chemist. He was also the son of Francis Wollaston, whose star catalog was involved in the discovery of Ceres (see *Discovery of the First Asteroid, Ceres*). The younger Wollaston published this handbill to establish the priority of his work in the discovery of the element palladium, in case French chemists beat him to publication. And he took the further step, in November 1803, of informing Banks at the Royal Society of his discovery (Griffith 2003).

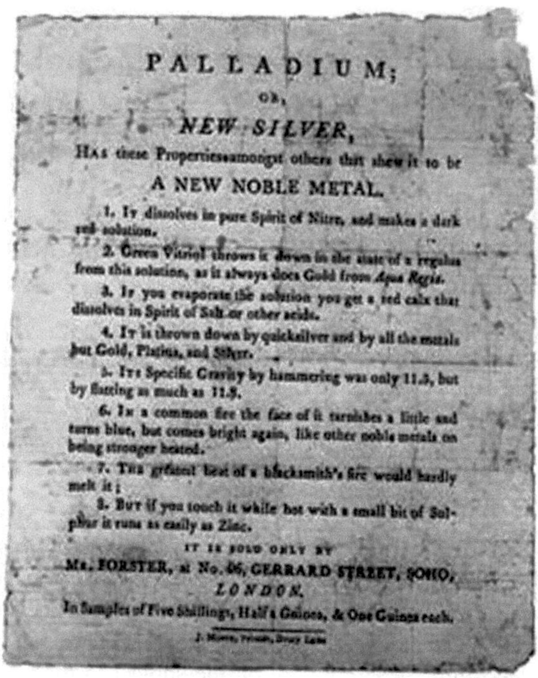

PALLADIUM;

OR,

*NEW SILVER,*

HAS these Properties amongst others that shew it to be

A NEW NOBLE METAL.

1. IT dissolves in pure Spirit of Nitre, and makes a dark red solution.

2. Green Vitriol throws it down in the state of a regulus from this solution, as it always does Gold from *Aqua Regia*.

3. IF you evaporate the solution you get a red calx that dissolves in Spirit of Salt or other acids.

4. IT is thrown down by quicksilver and by all the metals but Gold, Platina, and Silver.

5. ITS Specific Gravity by hammering was only 11.3, but by fusing as much as 11.8.

6. IN a common fire the face of it tarnishes a little and turns blue, but comes bright again, like other noble metals on being stronger heated.

7. THE greatest heat of a blacksmith's fire would hardly melt it;

8. BUT if you touch it while hot with a small bit of Sulphur it runs as easily as Zinc.

IT IS SOLD ONLY BY

MR. FORSTER, at No. 26, GERRARD STREET, SOHO,

*LONDON.*

In Samples of Five Shillings, Half a Guinea, & One Guinea each.

**Fig. 6.15** A handbill of 1803 issued for the discovery of palladium

**Fig. 6.16** William Hyde Wollaston

Wollaston first mentions the discovery in his notebook, now at Cambridge University. On the right-hand page he later wrote "The upper part of opposite page was written July 1802. I believe the C meant Ceresium a name which I once thought of giving to Palladium." (McDonald and Hunt 1982: 153) The name ceresium was inspired by the recently discovered asteroid Ceres. For the story of how the element Cerium got its name, see *Discovery of the First Asteroid, Ceres.* By August 1802 Wollaston had renamed his discovery palladium, as he described in a long sentence for the Transactions of the Royal Society of London three years later.

> *As I have already given an account of one product obtained from that ore* [platina], *which I considered as a new metallic substance, and denominated Thorium, I shall on the present occasion confine myself principally to those processes by which I originally detected, and subsequently obtained another metal, to which I gave the name of Palladium, from the planet that had been discovered nearly at the same time by Dr. Olbers* (Wollaston 1805: 316).

Palladium itself found a role in astronomy. "Wollaston used a palladium–gold alloy to make corrosion-resistant graduation plates for the "mural circle" erected by instrument maker Edward Troughton (1735–1835) in 1812 at the Royal Observatory in Greenwich (Griffith 2003: 181). From 1846 to 1860 the Wollaston medal of the Geological Society of London was made of palladium, and also from 1930 to the present (More on the subject in the next book of this series).

# References


Arabatzis, G.: 2004, Book review of The Seventh Sense by Peter Kivy. *Philosophical Inquiry* 26, 107-109.

Avery, C.B.: 1972, *The New Century Handbook of Greek Mythology and Legend.* Appleton-Century-Crofts.

Baxter, A.: 1745, Matho: Or, *The Cosmotheoria Puerilis, a Dialogue, Vol. 2* (translated and enlarged from a 1738 Latin edition). A. Millar, London.

Bjork, M.: 2017, Urania and Apollo: Myth and Identity in Tycho Brahe's Latin Poetry, *in The Classical Tradition in the Baltic Region: Perceptions and Adaptations of Greece and Rome* (A. Jonsson and G. Vogt-Spira, eds.). In preparation.

Boethius, Anicius Manlius Severinus: 1867, Boetii De institutione musica libri quinque. In Gottfried Friedlein. *Anicii Manlii Torquati Severini Boetii De institutione arithmetica libri duo: De institutione musica libri quinque. Accedit geometria quae fertur Boetii* (in Latin). in aedibus B.G. Teubneri. pp. 187.

Broadie, A.: 2012, *The Scottish Enlightenment.* Birlinn, Edinburgh.

Burtt, E.A.: 1932, *The Metaphysical Foundations of Modern Physical Science*, 2nd revised ed. Lund Humphries, London.

Campbell, T.: 1800, *The Pleasures of Hope*. William Durell, New York.

Donnellan, K.S.: 1970, *Proper Names and Identifying Descriptions*. *Synthese* 21, 335-358.

Dougall, J.: 1810, *The Modern Preceptor*, vol. 2. Vernor, Hood and Sharpe, London.

Englefield, H.: 1802. Letter to Thomas Young. *Nicholson's Journal*, 1802, pg. 197.

Ferguson, A.: 1767, *An Essay on the History of Civil Society*. Boulter Grierson, Dublin.

Griffith, W.P.: 2003, Bicententary of Four Platinum Group Metals, Part 1: Rhodium and Palladium. *Platinum Metals Review* 47(4), 175-183.

Goldstein, B.R.: 1985, *Theory and Observations in Ancient Medieval Astronomy*. Variorum Reprints, London.

Harrison, J.: 2005, *The Nightingales of Heraclitus: Shares Ideas of Art, Science, and Nature*. Prometheus, Peterborough, Canada.

Hawhee, D. and Holding, C.: 2010, Case Studies in Material Rhetoric: Joseph Priestley and Gilbert Austin. *Rhetorica* 28(3), 261-289.

Hutcheson, F.: 1753, *An Inquiry into the Original of our Ideas of Beauty and Virtue*, 5th ed., corrected. R. Ware, London.

Kant, I.: 1755, *Allgemeine Naturgeschichte und Theorie des Himmels (Universal Natural History and Theory of the Heavens)*. Johann Petersen, Koenigsberg and Leipzig.

Lalande, J.J.: 1789, Lettre sur le nom de la planete Herschel. *Journal des Sçavans*, 154-156.

Lalande, J.J.: 1804, History of Astronomy for 1804. *The Monthly Magazine* 20 (133), September issue, 130-133.

Laplace, P.S.: 1809, *The System of the World (J. Pond, tr.)*. Richard Phillips, London.

Leslie, J.: 1806, *A Dictionary of synonymous words, and technical terms, in the English language*. John Moir, Edinburgh.

Lofft, C.: 1800, Athenian Letters. *New London Review*, March issue, 226-228.

Mathieson, T. J.: 1999, *Apollo's Lyre*. University of Nebraska Press, Lincoln.

McDonald, D. and Hunt, L.: 1982, *A History of Platinum and its Allied Metals*. Johnson Matthey, London.

Olbers, W.: 1802. Manuscript archive number III 016, Bremen University, Bremen.

Olbers, W.: 1803. Manuscript archive number III 017, Bremen University, Bremen.

Ossendrijver, M.: 2012, *Babylonian Mathematical Astronomy: Procedure Texts.* Springer, New York.

Peterson, M.A.: 2011, *Galileo's Muse, Renaissance Mathematics and the Arts*. Harvard University Press, Cambridge.

Priestley, J.: 1777, *A Course of Lectures on Oratory and Criticism*. J. Johnson, London.

Rosenfeld, S.: 2001, *A Revolution in Language: The Problem of Signs in Late Eighteenth-Century France.* Stanford University Press, Stanford.

Schaffer, S.: 1981, Uranus and the Establishment of Herschel's Astronomy. *JHA* 12, 11-26. Letter quoted from Herschel to Lalande is RAS manuscript W.1/1, pg. 57-59.

Smith, A.: 1795, History of Astronomy, in *Essays on Philosophical Subjects* (D. Stewart, ed.). Wogan, Dublin.

Strawson, P.F.: 1959, *Individuals.* Metheun & Co., London.

Van Helden, A.: 1985, *Measuring the Universe: Cosmic Dimensions from Aristarchus to Halley.* University of Chicago Press, Chicago.

Wilczek, F.: 2015, *A Beautiful Question: Finding Nature's Deep Design.* Penguin Press, New York.

Wollaston, W.H.: 1805, On the Discovery of Palladium. *Phil. Trans.* 95, 316-330.

Young, T.: 1802, The Bakerian Lecture: On the Theory of Light and Colours. *Phil. Trans.* 92, 12-48, London.

Young, T.: 1804, Reply to the animadversions of the Edinburgh reviewers. In *Miscellaneous Works of the Late Thomas Young*, vol. 1, 1855 (George Peacock, ed.). John Murray, London.

# Chapter 7
# The Olbers Letters

## Introduction to the Letters

This chapter and the following two chapters present a corpus of correspondence in the Republic of Letters about Pallas over a span of four decades from 1803 to 1843. The unique value of such a collection for historical research was best expressed by the writer Thomas Birch (1705–1766; 1749: ix) when he said that in "original letters and papers, facts are represented in the most artless and undisguised manner, and in the order, in which they happened; and the secret springs, causes, and motives, which produced them, are opened to view." In our own time, Johnson (2004) likewise assigns them a special worth: "Letters bring a man before us more clearly than any other kind of document, a really intimate diary alone excepted."

## *Magnitudes of Ceres of Pallas*

One theme running through the letters in this book is the persistent problem of distinguishing the asteroids from the multitude of stars, "Those tremulous tapers of the skies," as the Italian poet Giambattista Marino (1569–1625) so elegantly termed them. The number of stars in the range of visual magnitudes these astronomers were searching through gives an idea of the task before them: mag. 6 (5946); 7 (17,765); 8 (51,094); 9 (140,062); 10 (404,194); 11 (1,196,690). Maskelyne (1802b) compiled a list of the magnitudes of Ceres and Pallas, covering the period February 3 to June 28, 1802, which suggests he was able to discern Pallas as faint as magnitude 11–12. The first portion from the unpublished manuscript is about Ceres, with some comparison of Pallas seen on the same night of April 22; the last half is of Pallas magnitudes only. The initials TF refer to Thomas Firminger (1775–1861), who assisted Maskelyne at the Royal Greenwich Observatory from 1799 to 1807.



C.J. Cunningham, *Studies of Pallas in the Early Nineteenth Century*,
DOI 10.1007/978-3-319-32848-5_7

George Gilpin (1754–1810) was clerk of the Royal Society and a friend of both Herschel and Maskelyne.

*1802 Observations of the app. Magnitudes or brightness of the two new planets*
*Feb. 3 Ceres 7.8 M..9..8..8..8..8 Mean 8 M fine night*
*March 4 Ceres 9 M or less*
*April 22 Ceres & Pallas appeared of equal brightness. On Apr. 22d & 23d they appeared the same to Mr. Gilpin; on 10th & 12th the French.*
*May 21 8 M 8 M T. F.*
*22 8 M..8 M..8 M T.F.*
*June 20 10 M. a very fine night*
*July 3 10 M.*

*April 15 Pallas 7 M by T.F. but he makes magnitudes too great, not being used to it*
*April 21 9 M..8..8*
*April 22 9 M..9*
*May 21 8 M. T.F.*
*22 8 M T.F.*
*May 17 9 M..9*
*June 11 9 M*
*June 18 11 M T.F.*
*June 20 11 M. 11.12 A very fine night*
*June 28 10 M A very fine night*

From these observations, Maskelyne arrived at a summary of magnitudes for the two objects

| *Ceres* | *Pallas* |
|---|---|
| *Feb. 3 8 M* | *Apr. 21 9 M* |
| *March 4 9 M or less* | *May 17 9 M* |
| *Apr. 22 9* | *June 11 9 M* |
| *May 17 9* | *June 20 11 M to 12 M* |
| *June 20 10 M* | *June 28 10 M* |
| *July 3 10 M* | |

We can see here how inaccurate magnitude estimates were before the days of photometers. In the case of Pallas, for example, its magnitude was said to range from 9th magnitude to as faint as 12th magnitude in the space of only nine days from June 11 to June 20. Pallas reached opposition on March 26, 1802, just two days before it was discovered. Its magnitude then was about 8, so a fading to 9 throughout April, May, and June was a reasonable estimate, but the sudden dip to 11 or 12, followed by a rise to 10 just eight days later, shows the unreliability of such visual measurements. This is even more remarkable when we consider Maskelyne's notes about the two nights of June 20 and 28. He regarded both as 'very fine,' and yet saw very different magnitudes. This was from an extremely experienced observer. For the observation of 15 April he derided his assistant Firminger for making magnitudes too great due to lack of experience, but it appears from this that experience counted for little.

## Mutual Orbital Properties of Ceres and Pallas

Another theme, and one that takes center stage in the letters between Bode and Olbers, is the mutual orbital properties of Ceres and Pallas. Only one English scholar in the early nineteenth century wrote about the orbital properties of the asteroids. Unfortunately we do not know who he was, as the title page to his book *Essay on the System of the Earth* is missing. Only one copy seems to exist, at Radcliffe Observatory in Oxford

> *The difference of mean distance from the sun of the two asteroids Ceres and Pallas, is stated in Vince's* Astronomy *to be as the difference between 276.74 and 276.759 or about .019, the distance of the earth being the standard at 100. Taking the real mean distance of the earth from the sun at about 95,000,000 miles, according to the table of Dr. Maskelyne before referred to, the distance between these asteroids must be about 18,000 miles. At this distance I cannot suppose it possible that they can permanently keep asunder, or that they could now be separate if this had been their original relative position. The difference of the inclinations of their orbits may for a time keep them separate when they are in the same position in longitude as to the sun, and if they are approaching to Jupiter with different velocities, it is possible that they may heretofore have been and may again become sufficiently separated to preserve their separate existence, before the chance of their being both in the nodes or points of mutual intersection of their respective orbits at the moment of their being also in conjunction or in the same meridian of longitude as to the sun, must otherwise bring them together, if within 18,000 miles of each other. If we suppose the present position of Ceres to be in part the effect of a late accession of a comet, this might afford a second possible explanation of the present separate existence of this asteroid. But we cannot suppose four comets all fixed so near to each other in the places of their new courses, as asteroids.*
>
> *It may also here be remarked that if these asteroids shall fall into Jupiter, as I have supposed to be presumable, they must be expected to affect the inclination as well as excentricity of the orbit of the increased planet in a proportion adequate to the difference of the two bodies; and if moving in an orbit at a considerable angle of inclination to the orbit of Jupiter, they must be expected also to produce a change in the position of the axis and of the poles of the aggregate planet. Thus if all the planets be aggregations of numerous asteroids, a second possible explanation is afforded of that phenomenon which we find in almost all the planets, of the axis not being perpendicular to the orbits.*
>
> *Before I proceed to consider other consequences of this view of the asteroids, I will here take the opportunity of remarking, that this explanation of their origin will account for the plane of the orbit of Ceres being at so much greater an angle to the plane of the sun's equator, than any of the aggregate planets.* (Anon 1811: 193–194)

In the past, there have been a series of close approaches (at less than 10 million km or 6.2 million miles) between Ceres and Pallas, around year–13,900, spaced by 4.6 y (just one orbital period). None will occur in the future before year 20,000 (Aldo Vitagliano, private communication). Orbital analysis (Michalak 2000) has shown the closest approach of Ceres and Pallas since their discovery occurred in 1825. This was 17.5 million miles, so the conclusions drawn by the author of 1811 from his calculation of a mere 18,000 miles are faulty. Vince's *Principles of*

**Fig. 7.1** August von Kotzebue

*Astronomy*, mentioned near the beginning of this passage, was a standard work in astronomy. It was published in three volumes between 1797 and 1808 by Samuel Vince (1749–1821), Plumian professor of astronomy at Cambridge.

## The Letters

A complete list of all the asteroid-related correspondence used in this series of books, together with their archival locations, can be found in the next book in this series: The Discovery and Early Study of Juno and Vesta.

Bode to Olbers Berlin March 5, 1803

*I was very surprised by the news of Pallas' rediscovery, probably no astronomer has expected this to happen so soon. I am thankful and when this moonlight has passed I am determined to look for the planet during nocturnal hours and find it with my telescopes. I announced without delay yours and Mr. Harding's news at the meeting of the* Akademie *and publicized it through local newspapers. And I also communicated extracts of the letters to Mr. von Kotzebue* (Fig. 7.1[1]) *for his journal in order to spread the news quickly.*

[1]August von Kotzebue (1761–1819) was a noted playwright and civil servant. He was active in Berlin from 1802 to 1806 as a journalist and member of the Academy of Sciences. His journal was named *Der Freimüthige*. Kotzebue wrote the text for the 1811 festival play *The Ruins of Athens*, in

*I mapped a sketch of the apparent orbit of Ceres and Pallas for this year in my chart according to which I will search for those henceforth, especially since Dr. Gauss's calculation is so correct. Actually, I have since long recognized Pallas as planetary body which differs only by its great inclination and eccentricity of its orbit from all other planets and why do we not recognize her consequently as a singular planet if the name comet should not be used. But the name is of course unimportant, but the comet of 1754 and maybe also the one of 1770 are singular planets and the orbit of Pallas intersects Ceres' orbit which is otherwise not the case with planets. All the same, dearest friend, you made an important astronomical discovery and immortalized your name. And also you found according to photometry the apparent diameter of Pallas so small? Should not be the fact that the texture of its surface reflects the light of the Sun only faintly the main reason of its small apparent size? The observation of the small Pallas will cause a lot of trouble, will it tolerate the illumination of the hairs of the wall quadrant and PI in the future? I doubt that I can catch her with my Dollond. But maybe it is not necessary that each and every astronomer chases her and thus neglects other important studies.*

Oriani to Olbers Milan April 9, 1803

*I believe to have seen the shy Pallas once. It is permanently overcast here which renders our instruments useless. On March 22 I believe to have observed her three times and only on April 1 was I able to see the same stars through light clouds, and it seemed to me that the one I designated as Pallas? was no longer at the same place; but since its RA is lesser by 2′, which it should not be according to Mr. Gauss's ephemeris, it could well be that I was mistaken. The observation of April 1 is incomplete & we must wait another eight or ten days when the Moon will be not over the horizon, to see this planetary atom again and rectify the observation of March 22.*

Bode to Olbers Berlin July 18, 1803

*I would like to express my sincere gratitude for the kindly communicated observations on Pallas, I have already printed them for the Yearbook 1806, and would like to ask humbly for future observations for the same volume. I recognized Ceres and Pallas several times by its motion, too, and tried to determine their position as good as the difficult circumstances permitted. I estimated Ceres of the $7^{th}$ order of magnitude at its present low position but Pallas seemed to have at its high position above the horizon hardly the $10^{th}$ order of magnitude. The latter will be easily followed in decreasing dusk for the next few months.*

*I wish to soon have tables of Pallas' path in order to announce it in advance in my Yearbook 1806. Until today I did not have the time to calculate those from Gauss's elements. Last night I observed Ceres again at the circular micrometer and compared with 119 Sagitarii. I will publish my observations on Ceres and Pallas in my* Yearbook *1806.*

Bode to Olbers Berlin Jan. 7, 1804

*I am admiring your tireless research on Pallas. I wish I received the formulae calculated according to the latest elements for calculating Ceres' and Pallas' tables more accurately than I have been able to (also including the perturbation formulae). I wanted to include those in my* Yearbook *1807 and asked Dr. Gauss already a couple of months ago. (I am trying to find Ceres and Pallas in the old star catalogs, but yet in vain.)*

(Footnote 1 continued)

which a German poet appears on the ruins of ancient Athens, where he invokes the goddess Pallas Athene.

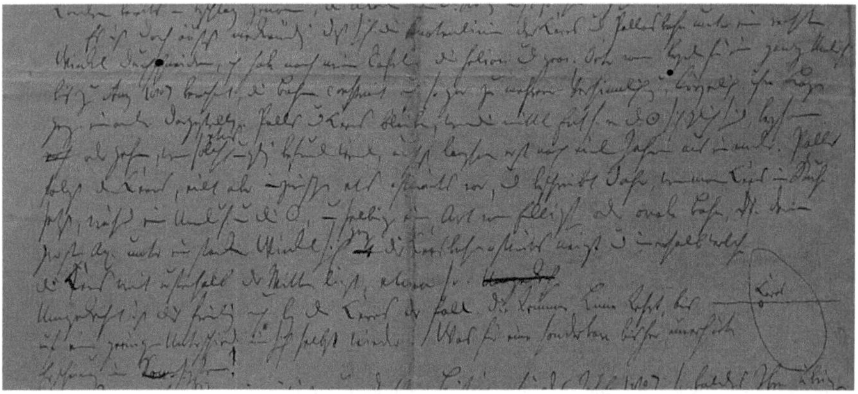

**Fig. 7.2** Detail from the April 21, 1804, letter by Bode showing his sketch of the orbit of Ceres

Bode to Olbers Berlin April 21, 1804

*It is very peculiar that the node lines of Ceres and Pallas intersect at a right angle. I calculated according to my tables the heliocentric and geocentric positions of both for an entire period until August 1807, constructed the orbits and tried to depict their position to each other. Pallas and Ceres are remaining, if the mean distances from the Sun are equal, close to each other, but will be moving, if those are rather different, extremely slowly and only after many years apart. Pallas is following Ceres, but is already moving slightly eastwards and describes thus, if Ceres is assumed at rest, during one revolution around the sun, around it a kind of ellipse or oval orbit whose major axis inclines eastwards at a great angle against Ceres' orbit and in which Ceres is located far from the center. Like this* (see Fig. 7.2)

*But this is also true for Ceres. The twisted line repeats itself except for a small difference. What a peculiar and unheard phenomenon in the Solar System!*

[The apparent motion of Ceres at the time of discovery was 5′ 19″, while that of Pallas was much faster at 22′ 22″.]

Olbers to Oriani Bremen June 11, 1804

*The prude*[2]*Pallas, as You called her in a letter to our common friend Mr. von Zach, had however the kindness to show herself to me at the beginning of May.*

*Here are my first three observations*

| *Mean time Bremen* | *app. RA* | *app. Dec.* |
|---|---|---|
| *1804 May 8 13 h 50′ 0″* | *334° 38′ 35″* | *8° 30′ 6″* |
| *Jun 1 12 4 50* | *338 27 27* | *10 10 32* |
| *3 11 53 25* | *338 41 25* | ——— |

[2]Olbers (in this French-language letter) uses the word 'prude' (= shy/prude). In the letter by Oriani to Zach referred to here, Oriani uses the word 'pudique,' which is different but the meaning is the same.

*Dr. Gauss's ephemeris of June gives the right ascension too small by 5' ½, the declination too great by a minute. The errors are quite pardonable, because Dr. Gauss has not yet considered the perturbations, but only determined the osculating ellipse*[3]. *It seems, the mean motion of Pallas will further approach that of Ceres – Have you not yet, Sir! entirely developed by Your profound research the perturbations of Pallas?*

Bode to Olbers Berlin August 25, 1804

*I am very grateful for your kindly communicated observations on Pallas, I saw this small planet several times and twice at the circular micrometer. I will try to catch it at the wall quadrant if its faint light tolerates the illumination of the hairs. I gave a talk the other day to the* Akademie *about the true and apparent orbit of Pallas and Ceres and their connexion. You will find it in my* Jahrbuch *1807 in excerpts. Heliocentrically Pallas is moving relatively around and against Ceres on a strangely crooked and itself repeating pear shaped line. But Pallas is far from the centre of this apparent orbit and also outside its plane. Pallas describes in the sky a very long narrow oval that goes from the Dragon* [Draco] *to Tucana, and bending up at the Dragon but making a node down at Tucana. Pallas comes there with the Sun in opposition and then at that time closest to Ceres' orbit. All this assuming that the periods of revolution are equally long. If you assume Pallas at rest, Ceres describes in a similar manner around it the apparent and relative orbit but* vice versa. *In general from this can be concluded that both have a close connexion and maybe are linked together by mutual attraction. I made sketches and calculated everything for the revolution from January 1 through August 10, 1807.*

Bessel to Olbers Lilienthal February 8, 1807

*Harding saw on January 9 16 h in the region of Pallas a* $10^{th}$ *-*$11^{th}$ *magnitude star, which he later was unable to notice; besides this star he also saw Pallas, but observed her not. I indeed searched on January 10 that region but did not note that star. If Harding did not mis-record, there is another wandering star in that region. I do not know for certain if I truly observed Pallas on January 10; in the moonlight there seemed to be one star amiss; but due to the brightness several were missing.*

*We had a clear morning, but I awoke only in the dawn due to the negligence of the warden. Today's morning, which maybe was clear also was lost to me because yesterday I did not ask to be woken due to yesterday's bad weather. If only the weather was more bearable, we will observe Pallas soon. Harding wrote to me on January 27 that the approaching evening promised to be beautiful but nothing of the comet. Shall he have forgotten it?*

*If you want me to free your observations from the refraction, I am asking for the diameter of your field and the distances of the stars from the center of the field.*

Bessel to Olbers Lilienthal February 22, 1807

*Revising my map, I missed your Pallas; but I convinced myself that I truly observed her on January 10, very close to the position prescribed by Gauss's ephemeris. It is strange that the entire region has become almost unrecognizable for me; but the duller weather might be the cause why I do not recognize several small stars in the neighborhood of La Lande's star, to which she was close. But I was only able to commit a certain amount of time to this research, since the comparison of my map keeps me extremely busy.*

---

[3]An osculating orbit is the truly elliptical orbit that a celestial body would follow if the perturbing forces of other bodies were to disappear so that it was subject only to the central gravitational field of the Sun (Collins Dictionary of Astronomy 2006).

| | | | | | | | | | |
|---|---|---|---|---|---|---|---|---|---|
| Juli 22. | 11ᵘ | 49′ | 50″ | . 226° | 21′ | 43″5 | . 20° | 42′ | 41″9 |
| „ 25. | 10 | 57 | 22 | . 226 | 39 | 26 ,3 | . . . . . . | | |
| „ „ | 11 | 12 | 54 | . . . . . . . . | | | . 20 | 13 | 5 ,8 |
| „ 27. | 11 | 45 | 34 | . 226 | 52 | 47 ,6 | . 19 | 52 | 47,1 |
| „ 28. | 10 | 25 | 55 | . 226 | 59 | 17 ,1 | . . . . . . | | |
| „ „ | 10 | 31 | 3 | . . . . . . , . . | | | . 19 | 43 | 5,8 |
| „ 29. | 11 | 42 | 31 | . 227 | 6 | 48 ,3 | . . . . . . | | |

**Fig. 7.3** Observations of Pallas by Olbers from July 23 to 29, 1807

Olbers to Bessel Bremen August 3, 1807

*I did not observe the eclipse of the fixed star but concentrated on Pallas. Here are my observations until now* (Fig. 7.3):

*The RA of the 25th are based on a star of the* Histoire celeste, *for which the time of passage is given 15 h 7' 3". I have corrected this to 15 h 7' 1."3 according to several comparisons to preceding and following stars. Harding's chart of Pallas attached to the July issue, is truly too late. The star with RA 226° 27' and 21° 20' Decl. N. is of 5th magnitude according to the* Histoire celeste *and Harding's chart is only of 8th magnitude.*

# References

Anon: 1811, *Essay on the System of the Earth.* Title page missing.
Birch, T.: 1749, *An Historical View of the Negotiations between the Courts of England, France and Brussels from the Year 1592 to 1617.* A. Millar, London.
Collins Dictionary of Astronomy, entry for Osculating Orbit. Market House Books (2006).
Johnson, P.: 2004, *Spectator*, October 9 issue, 296 (9192), 34.
Maskelyne, N.: 1802b. Manuscript in RGO 4/120. Cambridge University archives.
Michalak, G., 2000, Determination of asteroid masses I. (1) Ceres, 2 (Pallas) and (4) Vesta. *Astronomy and Astrophysics* 360, 363-374.

# Chapter 8
# The Gauss Letters

This chapter is dominated by letters between Gauss and Olbers, who kept up a lively correspondence for many years. Other correspondents included here are Bessel, Encke, Gerling, Hansen, Laplace, Maskelyne (Fig. 8.1), Nicolai and Zach. Previous correspondence about Pallas can be found in *Early Investigations of Ceres and the Discovery of Pallas*. We begin with a report by Gauss to England's Astronomer Royal on his study of the orbit of Pallas, and then the rediscovery of Pallas by Carl Harding in 1803.

Gauss to Maskelyne    Brunswick    30 January 1803

*Your great present of Taylors Tables I have received already in the month of December of the last year, for which inappreciable work I ought sooner to have acknowledged my gratitude. The cause of my delay was my wishing to give You at the same time some account of the last researches I have made on the orbit of Piazzi's planet. I have once more subjected the whole series of the observations which have been made till now, to a scrupulous examination, but the result has been, that it is not yet possible to give the elements a high degree of precision; indeed I have convinced myself that different elements may be assigned agreeing with each other to a few seconds, for all the observations from January 1, 1801 till August 5, 1802, but differing 10' or even more in longitude for the next opposition July 1 of this year. I have taken pains, to bring out such elements as make the calculated places coincide as nearly as possible with the observed places, and computed thereafter an ephemeris for 6 months which I have the honor to send You here enclosed. Though, by the above reasons, I cannot pretend to warrant this ephemeride to one single Minute, Yet I trust that the planet may be found out thereafter without any difficulty, and as soon as the error of the reckoning out the rate of its increase or variation have once been settled by good observations, the deviation will be no more of any consequence. The places of Ceres calculated by Mr. Triesnecker and Bode after my VII elements differ very little from these new ones, however I hoped that an ephemeris for Right Ascension and Declination would be more commodious for immediate application, than the longitudes and latitudes. The last column contains the value of 1/(dist. ⚳ a θ x dist. ⚳ a ♁)$^2$ which is considered as proportional to the apparent brightness (ie the illumination received by Ceres from the Sun, and that reflected from Ceres on to the earth, in each case decreases with the square of the distance between the bodies concerned). But to make a just estimation, it will be necessary to take also into consideration, that will start much lower in this year than in the two years past. The value of the above formula is*



C.J. Cunningham, *Studies of Pallas in the Early Nineteenth Century*,
DOI 10.1007/978-3-319-32848-5_8

**Fig. 8.1** Nevil Maskelyne by John Russell, circa 1776. National Maritime Museum, ZBA 4305

| | | |
|---|---|---|
| *for 1801* | *January 1* | *0.03637* |
| | *Feb. 11* | *0.02409* |
| | *Dec. 7* | *0.02494* |
| *1802* | *Feb. 26* | *0.05633* |
| | *May 17* | *0.03608* |
| | *Aug. 5* | *0.01543* |

*My researches upon this subject has led me to detect an error in one of your meridional observations Viz. the declination of May 11, which You make 16° 21′ 41″.8 ought to be about 2′ 30″ greater or nearly = 16° 15′ 12″ and therefore the longitude about 1′ smaller, the latitude about 2′+ greater. Perhaps You will be able to make out the cause of this difference, which comes out not only from the calculation but is confirmed also by two observations made on the same day by Cesaris at Milan and Mechain at Paris – At least this circumstance will be worth examination, as all your other Meridian observations are of an admirable exactness and agree among themselves to one or a few seconds.*

[Rough notes by Maskelyne contain a reference to a letter from Méchain dated January 27, 1802, in which Méchain said that his initial observation of Ceres on January 24 yielded a declination 17′ more than Zach's computation; RGO 4/120/xi].

*In the month of March I hope You will find out one of the planets if not both, by Your equatorial Sector. I flatter myself, that, when You have got some observations of them, You will honor me with the communication. I am completely prepared to convert any new observations into immediate use for correcting the orbit, and I have already experienced, what uncommon abridgement in calculation is derived from the excellent works Your bounty has furnished me with. The ephemeris for Pallas which I had the honor to send you in the month of November of the last year, I hope is come to your hands. Both planets are in an unfavorable position this year; Pallas for her great distance from earth and Sun, and the faintness of her light resulting thence; Ceres partly for the same reasons though in a more moderate degree, partly from her great south declination, which in our northern latitudes scarcely will permit her to emerge from the vapors of the horizon. Probably therefore the astronomers of Italy will sooner be able to find her out, than those of the more northern countries, and in St. Petersburg and Stockholm she will not be seen at all.*

[Appended to this is an ephemeris entitled "Geocentric motion of Ceres Ferdinandea in the year 1803," containing the equatorial co-ordinates (correct to the nearest minute of arc) and corresponding values of the 'force of light' derived from the equation quoted in the text of this letter, from February 4 to August 9, 1803]

No. 52 Olbers to Gauss Bremen February 22, 1803

*With great pleasure I'm hurriedly sending you the important news that Mr. Harding, astronomer at Lilienthal, has been very fortunate in rediscovering Pallas already on February 19 h 15. It appeared as a small 12th or 13th magnitude star just above No. 36 of Poniatowski's Bull (according to Mr. Bode) almost exactly at the place predicted by your computed ephemeris. He recognized it because, on the previous morning, he had examined No. 36 and its surrounding very closely, and thus found a small star at the indicated location which wasn't there 24 hours before. On February 21 in the morning, or on the 20th h. 15, he saw Pallas had moved away as required by theoretical considerations, and could compare it five times with No. 36 which it trailed by 55s at 15h 45m 17s mean time. He was kind enough to inform me immediately of his important discovery and I have hence also again seen and observed Pallas during the previous night. My observation is not very accurate due to the small planet's dimness. A preliminary reduction gives:*

*Feb. 21 17h 0m RA of Pallas 272° 57' Dec. North 7° 32'*

*Your ephemeris is thus still admirably correct. This accurate agreement with your calculation was all the more desirable for it otherwise would have been almost impossible to locate the presently so small Pallas among the innumerable multitude of similarly small telescopic stars.*

*I'll inform you immediately as soon as I make or receive an acceptable observation of Pallas. It's unfortunate that the observations must be made at such an inconvenient time of night.*

No. 53 Olbers to Gauss Bremen February 26, 1803

*The weather isn't at all favorable for observing Pallas. Since my last letter I've been able to observe it only once. The Dec. is somewhat suspect in both of my observations. Nevertheless I'm giving them here after appropriate reduction:*

| *Mean time in Bremen* | *RA* | *Decl.* | *Reference star* |
|---|---|---|---|
| *Feb. 21 17h 6m 10s* | *272° 56' 45"* | *7° 31' 14"* | *No. 36 Bode T.P.* |
| *23 15 24 36* | *273 28 39* | *7 46 1* | *No. 42 Bode T.P.* |

*No. 36 and No. 42 are adopted based on Poczobut's* [Marcin Poczobut, 1728–1810, professor at Vilnius University] *and Lalande's determinations respectively. The position of Pallas relative to the two stars was not favorable for the Dec. observations obtained with the circular micrometer (a hair-line micrometer could not be used because of the weak light). I consider the second RA to be quite good since the star was compared with No. 42 six times. The Dec. of Pallas could have been observed very well on the 24th and 25th because it crossed the parallel of No. 42.*

*I'm very curious about the comparison of these observations with your elements. One has to know Pallas' position very accurately and be able to recognize the telescopic stars beforehand in order to discover it among them. On the 23rd it stood between 4 other small telescopic stars from which I had to select it using only a few observations. Without your ephemeris it would again be lost due to moonlight and inclement weather.*

No. 54 Gauss to Olbers Brunswick March 1, 1803

*Many thanks for the important news concerning the rediscovered Pallas. The extent of my joy that this pet project of ours, whose just recent rediscovery this year seemed uncertain and which has again become visible so quickly, you can best judge against that of yours. During the year we'll be able to obtain a wonderful collection of observations of this planet, which will probably be visible until Michaelmas Day (September 29). Since the meridian observations presumably won't be of long duration, it is all the more essential that the off-the-meridian observations span the longest possible time interval. The good agreement among the previously computed positions with the observed ones is particularly pleasing to me because one can hope that the planet will, in time, be visible wherever there are astronomers, and because one needn't be concerned about losing and again having to laboriously search for it. The true elliptic elements for the present osculating section of the orbit will suffer only a small modification, and as soon as one has somewhat reliable observations, one will be able to give in advance with high accuracy the small correction to the ephemeris for this year.*

*The time of night can of course make the observation inconvenient. If you, nevertheless, devote an hour to Pallas during a clear night, then you'll of course at some time look around for its older companion, the low position of which is of course an additional difficulty. I'm enclosing herewith a small ephemeris which you'll find somewhat more convenient than that of Triesnecker* [Franz de Paula Triesnecker, 1745–1817, Director of Vienna Observatory] *and Bode, from which the RA are missing or from which the culmination time can be derived only in quarter degrees, – although I won't attempt to pass off the present ones as being more accurate, especially since the difference is quite negligible.*

*I would have sent you my tables for the perturbations of Ceres long ago if Zach hadn't written me that everything would appear in the February issue of the M.C. Since this letter won't have sufficient space, you'll receive them in my next one. I have also computed the remaining tables necessary for the motion of Ceres, 'leni calamo'* [with a fast pen] *just to the minute, actually only for my use, in order to examine the just recently obtained* Hist. Cel. *I have already glanced through it and have conducted several cursory attempts. I'm almost abandoning hope of finding Ceres therein, though I've located several days in which the difference in Dec. amounted to only a few degrees. I'll also send you this table shortly. Both tables are meanwhile provisional. At the conclusion of this year's observations we'll have much more accurate elements and then it will certainly be worth the effort to more accurately and completely calculate and incorporate the perturbations.*

*I've just received your second letter, and immediately calculated the following, using interpolation, from my ephemeris which is given to the second.* (Fig. 8.2)

[Enclosed is the geocentric path of Ceres for midnight at Seeberg calculated from March 3 to April 29, 1803].

| | AR calc. | Decl. calc. | Differenz AR | Differenz Dekl. |
|---|---|---|---|---|
| Febr. 21. | 272° 58′ 47″ | 7° 30′ 40″ | + 2′ 2″ | − 34″ |
| „ 23. | 273° 31′ 14″ | 7° 45′ 4″ | + 2′ 35″ | − 57″ |

**Fig. 8.2** Differences in RA and Dec. for February 21 and 23, 1803

| | |
|---|---|
| Epoche in Seeberg 1802 . . . . . . . . . . | 143° 28′ 17,2″ |
| „ „ „ 1803 . . . . . . . . . . | 221° 28′ 54,0″ |
| Tägl. trop. Bew. . . . . . . . . . . . . . | 769, 4161″ |
| Log. der halben Axe . . . . . . . . . . | 0, 442 616 0 |
| Excentricität . . . . . . . . sin 14° 13′ 6,3″ = | 0, 245 619 |
| Neigung . . . . . . . . . . . . . . | 34° 38′ 19,8″ |
| Sonnenferne } siderisch ruhend, für die Epoche | 301° 24′ 12,6″ |
| Knoten } von 1803 | 172° 28′ 7,9″ |

**Fig. 8.3** Elements of Pallas calculated by Gauss in February 1803

*Here also are the preliminary improved elements of Pallas* (Fig. 8.3):

No. 55 Olbers to Gauss Bremen March 4, 1803

*P.S. from March 5.*

*The previous night was clear. Pallas had to be on the parallel of No. 58. I observed 4 small stars on this parallel, and Pallas lay among these stars. It preceded No. 58 by 2m 33s and was only 0' 20" farther south. I still haven't been able to reduce the observations, but I'll send them to you before long.*

*It thus appears that Harding was the first to have again located Pallas.*

No. 56 Olbers to Gauss Bremen March 9, 1803

*I promised to send you the reduction of my last observation of Pallas on the morning of March 5. According to your ephemeris Pallas had to be near the parallel of No. 58 Tauri Poniat. According to Bode, and quite close to the small star from the* Hist. Cel. *p. 87 of June 17, 1794, for which the relevant data is:*

| | |
|---|---|
| *8.9 18h 23m 26s* | *zenith distance 39° 52' 5"* |
| *correction +12.0s* | *correction +1' 7"* |

*I initially let 2 small stars, one 12th and the other 9th magnitude, and No. 58 pass; but they were too westerly and too southerly to be Pallas or the star from the* Hist. Cel. *I then took 2 other small stars of similar magnitudes and luckily one was Pallas and the other was the star from the* Hist. Cel. *Here first is the situation of the 3 stars based on my observation, reduced to January 1, 1801:*

| | | |
|---|---|---|
| *12* | *275° 33′ 47″* | *8° 52′ 51″* |
| *9* | *275 35 47* | *8 53 9* |
| *8.9* | *275 59 29* | *8 57 22 * of the Hist. Cel.* |

*The observation of Pallas gives:*

| *Mean time* | *Apparent RA* | *Apparent Dec.* |
|---|---|---|
| *March 4 17h 11m 41s* | *275° 52′ 38″* | *8° 58′ 23″* |

*Although the observation was made only once, I still consider it to be good and in particular the Dec. was certainly not smaller, provided No. 58 is well determined. Because of the moon I won't be able to see Pallas again until after the 15th. From outside sources still nothing.*

No. 57 Gauss to Olbers Brunswick March 14, 1803

*Enclosed are my perturbation tables for Ceres. I am pleased that you also still found space in the M.C. alongside Regner's article.*

*Your observation of Pallas on March 4 agrees superbly with the new VI Elements. Here is the comparison. Since I don't know if I've already sent you the comparison of the first two observations with these elements, I'll also enclose it as well:*

| | | | | *Difference* |
|---|---|---|---|---|
| *1803* | *RA calc.* | *Dec. calc.* | *RA* | *Dec.* |
| *Feb. 21* | *272° 56′ 25.″0* | *7° 31′ 29.″8* | *−20.″0* | *+15.″8* |
| *23* | *273 28 46.8* | *7 45 52.3* | *+7.8* | *−7.7* |
| *Mar. 4* | *275 52 35.9* | *8 58 18.1* | *−2.1* | *−4.9* |

*The calculation is carried out very accurately and nothing is neglected except parallax. It seems that I've been fortunate as regards the speculation over your initial observations. The VI Elements now give 2½ minutes, on August 9th 4 min. 6 sec. less for the RA; and nearly 1 min. more for the Dec. than the V Elements. The ephemeris in the December issue of M.C. can accordingly be improved so accurately that one can always know the location of Pallas beforehand almost to the minute. It's a pity that Mr. Harding has entered into his chart the positions based not on this ephemeris, but rather on the older, much less appropriate, one from the October issue. I've already continued the ephemeris from August 9 to October 20 based on the VI Elements. This will follow before long, since I do not have sufficient time at the moment to copy it. This also applies to my small tables for Ceres and Pallas.*

[The ephemeris for Pallas from August 9 to October 20, 1803, follows].

No. 58 Olbers to Gauss Bremen March 23, 1803

*The observations of Pallas are still scanty. Indeed, it's very tiresome to pick out the small planet from among the large numbers of telescopic stars. Without your ephemeris it wouldn't even be possible. On March 21 I had, along with Pallas, 23 other stars in my telescope's field of view and already had mistakenly observed 3 stars as the planet before realizing it. Here are my observations:*

| | | RA | Dec. | Comp star |
|---|---|---|---|---|
| *Mar. 16* | *14h 10m 28s* | *278° 37′ 16″* | *10° 42′ 21″* | *No. 78 Bode* |
| *21* | *13 40 47* | *279 37 5* | *11 30 3* | ** Conn des Temps* |
| *22* | *13 14 22* | *279 48 10* | *11 38 45* | *No. 76 Bode* |

*Already on March 16 No. 76 Bode passed through my telescope's field of view, and I accordingly increased the RA given by Bode by 42″ that is also very nearly the case for the* Hist. Cel.

*Harding was truly the first, after the* Journal de Paris, *to have again located Pallas. Messier searched for it in vain. It doesn't appear that an observation was made of it in Paris before March 13.*

*I received Vol. 3 of* Mecanique Celeste *the other day. Pallas is mentioned merely in the preface and Ceres in the text itself. But the application of Laplace's formulae has been postponed to the next volume.*

Also on March 23, Zach wrote letters to Sniadecki in Poland and Banks in England, updating them about the improved elements of Pallas derived by Gauss.

Zach to Sniadecki  Seeberg  March 23, 1803

*I promised you in my last letter to send you the second sheet of my map of Pallas' path, and it is my pleasure to keep my word and to send you a small ephemeris of the path of this planet as well that will facilitate your search when observing at the meridian. The ephemerides of this little planet that I have sent you in my last letter are useful to refind without doubt Pallas, but according to the observations of this year and Dr. Gauss's improved elements you only need to subtract 2.5 min of all the RA's that I sent you lately, and one minute of all declinations. You will find these new orbital elements of this planet printed on those pages that I am sending you on the reverse. The last observation of March 4 thus agrees with these elements:*

| | |
|---|---|
| *RA Pallas – 275° 52′ 35″.9 calculated. Diff. with observation* | *– 2″.1* |
| *Declin. N. – 8° 58 18.1* | *– 4.9* |

*No news of our Ceres, the mists of the horizon prevent its discovery that would otherwise not be difficult, there can be no doubt that it will be at that position where Gauss's ephemerides indicate.*

Zach to Banks  Seeberg  March 23, 1803

*According to my last promise, I have the honor to send you here, the continuation of the celestial map, that represents the path of Pallas till Jun 28$^{th}$. Tho' this map has been performed long before the planet has been discovered again in Febr. of the present year, the elements of the orbit had been so well ascertained by Dr. Gauss, that the difference is but a trifle, not even to be represented upon the scale, on which the map is drawn. Nevertheless the nice astronomical observations will shew a little difference, but which Dr. Gauss has already corrected by mending the elements of the orbit, which he has calculated last year, upon the series of my observations of this planet, in conjunction with these made at the Royal Observatory in Greenwich. Only such most exact observations, strained to the highest pitch of astronomical exertion could produce elements so near to the truth after so long a period of near 7 months this little planet vanish'd out of sight. Dr. Gauss's old elements had been published in my* Monthly Correspondence *Decemb$^r$ 1802 p. 581. The* *new now corrected, upon a few observations, you will find in the present printed sheets, which make part of my Journal April 1803, which comes late to your hand. I only add that the last observation of this planet is thus represented by these elements.*

| | | *Diff. with observation* |
|---|---|---|
| *1803 March 4 Calculated RA ×* | *.... 275° 52′ 35″.9* | *−2″.1* |
| *——— Declin.×* | *.... 8° 58 18.1 N* | *−4.9* |

*The little printed ephemeris of the planet, which I had the honor of sending to you in my last, are calculated upon the obs elements quoted above, in order to reduce them to the present elements annexed here, and marked VI, it is only necessary to subtract in the beginning of every right ascension 2 ½ min. and avg up ye 9th...9′. 6″ in the middle time proportionally. From the declinations only subtract thoroughly 1 min. so this ephemeris shall be reduced to these new elements, and will exactly agree with the heaven, which by such a little and nice heavenly body is so much the more necessary, to avoid mistakes, & not to confound the planet with the multitude of little stars which might surround her.*

*Dr. Gauss has calculated since a new ephemeris for Pallas by his new (VI) Elements from August 9th down too* [sic] *Octobr 23rd. Which very likely will be the last term of visibility. I have the honor to submit here a copy.*

No. 59 Gauss to Olbers Brunswick March 29, 1803

*I've taken your new observations of Pallas, which you've kindly sent me, into account after the VI Elements. The result of the comparison appears a posteriori to demonstrate the great difficulty in observing this dim heavenly body. Here it is:*

| | | | | *Difference* |
|---|---|---|---|---|
| | *Calc. RA* | *Cal. Dec.* | *RA* | *Dec* |
| *Mar. 16* | *278° 37′ 10.″1* | *10° 43′ 18.″7* | −5.″9 | +57.″7 |
| 21 | *278 37 19.0* | *11 30 22.1* | +14.0 | +19.1 |
| 22 | *279 48 30.2* | *11 39 49.6* | +20.2 | +64.6 |

*The real and sole skill with which you find your way about the plethora of small stars and are able to locate the small planet evokes my greatest admiration. Let us hope you could soon help us in locating Ceres using the same skill.*

No. 60 Olbers to Gauss Bremen April 2, 1803

*You are quite right, one can tell from my observations, last compared by you, that the small faint planet is hard to observe. However, this reproach only applies to the decl. of March 16; for I marked the decl. of the 22nd as dubious. Much could be due to the uncertain location of the small stars, with which I must now compare Pallas.*

*On March 24, Pallas was compared with a star from the Conn. des Tems and I concluded from the observations:*

$$March\ 24\ \ 13h\ 2m\ 12s\ RA\ 280^\circ\ 10'\ 25''\ Dec.\ 11^\circ\ 58'\ 43''$$

*The observation was an average one.*

*On the 27th Pallas was much more conspicuous than I've ever seen it this year. It was positioned over the small star from the* Hist. Cel. *which appears on pg. 170 as: 18h 41m 36.5s zenith distance 36° 23' 0".*

*I found*

| | | | | | |
|---|---|---|---|---|---|
| *Mar. 27* | *12h 55m 23s* | *Pallas goes to the star forward* | *6."5s* | *and is north* | *2' 38"* |
| | *14 30 47* | | *4.0* | | *3 23* |

*the observation are good. I haven't reduced them yet due to lack of time.*

*On March 31, Pallas should have been just below No. 12 Aquilae according to Bode. Yet I couldn't at all recognize it by moonlight with the usual 45-times magnification of my most-used circle micrometer. I had to increase the magnification to 106 times, and with this I saw the small planet clearly. I compared it 7 times with No. 12 Aquilae. At 13h 24m 25s it trailed No. 12 by 3.5s and was only 0' 21" southwards.*

*I don't consider the RA as reliable as the Dec. which must be very accurate if No. 12 Aquilae, one of Mr. Lalande's stars, is correctly determined. The observations gives:*

*March 31 13h 24m 25s apparent RA 281° 20' 18" Dec. north 13° 8' 50"*

*Isn't it strange that the bright, weaker magnification didn't at all permit the planet to be glimpsed, which nevertheless could be found without effort using the darker, 106-times magnification. Well, I've again bidden farewell to Pallas for 12 or 14 days; it would serve no purpose to try to observe it during increasingly troublesome moonlight.* [The last part of this letter, about Ceres, is in *Discovery of the First Asteroid, Ceres*]

No. 62 Olbers to Gauss Bremen April 16, 1803

*Many thanks for the Ceres and Pallas tables you've kindly sent and for all of the other interesting news in your letter. I've already used the former to compute the position of Pallas for March 1, 1797. But even according to this calculation this position falls further south and outside of the area observed at that time. I'm now almost convinced that Pallas, as well as Ceres, has never been observed among and along with the 50,000 stars.*

*I've become very aware of the conjunction of Pallas with ε Aquilae. However, on moonlit April 10, I couldn't catch sight of it with any magnification whatever of my Dollond. The following day went better. Here are 3 observations to which I've at least applied extreme care.* [The Apr. 11 & 13 observations were published in the June 1803 issue of the *MC*, but the Apr. 12 data was not.]

| | | | |
|---|---|---|---|
| *Apr. 11* | *12h 10m 9s* | *RA 282° 43' 50"* | *Dec. 15° 0' 50"* |
| *12* | *12 12 18* | *282 49 44* | *15 11 17* |
| *13* | *12 13 41* | *282 55 15* | *15 21 18* |

*The RA is determined by ε Aquilae; for Dec. I've always chosen stars from the* Hist. Cel. *Whose parallel has very nearly coincided with that of Pallas. Pallas was also observed on April 15; the observation, however, isn't reduced yet.*

*According to a very friendly letter of Dr. Burckhardt's from Paris, in which he points out that the* National Institute *has awarded the discovery of Pallas in the previous year with a prize, courtesy of Lalande, it almost seems that this planet has thus far been rarely observed in Paris. Here are his own words:*

*'Your planet was observed here after the arrival of your observations; Mr. Mechain observed it on March 12 at 16h 59m mean time at RA 277° 46' 42" and Dec. 10° 6' estimated. The error of my elements is +1' 56" in RA according to this observation; according to yours of February 23 the error should be +2' 13". But Mr. Mechain is not very content with his observation, having only once been able to compare the planet with a star. Since then the preparation for the continuation of the meridian has occupied him quite a lot. Mr. Messier has not been as fortunate as Mr. Mechain in rediscovering it immediately, since the bad weather and the large number of small stars in the Milky Way has prevented him from doing so.'*

Gauss to Maskelyne Brunswick April 24, 1803

*Though, probably, you are already fully informed of every essential point concerning Mr. Harding's at Lilienthal finding out again Dr. Olbers' new planet: yet I hope, that perhaps some of the following particulars will be not without interest to you.*

*Mr. Harding saw Pallas, for the first time, again February 18 and became sure of his discovery the following night. But Dr. Olbers, the discoverer of the planet, who immediately had intelligence thereof and accordingly found out his planet Febr. 21, is, at least in Germany, the only one who makes due observations of it. He is only provided with a 5 feet refractor of Dollond (but an excellent one, 4 in clear aperture), the campus of which serves to him as micrometer. However so great is his skill in this kind of observations, that those made by him have a degree of precision, which hardly could be expected from this manner. The result of them is that my last ephemeris, which I had the honor to send you Nov. 18 (if I am right) of the last year, gave the Right Ascension too great by 2 ½ min., the declination too small by ½ or 1 min. This is to be understood for his first observations; at present the error of the Right Ascension has increased to 3 min. or some seconds more, which are to be subtracted from the place given in the above mentioned ephemeris; the error of the declination has remained nearly the same. With this correction the ephemeris serves to find out the planet at every time with all facility, that can be attained under the particular difficulties caused by the extremely faint light of the planet and the vast number of telescopic stars in this zone of the heavens. Probably you will have found out also the planet at present, and make yourself all the preceding remarks by your own observations.*

*As soon as I had received the two first observations of Dr. Olbers, I computed thereby the following new elements that I shall not touch anew, before the observations of this season shall be finished.*

| | |
|---|---|
| *Mean longit. 31 Dec. 1801) noon in the mer. of* | *(143° 28' 17."2* |
| *31 Dec. 1802) Seeberg* | *(221 28 54.0* |

*Therefore, for the meridian of Greenwich*

| | |
|---|---|
| *Epocha 1802* | *143° 28' 40." 1* |
| *1803* | *221 29 16.9* |

| | |
|---|---|
| *Aphelion 1803 (supposed to be at rest sidereally) 301° 24' 13"* | |
| *Ascending node, at supra* | *172 28 8* |
| *Mean diurnal tropical motion* | *769."4161* |
| *Logarithm of the semi axis major* | *0.4426160* |
| *Excentricity* | *0.245619* |
| *Inclination of the orbit* | *34° 38' 19."8* |

*With these elements I have compared all the hitherto-made observations of Dr. Olbers, which stand thus.* [Gauss writes here a table of observations from February 21 through April 13.]

*My ephemeris of the course of Pallas in this year expiring with August 9, I have continued it from the new elements till Oct. 20, which certainly will be more than sufficient, as the faintness of the planet's light will much sooner impose an end to the observations. Here I have the honor to transcribe the Continuation of the Ephemeris of Pallas.* [Gauss prints the ephemeris from Aug. 9 to Oct. 20.]

*As to the other new planet, Ceres Ferdinandea, I have not yet, till this hour, received any new observations. I am only informed that Baron* [Ferdinand] *Ende at Celle in company with Mr. Harding found the planet without difficulty March 21 and perceived its motion the following nights, but I could not get any precise observations of it. I am for this reason still ignorant, what corrections my ephemeris, which I sent you about the 31 of January last, may stand in need of.*

*I acknowledge with high pleasure that I am mostly indebted to the precision of your excellent observations of Pallas, for the exact agreement of its foretold course with the real, as likewise I am for the facility with which I now make all my calculations to your precious present of Taylor's Tables. I hope, you will farther grant your favor to my endeavors, to bring the theories of those remarkable celestial bodies nearer to perfection, by continuing to communicate to me your future observations of them.*

No. 64 Olbers to Gauss Bremen May 5, 1803

*Thanks for the compared observations of the $11^{th}$, $12^{th}$ and $13^{th}$. The calculated Dec. of the $11^{th}$ is 15° 0' 39."4 in your letter; I believe it ought to be read as 15° 0' 59."4.*

*Here are further observations of mine which, however, may well be somewhat less reliable because of the few accurately determined stars.* [These observations from Apr. 15 through 25 were published in the June issue of the *MC*.]

*Mr. v. Zach has sent me the following observations from the canon David in Prague.* [These observations were published in the June 1803 issue of the *MC*.] *Oriani has also observed Pallas on March 22 and April 1; the observations aren't reduced yet.*

No. 65 Gauss to Olbers Brunswick May 30, 1803

*Pardon me for the long delay in thanking you for your last letter of the $5^{th}$ of this month and the information therein. Of the observations of Pallas, I've only now calculated yours. Here is the result:* [This data was published in the July 1803 issue of the *MC*.]

*Your assumption concerning the Dec. of April 11 was correct; it was merely a slip of the pen in my letter. Thus far I still don't have news of any observation of Ceres in this year. I see from the* Connaissance des Tems An XIII, *which I'm receiving as a gift from the B. d. L. through von Zach, that Mr. Burckhardt has, in June of this year, set the longitude ¼ degree larger than that in my tables (assuming that the positions there are calculated at noon). Indeed, much uncertainty can still very well occur.*

No. 66 Olbers to Gauss Bremen June 8, 1803

*From your last letter, I see that my observation of Pallas up to the end of April still doesn't demonstrate a definite deviation of your osculating ellipse from the true orbit. I'm afraid the following few observations will also decide nothing because I've compared the planet, for the most part, with stars whose position has been indicated only by Flamsteed.* [See the July 1803 issue of the *MC.*] *The last observation was made not with the circle micrometer but rather with the cross-hairs micrometer.*

*You can see that I haven't been diligent. Much depended upon the weather. On May 10 I could not (for the first time this year) find Pallas. The reason was, I misidentified star No. 494 B. On May 25 I observed, instead of Pallas, a fixed star which wasn't far from Pallas and which I mistook for the planet.*

*I still haven't heard anything about Ceres. I'm regretting it more and more that I didn't follow your urging to search for this planet. For me, Ceres is now too low because some buildings are restricting my view. Haven't you heard anything from England about Ceres and Pallas? Hopefully we'll be receiving meridian observations of Pallas and Ceres.*

No. 67 Olbers to Gauss Bremen June 16, 1803

*I've never again seen Pallas. I also believe I'll be able to discontinue the – for me – very tedious observations because of its high position, for it will certainly be observed at several other places on the meridian.*

*Here is an observation from Messier (off the meridian), sent to me by Burckhardt:*

| |
|---|
| *May 28 10h 40m 6s true time in Paris* |
| *RA 281° 50' 34" Dec. 22° 5' 28"* |

*Ceres hadn't even been observed in Paris up to May 29.*

No. 73 Olbers to Gauss Bremen August 21, 1803#262

*I wanted to begin observing Pallas again after the middle of August, but until now have been prevented by the weather etc. Now I'll pay attention very carefully.*

*Here's the promised small list of Flamsteed's stars which, according to Lalande's assurance, are missing from the sky and each of which Flamsteed has definitely observed only once. They are the only ones of Flamsteed's stars among the 146, which Lalande has failed to see, of which this can be said.* [a table of six stars follows.] *Initially I hoped to find Pallas perhaps by No. 20 Eridani or No. 10 Canis maj., but the time just doesn't tally. I take No. 42 Scorpii and one of Mayer's stars of 8$^{th}$ magnitude to be one and the same.*

*When you go to Seeberg, it would be very desirable to obtain the times from Mayer's original observations, which are in safekeeping with Mr. v. Zach, when the following stars were observed: No. 78, 704, 860, 982, which Lalande failed to see, and 784 which according to Koch* [J. A. Koch in Danzig] *is missing from the sky. 860 and 982 could well have been Ceres and Pallas respectively.*

*According to Lalande several stars from La Caille's zodiac list are also missing. Yet I don't know how or where the times of their observations can be found.*

*Our speculation about the descending node of Pallas' orbit on Ceres' orbit now appears to be much more precarious, for I fear that this descending node is shifting very noticeably due to the attraction of Jupiter. Yet all of those missing stars make it all the more certain that there is still a multitude of asteroids and that we'll soon still discover some, particularly if our astronomical society becomes very active. I expect this especially from the well-known enthusiasm and perseverance of our worthy head appeal counsel v. Ende.*

*The reason I'm not really satisfied with the first observation is that I feel that the reduction of the small star from the* Hist. Cel. *is subject to a small uncertainty of 12". Moonlight interfered with the latest observation. In the third one the position for the determination of Dec. was not favorable.*

No. 76 Olbers to Gauss Bremen October 28, 1803

*I received two observations of our Pallas in October; the first one very good, the second incomplete. The weather, after many overcast days, finally cleared up on October 9. Pallas appeared like as exceedingly small star of 14^th^ magnitude very close to the most easterly of three small closely clustered stars that Harding recorded in his chart at 73° 50' RA and 6° 56' Dec. north. Just one of these small stars, precisely this most easterly one, appears on pg. 88 in the* Hist. Cel. *Harding, as noted by himself, has added the other two. The three small stars actually form a right-angled triangle whose base lies southerly and Lalande's star lies at the bases eastern acute angle. On pg. 88 of the* Hist. Cel. *this star appears as: 18h 14m 43s 41° 53' 12".*

*I compared Pallas with this star five times and found, by averaging the very much consistent results, that Pallas on October 9 at 7h 26m 7s mean time trailed this star by 4.2s and was 0' 40" southwards. On October 10 it was compared with the same star and at 7h 2m 10s now trailed it by 51.5s. The Dec. couldn't be noted on this day. As the basis for the star's reduction, I've used the position of No. 42 of Poniatowski's Bull as given, according to Lalande, by Bode and find:*

| | | *RA* | *Dec.* |
|---|---|---|---|
| *Oct. 9* | *7h 26m 7s* | *273° 51' 48"* | *6° 55' 26"* |
| *10* | *7 2 10* | *274 3 39* | —— |

*Pallas entered the tail of the cluster of stars above ν in Ophiuchus after October 10 and on the 11^th^ it was no longer possible to locate it in the throng of small stars. That is why I've given up further observations for this year. Please communicate this to Mr. Zach.*

*Since we still saw Pallas so close to its aphelion in October, it's quite certain that we'll be able to see it every year. Its rediscovery will be much easier in the coming year since it will, in general, present itself for better viewing. Just out of curiosity I've computed its position for every first day of the month from May onwards using the small tables that you so kindly gave me. We are, however, hoping for an ephemeris from you soon.*

No. 77 Gauss to Olbers Brunswick December 18, 1803

*I was not able to work on theoretical matters in Gotha. You might be interested in a study, which I, if I am not mistaken, mentioned already during my visit in Bremen, on the limits of geocentric motions of planets (also comets). The geocentric locations of every planet are, as far as its and the Earth's elliptical elements are regarded as constant, in a zone with particular limits; the exact determination of those is not without analytical elegance. I apply the results of this study to Ceres and Pallas; the calculations necessary for the first are almost finished, for Pallas. I will maybe first retouch the elements. I intend to publish an article about these studies in the* MC. *Our friend Harding leads me to expect that he will give us by and by a complete atlas of star maps for both zones which will be indispensable and invaluable for observing these two aloof goddesses.*

No. 78 Olbers to Gauss Bremen January 7, 1804

*I do not believe that you treat Pallas compared to Ceres shabbily although you calculated the perturbations of it already in the second year. For basically and seriously spoken, I cannot recommend to tackle this ample calculation of the perturbations at this point. The pure elliptical elements, which you will probably adjust some more to the latest observations, are sufficient enough to find Pallas again next June, and what else do we need for a start? Unfortunately, we did not find any older observations, for whose sake it would be desirable, to know the perturbations at least* quam proxime [approximately]. *It is really surprising that the observations of October still match your VI elements. – Hopefully you will give us soon an ephemeris of Pallas for the next year, which Harding needs very much for working on his map. I believe it sufficient if it is to start with May.*

*I look forward to your analytical determination of the geocentric limits of the latitude of Ceres and Pallas; for certainly I will have a lot to learn and admire, the more so, since this problem has induced me to some studies. But even if I assume the Earth's orbit to form a concentric circle with the Sun (which should always be permitted when only giving the limits for sketching maps), my formulae become quite complex and inconvenient.*

No. 79 Gauss to Olbers Brunswick February 7, 1804

*I have the pleasure to send you a transcript of those stars on Piazzi's list which lie within Pallas' zodiac from RA 327° to 345°, which therefore comprise the whole area within which this planet will be able to be observed this year. Kindly forward them to our friend Harding at your leisure. I've already sent him my computed ephemeris for Pallas several weeks ago, and presumably this same one will already have appeared in the February issue of MC. For the calculation of that ephemeris I used the following new elements. These eliminate the error of the latest observations without resulting in any particularly worse agreement with the former ones. Since these elements don't differ much from the VI Elements, thus in some places they will cause very noticeable differences between my ephemeris and Bode's (which is calculated according to these elements), probably because of the limited accuracy of the latter.* [These elements VII are printed in the *MC*, March 1804 issue.]

*Because of the <u>very</u> accurate agreement of these elements with the existing observations, I, as you, do not consider it necessary to take into consideration the variations in the elements, especially since, unfortunately, no good and complete meridian observations whatsoever have become available from last year.*

*My calculations concerning the zodiac of Ceres and Pallas have for the most part been completed, and I've sent Mr. Harding as many of the results as is necessary for him to prepare his proposed Atlas. In Dec. Ceres can reach 32° 52'N and 31° 52' S; Pallas however 26° 25'N and 33° 29' S. I have just a few more things to add about Pallas, and as soon as this is done I'll prepare an article for publication. Since you haven't considered this problem unworthy of your attention, then perhaps it might not be troublesome for you to briefly review my method beforehand. Clearly it depends only on the fact that for each heliocentric position of Pallas (or of any other planet or comet), one knows how to indicate the position of the Earth (or etc.) from which the former is seen as it appears within the boundaries of its zodiac. One can then take as many heliocentric positions of Pallas as one finds reliable as a basis from which to calculate the corresponding geocentric positions, and in this manner directly find as many points of the boundaries as one wishes. (I've taken 36 heliocentric positions of Pallas as a basis and, using them, have calculated 36 points of the northern and 36 of the southern boundary; half, even a third, of them would have been sufficient if accuracy only to the minute is required.) By using <u>appropriate</u> interpolation*

*e*, *e'* Excentricität der ⚴, ♁
*p*, *p'* halber Parameter der ⚴, ♁ $= \frac{1}{2}$ axis major $\times (1 - e\,e)$
*h*, *h'* Länge der Sonnenferne von ⚴, ♁ — Länge des ☊ von ⚴ auf der Bahn der ♁
*t*, *t'* Abstand in der Bahn vom ☊ der ⚴

$$\frac{p}{\cos t - e \cos h} = \frac{p'}{\cos t' - e' \cos h'}$$

oder zur Rechnung bequemer

$$\cos t' = \frac{p'}{p} \cos t - \frac{p' e}{p} \cos h + e' \cos h'$$

in Zahlen, die Elemente für 1803 zu Grunde gelegt,

$$\cos t' = \cos(\text{long. ♁} - 172^0\, 28'\, 14'') = 0{,}3841934 \cos t' + 0{,}0542514.$$

**Fig. 8.4** Gauss's formula for interpolation

*techniques one can then, if one wishes, interpolate between those points and thus obtain as many others as are desired at intervals (or longitudes) which increase, for example, by 5°. Indeed, for each given geocentric RA (or longitude) one could also immediately look for the corresponding Dec. (or latitude), but one would necessarily encounter fourth degree equations and this would give rise to much more work than the above-mentioned procedure.*

*Anyway, I've found the following simple formula* (Fig. 8.4) *for the solution to the above-mentioned task, merely by analysis,*

*From t', t I can then immediately calculate (without the interim task of finding longitude and latitude) the geocentric RA and Dec.*

*Only after I had already found the above formula analytically* [see Gauss's Works, Vol. 7, pg. 111 and 114] *did I notice that it could be made simpler by using geometric considerations. It can easily be shown that the essence of the problem must be the following:*

*'The tangents at the corresponding positions of Pallas and Earth must lie in a plane.' They must therefore intersect; or what amounts to the same thing, 'they must intersect the nodal axis at a point.' A simple calculation then gives that*

$$p/\cos t - e\cos h \quad p'/\cos t' - e'\cos h'$$

*are the distances of the points of intersection of the tangents and nodal axis from the Sun. Incidentally, the above formula obviously gives two values for t'. The one between 0° to 180° gives the northern boundary; the other between 180° to 360° the southern one.*

Gauss to Olbers Brunswick February 22, 1804

*Many thanks for your last kind letter and especially for the gentle way you pointed out the incorrectness of a method in the particular problem, which is indeed so defective that all calculations made according to it would be useless. Without doubt I must have caused you to assume by an expression not well chosen in the haste, that that method and idea were mine. But this is not the case. I have* [verb missing] *every hel. place of the Earth corresponding to every hel. place of Pallas, from where Pallas appears in its limits, i.e. where it has the greatest geocentric latitude,* verbi caussa *not that it* manenti eodem loco *helioc. Pallas has among all possible places of Earth, but that it among all possible combinations of Pallas and Earth positions, which give the identical geocentric longitude.*

*Actually, I believe your study no. 2 must yield the same results like mine, only that does not seem to lead to the goal as easily. In the complete development of my analysis there will be no misunderstanding about the actual sense of the limits. By the way, that your first way of treatment cannot be identical with mine, you can easily find, if you, for instance in the case of a circular orbit, compare the results of that with mine. – I have not yet finished the numerical calculations in the case of Pallas, but hopefully will be, then I will present the results to you. Then the composition of the analytical article will be done quickly. It is my forwardness to be sure of my results to several seconds, in order to (as I also did in my ephemerides) guarantee for the minutes. In the future any dissipation across the limits, or not touching those at the indicated time, will document the fact that the elements need improvement.*

*PS: Dearest friend, do I not abuse your patience if I come back once more to our problem? I cannot refrain from enclosing its solution geometrically, which I mentioned in my last letter since it is so easy and leads so easily to the final results.* (Fig. 8.5)

*I place ♁ (Earth) and ⚴ (Pallas) at arbitrary positions in their orbits, and through both a (by the way arbitrary, but her* verbi caussa*) plane supposedly perpendicular to the ecliptic. I give this ecliptic an ever parallel motion and ♁ and ⚴ always stay in it. The geoc. longitude of ⚴ stays thus the same, the latitude increases or decreases as long as it reaches a maximum or minimum. Then ♁ and ⚴ are in corresponding positions, we have a point from the limits and the differential of the latitude is = 0. (For brevity's sake I need infinitely small values, which you can easily translate into stricter terms.) Let ♁ ⚴ P be the position of the above plane in those corresponding points, where P ♁ × will be the geoc. latitude of Pallas, and ♁′ ⚴′ P′ the position after an infinitely small displacement. So the angle P′ ⚴′ ♁′ = P♁ ⚴, so the lines ♁ ′ ⚴′ and ♁ ⚴ parallel. From this follows without effort that the elements ♁ ′ ♁ and ⚴′ ⚴ and therefore the tangents to ⚴ and ♁ lie in the same plane and from this follows also the formula given in my last letter.*

*Olbers to Gauss Bremen February 29, 1804*

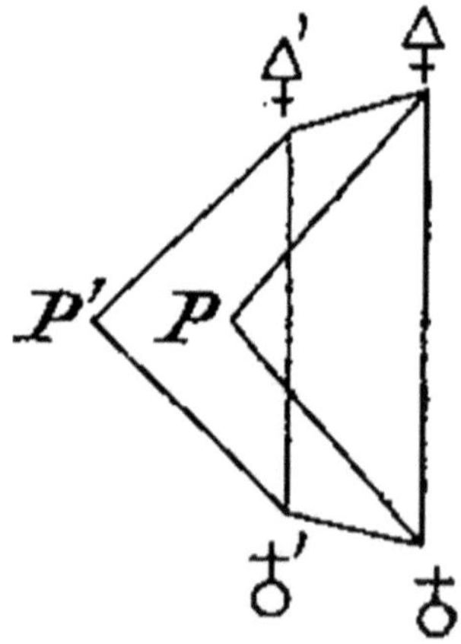


**Fig. 8.5** Diagram showing positions of Earth and Pallas

*Not you, but solely me, is to blame that I misunderstood your last letter. You expressed yourself completely clear and definite. Understandably, it only depends on the fact that one knows for every heliocentric position of Pallas the location of the Earth from which 'she appears in her limits'. Inadvertently, I understood the words 'in her limits' as the greatest possible geocentric latitude of the assumed position of Pallas, even more so, since I actually hit upon the idea initially, to determine by this erroneous way the limits. I see clearly now. The whole evident geometrical sentence gives the solution a great elegance and lightness. The equation*

$$\frac{p}{\cos t - e \cos h} = \frac{p'}{\cos t' - e' \cos h'}$$

*then follows, as you said, effortlessly. But the analytical way, through which you found this equation, must have had its difficulties, and I am very curious to see it. The problem seemed so complex initially that one should not have expected such a simple equation. And it struck me that the inclination was not taken into account. It is also strange that the Earth can see only on parts of its orbit (between* $2^s\ 3°\ 12'\ 22''$ *and* $3^s\ 18°\ 28'\ 31''$ *and between* $7^s\ 26°\ 27'\ 57''$ *and* $9^s\ 11°\ 44'\ 6''$*) Pallas in its limit, and that Pallas must touch twice a year its limits. – Shortly, I very much admire your wonderful solution of this problem, and thank you for your patience with which you considered my inconsiderate objections and disabused me.*

Gauss to Laplace Brunswick March 28, 1804

*I am hastening, Sir, to let the* Institut National *know, how much I was flattered by the title bestowed on me by this illustrious council, and how much I value such a distinction. For, were it not only the suffrages of enlightened judges to which righteous minds must aspire: The wish to deserve the approval of a tribunal of such legitimacy like the* Institut National de France *will be a further encouragement to indulge in the most difficult tasks. Would you please, Sir, convey my feelings of gratitude and devotion to that illustrious Society and my deepest respects which I give with all my heart to the merits of the great scientists of whom it consists.*

*I have just finished a research on the limits of the geocentric motion of the planets, of which the results might be of interest to the* Institut. *All planets and comets which move in a plane inclined to the ecliptic appear only twice during each revolution: for the rest they seem to sometimes to move away or approach, seen from the Earth. However, all of the possible positions a planet can have, seen from the Earth, are to be found in a zone with well defined boundaries: the exact determination of the limits of that zone, which I call the zodiac of the planet, is a rather curious problem, which has a simpler solution than one would expect even taking all the circumstances into account. I expound the one that showed itself to me in an article that will be printed in the astronomical journal of von Zach. One particular circumstance made me apply it in detail to the two recently discovered planets; it is my pleasure to enclose the results with this letter. Mr. Harding, known for his beautiful maps of Ceres' and Pallas' paths in 1803 and 1804, is fostering the project of a general atlas of those parts of the skies where these two planets can be found. This useful undertaking deserves all the more encouragement and recommendation since Mr. Harding volunteers to not only include every star of the* Histoire Celeste Francaise *on any of his new maps, but to verify with his own eyes the existence of each star he includes. The astronomers who have observed these new planets know how difficult it is to untangle those small twinkling dots from the surrounding telescopic stars: good and very detailed maps are indispensable for this purpose; and they will be grateful to Mr. Harding for his hard work which will satisfy a need which the maps of Mr. Bode could not fulfill.*

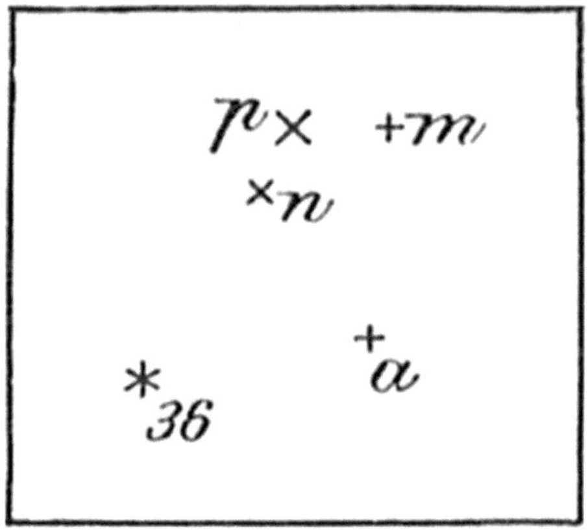


**Fig. 8.6** Finder chart for Pallas

No. 88 Olbers to Gauss Bremen May 16, 1804

*I believe I've already observed Pallas during the night from May 8 to 9, but until now I haven't been able to come to any definite conclusion about it. I have certainly seen it. That night was especially clear. I found 3 small stars over 36 Pegasi in an area where Pallas ought to have been, in the following configuration star a is a recognizable star of 10*$^{th}$ *magnitude* (Fig. 8.6). *Of the three smallest stars m, n, and p, the brightest was p and this one I took to be Pallas. In order to be able to orient myself with respect to these stars in the future, I noted that the angles 36 a p and a m p were 90°; however, p n 36 was colinear.*

*I've compared p three times with No. 36, still another star of Piazzi's and a star from the* Hist. Cel. *The observations agree very well over-all, and from them I've computed for p*

*May 8 13h 50m 0s apparent RA 334° 38′ 35″ apparent Dec. 8° 30′ 6″*

*On May 11 it wasn't very clear, and I saw neither p nor m and n. Last night (from the 15*$^{th}$ *to 16*$^{th}$*) it was very clear. However, dusk had become so intense by 1:30 am that such small stars weren't at all recognizable in a low position.*

*I caught brief glimpses of m, n was doubtful, and no trace of p. Subsequent observations of Pallas will soon reveal if I've been duped by a small fixed star.*

No. 89 Olbers to Gauss Bremen June 5, 1804

*I haven't again been able to find Pallas with certainty from May 8 to June 1. The eastern horizon was never completely clear; at other times moonlight and approaching dawn prevented me from seeing the still-dim planet. I finally observed it again on June 1 and 3. Here are all 3 observations, should you not have received my previous letter:*

| | *Mean time* | *RA* | *Dec* |
|---|---|---|---|
| *1804 May 8* | *13h 50m 0s* | *334° 38′ 35″* | *8° 30′ 6″ south* |
| *June 1* | *12 4 50* | *338 27 27* | *10 10 32* |
| *3* | *11 53 25* | *338 41 25* | ——— |

*I'm of the opinion that the first observation is the best; the last is the worst.*

No. 90 Gauss to Olbers Brunswick June 17, 1804

*I was very pleased by the news of your unexpected early rediscovery of your planet and your further observations on it. The good child, who has been until today – quiet and*

| | | 1803 | 1804 |
|---|---|---|---|
| Sonnenferne — | Knoten + $A$ = | 32° 15′ 6,4″ | 32° 15′ 50,7″ |
| „ | „ + $B$ = | 301° 25′ 54,5″ | 301° 26′ 44,1″ |
| „ | „ + $C$ = | 143° 19′ 30,5″ | 143° 17′ 47,2″ |
| | log sin $a$ = | 9,998 799 | 9,998 804 |
| | log sin $b$ = | 9,992 125 | 9,992 122 |
| | log sin $c$ = | 9,307 080 | 9,307 028 |

**Fig. 8.7** Constants for 1803 and 1804, as determined by Gauss

*obedient – easy to parent, comes now into the years when the influences of foreign powers must soon become visible, and when we must no longer ignore its inclinations and convergences towards these, if we do not want to be often confounded and lose our hitherto undeniable ascendant because of it. Without metaphor. Dear friend, I made another attempt if the present error of 5 min. of my last elements can be eliminated, disregarding the perturbations, and I actually found that it looks quite good and that the other earlier observations can be combined without much force; but the corrections of the elements are almost greater than I thought. Here are the new elements:*

| *Epoch 1803 Seeberg* | *221 31 12."8* |
|---|---|
| *Aphelion* | *300° 54' 59* |
| *Ascending node* | *172 28 9* |
| *Inclination* | *34 26 23.1* |
| *Eccentricity* | *0.2461016 = sin 14° 14' 49"* |
| *Log semimajor axis* | *0.4416223* |
| *Daily tropical movement* | *772."0597 (in 365$^d$ 78° 16' 41.8")* |

[Compare to elements VII in *MC*, March 1804.]

*For an easier calculation of the geoc. RA and decl., I determined the following constants in regards to my article in the* MC *May* (Fig. 8.7):

$$Sonnenferne = Aphelion\ Knoten = node$$

*Like in that article the angles A, B, C, had to be added to t, the argument of the latitude, the present are directly added to the true anomaly. The aphelion is still supposed to be at rest sidereally, and the diurnal motion of Pallas against it to be 771.9225" (log 2.8875737). The constants already contain the nutation*[1] *. These elements, based on the observation of June 1, which, if the calculation is correct, must be accurately represented by those, differ in the case of the two other*

| *May 8 RA –30"* | *Decl. –36"* |
|---|---|
| *June 3 RA +14.5"* | *Calculated Decl. 10° 17' 14.7"* |

[1]For a brief historical overview of the discovery of nutation and its role as a source of uncertainty, see Belenkiy and Vila-Echague 2016. It was d'Alembert who provided analytical proof of nutation's existence in 1749.

*The future places of Pallas this year result according to these elements differently by the following differences than according to the ephemeris:*

| | | |
|---|---|---|
| *1804 Aug. 4* | *RA greater 9' 17"* | *Decl. South 20" also less* |
| *Oct. 27* | *7 11* | *4' 18" also greater* |
| *1805 Jan. 19* | *10 53* | *3 48" also greater* |

*This calculation was only made hastily and needs revision. My intention calculating the new elements was actually mainly to be able to discuss the future observations of this year in a better and more reliable way. God give that the keeper of the larger observatories will be more eager this year than last year. Then also the perturbations shall be allowed for, for which I am planning to make the preparations. I already sent the article on the zodiac of the planets some time ago to General v. Zach* [Zach held a military rank of general] *and will probably be printed in the August or July issue of the MC. I only developed the analytical method and only hinted at the geometric one. The task still has some peculiar difficulties across which I came elaborating and which I only randomly scratched in this article.*

*PS: Is it not weird that the mean motion of Pallas, which until now appeared to be slower than Ceres, now appears to be faster and does not remain a possibility that the strange phenomenon of a complete identity may really occur?*

No. 91 Olbers to Gauss Bremen July 6, 1804

*I was also surprised by your new elements of Pallas' orbit. I had not expected the correction of the elements to be so considerable. It is a proof of the considerable influence of the perturbations on the motion of this planet. Here are three more obs. on Pallas:* (Fig. 8.8)

*Although Pallas is quite easily visible in the telescope right now, the weather was nasty and I am not quite satisfied with the obs. of July 1 and 4. But the RA of July 5 is as well determined as was possible at the circular micrometer. Four obs., all four marked as good, corresponded up to ½ second of time.*

*I will not be wasting my time right now on obs. on Pallas. Their only purpose could be to correct the elements so that the ephemerides can be corrected for an easy meridian observation. This purpose, as I believe, has been achieved sufficiently.*

*You probably know Messier's obs. from public papers. Unnecessarily, I am giving them here:*

| | | |
|---|---|---|
| *June 2 13h 5m 36s true time* | *338° 36' —* | *10° 14' 22"* |
| *17 12 41 56* | *339 56 30"* | *10 55 23* |

*Also Messier complains about the difficulty observing the faint planet, and the obs., I believe, will show these difficulties.*

| | Mittl. Br. Zeit | Scheinb. AR | Scheinb. Dekl. | Vergl. Stern |
|---|---|---|---|---|
| Juli 1. | $11^h 37^m 10^s$ | 340° 22' 3" | 11° 6' 52" | ζ *Pegasi* |
| „ 4. | $12^h 33^m 6^s$ | 340° 21' 11" | 11° 5' 58" | |
| „ 5. | $11^h 30^m 31^s$ | 340° 20' 18" | — — — | |

**Fig. 8.8** Observations of Pallas by Olbers on June 1, 4, and 5. Columns from left to right are: *Mean time Bremen; Appar. RA; Appar. Decl.; Comparison Star*

*Harding has probably told you his obs. himself. He sent me the following:* [see Harding to Gauss, June 21 letter] *The RA is probably more reliable than the decl. with which friend Harding is not always fortunate.*

*Do you consider it useful in any way to continue the observations at the circular micrometer, since soon we can expect meridian obs.? After having finished the meridian obs. I will indeed try to follow Pallas if possible.*

No. 92 Gauss to Olbers Brunswick July 24, 1804

*You are right, the influence of the perturbations on the motion of our Pallas, seems to become more and more apparent. I compared my last elements with your latest obs. and was surprised to find the following rapidly growing differences; I am also giving the older ones here, since in the one of May 8 occurred an error in calculation regarding the place of the Sun.*

*Error of the calculation.*

| *May 8* | *RA - 20″* | *Decl. –36″* |
|---|---|---|
| *June 1* | *0″* | *0″* |
| *3* | *+ 14.5″* | .... |
| *July 1* | *+1′ 24″* | *–37.8″* |
| *4* | *+1′ 11″* | *–53″* |
| *5* | *+1′ 13″* | .... |

*If no errors have crept in my rather hasty calculations, I must doubt if all obs. can be fairly described by purely elliptical elements. If only soon meridian obs. were made and publicized! From Gotha and Paris is probably not much to be expected since Zach and Méchain are still busy with the Earth.*

*The other day I calculated all of Buerg's* (Fig. 8.9), *Zach's and Oriani's meridian obs. on Ceres of last year according to my latest elements, which were only corrected according to the three much earlier of Palermo, and the correspondence was quite good. Consequently, I am awaiting longingly how these elements correspond to this year's observations. I would be very pleased if you could snatch one or the other before your departure to Rehberg. I only know from hearsay that the National Institute Paris has put a prize on the theory of the perturbations of Pallas, since I skim through the newspapers only. If the program is given in extenso in the* Moniteur *or anywhere else, please give me the number. I almost believe such an extensive work too premature. If, on the other hand, quite good meridian observations will be made this year and some of the last year become known (without which the inclination of the orbit and other things can only be determined preliminarily) – maybe in the* Conn. des tems XIV, *which I have not seen yet, can some be found by [Alexis] Bouvard, or of the* Observat. de l' ecole militaire, *by Burckhardt or Le Francais – I am possibly inclined to tackle such a task.*

Johann Tobias Buerg (1766–1834; Fig. 8.9) was an assistant at Vienna Observatory (from 1792) and professor of mathematics and astronomy at the University of Vienna beginning in 1806. He was one of the first people to see Pallas. He did so in the presence of Zach on April 4, 1802, as related in Chap. 11. Buerg resided with Zach for 10 months as a calculator at Seeberg Observatory, during which time (1801/02) he observed Ceres and Pallas.

**Fig. 8.9** Tobias Buerg. On 7 July 1801 Zach wrote to Oriani that he and Buerg were "chasing a small planet." Buerg became one of the first to see Ceres and Pallas (Vargha, 2005: 55)

No. 93 Olbers to Gauss Bremen July 27, 1804

*I consider the great deviations of the observations of July very remarkable, but I admit, your severe corrections which your elements suffered in your last calculation are no less so. This quick deviation of the elliptical elements can hardly be explained by the perturbations. Did not rather the major axis become too small and the mean motion thus too great by the unreliability of my observation of June 9 than by an error in calculating? – Harding's observation of the RA of June 1 agrees with mine, as far as I remember, but for 15." (I looked it up, Harding gave June 1, 12h 51m 44s mean time at Lilienthal app. RA Pallas 338° 27' 10.4" the declination completely wrong 10° 13' 22"). His observation of July 1 agrees with mine in RA but for 1". But the RA of July 1 is not as reliable as that of July 5. – Also in the case of May 8 I do not believe to be wrong in the RA by 20". Four observations gave the difference of no. 36 in time 0m 48s, 0m 47.5s, 0m 48s, 0m 48s. – The second and third are identical, only immediately determined by Piazzi by no. 36 and another star. Weather and moonshine give me little hope to see Ceres now, but I will make an effort.*

No. 95 Olbers to Gauss Bremen September 1, 1804

*I read, studied and admired your excellent article on the zodiacs of planets with great pleasure. It occurred to me if, after your calculations, it might not have been easy for you also to give the day of the year for each of the 144 points when Pallas or Ceres might be locatable within these boundaries. This would tremendously simplify any searches of missing stars in the* Hist. Cel. *or in Bradley, Flamsteed, etc., if they perhaps could have been Ceres or Pallas. For one could then approximately estimate for every other day the Dec. which the planet could have for a given RA.*

*I observed both planets on August 29 very well, at least as far as the RA is concerned, but haven't yet had time to reduce them. Bode has sent me several helpful comments not only concerning their apparent relative motion during a revolution, but also concerning the motion of each of them as seen by the other. Relative to Ceres, Pallas describes a long narrow oval against the sky. This oval stretches from the Dragon (Kite) to the Tucan (Pepper Eater), curves itself in a regular manner at the top by the Dragon, but ties up into a knot at the bottom by the Tucan. Pallas would never come in opposition or upper conjunction with the Sun there, but only once the lower conjunction and is then the nearest to etc.*

[The Sept. 7, 1804, letter dealing with the Paris Prize question was given in Chap. 3.]

[The discovery of the third asteroid, Juno, elicited a flurry of correspondence between Olbers and Gauss at this juncture. These will be published in a future book in this series. Only sections of the letters dealing with Pallas are printed below.]

No. 106 Olbers to Gauss Bremen October 10, 1804

*Here is also one observation of Pallas:*

*Oct. 9 7h 2m 27s 327° 42′ 28″ 3° 10′ 28″ south*

*Unfortunately, I can't find any observations of Pallas in the most recent volume of the* Conn. des tems *apart from my own. I've urgently requested Burckhardt to send those of the previous year.*

Gauss to Maskelyne Brunswick October 16, 1804

[The first part of this letter, dealing with Juno, will be printed in a future book in this series.]

[Gauss includes here the ephemerides of Ceres, Pallas and Juno for the period Oct. 18 to Dec. 29.]

*Nota*

1) *The declinations are all south.*
2) *The hour for all positions is 12h mean time at Seeberg.*
3) *The ephemeris of Pallas gave on October 9 the RA 7′ too small the Decl. 2′ too small.*
4) *The ephemeris of Ceres gave on Sept. 15 the RA 8′ too small the Decl. 4′ too great.*
5) *The errors of Juno I cannot yet guess at.*

*Ceres still is pretty bright and you will find it without pain, but Pallas is already so faint, and those parts of the heavens in which it moves this year so rich of small stars, that it is very necessary, to be very attentive to the small stars through which it passes, in order to mark its motion from one evening to the other. But the more precious will be exact observations of this planet. About September 18 Ceres and Juno will be very near each other. I wished Dr. Herschell might profit of this favorable occurrence, to compare their apparent diameters and brightness – I hope, dear Sir, you will make a rich series of observations of all the three planets, the kind communication of which shall highly oblige.*

No. 109 Olbers to Gauss Bremen October 31, 1804

*Here also are a couple of good observations of Pallas, assuming No. 109 Bode has been correctly given by Lalande:*

| | | | |
|---|---|---|---|
| *Oct. 23* | *7h 9m 36s* | *327° 30′ 49″* | *5° 28′ 57″ S* |
| *24* | *6 39 36* | *327 32 34* | *5 37 23* |

No. 111 Olbers to Gauss Bremen November 7, 1804

*Here are also two good observations of Pallas that have been compared with No. 30 Aquarii according to Piazzi.*

| | | | |
|---|---|---|---|
| *Nov. 5* | *6h 6m 21s* | *328° 15′ 20″* | *7° 5′ 17″S* |
| *6* | *6 13 16* | *328 20 47* | *7 11 32* |

*I'm curious whether these observations, obtained almost 6 months after my first ones this year, will still confirm the strong variations which you discovered in their elements and particularly in the mean motion.*

*Pallas appeared on these two nights again comparatively very bright, and seemed almost brighter than on the 23rd and 24th of October. Perhaps this was due to the increased clarity of the atmosphere.*

*Ceres, on both nights, was very close to a recognizable star of the Whale and exceedingly easy to observe, but I didn't have the time to do so.*

No. 112 Olbers to Gauss Bremen November 17, 1804

*In 1802 I estimated Pallas' diameter to be 0.7 that of Ceres' diameter with high probability. For on March 28, 1802 one had to compare Ceres and Pallas very accurately in order to notice that Ceres, having higher altitude, had in fact a somewhat stronger brightness than Pallas. However I don't think that the ratio was greater than 5:6. Had both brightnesses at that time been identical then Pallas' diameter would be 0.77. Granted, Ceres was brighter in the ratio of 6:5, hence Pallas' diameter will be 0.70.*

No. 115 Olbers to Gauss Bremen December 5, 1804

*I've observed mainly Pallas since my last letter.*

*Here are four observations of it, of which especially the last three are good:*

*As regards ν and ρ Aquarii, Zach and Piazzi are in excellent agreement* (Fig. 8.10).

Gauss to Maskelyne Brunswick December 7, 1804

*For the commodity of the observers, I have calculated a new ephemeris which I have the honor to communicate to you. The ephemeris of Ceres, sent you in my last letter, gave on Dec. 4 the Right Ascension too small 6′, the declination too great almost 4′. Therefore Juno and Ceres will on Dec. 21 come so near one another, that the distance is less than ¼ degree. Juno now is considerably fainter in light than Ceres, though still brighter than Pallas, which Dec. 3 was very near ζ Aquarii.* [Gauss prints here an ephemeris from Dec. 5, 1804, to Jan. 31, 1805.]

| | | | |
|---|---|---|---|
| Nov. 20. 8h 21m 20s | 330° 7′ 0″ | Dekl. 8° 22′ 52″ | Stern nach Bode. |
| „ 25. 5h 49m 26s | 330° 55′ 47″ | 8° 40′ 26″ | ϑ Aquarii |
| „ 27. 5h 54m 59s | 331° 17′ 22″ | 8° 46′ 27″ | |
| Dec. 2. 6h 56m 36s | 332° 15′ 20″ | 8° 58′·52″ | ϱ Aquarii |

**Fig. 8.10** Observations of Pallas by Olbers from November 20 to December 2, 1804

*I expect with impatience the continuation of your observations of Juno, as those which you shall have made of Ceres and Pallas: and I shall be happy, to send you the further results of my investigations on these three remarkable planets.*

No 1 Gauss to Bessel Brunswick December 21, 1804

*Our common friend Dr. Olbers told me that you were so kind as to support me in calculating the geocentric orbit of the three new planets and to take on the task of calculating the positions of the Sun. I have almost finished my work on new elements of Ceres and thus will soon need the positions of the Sun for Ceres' ephemerides and take the liberty of asking you to take on this mechanic and more useful than interesting work. I would like to start Ceres' ephemerides with July 28$^{th}$ at 12 o'clock mean time at Seeberg, because Ceres will probably be visible around the beginning of August 1805 and since Juno will be visible until July 1806, if its faint light will be no hindrance, the positions of the Sun should comprise one year. But I only need them every twelve days what amounts to 30 or 31 positions. I would prefer the positions of the Sun exact to 0".1 and the logarithms of the distances to the seventh place, the coordinates, however, are sufficiently exact if they are accurate to 0.00001. Regarding the Sun's longitude and distance I only need the equations dependent on I, II, III, V and Ω of the many equations of Mr. von Zach's new Sun-tables; in I, the equation of the center, I also look for the decrease in eccentricity and in the interpolation for the second differences. The enclosed scheme will explain this further.*

No. 120 Olbers to Gauss Bremen January 16, 1805

*Here are two observations of Pallas, perhaps the final ones before it disappears this time. Pallas was compared with 232 Aquarii according to Bode's list. Bode attributes the star's position to Piazzi; however, Piazzi hasn't included it in his extensive catalogue. It also doesn't appear in the* Hist. Cel. *I've adopted the star's apparent position to be*

*RA 339° 33' 4.3" Dec. 9° 7' 9.2" and thereby found for Pallas*

| | | | |
|---|---|---|---|
| *1804* | *Dec. 30 6h 31m 30s* | *339° 1' 44"* | *Dec. 9° 9' 14"* |
| | *Dec. 31 5 47 24* | *339 17 56* | *9 8 4* |

*Schroeter writes me that his treatise on the small planets, which should soon be published and constitutes a complete volume, is almost finished. I quite candidly brought my doubts about the accuracy of his measurements to his attention. He however stands by his conviction. Pallas is therefore, according to him, the largest and as large as our Moon!!*

No. 121 Olbers to Gauss Bremen January 20, 1805

[The text of this letter dealing with Vidal's planet is published in Chap. 5 of this book.]

*I've had to discontinue observations of Pallas because of its low position.*

No. 122 Gauss to Olbers Brunswick January 25, 1805

[The first part of this letter, about Ceres, was in *Discovery of the First Asteroid, Ceres.*]

*Next I'll apply myself to the calculation of the Milan and Seeberg observations of Pallas in order to improve the elements anew using that calculation and yours. I believe that those forwarded to you in 1804 won't differ much from these observations. I'd give much to have several reliable Dec.'s from June or July 1803, or indeed only generally from that year 1803, without which each improvement must still turn out to be precarious. I therefore consider it very superfluous to make any allowances for the perturbations, the more so as next year Pallas will be so bright that there cannot be any difficulty in locating it, especially in the southerly regions. An incomplete calculation of the perturbations doesn't help any more than none at all, as was demonstrated by Ceres last year.*

No. 123 Olbers to Gauss Bremen January 31, 1805

*I've had to totally give up the observations of Pallas because of the inclement weather. Since you are exchanging letters with Piazzi, I'm beseeching you to remind him to observe this planet in the coming year, and for that reason to send him the ephemeris sufficiently early. We certainly won't be able to make any observations here around the time of opposition.*

*I would still wish to experience the year 1811 because of Pallas, when it will probably be visible to the naked eye as a 5$^{th}$ magnitude star.*

*I hope you'll share with us in the* M.C. *something more detailed about your latest investigations into Ceres' orbit. This must be very interesting.*

*I'm now getting Le Caille's* Coelum Australe *in order to see whether it perhaps contains Pallas. I fear, however, that the observations start too late. In the winter of 1751 it appeared especially bright with a considerable southerly declination.*

No. 124 Olbers to Gauss Bremen February 7, 1805

*My hope to locate Pallas among La Caille's stars is shattered. La Caille started his observations only on August 6, 1751.*

No. 125 Gauss to Olbers Brunswick February 12, 1805

*I am now occupied with Pallas. I've compared all of the Milan and Seeberg observations of 1804 with the elements very precisely; I found them from your first observations and have communicated them to you. The results follow:*

| | | |
|---|---|---|
| *Aug. 27* | *+66.4″* | *–43.0″* |
| *28* | *+64.2* | *–47.9* |
| *29* | *+60.9* | *–49.3* |
| *30* | *+58.7* | *–51.4* |
| *31* | *+54.6* | *–51.0* |
| *Sep 6* | *+43.9* | |
| *7* | *+44.1* | |
| *9* | *+34.7* | |
| *10* | *+36.9* | |
| *12* | *+30.9* | |
| *13* | *+36.1* | |
| *14* | *+35.3* | *–75.2* |
| *15* | *+25.5* | |
| *17* | *+26.8* | |
| *28* | *+16.3* | *–98.9* |
| *30* | *+20.3* | *–106.4* |
| *Oct. 2* | *–50.2* | *–239.4* |
| *4* | *+18.3* | |

*An error or perhaps a mix-up must have occurred with the observation of October 2. I find the opposition as follows:*

*Aug. 30 5h 1m 40s Seeberg mean time 337° 0' 36.1" +15° 1' 49.8"*

*I'm now up to the observations that were made at around the time of opposition 1803. Unfortunately, I can't trust either Oriani's or David's, the difference in both calculations of the opposition is already a bad omen. As soon as I've finished the ephemeris, I'll urgently request Piazzi to observe Pallas next year. Since it will be much brighter than in the previous years, he'll be able to observe it so much the better. Oriani's observations of 1803 must either be very bad or he calculated them incorrectly, for the VI Elements couldn't have differed by 3' in June 1803.*

*I can't rely too much on v. Lindenau's meridian observations of Pallas in January 1805. He appears, speaking confidentially, not to possess sufficient practical proficiency. He has, at my request, sent me several positions of fixed stars that are certainly not reliable.*

No. 126 Gauss to Olbers Brunswick February 26, 1805

*I'll soon be finished first of all with Pallas. I do believe that I've worked out the opposition of 1803 reasonably well. I've reduced Bittner's observations from scratch* [Adam Bittner, 1777-1844, astronomer in Prague]. *446 Bode appears, according to the* Hist. Cel., *to have been correctly given by Bode. I've therefore again added the 16" to the RA which was removed by Bittner while increasing the Dec. by 6 ½". I thus found Bittner's and Oriani's observations to be in close agreement, but precisely those observations, on which both oppositions depended, aren't worth much. Bittner's Dec. is too small, Oriani's too large. My result is*

*June 30, 1803 0h 31m 12s Seeberg mean time 277° 39' 26" 46° 26' 49"*

*And lies between the two.*

*Using the 5 observations of*

April 1802
May 1802
June 1803 opposition
Aug. 1804 opposition
Nov. 1804

*determined according to your latest observations and by means of the differential method, I have thus for the first time improved the elliptical elements of Pallas to such an extent that the differences are as small as possible, and have found the following elements* [published as VIII Elements of Pallas in the *MC* Vol. XI p. 377; *Gauss Werke*, Vol. VI, p. 263]. *The difference of the observations are as follows:*

| | *Long* | *Latitude* |
|---|---|---|
| *1802 4 April* | *+18.9"* | *–6.2"* |
| *16 May* | *+3.6* | *+11.3* |
| *1803 30 June* | *+7.3* | *–6.2* |
| *1804 30 Aug.* | *+21.7* | *+15.7* |
| *28 Nov.* | *–14.9* | *–33.1* |

*The use of a pure ellipse wouldn't account for the observations appreciably better. With the calculation of Pallas' ephemeris I have already made a start, and hope to be able to send it to Harding on the next postal day.*

No. 128 Gauss to Olbers Brunswick March 25, 1805

*I've now begun to occupy myself with the perturbations of Ceres but there certainly won't be any quick progress. This is, so to speak, only a trial for which I'm using a specific method. Someday I might be able to apply it to Pallas and Juno if it fulfills my requirements. There is indeed much work to it, but the way in which I've approached the problem will not neglect the $11^{th}$ power of the eccentricity and inclination. Meanwhile, the by far longest part of the calculation is so routine that I'll certainly be able to use outside help with it. Perhaps in future I'll take advantage of Mr. Bessel's readiness to oblige. He has recently sent me his results concerning the comets of 1618 which, along with his skill with calculus, are continually increasing my grasp of the problem.*[2]

No. 129 Olbers to Gauss Bremen April 4, 1805

*A particular fate appears to have befallen all older observations of the 3 new planets. I have now received La Caille's* Coelum astrale Stelliferum. *At the beginning of 1751 Pallas was still very conspicuous on the other side of the Tropic of Capricorn. But La Caille started his observations only in the summer of 1751, when Pallas was already approaching the equator.*

*I'm anxiously awaiting the results of your perturbation calculation for Ceres, for which you're using a specific method.*

*Schroeter has, as he informs me, changed much in his work concerning the new planets based on ideas I had pointed out; I thus hope that you will no longer consider the calculation of the masses, densities, and gravitation at the surface of these small heavenly bodies. The determination of these details rests upon a totally erroneous application of an unprovable statement of Melanderhjelm.* [Daniel Melanderhjelm, 1726–1810] *He had adopted the hypothesis that the planet's atmospheric density at the surface varies as the square of the gravitational force at the surface. Schroeter believed he could conclude the reverse, that the atmospheric density at the surface varied as the height of the visible portion of the atmosphere. For our Earth he adopted, along with La Hire* [Philippe de la Hire, 1640–1718], *a height of 38,000 Toisen. Since his telescopic observations gave him the heights of Pallas' and Ceres' atmospheres from 100 to 150 miles, he thus decided on a high atmospheric density at the surface of both planets, and this the same for the gravitational force and density. The result is, e.g., that the density of Ceres is 4½ to 5½ times that of gold, etc.– I pointed out to him (1) that Melanderhjelm's so-called theory merely entails the somewhat strangely expressed theory that the ratio of the mass of the atmosphere of every planet to its total mass is always the same, and thus with every planet it would be about 1/800000 of its mass; (2) that this hypothesis, in itself very improbable, is refuted precisely by his observations of such large atmospheres surrounding such small heavenly bodies; and (3) that the heights of the visible atmospheres could by no means vary just like their density at the surface, etc. Just between us, I can't at all believe that Ceres and Pallas have these large atmospheres. Rather, I assume them to be due to irradiation in the telescope.*

[For more on how irradiation was explained by Leonardo, Galileo and others, see Dupré 2003.]

[2]Three notable comets appeared in the year 1618. See Drake and O'Malley (1960)

No 6 Gauss to Bessel Brunswick September 3, 1805

*You once had the kindness to take on a tedious calculation for me and today I would like to impose a new one on you. I have been working for some time on calculating perturbations. I have developed a special method for such planets like Pallas and Juno that show a considerable eccentricity or inclination. This method is much, very much work but I do not consider this a disadvantage since in my opinion all known methods are deficient. A characteristic of this method is that the development of coefficients of such an expression*

$$(aa+a'a'-2\,a\,a'\cos\varphi)^{-1/2} = \tfrac{1}{2} A^o + A'\cos\varphi + A''\cos 2\varphi + A'''\cos 3\varphi \text{ etc.}$$

*is requested – like it is the case in all hitherto used methods–not only for one value a, a' but for a large number of different values. For example in a calculation I made for Ceres I needed 50, for calculating Pallas and Juno I will need several more. But I am able to determine those coefficients at high speed because I know special dodges, partly based on seeming heterogeneous studies. But the large quantity renders the work tedious and since the work has to be done for one and the same planet several times (with the successively improved elements) I am planning on calculating an independently valid table. With the help of which the needed coefficient would be given without much effort. And I am so immodest as to ask you for support. Naturally the table should not state $A^o$, A', because then there would be doubled entries. But one needs only one entry for a $A^o$, a A', a A'' etc or a'$A^o$, a'A', a'A'' etc because the values are only functions of a/a'. It is not difficult to develop the following series, a/a'= f (on the premise a < a')* (Fig. 8.11) *It is known furthermore, that the following equations are true* (Fig. 8.12).

*Normally $A^o$ and A' are calculated from above series and then A'', A''', $A^{IV}$ etc. But this procedure is not applicable for my intention, for if $A^o$ or A' are slightly inaccurate, ie by 1 in one of the decimal places, the errors in A'', A''' are increasing, almost in geometric progression and the more so the smaller f is and one can possibly obtain for more distant coefficients, for example for $A^x$, which is probably necessary for Pallas, completely wrong values. This fact has not been mentioned yet, as far as I know, but it is nevertheless very important. Therefore I thought of a different method where I can obtain the logarithms even of $A^x$ to the $6^{th}$ decimal place with a normal log table.*

$$\tfrac{1}{2}a'A^0 = 1 + \tfrac{1}{4}ff + \tfrac{1}{4}\cdot\tfrac{9}{16}f^4 + \tfrac{1}{4}\cdot\tfrac{9}{16}\cdot\tfrac{25}{36}f^6 + \text{etc.}$$

$$= 1 + \tfrac{1}{2}\cdot\tfrac{1}{2}ff + \tfrac{1\cdot3}{2\cdot4}\cdot\tfrac{1\cdot3}{2\cdot4}f^4 + \tfrac{1\cdot3\cdot5}{2\cdot4\cdot6}\cdot\tfrac{1\cdot3\cdot5}{2\cdot4\cdot6}f^6 + \text{etc.}$$

$$\tfrac{1}{2}a'A' = \tfrac{1}{2}f\left(1 + \tfrac{1}{2}\cdot\tfrac{3}{4}ff + \tfrac{1\cdot3}{2\cdot4}\cdot\tfrac{3\cdot5}{4\cdot6}f^4 + \tfrac{1\cdot3\cdot5}{2\cdot4\cdot6}\cdot\tfrac{3\cdot5\cdot7}{4\cdot6\cdot8}f^6 + \text{etc.}\right)$$

$$\tfrac{1}{2}a'A'' = \tfrac{1}{2}\cdot\tfrac{3}{4}ff\left(1 + \tfrac{1}{2}\cdot\tfrac{5}{6}ff + \tfrac{1\cdot3}{2\cdot4}\cdot\tfrac{5\cdot7}{6\cdot8}f^4 + \tfrac{1\cdot3\cdot5}{2\cdot4\cdot6}\cdot\tfrac{5\cdot7\cdot9}{6\cdot8\cdot10}f^6 + \text{etc.}\right)$$

etc.

**Fig. 8.11** A series calculation developed by Gauss in perturbation work

**Fig. 8.12** Further series equations

$$A^0 - 2\left(f + \frac{1}{f}\right)A' + 3A'' = 0$$

$$3A' - 4\left(f + \frac{1}{f}\right)A'' + 5A''' = 0$$

$$5A'' - 6\left(f + \frac{1}{f}\right)A''' + 7A^{IV} = 0 \text{ etc.}$$

No. 134 Olbers to Gauss Bremen September 8, 1805

*The weather this year could drive astronomers to despair. I haven't been able to observe Pallas at all, but I was finally able to get an observation of Ceres. On the 5th I compared it with 2 small stars from the* Hist. Cel.*, both observed on two hairlines and in excellent agreement. I could only make two comparisons, both however gave the same RA. The Dec. is based on only one comparison and is hence less reliable.*

*Sept. 5 13h 8m 47s mean time apparent RA 95° 59′ 25″ Dec. 22° 21′ 25″.*

*The RA seems to differ from the ephemeris still by about 4 minutes.*

*I found the comments in your last letter concerning the perturbation calculations of the small planets, particularly of Ceres, extremely interesting. How laborious the final correction to Pallas' orbit will be! Laplace doesn't appear to want to concern himself with it at all. In the introduction to the 3rd volume, he had promised the perturbation theory of these planets, yet in the 4th there's no mention of it whatever. Moreover, he now says everything that he wanted to reveal concerning the mechanics of the heavens has ended, and there remains merely the story of the discoveries which he will present in the 5th volume. – I'll write more about this 4th volume before long.*

*I've heard nothing more concerning Legendre's new work on the planets. I've carefully preserved your small article on the determination of planetary orbits – an invaluable present. Will you allow me to make a copy of it before returning it to you and also show it to Bessel?* (Fig. 8.13) *Thus far it hasn't been out of my sight.*

**Fig. 8.13** A bust of Bessel in the German city of Bremen. (Photo by the author, 2004)

Zach to Gauss Seeberg Nov. 25, 1805

*I have not neglected your newly discovered planets and derived some pleasure from being reacquainted with them. It was Lindenau's belief that Pallas would never be seen as far North as to come before your eyes again (*MC *XI April 1805, pg. 383). If the sky was not overcast, I could observe the planets from my present position, and found it possible to document their position. Until then, I am sending you the few results I could obtain in the South, in the hope that they may be of help in improving the accuracy of your orbital calculations.*

No. 139 Gauss to Olbers Brunswick January 3, 1806

*Of our asteroids I observed only Ceres a few times in November. Apart from these and some of Harding's, I've still not seen any other observations of Ceres. Zach sent me a few observations of Pallas among which was nevertheless only one Dec. This one agrees well with the ephemeris, the RA differ by 13'.*

No. 17 Bessel to Gauss Lilienthal May 28, 1806

*I take the liberty, my dearest friend, to send you some observations of Ceres and Juno.*

| *Ceres* | | | |
|---|---|---|---|
| *May 16* | *10h 8m 41s mean time* | *119° 1' 11"* | *28° 37' 18"* |
| *17* | *10 52 25* | *119 26 57.5* | *28 32 27.5* |
| *19* | *10 3 17* | *120 15 16* | *28 22 55.0* |
| *23* | *10 24 16* | *121 56 7* | *28 2 45.0* |

[The remainder of the letter, dealing with Juno, will be published in a future book in this series.]

Gauss to Bessel Brunswick June 16, 1806

*Thank you very much for the kindly communicated observations on Ceres and Juno. I received a wonderful suite of meridian observations on the latter planet from Milan. What an excellent instrument their mural quadrant by Ramsden must be at which the faint planet could be observed this year with illumination!*

No. 147 Gauss to Olbers Brunswick July 30, 1806

*I've had nothing to do with Pallas for a long time; the Milan observations served me well (Zach's observations were, as I already judged before, totally wrong). One can still bring all observations of 1802, 3, 4, 5, 6 into considerably good agreement without taking perturbations into account. Now there only remains Ceres, of which I, however, don't have any observations other than those of Harding and Pasquich; I'm hoping to obtain still better ones.*

[Janos Pasquich (1754–1829), professor of mathematics and director of Buda Observatory]

No 23 Gauss to Bessel Brunswick January 5, 1807

*And I had just as little time to search for Ceres, Pallas and Juno. If you should be more successful observing one of these planets, please let me know as soon as possible. Of the above-mentioned observations on Ceres, which I use as examples in my book, the one of May 23, 1806, is yours.*

No. 157 Olbers to Gauss Bremen March 12, 1807

*Sincere thanks for your observations of Pallas. I still haven't observed it but saw it again on March 5 when it was in close conjunction with a small star from the* Hist. Cel. *A star in its vicinity, seen by Harding on January 10 but which he later lost sight of, kept Bessel and*

**Fig. 8.14** Jean Baptiste Joseph Delambre

*also me busy trying to locate it. It could very well be an asteroid. Our efforts were futile. – Already a few months ago I had a circle micrometer manufactured. Included is a small plate that accurately gives the circle's diameter. Of course I must now make an observation at an additional moment of time during the transit, but I find the differences in Dec.'s to be much more accurate. I thus hope soon to be able to furnish you with acceptably good observations of Pallas. I'm having the* Conn. des tems *for 1808 sent from Hamburg. Delambre* (Fig. 8.14[3]) *is the editor now. It contains Bouvard's original 1804 observations. Among these are also several observations of Pallas which seem to be very accurate and occur at the transit instrument and quadrants. If it's of any use to you, Bessel would gladly reduce them for you.*

No. 158 Gauss to Olbers Brunswick March 24, 1807

*I am very pleased by the hope you give me to send obs. on Pallas. You will probably do everything in your power to give us some obs. on Juno with your wonderful Dollond, which will probably elude most astronomers due to its faintness this year. I will not even try with the local Short, which has been deteriorating over the past year – probably in maladroit hands.*[4]

No. 165 Olbers to Gauss Bremen April 18, 1807

*Soon you'll also receive several observations of Pallas, which I'm not totally neglecting. I still observed it twice this evening because the region where Vesta stood always remained*

[3]Jean Baptiste Joseph Delambre (1749–1822) was appointed director of Paris Observatory in 1804. He was one of the early observers of Pallas, and wrote about Bode's law (see Chapter 11 in this book)
[4]Scottish instrument maker James Short made many telescopes used in Germany.

*cloudy. However, my observations are not yet reduced.* [Remainder of this letter is in future book in this series.]

No. 28 Bessel to Gauss Lilienthal July 15, 1807

[In this letter Bessel sends Gauss the observations of Ceres and Pallas made in Milan during April and May of 1807. These were published in the *MC*.]

No. 184 Olbers to Gauss Bremen July 25, 1807

*I searched again for Pallas which is now very faint, less bright than a star of 10th magnitude. So far, the constant thunderstorms permitted only one observation. It was compared twice with one of Lalande's stars, and this one was reduced by Piazzi's stars.*

$$\text{July } 22\ 11h\ 49m\ 50s\ 226^{\circ}\ 21'\ 43.5''\quad 20^{\circ}\ 42'\ 41.9''$$

*Here, too, we have continuously an intolerable heat with many thunderstorms. On the 23rd, the highest mark on the thermometer was 22.8°C.*

No. 185 Olbers to Gauss Bremen August 5, 1807
*Here are some reports on observations on Pallas:*

| | | | |
|---|---|---|---|
| *July 22* | *11h 49m 50s* | *226° 21' 43."5* | *20° 42' 41."9* |
| 25 | *10 57 22* | *226 39 26.3* | ——— |
| 26 | *11 12 54* | ——— | *20 13 6.4* |
| 27 | *11 45 34* | *226 52 47.6* | *19 52 47.1* |
| 28 | *10 25 55* | *226 59 17.1* | ——— |
| 28 | *10 31 3* | ——— | *19 43 5.8* |
| 29 | *11 42 31* | *227 6 48.3* | ——— |

*The declination on the 22th is less reliable, as the star by which it was determined didn't have a suitable position. For the remaining declinations, I could use stars that were almost on the parallel of Pallas, and the observation error might not be 5", if Lalande's Zenith Distance (Z.D.) are correct. Since the 25th, the RAs are based mainly on the star, listed on p. 75, of 8th/9th magnitude: 15h 7m 3s Z.D. 28° 53' 27". After several comparisons with preceding and following stars, I had to correct the transit period through the center filament by 1.7s so that instead of 15h 7m 3s it now reads 15h 7m 1.3s. However, should Lalande's data be correct, then all the RAs of Pallas since the 25th would have to be increased by 25.5".*

*If the weather is still promising, I plan to provide some further observations of Pallas. However, it will be in a favorable position against Piazzi's stars only at a time when the moonlight will be hindering the observation of this faint planet. – Harding's map did appear rather late in the recent M.C. With respect to its use during observations of the small planets, I do not consider it very convenient that Mr. Harding has entered stars which exist neither in the* M.C. *nor with Piazzi, without distinguishing them from the rest.*

No. 210 Olbers to Gauss Bremen July 12, 1808

*Harding kindly donated to me a map of this year's orbit of Pallas, for which I am very grateful. It has been extremely helpful. Please convey to our dear friend my preliminary gratitude, as I do not have the time today to write to him.*

*I have good observations of Ceres made in Seeberg, also of Pallas, but for the latter only the RA. Unfortunately, the few observations of Pallas carried out here with the circular micrometer are dubious with respect to the declination. Recently, I submitted a treatise to the Royal Society in which I am dealing with that investigation which I repeatedly have told you about. It is the same one which has teased me for 4 years and which I could finish only in the year 1805. You shall find a short report on it in No. 151 of the academic announcements* Gel. Anzeigen; *as soon as the treatise itself is printed, I shall offer you a copy.*

No. 37 Gauss to Bessel Goettingen July 25, 1808

*We have only been able to observe Juno, Vesta and Pallas of the new planets and only at the circular micrometer; our quadrant will most likely be only useful for observing Vesta. We are not fortunate in the case of Pallas. It is too high in the sky for the Herschel telescope and too faint for the small Dollond, thus the observations cannot be accurate and reliable, especially the declinations. Hopefully you have been more successful with your splendid Herschel and made some nice observations. Lindenau has been observing Ceres for three weeks now in the meridian and Pallas and Juno too, but I assume the latter two only at the passage instrument. Here are our observations.*

*Pallas*

| | | | |
|---|---|---|---|
| *July 13* | *12h 3m 17s* | *300° 21' 49"* | *18° 41' 44"N* |
| *15* | *12 1 28* | *299 57 3* | *18 32 40* |

No. 231 Gauss to Olbers Goettingen August 6, 1810

*I have started to write an article*[5] *for the* Societät *about my calculations concerning Pallas, describing besides the numerical results my method to determine a planetary orbit from 4 oppositions (actually from 4 longitudes of the orbit the elliptical elements) and those shortcuts I use with the method of the least squares, what might be interesting for you. If the deadline of the Parisian prize question were 4 months later, I would probably apply; but there is so little time and for 10 months I have been lacking energy and mind to start any complex work. But nevertheless, I intend to work out all parts of the method according to which I will be able to reasonably calculate Pallas' perturbations in 4 or 5 single memoirs for the* Societät.

No. 233 Gauss to Olbers Goettingen October 24, 1810

*Soon I will give a lecture to the* Societät *on my work on Pallas, already mentioned to you. You then will find a complicated report about it in the papers. It also deals with theoretical matters: the development of my method to determine an orbit from 4 oppositions I have been employing for years, a supplement to the second section of the second volume of my* Theoria; *the explanation of a practical dodge of considerably shortening the otherwise tedious task of eliminating in the method of the least squares.*

*Thus, soon I will start to calculate the perturbations of a new planet in greater detail. If I knew that the institute would prorogue the prize question I would not be reluctant and choose Pallas otherwise I have to prefer Ceres which will be however more satisfying since soon the opposition is observable. In the case of Pallas I would start with considering the elements as variable and determine the perturbations during 7 years by quadratures for what I designed the formulae that I consider more convenient than those of Laplace. If then*

[5]See Vol. VI, p. 1 of Gauss Werke; see Brendel and Schlesinger.

*the combination of hitherto observed 6 oppositions succeeded (what is likely to be expected) which cannot be combined into an ellipsis, I would – according to my method developed 4 or 5 years ago – calculate the perturbations in the common way as periodic perturbations in longitude, latitude and radius vector or even better in its logarithm.*

No. 235 Gauss to Olbers Goettingen November 26, 1810

*I believe I have already informed you in my last letter that I was about to start a work on the perturbations on Pallas by Jupiter, but not to develop a general theory of them but only the amount from 1803 to 1811. The calculation of three systems of elliptical elements from the oppositions of 1803, 1804, 1805, 1807 – 1804, 1805, 1807, 1808 – 1805, 1807, 1808, 1809 showed the strong impact of the perturbations; judge yourself according the following samples:*

| *Perihelion 1803* | *121° 3′ 11″* | *121° 5′ 22″* | *120° 58′ 5″* |
|---|---|---|---|
| $\varphi$ | *14 10 59* | *14 10 4* | *14 9 37* |
| *Daily mean movement* | *770.214″* | *770.447* | *770.926* |

*I was curious whether these six oppositions would be combined much more easily if the perturbations by Jupiter were taken into account. I admit, that such an unpleasant work holds little interest to me and that even the prize of 6,000 Francs could hardly have made it more tempting as long as all observations were represented by one ellipsis. But now, however, I have started this task bravely and the main work is done because I committed most of my spare time during the last 4 weeks and I hope to harvest the fruit of it soon if there are no major errors in my calculation. I considered it most useful to regard the elements as variable and to calculate its daily change through Jupiter every 50 years over a period of more than 8 years. I started to develop my own formulae or rather to transform those developed by me several years ago into more adaptable ones; because I considered those of Laplace much too tedious and I would have had to commit certainly twice or three times the amount of time employing them instead of mine. I calculated Jupiter's locations according to Bouvard's tables (by using a special dodge) and Pallas' locations according to a 4$^{th}$ system of elements, fitting to all oppositions as accurately as possible. Finally, I finished today those 61 daily changes of the single six elements so that there is only the mechanic integration (merely an addition) and the new calculation of the basic elements, representing (together with the perturbations) the oppositions best. These perturbations are far more considerable than I thought and especially strong in 1809 when Jupiter and Pallas were at conjunction; I thus believe my ephemerides will deviate considerably in the next opposition. Recently Harding has searched for Pallas but is still uncertain whether he found it. If you observe it – please inform me as soon as possible. The daily motion is probably now by 2″ larger than in 1803; the angle $\varphi$, measured continuously, probably smaller by 13′, the perihelion decreased by 20′ or 30′ etc. Should the correspondence to the six hitherto oppositions and the 7$^{th}$ of next year turn out satisfactory, I may take on the general theory, although the tremendous amount of mechanic calculations is very deterring. The perturbations by Jupiter are so predominant that one can easily neglect those by other planets for the time being.*

*Should you succeed in arranging a second extension of the prize question in Paris I would thereby have one more reason to tackle the above-mentioned work. The success of my current work however must be the first precondition. In 14 days I will know for certain – at least regarding the hitherto observations. And then I will be twice as curious about the next, when the perturbations suffered during the conjunction will show in full.*

*PS: Before having finished this letter I completed the integration of the perturbation of the inclination that is the least considerable. If I set it in the 1st opposition thus it is*

| | |
|---|---|
| *1803* | *34° 37' 47"* |
| *1804* | *34 37 36* |
| *1805* | *34 37 12* |
| *1807* | *34 37 25* |
| *1808* | *34 37 31* |
| *1809* | *34 36 40* |
| *1811* | *34 35 5* |

No. 236 Gauss to Olbers Goettingen November 30, 1810

*I cannot help but write you at least a few words and to tell you that I have almost finished my work about which I wrote you in my last letter and that it will turn out, if I am not mistaken, satisfactory beyond my expectations. Yesterday I calculated the normal elements from the first four oppositions that represented those exactly in combination with the previously integrated perturbations and these differed to my great horror from the 5th opposition by 4' and from the 6th by 20'. I thus discovered that an error has sneaked into my calculation in so far as the main part of the perturbation of the mean motion showed the wrong sign. Today I redressed everything and according to the preliminary normal elements – calculated leni calamo* [with a fast pen] *only with the minor tables – those will deviate as follows*

| | | | | | |
|---|---|---|---|---|---|
| *6 longitudes 8"* | *1* | *8* | *0* | *6* | *5* |
| *6 latitudes 2"* | *5* | *6* | *5* | *17* | *3* |

*The polishing up has yet to be done (the latitude of the 5th opposition was as is well known badly observed and I believed an error of 30" possible). The elliptical elements corresponding best differed*

| | | | | | |
|---|---|---|---|---|---|
| *in longitude –111"* | *+59* | *+20* | *+86* | *+137* | *–217* |
| *in latitude –8"* | *–37* | *0* | *+25* | *+29* | *+83* |

*In order to give you an idea of the perturbations of the elements, here are some:*

| *Opposition* | | | | | | |
|---|---|---|---|---|---|---|
| *1803* | *1804* | *1805* | *1807* | *1808* | *1809* | *1811* |
| *Diurnal motion* | | | | | | |
| *770."721* | *770.389* | *770.133* | *770.497* | *770.154* | *768.293* | *768.860* |
| *φ 14°12' 47"* | *11' 33"* | *10' 59"* | *10' 10"* | *8' 10"* | *1' 35"* | *13° 58' 57"* |

No. 237 Gauss to Olbers Goettingen December 13, 1810

*My last two letters crossed with yours. Since then, I finished my work on the perturbations of Pallas by Jupiter and found out an agreement that surpasses my most daring expectations. Please note the differences here:*

| | *1803* | *1804* | *1805* | *1807* | *1808* | *1809* |
|---|---|---|---|---|---|---|
| *Mean longitude* | *+1.3"* | *−3.8* | *+3.9* | *−3.3* | *+3.2* | *−1.4* |
| *Heliocentric latitude* | *−1.0"* | *+41* | *+6.0* | *+3.9* | *−16* | *−3.9* |

*When calculating the perturbations, I based them on the elements which you find on the enclosed sheets 1) p. 1971 as No. IV; I have started to repeat the complete calculation once again by dividing the entire time from 1803 – 1811 into 7 or 8 periods and taking separate elements, as obtained from my first calculation, as a basis for each of them. But will you join me in being very surprised when I tell you that my ephemeris in the October issue of the* M.C. *2) now is wrong by more than 1°, if I can rely on the results of the perturbation calculations? I do believe to be on sure grounds; but not sooner shall I rest until I have found Pallas again, and I am now impatiently hoping for favorable weather. In case you are luckier in Bremen, then please do look around. According to the perturbed elements, the RA should be smaller by 1° 10' on December 19, by 1° 40' on January 10, the southern declination should be larger by 12' ... 10' than the ephemeris. In the opposition of 1809 with Jupiter, Pallas is far too much perturbed. If my calculation here, too, is confirmed to my satisfaction, then I might perhaps deal with Juno and Vesta in a similar manner. The article, the announcement of which I herewith enclose, shall be published soon.*

1) *Goettinger Gelehrten Anzeiger, issue 198, Dec. 13, 1810, pp. 1969-1973. Gauss's works, vol. VI, pp. 61 ff. Sch.*
2) M.C., *vol. XXII, p. 401 ff, Oct. 1810; Gauss's works, vol. VI, p. 320, though the ephemeris is not printed there.*

No. 238 Olbers to Gauss Bremen December 19, 1810

*Though only with a few lines, I have to thank you heartily for your most interesting communications; and my congratulation for the admirable success by which you have overcome the tremendous difficulties which were opposed to the determination of the perturbation of Pallas. Truly, such an agreement among the 6 oppositions one could not, or should not, expect. How right your judgment has been, too, that all other perturbations of this small planet are quite inconsiderable as opposed to those originating from Jupiter. I am most curious how you now find – according to probability – the real, always invariable large axis of the Pallas orbit, and whether this is still similar to the one of the Ceres orbit? – The almost unbelievable deviation of your ephemeris from the finally determined perturbed elements stimulates my longing for finding Pallas again, and I certainly will try to pay attention to it. Unfortunately, the weather is so unfavorable, and the constellation of the Cat* [the obsolete constellation Felis, created by Lalande in 1799; it was between Antilia and Hydra] *is still rising very late. As soon as the weather settles, I will nevertheless risk a few nights. I hope to find soon, possibly still this week, a reason for writing to Delambre and will try – of course without compromising you in the least – to find out if the prize deadline for the Pallas perturbations could not be extended. Forgive me today's short and hasty letter, dear Gauss. Today, we have received news from Paris that no longer permit any doubts about the unification of our small republic with the French empire. Fuimus Troes!* [True Trojans!] *With all my philosophy of life, this blow afflicts and moves me too much! You do not know how much I cling to our present fortunate constitution. All my previous plans for my children's future are destroyed, too!*

Gauss to Schumacher Goettingen January 6, 1811

*In case the Parisian institute poses Pallas' perturbations for the fourth time as a prize question I am not averse to reflecting on that because the matter has become interesting to me. Since you read the* Moniteur *there yourself I would appreciate it very much if you kindly informed me if you read something about it.*

No. 239 Olbers to Gauss Bremen January 26, 1811

*With the greatest pleasure I have seen in the* Moniteur *that the prize for the perturbation of Pallas at 6000 franc has been prorogated until October 1, 1816, in the same measure as it will be given to the first paper which is submitted during this period and meets the intentions and requirements of the competition. I don't have any doubts now, my dearest friend, that you undertake the complete solution of this problem according to your remark which pleased me so much; and thus, you shall complete the education of this hard-to-master goddess which already caused you so much effort and did you so much credit. According to a message obtained from Mr. von Lindenau, the sky completely confirms your calculation of the perturbation of Pallas. I have not yet been able to convince myself, notwithstanding the many clear nights; a very stubborn and bad cough still prohibits me from doing any astronomic observations. It costs me quite some effort to hold myself back.*

No. 31 Schumacher to Gauss Altona February 26, 1811

*Dearest friend, here are my meridian observations of Pallas. These are the first from me you are calculating and I hope you will not be deterred. On the 23$^{rd}$ I saw Pallas entering but a small cloud prevented the observation. The declinations are based on a pole height 53° 32′ 51″.5. Hardly any other observatory is able to state the declinations so exact.*

| | *RA in time* | *RA in Bogen* | *South Declination* |
|---|---|---|---|
| *Feb. 19* | *9h 46 29.0* | *146° 37′ 15.0* | *12° 42′ 45.3″* |
| *20* | *– 45 47.5* | *– 26 52.5* | *12 18 32.7* |
| *21* | *– 45 6.5* | *– 16 37.5* | *11 53 47.7* |
| *22* | *– 44 26.4* | *– 6 36.0* | *11 28 53.5* |

*In case you would like to reduce the declinations yourself, I am adding the observed zenith distances.*

| | *Zenith distance* | *Reaum. Temp.* | *Paris zoll* |
|---|---|---|---|
| *Feb. 19* | *66° 13 27* | *−3°* | *28 6.0* |
| *20* | *65 49 18* | *−5* | *28 0.3* |
| *21* | *65 24 38* | *−3* | *27 9.9* |
| *22* | *64 59 48* | *−2* | *27 7.0* |
| *collim. = −11.2″* | | | |

[A zoll is approximately one inch, or 2.6 cm.]

*I could not find Ceres according to Bode. The continuous nights spent awake have weakened my health. The day is wasted with sleeping and relaxing. Please, be so kind and continue your corrected ephemerides of Pallas, which I have only until March 3$^{rd}$ and keep me informed. It is useful for finding Pallas immediately again after a period of dark and bad weather.*

No. 32 Gauss to Schumacher Goettingen March 10, 1811

*I am much obliged to you for sending me your observations of Pallas. I hope to calculate them soon. Further observations will probably not be possible this year, but I would much appreciate it if you could send me some observations of Vesta and Juno.*

Schumacher to Gauss Hamburg June 30, 1811

*Only now I had the chance to read the* Moniteur *of June 9, 1811 and I rush to inform you – in case you do not know yet – about the details of the prizes. Regarding Pallas' perturbations two treatises were submitted – one has been neglected while the other shows a detailed understanding of analysis but was nevertheless insufficient. Thus the deadline has been extended by 5 years until June 1816 and the institute will crown the first treatise submitted within this time that fulfills the requirements. The prize is still the same, namely 6,000 Francs.*

No. 243 Gauss to Olbers Goettingen August 12, 1811

*I did not write earlier in order to be able to write some more about your Pallas. I am ignoring some information that you will find in one of the next issues (I believe of Aug. 17<sup>th</sup>) of our* Gel. Anz. *and just mention the general results of perturbation. I completed the perturbations of the inclination and longitude of the line of nodes (being the least arduous) in a first calculation, also checked the first with the 7 oppositions (I obtained the elements from my previous calculations). At first a bad correspondence, until I found out that 2 equations showed the wrong sign. Now they correspond to my satisfaction. In the next few days I will finish the checking on the node and then I am planning to work on the eccentricity, which will probably take one month. In order to give you an idea of the results I enclose 40 perturbation equations for the inclination; I only used what exceeded one second; and there might very well be some small equations which were dependent on the nine- or tenfold longitude of Jupiter.* [Gauss includes a table of perturbation equations]

*The 80 equations for the inclination and node could be summarized into 40 for the latitude, but in my opinion nothing is won by doing so because if one lets the elements themselves perturb thus one can without hesitation regard perturbed elements as valid for several months and consequently only needs to calculate only once every year the perturbations for 6 elements (amounting to maybe 300-400 equations); whereas, if only secularly changes are applied to the elements and the periodic to latitude, longitude and radius vector (together about 200) for one year at least 6 locations would have to be calculated in order to interpolate. But that might be done as one pleases, if only all perturbations are available in some form.*

Gauss to Olbers Brusnwick Mar. 20 & 31, 1812

[For the text of these letters dealing with the Gauss anagram, see Appendix A in this book.]

No. 259 Olbers to Gauss Bremen April 5, 1812

*With pleasure I am sending you the two first observations on Pallas which, at least, correspond excellently with the ephemeris. I took advantage of the first reasonably clear evenings to search for the small planet of the 9<sup>th</sup> or 10<sup>th</sup> order of magnitude. I would consider the obs. good if only they were not based on stars of the* Hist. Cel. *reduced by a star of Piazzi. I added to the RA of Piazzi's star 4".*

| | *Mean time* | | |
|---|---|---|---|
| *1812 Apr. 3* | *11*h 26m *44*s | *268° 42' 23."1* | *15° 25' 4."9 north* |
| *4* | *12 2 33* | *268 47 12.5* | *15 37 41.1* |

*I am delighted by the most peculiar results of your profound studies of Pallas' orbit that you are communicating. But I consider your discovery that 18 Jupiter was = 7 Pallas one of the most peculiar that was made in our Solar System for long time. – But what about Ceres? Has Ceres an equation considerably dependent on 18 Jupiter –7 Ceres? Will it not be possible to find for the other small planets as well equations of that kind m Jupiter = n Ceres etc?*

Gauss to Bessel Brunswick May 5, 1812

[See the Gauss anagram chapter in this book for this text.]

No. 260 Olbers to Gauss Paris May 12, 1812

*Laplace eagerly inquires after you and about your work on Pallas. He appears to be very curious about your new method, but I was neither willing nor able to tell him anything about it.*

No. 262 Olbers to Gauss Paris July 18, 1812

*I am using the otherwise extremely unpleasant departure of my friend Palm to send you as soon as possible the second part of Laplace's* Théorie des probabilités, *that he gave me for you. Unfortunately, we are still here, and our departure depends entirely on a special permission of His Majesty, which might, under the actual circumstances, take some time, because in such times the Emperor* [Napoleon] *hardly has time to think of such trifles. I have not yet received any obs. of Pallas from the* Observatoire Impérial, *where they have only started to observe Pallas since July 12th. They justify it with trying to obtain more of them. – Burckhardt gave me two additional obs.*

*– Meridian of the Ecole militaire*

*Counting the day from midnight on 1812, June 13: 23h – mean time 17h 26m 11.28s sid. t. decl. 25° 6' 41.5"*

*We have only obtained the passage of one single thread, having taken a small star that preceded for the planet*

*July 1st 17h 11m 53.28s 24° 14' 45.4"*

[More of this letter, dealing with the asteroid explosion hypothesis, is in Chap. 2 of this book.]

No. 266 Gauss to Olbers Goettingen April 8, 1813

*I am almost done with the perturbations on Pallas; there is only one element missing, the epoch. Regarding the perihelion there are 183 equations (caused by Jupiter) exceeding 0".1. When the epoch is done, I will at least make one probe with the tables; although 800 equations are deterring but the fact that 7 Pallas= 18 Jupiter (quam proxime; approximately) allows a major shortening of the calculations. In the meantime Mr. Nicolai is probably working on the perturbations by Saturn and maybe the very skilled Mr Encke, another of my students, is going to work on the perturbations by Mars.* [Johann Franz Encke, 1791–1865] *Your time permitting, please help to search for Pallas in the near future.*

No. 267 Gauss to Olbers Goettingen July 2, 1813

*The perturbations on Pallas by Jupiter, as far as I intended to calculate them, are almost completed. But the numerical calculations for the hitherto oppositions are lagging behind.*

Nicolai to Gauss Seeberg March 19, 1814

*I have been working again for some time now on the perturbations on Pallas by Saturn and I am pleased to send you today three elements. You are probably surprised at the fair number of equations that astonished me at first as well; but most of it is codswallop having probably no effect on the results. I did not neglect an equation being in* log a *more than unity and in the other elements more than 0.1", just like you did it in the case of Jupiter. Furthermore I can vouch for the correctness of every equation, because first I calculated them with great carefulness and eagerness and then I looked them over. I believed I had to do so because there is no check in the entire calculation, apart from the interpolations... And I have to say that I assumed Saturn's mass = 1/3512.08 according to newer studies, as Laplace mentioned in the latest, 4th, edition of his* Exposition du Système du monde *that I received several weeks ago.– The calculation according to this other method appears to me easier and more convenient than to the first; and without doubt less errors are possible; I am only a bit afraid of the second part of the epoch. – The secular changes for inclination and node as well as the larger equations correspond quite well to my first calculation. – I should not think that the assumed number of twelve positions of Pallas will be too small a number because of the perihelion and the 2nd part of the epoch; but if I compare the number of equations for these two elements according to my previous calculation to that of the other elements according to the same calculation and then look at the number of equations for the latter which was derived from the new calculation, I am almost afraid that might be the case. Because of this I now wish I had used 18 positions of Pallas. The further course of the calculation will tell. The numeric calculation of all equations for the 9 oppositions is actually quite a bit of work* (Figs. 8.15 and 8.16).

*The left term gives the perturbations of log a in units of the 7th decimal, the coefficients on the right-hand side give the perturbations $\delta_{\int}$ n d t combined with the corresponding cosine.*

*Footnote: The perturbations on Pallas by Saturn were calculated by Nicolai mainly in the same way like those by Jupiter, by developing the perturbing forces according to the interpolation method in series with the arguments M and M'. The three values T, V, W were used as components of the perturbing forces but these were divided into three right-angled components of which the first tends towards the direction of Pallas' line of apsides, the second lies perpendicularly to it within Pallas' orbital plane and the third stands vertically to this orbital plane. Thus x, y, z, x', y', z' are the right-angled coordinates of both planets in respect of the above mentioned axes, and thus the components are:* (Fig. 8.17)

*and their relations to the ones used formerly are:* (Fig. 8.18)

*The perturbations of the elements are then determined by the equations* (Fig. 8.19)

*that can easily be derived from the articles 8-12 of the* Exposition d'une nouvelle méthode (Exposition of a new method).

Nicolai to Gauss Seeberg Good Friday 1814

*I am pleased to be again able to give you two elements and by the end of next week I hope to be done with this work* (Fig. 8.20). *My worries about the paucity of Pallas' positions have turned out to be unfounded. The equations 14, 19, 20, 25 (These are the equations with the arguments:* 4⚴ - 2♄, 4⚴ - 3♄, 5⚴□ - 3♄, 4⚴ - 4♄*) of the perihelion might be slightly affected by the inaccuracy caused by the small number but the coefficients of these equations are so small that in general there is no possibility for inaccuracy.*

Nicolai to Gauss Seeberg April 22, 1814

*You are probably in receipt of my letter of April 8th regarding the perturbation formulae for eccentricity and perihelion; today I am pleased to send you the rest of the work now finished, namely the equations for the 2nd part of the epoch.*

**Log *a* und erster Theil der Epoche[*)].**

| | |
|---|---|
| 5 sin( ♀ + 79° 15′) | 0″,39 |
| + 7 sin(2♀ + 93 48) | 0,24 |
| + 2 sin(3♀ + 341 19) | 0,04 |
| + 2 sin( ♀ + ♄ + 87 31) | 0,15 |
| + 2 sin( ♀ − ♄ + 92 48) | 0,20 |
| + 2 sin(2♀ − ♄ + 77 14) | 0,07 |
| + 2 sin(3♀ − ♄ + 88 13) | 0,05 |
| + 1 sin( ♀ + 2♄ + 140 23) | 0,06 |
| + 38 sin( ♀ − 2♄ + 34 30) | 3,92 |
| + 39 sin(2♀ − 2♄ + 88 29) | 1,65 |
| + 9 sin(3♀ − 2♄ + 331 1) | 0,23 |
| + 2 sin(4♀ − 2♄ + 190 0) | 0,03 |
| + 10 sin( ♀ − 3♄ + 84 44) | 1,37 |
| + 6 sin(2♀ − 3♄ + 84 30) | 0,27 |

**Fig. 8.15** Log a and first part of the epoch

*I applied a slight transformation in the analytic term for dL, which I would like to give you for your information. If the value 1 + cos v cos E = u, thus*

$$d\varphi = -m\ a\ a \sin v \cos E\ .X + m\ a\ a\ u\ Y$$

$$d\varpi = A - m\ a\ a \operatorname{cotang} \varphi.(3-u)\ .X + (m\ a\ a/e) \cos v \sin E.Y.$$

*If the latter is multiplied, excluding A, with 1 − (rr/a a cos φ)], the first with −(cos φ + (r/a cos φ)), sin E, thus.* (Fig. 8.21)

*I did not try to simplify the bracketed factors further since they appeared good enough for a numerical calculation and the more so in combination with your auxiliary tables. The transmutation and introduction of the value u is caused by the relation between the true and the eccentric anomaly, which I accidentally came across only recently. If you imagine the complements of v and E to 90° (or 450°) as both sides of a spherical triangle, meeting in φ so the third side of the triangle – opposite to φ constant and also = φ. Also the two other angles of this triangle have peculiar characteristics; the one opposite to the side v is always equal to that side and the other, opposite to the side E, is the complement of this side to 180° or = 90° + E.*

| | |
|---|---|
| + 9 sin(3♀ − 3♄ + 81° 4′) | 0″,24 |
| + 3 sin(4♀ − 3♄ + 325 50) | 0,06 |
| + 1 sin( ♀ − 4♄ + 113 8) | 0,28 |
| + 2 sin(2♀ − 4♄ + 81 57) | 0,09 |
| + 2 sin(3♀ − 4♄ + 82 45) | 0,05 |
| + 2 sin(4♀ − 4♄ + 72 40) | 0,04. |

Neigung der Bahn.

Jährliche Änderung = + 0″,16950

| | |
|---|---|
| 0″,42 sin( ♀ + 42° 58′) | + 0″,11 sin( ♀ + 2♄ + 129° 41′) |
| + 0,24 sin(2♀ + 98 47) | + 0,56 sin( ♀ − 2♄ + 243 21) |
| + 1,02 sin( ♄ + 37 58) | + 0,29 sin(2♀ − 2♄ + 275 8) |
| + 0,18 sin(♀ + ♄ + 89 3) | + 0,23 sin( 3♄ + 83 5) |
| + 0,11 sin(♀ − ♄ + 355 47) | + 0,52 sin( ♀ − 3♄ + 260 41) |
| + 2,27 sin( 2♄ + 105 16) | + 0,13 sin( ♀ − 4♄ + 273 9). |

Länge des aufsteigenden Knotens.

Jährliche Änderung = − 1″,47934

| | |
|---|---|
| 0″,89 sin( ♀ + 281° 36′) | + 1″,22 sin( ♀ − 2♄ + 111° 22′) |
| + 0,39 sin(2♀ + 11 13) | + 0,59 sin(2♀ − 2♄ + 184 51) |
| + 1,79 sin( ♄ + 246 49) | + 0,31 sin( 3♄ + 327 21) |
| + 0,17 sin( ♀ + ♄ + 350 51) | + 0,36 sin( ♀ − 3♄ + 3 31) |
| + 0,75 sin( ♀ − ♄ + 186 33) | + 0,14 sin(2♀ − 3♄ + 143 10) |
| + 0,13 sin(2♀ − ♄ + 301 7) | + 0,11 sin(3♀ − 3♄ + 180 41) |
| + 3,27 sin( 2♄ + 2 51) | + 0,13 sin( ♀ − 4♄ + 25 2).} |
| + 0,15 sin( ♀ + 2♄ + 37 19) | |

**Fig. 8.16** Headings in the column read: *Inclination of the orbit. Annual change = 0″.16950; Longitude of the ascending node. Annual change = −1″.47934*

**Fig. 8.17** Equation

$$X = \mu\left(\frac{x'}{r'^3} - \frac{x'-x}{\rho^3}\right)$$

$$Y = \mu\left(\frac{y'}{r'^3} - \frac{y'-y}{\rho^3}\right)$$

$$Z = \mu\left(\frac{z'}{r'^3} - \frac{z'-z}{\rho^3}\right) \qquad (z = 0)$$

**Fig. 8.18** Equation

$$\frac{ar}{\cos\varphi}\left(Y\cos(v-\tilde{\omega})-X\sin(v-\tilde{\omega})\right)=V$$

$$\frac{a}{\cos\varphi}\left(Y\sin(v-\tilde{\omega})+X\cos(v-\tilde{\omega})\right)=T$$

$$-\frac{ar}{\cos\varphi}Z=W.$$

$$\frac{\mathrm{d}i}{\mathrm{d}M}=-\frac{ar\cos(v-\Omega)}{\cos\varphi}Z$$

$$\frac{\mathrm{d}\Omega}{\mathrm{d}M}=-\frac{ar\sin(v-\Omega)}{\cos\varphi\sin i}Z$$

$$\frac{\mathrm{d}\varphi}{\mathrm{d}M}=aa\sin(v-\tilde{\omega}).\cos E.X-aa(1+\cos(v-\tilde{\omega}).\cos E)\,Y$$

$$\frac{\mathrm{d}\tilde{\omega}}{\mathrm{d}M}=\frac{aa}{e}(\cos\varphi+\sin(v-\tilde{\omega}).\sin E)\,X-\frac{aa\cos(v-\tilde{\omega})\sin E}{e}Y+A$$

$$\frac{\mathrm{d}n}{\mathrm{d}M}=\frac{3a^3n\cos\varphi\cos E}{r}Y-\frac{3a^3n\sin E}{r}X$$

$$\frac{\mathrm{d}\varepsilon}{\mathrm{d}M}=\frac{\mathrm{d}\tilde{\omega}}{\mathrm{d}M}-\frac{rr\left(\frac{\mathrm{d}\tilde{\omega}}{\mathrm{d}M}-A\right)}{aa\cos\varphi}-\frac{\sin E(2-ee-e\cos E)}{\cos\varphi}\frac{\mathrm{d}\varphi}{\mathrm{d}M}$$

$$A=-\frac{ar\,\mathrm{tg}\,\tfrac{1}{2}i\sin(v-\Omega)}{\cos\varphi}Z,$$

**Fig. 8.19** Equation

*Should our factors of the perturbing forces not tally exactly, this might probably be caused by your use of the newest mean elements and I on the other hand used the ones derived from the very first calculation. Second part of the epoch.* (Fig. 8.22)

No. 278 Gauss to Olbers Goettingen April 23, 1814

*The* M.C. *is discontinued this December. However, as Lindenau shall return soon, I hope that it will soon start again. It would be a great loss, indeed, if we have to live without such a journal for a long time. I have already written to you how precisely the 9 oppositions of Pallas agree with my perturbation calculations. Yet, I still find the differences too large. Perhaps, the perturbations by Saturn and Mars might still bring them down a bit, but not too much according to an estimate. It seems doubtful, though, to attach importance to results which to find required the numerical calculation of about 1,000 equations for each opposition. As all these 9,000 equations have been calculated twice, and as I did the calculations with the greatest care, I hardly doubt that my mass of Jupiter needs any correction.*

Excentricitätswinkel.

Jährliche Änderung = − 0″,30592

| | |
|---|---|
| 0″,78 sin (♀ + 191° 38′) | + 0″,25 sin (3♀ − 2♄ + 320° 12′) |
| + 1,52 sin ( ♄ + 187 49) | + 0,60 sin ( 3♄ + 161 31) |
| + 0,24 sin ( ♀ − ♄ + 202 49) | + 0,76 sin ( ♀ − 3♄ + 206 23) |
| + 0,14 sin (2♀ − ♄ + 213 21) | + 0,55 sin (2♀ − 3♄ + 208 25) |
| + 3,52 sin ( 2♄ + 206 12) | + 0,30 sin ( ♀ − 4♄ + 199 50) |
| + 0,16 sin ( ♀ + 2♄ + 176 47) | + 0,12 sin (2♀ − 4♄ + 206 26) |
| + 4,08 sin ( ♀ − 2♄ + 206 13) | + 0,10 sin (3♀ − 4♄ + 205 41). |
| + 0,13 sin (2♀ − 2♄ + 256 43) | |

Länge des Periheliums.

Jährliche Änderung = + 0″,01491

| | |
|---|---|
| 2″,04 sin ( ♀ + 345° 24′) | + 2″,18 sin ( 3♄ + 61° 21′) |
| + 0,31 sin (2♀ + 210 31) | + 2,52 sin ( ♀ − 3♄ + 296 44) |
| + 0,15 sin (3♀ + 244 31) | + 2,06 sin (2♀ − 3♄ + 293 1) |
| + 3,68 sin ( ♄ + 336 2) | + 0,63 sin (3♀ − 3♄ + 191 7) |
| + 0,23 sin ( ♀ + ♄ + 319 18) | + 0,18 sin (4♀ − 3♄ + 232 6) |
| + 0,33 sin ( ♀ − ♄ + 250 31) | + 0,11 sin (5♀ − 3♄ + 120 31) |
| + 0,52 sin (2♀ − ♄ + 302 45) | + 0,19 sin ( 4♄ + 40 24) |
| + 0,16 sin (3♀ − ♄ + 200 37) | + 1,09 sin ( ♀ − 4♄ + 295 0) |
| + 12,74 sin ( 2♄ + 109 32) | + 0,42 sin (2♀ − 4♄ + 294 3) |
| + 0,52 sin ( ♀ + 2♄ + 68 51) | + 0,39 sin (3♀ − 4♄ + 283 14) |
| + 13,77 sin ( ♀ − 2♄ + 304 12) | + 0,18 sin (4♀ − 4♄ + 189 33) |
| + 2,47 sin (2♀ − 2♄ + 182 45) | + 0,34 sin ( ♀ − 5♄ + 303 19) |
| + 0,90 sin (3♀ − 2♄ + 244 15) | + 0,15 sin (2♀ − 5♄ + 292 20) |
| + 0,36 sin (4♀ − 2♄ + 126 6) | + 0,26 sin ( ♀ − 6♄ + 298 41).} |

**Fig. 8.20** Headers read: *Angle of eccentricity. Annual change = −0″.30592; Longitude of perihelion. Annual change = + 0″.01491*

$$\mathrm{d}L = A + \frac{m}{e}\{(rr - aa\cos\varphi)(3-u) + ae(p+r)(2-u)\cos E\}X$$
$$- \frac{m\sin E}{e\cos\varphi}\{(rr - aa\cos\varphi)\cos v + ae(p+r)u\}\,Y.$$

**Fig. 8.21** Equation

Jährliche Änderung = — 1″,31973

| | |
|---|---|
| 0″,45 sin( ⚴ + 180° 35′) | + 0″,27 sin(3⚴ — 2♄ + 65° 10′) |
| + 0,29 sin(2⚴ + 181 47) | + 0,30 sin( 3♄ + 207 41) |
| + 1,78 sin( ♄ + 140 25) | + 1,05 sin( ⚴ — 3♄ + 163 1) |
| + 0,32 sin( ⚴ + ♄ + 181 12) | + 0,45 sin(2⚴ — 3♄ + 154 50) |
| + 0,41 sin( ⚴ — ♄ + 178 5) | + 0,42 sin(3⚴ — 3♄ + 173 25) |
| + 0,11 sin(2⚴ — ♄ + 158 28) | + 0,21 sin( ⚴ — 4♄ + 166 29) |
| + 2,01 sin( 2♄ + 221 24) | + 0,15 sin(2⚴ — 4♄ + 159 55) |
| + 3,35 sin( ⚴ — 2♄ + 124 44) | + 0,11 sin(3⚴ — 4♄ + 156 39).} |
| + 2,04 sin(2⚴ — 2♄ + 178 31) | |

**Fig. 8.22** Text reads: Annual change = −1″.31973

*To find the real value of this correction, necessarily requires the prior screening of the perturbations by Saturn and Mars; Nicolai has almost finished the former, and I have started the latter. Yet, I could not resist my impatience to see as to how much those differences can be brought down (still without regard to the perturbations by Saturn and Mars), if the mass of Jupiter itself is dealt with as an unknown quantity. With great surprise I found that the agreement is thereby improved exceptionally well. See here the longitudinal differences:*

*With the old Jupiter mass*

*–0.5″ –21.0″ –13.2″ +19.6″ –25.2″ +16.6″ + 4.5″ –27.7″ –3.5″*

*With the corrected one*

*+3.5″ –3.1″ + 2.3″ –0.5″ + 1.3″ –5.9″ + 1.9″ –7.1″ +7.5″*

*Mean error without correction of the mass 17.3″*

*″ ″ with ″ ″ ″ 4.3″*

*I call mean error: the square root of the mean value of the error's square, i.e., of the sum of the squares divided by the number. The sum of the squares is:*

| | | |
|---|---|---|
| *Without correction* | | *2684* |
| *With* | *″* | *167* |

*Also the latitudes, which by the way agreed quite well even without the correction, do agree a bit better still with the same:*

| | | | |
|---|---|---|---|
| *Without correction,* | *mean* | *error* | *5.5″* |
| *With* | *″* | *″* | *″ 4.0″* |

*The correction of the mass is* $^{1}/_{43}$ *of the old, that is enlargement. Of course, this result is still precarious, yet I assume that it will not be smaller than* $^{1}/_{60}$ *after Saturn and Mars have been included. I hear that Laplace himself – in the 4th edition of* Expos. *(which I have not read yet) – has slightly increased the old mass according to the observations of Saturn, although only very little; I believe Pallas should yield more reliable results here (especially when still several more oppositions can be used). By the way, I wish that the above communication remains between you and me. The old determination was only based on the 4th satellite of Jupiter the distance of which Laplace gave as 8' 16"; Triesnecker found 1 ½" more, and one had to take still 1 ½" more to find my mass. If Wurm (*M.C. *1802) when calculating Triesnecker's observations wanted to take not the radius of Jupiter that was given by Triesnecker's own measurements, but to prefer the one found by Schroeter, it seems to me he is wrong. Without injuring Schroeter's measurements, it seems to me that not only the absolute right value must be used here, but rather the one that Triesnecker observed with his eyes and his telescope.*

*This winter I had here a young Saxon, Moebius* [August Ferdinand Moebius, 1790–1868; director of the observatory at Leipzig University from 1848], *whose skill has been of much value to me. After he is gone, I am for a while without any help. Perhaps, however, Mr. Encke returns this summer for some time.*

No. 280 Gauss to Olbers Goettingen June 15, 1814

*It is strange with Pallas. Nicolai is done with the calculation of Saturn's perturbations for 9 oppositions; but the former nice correspondence is deteriorating slightly thereby. But the necessity to increase Jupiter's mass remains, namely by $^1/_{40}$. Without an improvement of Jupiter's mass the best correspondence of the longitudes possible is as follows:*

*–6.7"*
*–27.7*
*–4.2*
*+28.7*
*+31.0*
*–18.7*
*–6.5*
*–33.3*
*+0.1*

*Sum of the squares= 4110"*

*Mean error= 21"*

*With an increase in Jupiter's mass by $^1/_{40}$ on the other hand:*

*–2.3"*
*–9.8*
*+11.3*
*+8.5*
*+5.1*
*–8.8*
*–5.7*
*–5.7*
*+7.4*

*Sum of the squares = 525, which can be reduced to 503 by polishing so that the mean error = 7".5. The latitudes in contrast correspond very well and even better than before having taken into account the perturbations of Jupiter, namely:*

*+5.8"*
*–0.1*

*–2.0*
*–3.2*
*–8.5*
*–1.7*
*–8.2*
*+ 1.2*
*–4.3*

*Sum of the squares is 210"*

*Mean error is 4.8"*

*Whether an error in one of the elements might have sneaked into Mr. Nicolai's calculation is not for me to decide; it is strange however that if all perturbations of the epoch are given the opposite sign, the correspondence improves and the sum of the error's squares amounts only to 200; I showed Mr. Nicolai a method to check on that. Mars' perturbations will hardly be completed before six months, however slight they are; my objective is to deal with them that they leave nothing to be desired.*

Nicolai to Gauss Seeberg August 6, 1814

*Regarding the perturbations in Pallas' orbit by Saturn, I have not yet made any checks by calculating various special perturbations (because since my return I have not worked on numerical calculations but only theoretical things) but looked over the entire calculation once again but I could not find any errors, but I had not expected any anyway due to the extreme carefulness I applied to the entire calculation. Also my assumed forces were correct and I – and I cannot remember the reason why – assumed also –; hopefully, your coefficients of all elements will correspond to mine. Another reason why I am reassured of the accuracy of my equations is that the larger ones correspond fairly well to my previous calculation applying another method. Shortly I am planning to calculate several special perturbations from one opposition to the other and I hope to find a nice correspondence. Should this not be the case I would be more inclined to blame it on the numerical calculation of the equations than on themselves. Who knows whether the perturbations on Mars will intensify the correspondence of the hitherto observed oppositions of Pallas, however slight it might be?*

No. 285 Gauss to Olbers Goettingen September 25, 1814

*I have not been working on Pallas' perturbations for some time now but soon I will continue and develop the theoretical part. The forthcoming opposition seems to fit quite well. I am curious whether Saturn will disturb the harmony again after having added this opposition.*

No. 63 Schumacher to Gauss Mannheim December 7, 1814

*I have been able to observe Pallas only once since the opposition that I would like to send you because there is not much hope to add another due to the fog. Thus, you can assume the quadrant's position unchanged as it could be derived from a comparison of the former zenith distances and I also added the stars of Capricorn and Fomalhaut for comparison.* (Fig. 8.23)

*Using Fomalhaut the observations always show differences as follows:*

| *96* | *90* |
|---|---|
| *Nov. 25 53".1* | *59".5* |
| *27 50.5* | *59.0* |

| | Bar. | Réaum. | Fahr. | | 96 Eintheil. | 90 Eintheil. |
|---|---|---|---|---|---|---|
| Dec.3 | 27 9,3 | + 5,3 | 44,0 | ☉ | 71°14′ 36″9 | 71° 14′ 38″3 |
| | 27 9,6 | + 3,9 | 40,9 | α Cygni | 4 51 20,0 | 4 51 21,3 |
| | 27 9,9 | + 3,0 | 38,9 | γ Capr. | 66 56 23,9 | 66 56 24,8 |
| | idem | idem | idem | δ Capr. | 66 24 35,0 | 66 24 35,0 |
| | 27 10,0 | + 2,8 | | Fomalh. | 79 59 48,5 | 79 59 53,8 |
| | 27 10,2 | + 1,4 | 35,5 | Pallas | | 77 9 9,9 |
| | | | | Stellula sequens. | | 77 14 25 |

**Fig. 8.23** Observations of Pallas by Schumacher on the night of December 3, 1814. Temperature is given in Réamur and Fahrenheit

*I was not able to observe Pallas' RA since I was occupied to arrange the light in such a way that I was able to see both star and wire and I only read the inner division in order to determine the following star (about 2 or 1.5' after Pallas). But the measurement could only take place when it had already passed the third wire. Please be so kind as to send me that part of the* G. Anz. *where the results of these observations are listed. Should the weather clear up against all expectations I will send you Ceres' opposition. At the moment there is no hope at all.*

No. 287 Gauss to Olbers Goettingen December 31, 1814

*You probably read about the Pallas' recent opposition in the* Gel. Anz. *Since then I have been fitting the elements and find the following differences:*

| | *Old Jupiter mass* | | | *Jupiter mass increased by 1/54* |
|---|---|---|---|---|
| | *Mean longitude* | *Heliocentic latitude* | *Mean longitude* | *Heliocentic latitude* |
| *I.* | *–2.6"* | *–5.1"* | *+4.4"* | *–3.0"* |
| *II.* | *–11.2* | *–12.2* | *+1.8* | *–8.5* |
| *III.* | *–17.3* | *–0.5* | *–5.9* | *–3.3* |
| *IV.* | *+ 3.2* | *–6.7* | *–10.8* | *–6.1* |
| *V.* | *+22.8* | *+9.0* | *+3.3* | *+11.5* |
| *VI.* | *+23.8* | *–3.4* | *+1.1* | *–6.5* |
| *VII.* | *+16.8* | *–7.2* | *+13.2* | *–5.4* |
| *VIII.* | *–17.9* | *+4.5* | *–1.2* | *+2.2* |
| *IX.* | *–9.6* | *–0.1* | *–4.0* | *–0.7* |
| *X.* | *–7.7* | *+2.2* | *–2.0* | *+5.5* |
| *Sum of* | | | | |
| *Squares* | *2284* | *391* | *381* | *325* |

*This is based on Jupiter's perturbations according to that calculation which I call the " special" one (continued from one year to the other by mechanical quadrature). The calculation according to my general formulae as well as to that of Saturn's perturbations is not yet finished. It normally takes some months. Now I will soon work out my method and make it public this or the other way to get rid of this worthwhile and successful work though exceedingly tiresome work.*

No. 291 Olbers to Gauss Bremen January 25, 1815

*Has the difference been clarified – according to your last letter Saturn's perturbations upset the perfect harmony among Pallas' oppositions with an increase in Jupiter's mass? – I do hope that you will notify the Parisian institute of your method. Just because of you the prize has not been awarded yet. No news from Paris.*

Nicolai to Gauss Seeberg July 17, 1815

*I spent yesterday afternoon working on both parts of the epoch perturbations of Pallas by Saturn and I am eager to inform you about the equations found. Those equations of the first part of the epoch, to which no corresponding could be found in the second part were all below 0".1 so that I did not mention them among the final equations for the epoch. This will be the case especially with Jupiter's perturbations. – In the course of this work I noticed that if the first part of the epoch perturbation is brought through addition of 90° to the constant angle, and thus is brought – just like the second – to the expression a* $\sin(\varphi + A)$, *those constant angles do not differ much, especially in the larger equations. Is this fact mere coincidence or is there a theoretical reason? – and does something similar happen in the case of Jupiter's perturbations?*

No. 83 Bessel to Gauss Koenigsberg December 24, 1815

*Here are my observations on Ceres and Pallas, which, as I hope, might be useful to you, although Pallas' declinations might have been better, if the planet had been observed at both rims of the hair, which was impossible as its diameter was smaller than that of a tiny hair in the telescope of the circle. But probably the resulting uncertainty is not considerable.* (Fig. 8.24)

No. 315 Gauss to Olbers Goettingen January 8, 1816

*Should the Pallas prize be awarded or withdrawn now I would publish my theory piece by piece in the* Comm. *or in any other special paper.*

No. 318 Gauss to Olbers Goettingen February 16, 1816

*I much appreciated your articles from the* Conn. des tems *1818. I let rest Laplace's calculations of the reliability of Jupiter's mass because it depends on his perturbation theory which produces results of precarious accuracy. The new planets will be the best teachers of Jupiter's mass in the future; after centuries Halley's and your comet will make contributions, too.*

No. 320 Olbers to Gauss Bremen March 7, 1816

[About the prize, see Chap. 3 in this book for the text of this letter.]

Lindenau to David Seeberg May 4, 1816

*Did you possibly receive the oppositions of Pallas and Ceres? I made here some good observations of these two planets; Pallas was quite luminous compared to previous oppositions, but nonetheless hardly discernible in my quadrant.*

No. 327 Gauss to Olbers Goettingen July 24, 1816

*Bearer of this letter is Dr. Tittel* [Peter Paul Tittel, 1784–1831] *from Erlau, an eager student of mine who is going to Pyrmont mainly to acquaint himself with you personally. It is sad that so far you had such a bad weather at your bathing-cure. Here, too, spring was rainy and cloudy almost throughout; in no other year did I do so few observations as in this one. I obtained only two observations in June that I communicate to you as follows:*

| | Ceres. | | |
|---|---|---|---|
| März 26 | 13h 22m 58s,7 | 205° 2′ 31″,8 | 6° 8′ 41″,3 |
| » 28 | 13 13 35,6 | 204 39 38,6 | 6 17 50,0 |
| April 1 | 12 54 40,4 | 203 51 36,3 | 6 35 1,7 |
| » 2 | 12 49 54,8 | 203 39 9,9 | . . . . . . . |
| » 3 | 12 45 8,8 | 203 26 35,7 | 6 43 58,5 |
| » 4 | 12 40 22,2 | 203 13 54,5 | 6 46 44,3 |
| » 9 | 12 16 24,7 | 202 9 13,9 | 7 3 32,3 |
| » 10 | 12 11 36,2 | 201 56 4,1 | 7 6 25,2 |
| » 12 | 12 2 0,3 | 201 29 57,6 | 7 11 44,4 |
| » 14 | 11 52 24,3 | 201 3 50,2 | 7 16 23,2 |
| » 16 | 11 42 49,2 | 200 37 57,0 | 7 20 21,5 |
| » 18 | 11 33 14,9 | 200 12 16,2 | 7 23 34,3 |
| | Pallas. | | |
| März 26 | 13h 47m 32s,2 | 211° 11′ 54″,0 | 14° 10′ 2″,4 |
| » 28 | 13 38 29,5 | 210 54 7,8 | 14 49 26,3 |
| April 1 | 13 20 11,8 | 210 15 30,2 | 16 5 49,5 |
| » 2 | 13 15 35,0 | 210 5 16,3 | . . . . . . . |
| » 3 | 13 10 57,4 | 209 54 49,1 | 16 42 39,4 |
| » 4 | 13 6 19,2 | 209 44 12,0 | 17 0 38,3 |
| » 9 | 12 42 58,7 | 208 48 50,1 | 18 26 17,5 |
| » 10 | 12 38 17,0 | 208 37 22,5 | 18 42 25,6 |
| » 12 | 12 28 53,2 | 208 14 17,1 | 19 13 38,9 |
| » 14 | 12 19 28,5 | 207 51 0,2 | 19 43 27,6 |
| » 16 | 12 10 3,6 | 207 27 40,6 | 20 11 45,9 |
| » 18 | 12 0 38,5 | 207 4 17,3 | 20 38 30,9 |

**Fig. 8.24** Observations of Ceres and Pallas by Bessel from March 26 to April 18, 1815

| | | | |
|---|---|---|---|
| *1816* | *June 12 10h 47m 27.5s* | *251° 46′ 8.0″* | *–3° 42′ 4.7″* |
| | *June 13 10 27 34.7* | *251 33 47.6* | *–3 41 11.7* |

*From this, I derived the opposition as well as I could, and the helpful Posselt* [Johann Friedrich Posselt, 1794–1823] *who is now here is calculating the ephemeris for next year. The present opposition of Pallas (XI) still agrees well with my theory and confirms the necessity of an increase of Jupiter's mass. I now find Sun/Jupiter = 1050 and the likely uncertainty of this number very close to = 1. Laplace finds from the motion of Saturn 1070 with a slightly larger uncertainty. This difference, thus, is enormous; yet, I do not believe that this permits one to conclude a different* affinitas chemica [chemical affinity], *as Laplace's result is based on his Saturn theory that has been developed according to a method the reliability of which can be questioned. My determination of the Jupiter mass I do not yet wish to become public. Each newly added Pallas opposition shall increase the accuracy. Though it is an enormous amount of work, as the calculation for one opposition cannot be done in 14 days. I am very curious what type of results Juno will yield soon.*

No. 329 Gauss to Olbers Goettingen February 15, 1817

*I, too, have started to calculate an auxiliary table for Pallas' perturbations, a work of approximately ½ a million numbers that I could not have undertaken without the help of several young men, especially Mr Westphal* [Johann Heinrich Westphal, 1794–1831]. *More than the first half is already finished. After having finished my work on the table I will once again calculate most carefully all oppositions and take into account even the observations of 1802 that have been neglected. The application of the theory of probability gives my determination of Jupiter's mass already the same reliability as Bouvard's gives Laplace's calculation. The last opposition has not yet been taken into account and it will change the value only insignificantly but increase the reliability considerably. And the next opposition even more so. Please look for Pallas as soon as possible.*

No. 340 Gauss to Olbers Goettingen December 2, 1817

*I believe I have already mentioned in several previous letters to you the tables for a much easier calculation of the particular value of Pallas' perturbations. Since I was without Mr Westphal's help I was able to work only slowly but they are finished for some time now. I did not consider it necessary to compare all of the hitherto observed 12 oppositions again with help of these tables to normal elements and to develop the resulting 23 condition equations (the latitude of one opposition is worthless because only David observed very bad declinations, No V of 1808) and to determine from this the 7 improvements (including that of Jupiter's mass). This tedious work is now finished – at least for the time being (some polishing is still needed). The correspondence is admirable, the more so since I did not take Saturn's perturbations into account, which were calculated by Mr. Westphal and would have made the correspondence worse. The correction of Jupiter's mass (+1/46) appears to be confirmed, the sum of the squares of the 12 differences in latitude, which without this improvement amounts to 4064" and with the improvement is brought down to 241", and according to my theory of probability the determination of Jupiter's mass from Pallas' perturbations is three times as exact as derived from the observations of Saturn from which – as it is well known – Laplace derived his result which is contradictory to my own. By the way, the consideration of Nicolai's perturbations of Saturn would not alter the result of the improvement of Saturn's mass. Have a look at the remaining differences, if you please:*

| | *Heliocentr. longitude* | *Geocentr. latitude* | *Difference according to Carlini's determination of the oppositions* | |
|---|---|---|---|---|
| *I. 1803* | *+0.8"* | *–1.2"* | *–5.2"* | *–30.2"* |
| *II. 1804* | *+4.0* | *–11.8* | *+4.5* | *–9.8* |
| *III. 1805* | *–2.1* | *–3.8* | *–3.7* | *+3.9* |
| *IV. 1807* | *–4.6* | *–2.3* | *+1.1* | *–4.5* |
| *V. 1808* | *–5.0* | *–16.5* | | |
| *VI. 1809* | *+0.7* | *–5.1* | | |
| *VII. 1811* | *+2.4* | *–6.6* | *+2.9* | *+6.0* |

(continued)

(continued)

| | *Heliocentr. longitude* | *Geocentr. latitude* | *Difference according to Carlini's determination of the oppositions* | |
|---|---|---|---|---|
| *VIII. 1812* | *+10.7* | *+4.8* | | |
| *IX. 1813* | *−0.7* | *–0.8* | | |
| *X. 1814* | *+1.8* | *–1.1* | | |
| *XI. 1816* | *–2.7* | *–8.5* | | |
| *XII. 1817* | *–6.5* | *+5.0* | | |

*These differences are the more insignificant since the determination of several oppositions made by other astronomers, e.g., Carlini, differs from mine more often and more significantly than the above differences are, and since the 'appetite comes while eating' I thus think about letting the oppositions II and VIII be calculated again from the observations and taking into account the osculating elements. And I would also like to see the observations of 1802 being discussed. Now the longitudes correspond even better than the latitudes which is not surprising at all since these are heliocentric and those geocentric.*

Gauss to Encke Goettingen March 25, 1818

*I dare once more take advantage of your often proven willingness to do me a favor and help me with your talent for calculating – hoping the subject holds some interest for you, too. My table of Pallas' perturbations is now completed and this asset leaves nothing to be desired. I compared all hitherto observed 12 oppositions of 1803...1817 anew to the elements using the table, polished them and reached a marvelous correspondence. Here the numerous observations of 1802 starting only after the opposition have not been taken into account, and I am very eager to know in what kind of correspondence my elements would result; maybe it would be better to thus obtain the opposition – because a comparison can be made with elements that have to come so close to the truth that also more distant observations could be used. So will you please be so kind: 'to take on the task of comparing all meridian observations of Pallas from April and May 1802 with the osculating elements?'*

*The elements themselves are according to my tables as follows; node and perihelion are referring to the mean equinox of the epoch and are to be regarded as sidereal stationary:*

| | *1802 April 19.49212 Goettingen Time* |
|---|---|
| *Mean longitude* | *166° 44' 30".43* |
| *Perihelion* | *121 12 17.28* |
| *Ascending node* | *172 27 35.19* |
| *Eccentricity* | *14 13 27.83* |
| *Inclination* | *34 37 47.88* |
| *Daily mean tropical* | *770 ".47396* |
| *Log semi-major axis* | *0.4422173* |

*You might improve the epoch of the longitude.*

No. 346 Gauss to Olbers Goettingen March 31, 1818
*Now that my tables of the perturbations that Pallas sustains from Jupiter's side are completed and the elements correspond to all 12 hitherto oppositions as accurately as*

| | | | | | |
|---|---|---|---|---|---|
| Mai | 8 | 13^h 28^m 38^s,5 | 248° 38′ 30″,6 | 24° 10′ 44″,4 | |
| | 16 | 12 50 52,7 | 247 3 37,2 | 25 11 51,2 | |
| | 22 | 12 22 14,5 | 245 47 43,65 | 25 44 6,7 | |
| | 23 | 12 17 27,3 | 245 34 52,65 | 25 48 21,4 | dunstig |
| | 28 | 11 53 31,7 | 244 30 40,35 | 26 4 12,7 | |
| | 31 | 11 39 11,9 | 243 52 34,2 | 26 9 40,8 | |
| Juni | 1 | 11 34 26,1 | 243 40 2,4 | 26 10 49,7 | |
| | 2 | 11 29 40,4 | 243 27 34,95 | 26 11 37,5 | |
| | 3 | 11 24 54,9 | 243 15 8,7 | 26 12 4,7 | |

**Fig. 8.25** Observations of Pallas by Bessel from May 8 to June 3, 1821. dunstig = hazy

*possible, I asked Mr. Encke to calculate all observations of 1802 according to these elements. Although you discovered the planet after the opposition it might be possible because the elements are so accurately determined to derive the opposition from the observations.*

Bessel to Gauss Koenigsberg June 18, 1821

*Since you are probably separated from your observatory during the summer, you are maybe pleased by my observations of Pallas' opposition; so I am communicating them here:* (Fig. 8.25)

Gauss to Schumacher Goettingen March 17, 1822

*As far as Pallas' tables and the latest elements are concerned all papers dealing with this matter are so scattered that it is simply impossible for me to immerse myself instantaneously and be able to give instructions for a reliable calculation. As long as there are no hitches suspending the continuation of my measurements this year, the astronomers have to manage without me.*

No. 552 Gauss to Olbers Goettingen March 20, 1825

*I have been eagerly observing Pallas and Ceres this month. The observations are peculiar, since in a way they form the beginning of a second period, within 23 years both planets complete almost 5 revolutions and are thus close to the positions where they were in 1802. Both planets are very bright. My last observations were:*

| | *Ceres* | | *Pallas* | |
|---|---|---|---|---|
| *March 18* | *180° 46′ 26″.4* | *+18° 24′ 4″.4* | *175° 1′ 15″.0* | *+5° 44′ 59″.3* |
| *19* | *180 33 28.9* | *+18 28 34.0* | *174 50 59.1* | *+6 8 56.3* |

*The information on Ceres and Pallas was much appreciated. I was looking for the latter a couple of times in vain, since I was searching at a different location. It seems that only you are able to indicate Pallas' position in advance with certainty. – But why do we not yet have tables for Ceres at least of the quality of those by Daussy* [Pierre Daussy, 1792–1860] *for Vesta?* [For details on Daussy's *Tables*, see a future book in this series.]

Gauss to Bessel Goettingen March 21, 1825

*I am sorry that I could not find the time to calculate an ephemeris for Pallas. That of Caturegli is no good. I only calculated some positions for a first rediscovery and then followed the planet according to the observations. The observations on Pallas and Ceres of this year are peculiar insofar as the beginning of a second period, for within 23 years they complete almost 5 revolutions and consequently, they are in the same region as on 1802. My last two observations were:*

| | | |
|---|---|---|
| | | *Pallas* |
| *1825 March 19* | *11h 39m 23.s94* | *+6° 8′ 56″.3* |
| *20* | *11 38 43. 08* | *+6 32 41.4* |
| | | *Ceres* |
| *March 19* | *12 2 13. 93* | *+18 28 34.0* |
| *20* | *12 1 22. 15* | *+18 32 50.0* |

No. 554 Olbers to Gauss Bremen March 26, 1825

*I was very pleased by the information on Ceres and Pallas. I have been looking for the latter several times unsuccessfully, because I was looking at a different place. It seems, that only you are able to indicate Pallas' position accurately in advance. – But why do we not yet have tables for Ceres, at least of the quality like Daussy's tables for Vesta?*

Gauss to Gerling Goettingen February 8, 1834

*In the next few days (after several years' interruption) I will start to observe Pallas that is close to opposition where Encke's ephemerides is wrong by 5 min. It is a painful thought that my work on Pallas' perturbations, made 20 years ago, has remained discontinued, undeveloped and unpublicized ever since and will – like many a thing – perish with me. You cannot imagine how hard it is for me to carry out scientific work because of a fragmentation of my time and the pressure of certain circumstances.*

[Christian Ludwig Gerling (1788–1864). In 1812 he published a book on solar eclipses. See Schrimpf (2014).]

Encke to Gauss Berlin October 4, 1834

*The method of partial perturbations, which you were kind enough to show me in the year 1811, was explained to me only with the remark that you did not wish to see it publicized since you yourself would like to do so. I have been using it ever since and am now forced by the Yearbook to apply it to the four small planets and even to the comet with short period constantly. The death of Heiligenstein* [Anton von Heiligenstein, 1805–1834] *made it necessary for Ceres, too, as it happened during these holidays. And I foresee that I alone will no longer be able and that even if one or another of your students will take on one or the other planets this would be no help for me because most likely I am even forced to hand over the work on the comet. I would like to ask you therefore to allow me to introduce everyone to your method who I entrust with such a task because work without knowledge is not possible and I would appreciate it even more if you allowed me to describe the method in detail in one of the supplements of the yearbook.*

Gauss to Encke Goettingen October 13, 1834

*Regarding my method of the special perturbation calculation, I am pleased to hear you want to publicize it* [footnote] *because at the moment I do not know when or whether I will have time for that and thus the indiscrete abuse of private communication might be opposed. I will retrieve my figures for the previous perturbations of Pallas for you as well as the amounts of the corresponding element perturbations: but I have not touched these papers for a long time.*

**Fig. 8.26** Peter Andreas Hansen

[See Encke's essays in the Berlin Astronomical Yearbook of 1837 and 1838, especially the Footnote: on p. 251 of the first.]

Peter Andreas Hansen (1795–1874; Fig. 8.26) was a Danish German astronomer and director of Seeberg Observatory. He was an expert on perturbations, having won a prize offered by the Berlin Academy in 1830 for the mutual perturbations of Jupiter and Saturn. The early work on comets he mentions here garnered him the prize of the Paris Academy in 1850.

Hansen to Gauss Gotha February 7, 1843

*I am very pleased to be able to send you a summary of a paper where I explain a method to calculate the absolute perturbations (i.e., the perturbations for an unknown time t) of celestial bodies moving on orbits of arbitrary elliptical eccentricity and inclination. As is generally known this problem is still unsolved and it is a problem whose solvability has been doubted now and then. First example of an application of my method is the calculation of the perturbing influence Saturn has on Encke's comet; further examples will be in my treatise that is ready for print. The method is not applicable to orbits with great eccentricity but the ellipse's eccentricity can be arbitrary. And indeed the calculating of perturbations on Pallas by Jupiter is according to my method very easy and these perturbations will consist of a moderate number of elements, fewer than the Moon's perturbations. Thus, I gave an applicable solution to the question that was raised by the Parisian academy in 1811 and rewarded by a twofold prize.*

Gauss to Hansen Goettingen March 11, 1843

*I restrict myself to express my pleasure that you proceed in calculating the perturbations in a similar way as I did some 30 years ago while calculating Pallas' perturbations in so far as you omit the use of series according to eccentricities and inclinations. Naturally you had to make special studies for the comet perturbations for which there had been no reason in the case of Pallas. In calculating the perturbations on Pallas by Jupiter I intentionally applied the method of variable elements for although I thus had to deal with 6 elements instead of only 3 with the other method, I preferred for practical usage the above mentioned method; for each year one only needs to make one calculation (for one day), for which I invented a special auxiliary table by means of which I was able to finish the calculation in comparatively short time, although 801 equations (1,602 if you want to count the sine and cosine parts of one and the same argument separately; every single one with a coefficient above 0".1 is mentioned) had to be taken into account.* By very simple means the result for the entire season of observations can be made sufficient without neglecting the accuracy. Also the perturbation by Saturn was calculated and that by Mars was started according to another method but not finished. Other – widely branching – affairs had not allowed me later on to return to those studies and I leave it open whether I will have at my age leisure and pleasure to do so, but there are many other different things I would only regretfully abandon and let perish. You are in the fortunate situation that you do not have to shatter your time and anyway it is a comforting reassurance that this branch of astronomy is in your good and capable hands. Whether I used something related to your mentioned way of integration I cannot say for certain, now, after 25 years I have become estranged from the whole matter but I would rather doubt it.*

**) I dare say it is unnecessary to tell you that the work I am talking about is completely different from that calculation using quadratures that I made simultaneously between 1802 and 1818 or 1820; this I call "special" that one "general" calculation and the latter found in the first the control necessary in such extensive calculations.*

Gauss to Bessel Goettingen March 21, 1843

*The first treatise (of Bessel) on Jupiter's mass has awoken a memory, one that is always painful to me, namely my old work on Pallas' perturbations. It has become so unfamiliar to me during the last 25 years that it is very difficult to make sense of the papers available. The necessity of increasing Laplace's mass of Jupiter was immediately recognized and at first I assumed it 48/47 without being able at the moment to find out when I made this determination; but later on I found that from the first 10 oppositions – 2376 days... +1759.5 days (reckoned from the beginning of 1810 as epoch) the elements have been improved radically and among others the above mentioned fraction was increased to 43/42. The sum of the squares of the remaining errors was (PS: without taking into account other perturbations, namely that of Saturn):*

| | | |
|---|---|---|
| *Laplace's mass 1/1067.09 (Hel.)* | *Long. 3175.69* | *(Hel.) latitude 348.54* |
| *Increased by 1/47* | *963.78* | *318.47* |
| *Increased by 1/42* | *918.46* | *322.66* |

*Alone in those papers I am able to find right now there is a new improved calculation of all elements based on the first 12 oppositions which has not been finished. While searching I found some older papers with a similar calculation from the first 9 oppositions without* *being able to say whether the correction of the mass had not been found earlier. But at the same time I realize that that written above from memory is wrong: Saturn's perturbing forces were taken into account. It is probably not necessary to remark that all those calculations are based on that perturbation theory that I call the general one since according to it the value of a perturbation can be found for any time; that makes altogether more than 800 equations (if you combine those containing sine and cosine of the same angle, without that more than 1,600) and this theory was more than once worked through – what was necessary to obtain first of all the mean elements themselves. Besides this general theory the calculation by quadratures was continuously made and that was an indispensable control. – The last thing I retrieved is the calculation and comparison of the 14$^{th}$ opposition of January 6, 1820. Whether and when I will find the time to dive into these studies, edit and publish them remains to be seen; without any change in my personal situation and being able to decide freely what to do with my time it will hardly happen especially since there are so many other things which I would just as unwillingly let perish, or better to which I attach greater value.*

[And so it happened that this was left to perish. Apparently this letter of 1843 is the last Gauss ever wrote on the subject.]

Bessel to Gauss Lilienthal October 1843

*The kind communication of the result of your former studies regarding the influence of Pallas' motion on Jupiter's mass, leaves me doubting whether La Place's mass improved in the ratio 43/42 is the former 1/1067.09 or the later on assumed 1/1070.5 according to Bouvard's studies. The latter would give 1/1045.6, differing from my determination, based on the satellites, only slightly in the same sense in which it deviates from that based by von Encke on the motion of Juno. I believe that one can only conclude from the newer determinations, starting with yours, that all are as far in accordance as can be expected from the mean values on which they are based. – Too bad, that such a great work as yours on Pallas, which, despite its results referring to this planet, must have been nothing else but momentous, has never seen the light of day! Your remarks regarding this matter leave little room for hope that you will publicize it.*

The Olbers-Gauss letters were published in German by Schilling (1900). The numbers associated with the letters from that book have been retained in this chapter.

## References

Belenkiy, A. and Vila-Echague, E.: 2016, Groping Toward Linear Regression Analysis: Newton's Analysis of Hipparchus' Equinox Observations. *Observatory* 136(1250), 1-22.

Drake, S. and O'Malley, C.: 1960, *The Controversy on the Comets of 1618*. Univ. of Pennsylvania Press, Philadelphia.

Dupré, S.: 2003, Galileo's Telescope and Celestial Light. *JHA* 34 (4), 369-399.

Schilling, C.: 1900, *Briefwechsel zwischen Olbers und Gauss*, vol. 1 of 2. Springer, Berlin.

Schrimpf, A.: 2014, An international campaign of the 19th century to determine the solar parallax. *European Physics Journal H* 39, 225-244.

Vargha, M.: 2005, Franz Xaver von Zach (1754–1832) His Life and Times. Konkoly Observatory, Budapest.

# Chapter 9
# The Harding Letters

This chapter includes all the Pallas-related correspondence between Carl Harding (Fig. 9.1) and Carl Gauss, in 1803 and 1804. Pallas-related letters by Zach, Maskelyne, and Banks from this period are also included.

Mapmaking is a major element of these letters. Finding Pallas in the sky relied heavily on accurate star charts, and there was a simultaneous effort underway on Earth. On April 10, 1802, just a few days after Pallas had been discovered, "the Great Survey of India was launched. It lasted sixty years…the British carefully mapped the whole of India, marking borders, measuring distances." (Harari 2014: 297) The effort to map the heavens that received an impulse from the discovery of the asteroids likewise led to a decades-long effort that resulted in the 1859 publication of the Bonner Durchmusterung, which gave the position and brightness of 324,000 stars. Harding was keenly aware of the need for a better atlas, as we read in several letters, including the final one given here, to Lalande in 1804.

Harding to Gauss     Lilienthal     February 22, 1803

*Let me communicate the fortunate news that I was able to find Pallas Olbersiana again on the 13$^{th}$ of this month at 14h 50'. As difficult it would be to find such a little star among the millions of telescopic stars of Poniatowski's Taurus, as simple this task became by your ephemeris calculated with exceptional accuracy for which I would like to thank you. I am sorry that I have not yet calculated my hitherto observations so you cannot see that they match your calculation to a very high degree. I have observed her already four times and always found her where expected. I immediately informed Dr. Olbers where he could find his Pallas and I do not doubt he found and observed her last night. According to my estimate Pallas appears as a star of the 12-14$^{th}$ order of magnitude. Such estimates may be not reliable but I do not believe to be wrong in the case of Pallas. With an excellent achromatic telescope of 3f., with which stars up to the 11$^{th}$ magnitude can be distinctly seen, she cannot be seen and thus for her observation I have to employ a telescope in which she always shows more luminosity than any other telescopic star. Without doubt Pallas has been seen and accurately observed in many places, because since the 18th there has been at least one clear hour every night, at least here, close to the North Sea and so I am hoping*

 
C.J. Cunningham, *Studies of Pallas in the Early Nineteenth Century*,
DOI 10.1007/978-3-319-32848-5_9

**Fig. 9.1** Carl Harding

*that you will soon have more accurate observations on Pallas than I was able to deliver since I can only use the circle for determining the positions of the stars.*

Zach wrote the following letter in English. His idiosyncratic spelling is retained.

Zach to Banks Seeberg Observatory March 1, 1803

*I take the liberty to inform you, of the appearance of the new Planet Pallas. Mr. Harding in Lilienthal saw it first Febr.$^{y}$ 19$^{th}$ at 15 o'clock near the star N$^{o}$. 36 of Taurus Poniatowski of Mr. Bode's Catalogue. This little Planet is very difficult to be seen, and not visible by a refractor of 3 feet. Mr. Harding discovered it with a reflector of 7 feet, and esteems the magnitude to be, like a star of the 13$^{th}$ magn. I saw this heavenly body with my refractor of 10 feet, and I can assign no magnitude, methinks the Planet appears quite as a telescopick* [sic] *star. Dr. Olbers looked at Pallas, with his 5 feet refractor, and saw it perfectly, to his opinion the Planet looks quite as the 4$^{th}$ satellite of Saturn. The reason and the possibility of discovering this little Planet so soon, is to be ascribed to the exactness of the elements of the orbit calculated by Dr. Gauss, and to the precision of the ephemeris given in the* Monatliche Correspondence, *Octobre* [sic] *1802, which placed the Planet exactly upon the spot, where Mr. Harding found it. There is therefore no difficulty more to find this Planet, whoever pleases to look for; he has only to follow the ephemeris, of which I have the honnor* [sic] *to join here a copy. I have also had engraved a map for my* Month. Corresp. *upon which the supposed path of Pallas has been described, in order to facilitate the search for this Planet, but beyond every expectation this map is become now an exact representation of the way, which this heavenly body takes in reality. I take therefore the liberty to send it to you annexed. It representes* [sic] *the motion from 3 to 3 days of the Planet from Febr. ye 4$^{th}$ till April ye 17$^{th}$. In my next letter, I shall have the honor to send to you the* *second part of this map, which contains the track of this Planet from April 5th till June 28th. By these means it is impossible to miss this nearly imperceptible point, which allways* [sic] *will be traced by these maps, and by the ephemeris.*

*Good observations are yet very difficult to be made considering the meanness of this little body. In the meantime some course observations have been made in order to evince, that this heavenly body is realy* [sic] *our Pallas.*

*1803 Febr. 20 at 15h 30' 13" mean time in Lilienthal*

| | | | |
|---|---|---|---|
| *AR ⚴ =* | *272° 38' 27"* | *Declin. ⚴ ——* | |
| *Febr. 21 at 17. 0 11* | *272 36 29* | *——* | *7° 32'N* |

*No tidings yet from Ceres. But there is no doubt she will presontly* [sic] *make her appearance upon the spot assigned by the ephemeris of this planet, I add here. Her very low position, and great south declination is probably the reason, she has not been seen yet, in the vapors of the horizon.*

## Zach to Jan Sniadecki Seeberg March 2, 1803

*I make this letter fly in order to tell you about the reappearance of Pallas, which was, as you know, a little questioned because of the extreme smallness and distance of this small planet in the part of the orbit it went through this year. Actually, right now this star appears as a 13th magnitude star and it is not visible with an achromatic telescope of three feet. It was Mr. Harding, Mr. Schroeter's assistant at Lilienthal, who saw it first on February 19 at 3 pm with a 7-foot-telescope. You will ask me how he could find an almost imperceptible spot among the myriads of these spots. And this is the miracle that Pallas was exactly at that position where Dr. Gauss's orbital elements and the ephemerides that I gave in the October issue of 1802 of my journal, indicated. You only have to follow these ephemerides and the enclosed map and you will find the planet. I had this map engraved in January in order to make the search for this small star easier, but now we have a very accurate map of the way Pallas will take. But let me repeat it again: this star is very hard to see. I saw it with my achromatic telescope of ten feet and it is impossible for me to assign a magnitude to it and thus I'd rather call it a star that is invisible to the naked eye. Mr. Olbers saw it with his 5 foot achromat and told me the planet appeared to him as the 4th satellite of Saturn. Here are some positions that prove that this moving star certainly is our Pallas and that it follows the path assigned by the capable Dr. Gauss.*

| *Mean time Lilienthal* | *RA Pallas* | *Decl. Pallas N* |
|---|---|---|
| *Febuary 20 15h 30' 13"* | *272° 38' 27"* | *———* |
| *21 17 0 11* | *272 56 29* | *7° 32'* |

*On February 23 it preceded the star no 421 of Bode's catalogue by 2' 15".5 in time at 3h 19' 24"pm. Thus you can see there is no doubt that this star is our Pallas Olbersiana. This planet courts the Polish since it traverses Taurus Poniatowski* [a constellation named after the King of Poland]. *I am asking you to inform Mr. Poczobut about this discovery with my compliments. I am enclosing a second copy of the map for him.*

*No news from Ceres. But it is supposed to arrive all the same since there is no doubt it is to be found at the named position. Its southerly declination, and the vapors of the horizon will delay its rediscovery a bit which is not subject to other difficulties. As soon as it will take place I won't hesitate to inform you.*

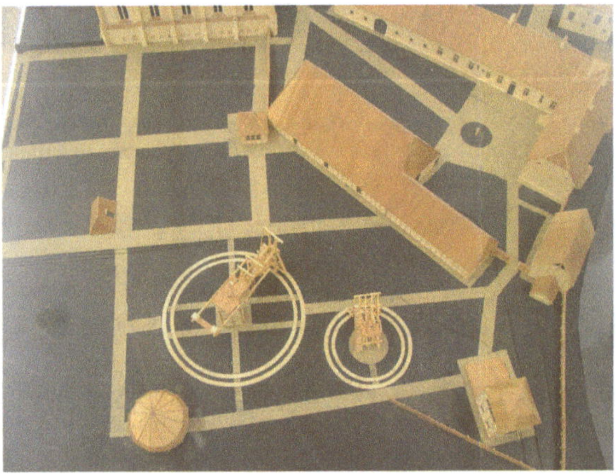

**Fig. 9.2** A model of the observatory in Lilienthal, at which Harding and Schroeter studied the asteroids. The two large telescopes are set up in the circular rings, which allowed them to move from one area of the sky to another. In 2015, a full-size replica was unveiled in Lilienthal of the large Newtonian telescope. A photo of it will be in a future book in this series, which surveys all the observatories involved in asteroid research

*Mr. Herschel is angry because I fought his asteroids in my journal (and as everybody believes quite successfully). But Senator La Place is on our side. He has just sent me the third volume of his Mec. Celeste. He keeps calling Ceres a planet. Mr. Piazzi made a nice joke of it and calls the asteroids Herschel's dynastie and he writes: presto vedremo nel cielo dei Duchi, Comti, et Marchesi.*

Harding to Gauss    Lilienthal    March 17, 1803

*I believed to not be able to thank you in any better way for your shared ephemeris of Ceres than sending you some observations on this planet that I hoped to find according to your instructions before the bad moon would come too close. Unfortunately, it has constantly been overcast and only on the 4th and 12th of this month it cleared up when I started eagerly searching. In the restricted southern horizon, which is hidden by a wood, barely 100 feet from the observatory, I was unable to see her on the 4th since dawn broke earlier than Ceres appeared from behind the woods. On the 12th I was certain to have found her; I observed a star of about 8th magnitude, west of no. 28 Flamsteed of Sagittarius, the only one I could discern in that region where Ceres was supposed to be due to brightness and the beginning dawn. But I soon found that it was not Ceres since the RA was too great by about 9' than it could have been according to your ephemeris. This seems to prove that Ceres is not as bright as an 8th magnitude star. I used our wonderful 10f. Dollond. Last night was completely clear but I was in Bremen and unable to observe either Ceres or Pallas. Our excellent Dr. Olbers told me he had observed Pallas well. This night is going to be clear again and I hope to see both planets* [at Lilienthal Observatory: Fig. 9.2].

*The clock used for the observations on Pallas lost 25" and that used for Ceres 10' 22" compared to Hubert's clock.*

*My observation of Ceres on June 26 is*

| | | | |
|---|---|---|---|
| *Immersion Ceres* | *12h 19'41"* | *Mars* | *12h 26' 9"* |
| *Emersion* | *21 53* | | *28 17* |
| | *12 20 47* | | *12 27 13* |
| | *20 47* | | |

*Difference RA = 6' 26"*

*Last night the sky was completely overcast. I hope to be able to observe both planets tonight if the awful dry fog will allow my doing so. I was able to observe Pallas only hazily because the sky put on an angry expression.*

| | | | |
|---|---|---|---|
| *Pallas* | *10h 57' 36"* | *Nr. 464 Herc* | *10 59 12* |
| | *11 0 48* | | *11 2 55* |
| | *10 59 42* | *differ 1' 41."5* | *11 1 3.5* |
| | *11 5 56* | | *11 6 40* |
| | *9 14.78* | | *11 53 00* |
| | *11 7 39.37* | *differ 1' 41."2* | *11 9 16.5* |
| | *11 12 1* | | *11 13 59* |
| | *15 13* | | *17 7* |
| | *11 13 52* | *differ 1'41"* | *11 15 33* |

*The clock lost compared to mean time 36", and I have just realized that these observations turned out better than I thought. Pallas' emersions that seem to have least luminosity were hardly discernible in dawn and the foggy sky.*

Blagden was the eyes and ears of Banks in Paris, which was quite useful as Banks (Fig. 9.3) could not read French. Blagden went to Paris even before the peace treaty between France and England had been ratified, the war having disrupted travel for several years. As early as April 1802, he told Banks how well he was being received by Laplace, Lalande, Lagrange and even First Consul Napoleon Bonaparte, who invited him to dinner in May. (O'Brian 1987: 274) On March 19, 1803, he related to Banks that Pallas had been spotted from Paris:

*M. Laplace desires me to inform you, that the planet Pallas has been seen here, nearly in the place calculated from Gauss's last elements, as laid down in Zach's journal; so that no doubt now remains as to the general form & situation of its orbit.*

**Fig. 9.3** Joseph Banks

On March 22, 1803, Banks wrote to Herschel about Pallas:

*By letters received this evening I learn that Pallas was rediscovered by Harding at Lilienthal on the 19th of Feb. & has since been seen at Paris nearly in the place where according to Gauss's calculation she was expected. No news has been received of Ceres she is still very faint.*

Maskelyne sent Banks a letter about Pallas on April 13, 1803. It shows that they were sharing information from Zach:

*I thank you for the communication of Dr. Zach's letter, ephemeris of Pallas's place from Aug. 9 to Oct. 25 last, & the celestial chart representing its motion. I have copied what I wanted, and now return the papers, except that the ephemeris is only a copy, but an exact one, of what you sent, the original having been mislaid. Yesterday morning & this morning I had very good observations of Pallas as follows*

| *1803* | *Mean time* | *Right ascension* | *Declination North* |
|---|---|---|---|
| *April 11* | *15h 26′ 41″* | *282° 44′ 12″* | *15° 3′ 38* |
| *April 12* | *15 31 5* | *283 2 30* | *15 13 47* |

*The correction of the Ephemeris sent before, and serviceable as it makes the observations agree almost exactly with it. Thus this morning the right ascension agreed exactly & the declination differed only 13" which is not to be reckoned anything, as the ephemeris was only set down to minutes. The observation made here on the 1st instant was only taken by the divisions of the instrument & therefore less exact, but agreed nearly with the ephemeris corrected. Thus it seems that Dr. Gauss's 6th sett [sic] of elements of the planet's orbit is pretty exact, & likely to represent the planet's place, probably within a minute during the remainder of the year. Astronomers are much obliged to him for his indefatigable labors, which are of great service to ascertain whether we observe the planet or a fixt star, as so many are scattered throughout the heavens. We shall, I think, have no difficulty about finding it in future. I am glad to learn that such grand operations are carrying on in the mensuration way in the North of Germany. When you write to Dr. Zach, be so good as to make my complements to him, and impart to him my two observations, & inform him I used the star of 6 Magnitude No. 170 of Aquila according to Bode's Catalogue, & took its Right ascension for the present time 18h 46' 30" of time, and its declination 14° 9' 13" North.*

Harding to Gauss     Lilienthal     June 30, 1803

*My dearest Doctor, please receive some observations on Pallas that I was able to make this evening because of clearest skies.*

| | | | |
|---|---|---|---|
| *Pallas* | *11h 2' 41"* | *Nr. 464 Herc* | *11 4 38* |
| | *11 6 22* | | *11 8 32* |
| | *11 4 35.5* | *differ 2' 3."5* | *11 6 35* |
| | *11 10 29.5* | | *11 12 3* |
| | *13 10* | | *16 12* |
| | *12 5.25* | *differ 2'2."25* | *14 7.50* |
| | | | *13 2.25* |
| | *11 17 36* | | *11 19 26* |
| | *21 9* | | *23 25* |
| | *10 22.5* | *differ 1' 41"* | *21 25.5* |
| | | | *19 22.5* |
| | | | *2 2.0* |

*The differences are not outstanding but it was hard to notice the emersions of Pallas since the telescope I used was not completely aligned and thus does not show distinctly around the edge. Pallas was very faint and did not change much within 1.5h. I noticed this because I have noticed almost every evening some kind of change in light within a shorter period of time, from which I felt enabled to draw the conclusion of a rotational period of 23 hours. Today's observations might as well go together with this hypothesis, for how long her least brightness lasts I do not know yet. I do not have enough observations to conclude anything certain regarding this matter.*

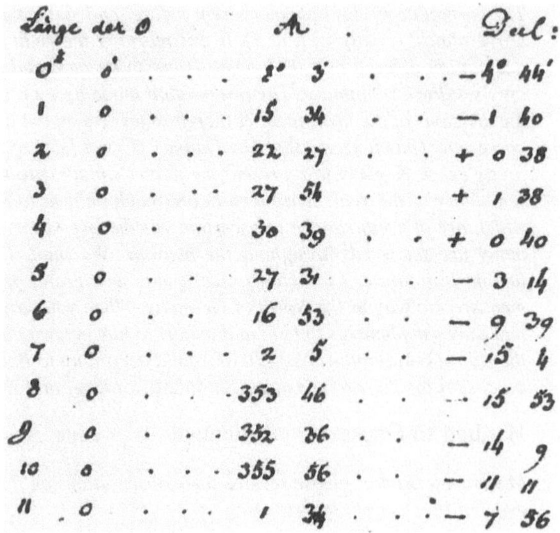

| Länge der ☉ | AR | Decl: |
|---|---|---|
| 0ˢ 0° | 8° 3' | −4° 44' |
| 1 0 | 15 34 | −1 40 |
| 2 0 | 22 27 | + 0 38 |
| 3 0 | 27 54 | + 1 38 |
| 4 0 | 30 39 | + 0 40 |
| 5 0 | 27 31 | − 3 14 |
| 6 0 | 16 53 | − 9 39 |
| 7 0 | 2 5 | − 15 4 |
| 8 0 | 353 46 | − 15 53 |
| 9 0 | 352 36 | − 14 9 |
| 10 0 | 355 56 | − 11 11 |
| 11 0 | 1 34 | − 7 56 |

**Fig. 9.4** Geocentric position of the ascending node of Pallas' orbit on Ceres' orbit

Harding to Gauss Lilienthal July 17, 1803

*Already in August 1802 I missed an 8th magnitude star which is mentioned in the* Hist. Cel. Fr. *vol. 1 and which drew my attention because of our current studies since it was in the region where the square position of Pallas' ascending node meets Ceres' orbit* (Fig. 9.4). *Last night I studied this region thoroughly and convinced myself again that it really is missing. You find this star's position indicated by La Lande:* Hist. Cel. Fr. *vol. 1, p. 191.*

*23h 35 18."5 zenith dist. 63° 17′ 9″*

*correction = + 26.3* *+ 1 47*

*23 35 44.8* *6*

*Accordingly, on September 17, 1795, the star's app. RA 353° 56′ 12″ decl. south 14° 29′ 41″.8. I do not believe an erratum of the* Hist. Cel *created this star, because 18″ before and also the following minute a star was observed and there is no star of this magnitude in that zone. It can only be concluded that that star was either an asteroid or had intense changes in light that has made it invisible since August 7, 1802. I have never seen a star becoming completely invisible but there is no end to the changes in light in the smaller stars, I am continuously collecting data in this regard. Please do not forget to send the calculation of Ceres' orbit for the stars in question, so that I will able to pick up the trail soon. It would be quite fun to find another small splinter.*

Harding to Gauss Lilienthal August 2, 1803

*What you kindly communicated regarding a planetary orbit is not very encouraging. But it seems very strange that you do not consider the identity of the two stars in question to be impossible. I admit, that I did not expect any other news from you than that my observations on the missed star do not correspond to La Lande's. If Dr. Olbers' splinter theory is correct we can assume there are several more fragments of a previously larger main planet and then it would still be an unthinkable* [word missing]. *Perhaps La Lande and I saw the very same piece. One observation on Pallas of May 1802 and one of Ceres of June 1803 would maybe give similar elements like those that you found from those two observations. I only want to say that I am not tiring to search the skies and that I like Dr. Olbers' hypothesis a lot. Since Pegasus and Cassiopeia are good to see I am going to search this celestial region thoroughly but the amount of stars in it is almost deterring.*

*Dr. Olbers' calculation for the descending node of Pallas and Ceres orbit is in my hands and I am enclosing a copy you asked for. I am already eagerly working on a chart of the region of its geoc. positions and will send you a copy of it too.*

## Harding to Gauss Lilienthal November 1, 1803

*My chart of the corrected orbit of the ascending node of Ceres is already finished and I gave Dr. Olbers one copy of it. You, too, will certainly receive a similar and since you reassured me that even von Zach would gladly accept one, I am going to send him one tomorrow as well. I am in doubt whether it is suitable for the MC since it is intended for a perennial use since according to Dr. Olbers' hypothesis we want to search for further fragments of the once exploded planet and secondly its size seems to be too unhandy for the journal because it is 30 × 16 inches, although the single degrees, of which it shows 43° in RA, are by far smaller than those on the chart of Ceres' orbit. I could not make this one as big since it would have become too unhandy to use. According to my opinion this fact does no harm because the Whale is not Poniatoiwski's Taurus.*

*But I have already begun working on a chart which shows the orbit of Ceres for 1804 according to your ephemeris and which I will offer Baron von Zach for his MC. I promise to send it with the next post. I made it, without taking away from the clearness, so small that it can easily see the light of day through this magazine. I have high hopes that we will reach our goal and find a sisterly splinter of Ceres next year if existent. Ceres and Pallas keep pace with each other in RA and will be in the region of the descending node next year. So if we man this pass it will not elude us: for there can hardly be doubt that it must have almost the same orbit as Ceres.*

*If my chart is to be published I hope that Mr. von Zach will communicate also the positions of those stars that are necessary to make useful observations or rather to make the good observations useful. An ephemeris of Ceres' orbit will make a wonderful gift for me and I promise to deliver a chart in return. The next post will bring you the promised positions of Ceres' orbit, I will send those to Gotha. If you have left already Baron von Zach can keep it for his* MC *anyway. If it does not prove suitable it can be forwarded to you.*

## Harding to Gauss Lilienthal November 5, 1803

*Please find enclosed, my dearest friend, the promised chart of Ceres' orbit, which I am asking to give Mr. von Zach with my humble regards. There are maybe 13 stars whose existence I have not checked. I tried to determine some of those at the circular micrometer because they are neither mentioned by La Lande nor Bode, others which I did not find in the skies are left out, namely those mentioned in my last letter. Last night I convinced myself again of the fact that Bode's no. 72 Ceti is not there. La Lande puts a 9-10 magnitude star in its place which really exists. But nevertheless Bode claims to have borrowed it from Piazzi. Please be so kind as to check Piazzi's catalogue whether it is really mentioned and how often he observed it. Too bad, P. did not use the space of his splendid folio at the time of his observations.*

*I made two mistakes on my enclosed chart, which can be easily corrected by the engraver. No. 30 Piscium and 28 Ceti Flamst. should be of $5^{th}$ magnitude but I assigned them, misled by the* Hist. Cel. Fr. *the $6^{th}$ magnitude, but they do not deserve this degradation. If future errors show please kindly inform me. Maybe I should omit it anyway but then the margin must be calibrated as I depicted the equator. But I leave this to Baron von Zach's judgment.*

Harding to Gauss Lilienthal December 1, 1803

*You probably have received my chart of Ceres' orbit and I wish for your and von Zach's approval. Pray, let me know what you think so I can improve further similar works. Do you consider the grid necessary? I do not believe the outer appearance gains anything but it facilitates finding each star's position immensely, if to several minutes only. Tell me whether I should leave it on the chart of Pallas' orbit which I am planning on designing. I only give the borders but not the contours of the constellations, so the chart is not too overloaded and actually I consider those dispensable. If you want me change it, I can easily add those. I happily take your advice since my wish is this little chart is of use to the practical astronomer. My plan is to depict in such a way the entire zone of the celestial sphere in which Ceres and Pallas can show geocentrically because I consider these charts almost indispensable for observations of both planets. This would be quite an opus but I am not afraid of tackling it if I only knew it is of use. Please let me know what you think. I have not yet begun to sketch the sheet of the geoc. position of Pallas node for you, because I wanted to finish the chart of Pallas' orbit first. Please excuse my negligence. But I will send it and if Mr. von Zach is content, I am happy. Dr. Olbers is pleased as well.*

Harding to Gauss Lilienthal January 6, 1804

*That you approve of my plan for the celestial charts is a real encouragement to tackle those. From the kindly communicated extreme points of Ceres' orbit I can already judge the region which needs to be worked on but when you have sent me those for Pallas' orbit as well I will know more since those determine the size of the charts. Since the zone of Ceres' orbit is always almost parallel to the ecliptic it might be most convenient and instructive to have charts designed according to longitude and latitude because much space could be gained this way. But by doing so the practical advantage, the depiction of the declination and parallel circles, who's keeping Dr. Olbers favors, would be sacrificed. The boundaries of Pallas' orbit are so distant from the ecliptic that here the equator and its parallels must form the basis and thus this way will probably be preferred. Consequently, I would much appreciate it if you could kindly send me some positions of the orbit of this planet as well. The chart of Pallas' orbit of this year is already finished but for eight weeks I have not been able to compare it with the sky due to bad weather. During the whole of December there was not one single clear night. It is high time it cleared up since Pegasus is approaching the Sun. I am afraid in the end I have to send the chart without knowing whether it shows a true image of the sky or not. Many a star of Bode has probably to be left out. I am eager for your ephemeris of Pallas, for I do not want to map Bode's calculation: it is erroneous for several times and results in an undulating line on the chart which is hardly described by the planet.*

Harding to Gauss Lilienthal January 19, 1804

*Thank you for your kind letter. I much appreciated Ceres' zodiac and am working eagerly on its depiction. But for now I am not sure how this one and that of Pallas can be combined since in some regions both are rather distant and in others rather close. If I intended to show both on the same sheets, some must have almost 60° in decl. others only about 30°; separated the work would be half as much but the number of sheets would be doubled. But maybe there is a way out. The other day I had the opportunity on two clear evenings to compare my sketch of Pallas' orbit with the sky, but I had to limit myself to the region where the planet will be. A dense fog covered the horizon up to 15° so that I could discern stars only to the equator.*

*Farewell, and if you happen to have calculated an ephemeris of Pallas – kindly remember your admirer Harding.*

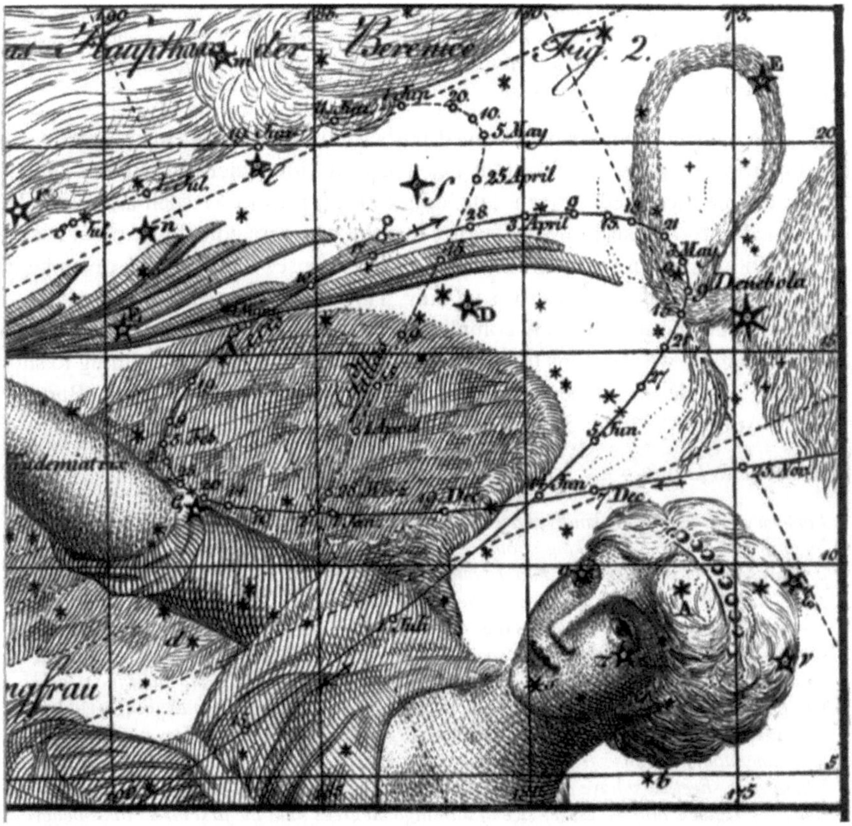


**Fig. 9.5** An example of the type of inadequate star chart Harding refers to. The paths of Ceres and Pallas in 1802 are overlaid on this chart that appeared in the *Berlin Astronomical Yearbook* for 1805. While graphically beautiful, such a chart was useless in actually finding the faint asteroids in a telescope

Harding to Gauss  Lilienthal  January 28, 1804

*Thank you very much for your kindly shared ephemeris of Pallas. The chart is to be sent to Seeberg within the next couple of days. If Pallas is only on June 5 as bright as it was on February 19, 1803, I consider it impossible to see her in March; twilight will hide her. But we will try.*

*How much Bode's ephemeris differs from yours! In October the difference in RA is nothing less than ½ degree. We are grateful for your calculation because Bode would have only misled us and robbed us of seeing Pallas.*

*Wherever you look in the sky, you notice how imperfect all our catalogues and illustrations of the celestial sky are* (Fig. 9.5). *I therefore wish to give my charts, which I am planning on sketching, more accuracy, for otherwise all of my efforts would have been in vain and I must ask myself what the raven of the Roman cobbler blabbed in his innocence* [see footnote below]. *I therefore will not list one single star which I did not at least see once for myself. For I believe that nothing is gained if you know what stars should be in the skies, one has to know which are really there. The first is also given by Bode there is an issue with at least 1,000 stars of his 17,000 if one really checks. I will soon write Baron von Zach about my plan and ask for his opinion.*

[FN: In ancient Rome people trained birds to make complimentary greetings to the emperor. Augustus would often buy the birds for generous sums of money. A poor cobbler acquired a raven, intending to train it to make such a remark. The bird turned out to be such a slow learner that the exasperated cobbler often used to say it, "Nothing to show for all the trouble and expense." One day the raven began to repeat its lesson as Augustus was passing. This time the emperor declined to buy, saying, "I get enough of such compliments at home." The bird, however, also remembering the words of his trainer's usual complaint, went on, "Nothing to show for all the trouble and expense." Amused, Augustus bought the raven (Bartlett's Book of Anecdotes 2000: 30).]

Harding to Gauss Lilienthal February 14, 1804

*I am very grateful for the kindly shared positions of Pallas' orbit which are sufficient for my intentions. I have been thinking about a plan for my project which is still shapeless due to concerns that have to be taken into account. To combine the zodiacs of both planets on the same pages is impossible because of their difference and I am aware of the necessity to separate them according to your suggestion. The difficulty regarding Pallas' zodiac is as follows. Firstly, it takes up so much space that the maps must have 36° to 43° in declin. if I wanted to give each only 30′ RA. This demands a format of Bode's atlas and the degrees can only increase by 1/3. From this you can see that such a format is as disagreeable as the fineness of the degrees. To take a format larger than that of Bode does not seem to be recommendable because it would become too unhandy. There are only three divisions where each sheet must have only an extension of 43°, the rest do not require more than 36° and thus 9 sheets allow a larger size of the single degrees than the first three. But I believe different scales of the pages would not be a good idea. A lot of work and space could be spared if the boundaries of Pallas' zodiac would be parallel to the margin of the pages. But this would give the declinations and parallel circles such an inclined positioning that I would rather omit them entirely if not their handiness advised their keeping. I do know that the handiness is to be preferred to the beautiful even more so since the maps neither shall nor can be works of art. I am planning on sketching a sheet according to both methods that I will show you for examination. Thus it can be easily determined which method is best.*

*Already eight days ago I sent Mr. von Zach the chart of Pallas' orbit. Hopefully he will approve. Whether it matches the sky I cannot say since the weather made any comparison impossible. I missed ten stars of Bode in a small region. Strange, that amongst those are three, of which he observed one himself, and he claims to have borrowed the second from Count Bruhl and the third even from Maskelyne. If his numbers are correct, the stars must have a change in light for I searched for them two evenings with our wonderful comet searcher, which you know, and which distinctly shows stars from the 9$^{th}$ magnitude on, but in vain. The stars are no. 139 and 141 Pegasi according to Bode's catalogue. June will tell if I was mistaken. One and a half years ago I missed for a long time an 8$^{th}$ magnitude star of La Lande in Aquarius.*

Harding to Gauss Lilienthal March 9, 1804

*Please find enclosed, most admirable friend, two draft sheets of my projected small atlas and am asking humbly for your opinion. No. 1 seems to have only few advantages looking at the huge dimension that it has to have plus the smallness of the single degrees, if the format is not to be extended to an utterly unhandy size. Its height cannot be smaller though because Pallas' zodiac has at some points a latitude of 43°. No 2 is not as beautiful with the inclined declination and parallel circles, but by this the format is handier and the single degrees can be made larger. I prefer the latter, due to this fact alone, since I believe that this size for special maps, like these are, is still slightly too small and I would like to have such degrees as those on the British maps of the orbit of Ceres. But I do not want to anticipate your decision but ask for your honest opinion on both since your discretion is to determine my well-being and I will be immune against any mistake. You see, both charts are merely a sketch, not even corrected, and thus have different variants. I only filled in some of the regions richest in stars to be able to see whether the degrees have the right size regarding the amount of stars. This does not seem to be the case since the region they depict is by far not the richest in stars, as you know for yourself, there are regions that are even more crowded with stars. I do not know if I could dare send these charts to Mr. von Zach and ask for his opinion, too. It is my wish, of course, but I am afraid to trouble him. If you believe He will not take it amiss, please send him my charts since you are in correspondence with Him. It would be less hassle, if He sent his opinion in a letter to you that you could kindly communicate some time, than Him writing a letter to me. Mr. von Zach was so kind as to send me some copies of the Ceres map I am enclosing one copy for you and one for Baron von Ende, if I may. You will find some corrections in red. Please show Mr. von Ende the above mentioned two maps as well.*

## Harding to Gauss Lilienthal April 22, 1804

*Soon I will start hunting Ceres and Pallas, but I am afraid the moon might be a hindrance. You will find Ceres immediately by means of my map since all stars to the 9$^{th}$ order of magnitude are mapped. The bright stars in Cetus and Pisces lead you to the location where she will be on April 30. Pallas, on the other hand, is more difficult to find because the region where she is only now, contains a lot of small stars which are not all shown on the map.*

*Dr. Olbers gave me the limits of Pallas' orbit and I am diligently working on it. I am very grateful for your remarks on the draft sheets I sent. I am glad that you also do not take exception to the inclined position of the parallel and declination circles. I plan on adding a list of all depicted stars to the charts which will be taken from the best catalogues available, namely Piazzi, Zach, Maskelyne, Cagnoli,[1] Wollaston, and Mayer. By doing so I hope to make them as useful as possible since these catalogues are not available to everyone.*

## Harding to Gauss Lilienthal May 12, 1804

*I have searched for the two new planets several times, but the twilight makes them completely invisible. I believe to be able to recognize Ceres at first sight but not the smaller Pallas since I examined her region with a telescope showing stars up to 10$^{th}$ order of magnitude only. I have been examining Leo for several weeks now and missed 4 of La Lande's stars of the 4$^{th}$ magnitude. You will find them in* Hist. Cel. Fr., *vol. 1, p. 147. The second among those seems to be suspicious because no star of that region matches its*

[1]Antonio Cagnoli (1743–1816), professor of mathematics at Modena and president of the Italian Society of Sciences, founded in 1782. In a roster of the society in 1802, Piazzi is not listed as a member, but he joined later. Herschel, Banks, and Maskelyne (in England) and Laplace and Lalande (in France) were all foreign members, and Zach joined them after 1802. Oriani and Lagrange were emeritus members (Cagnoli 1802a:xxii–xxvi). For his catalog of 501 stars, see Cagnoli (1802b).

*position. Could it have been Ceres or Pallas? But Ceres cannot move so far towards the equator. I will follow your advice and omit the names of Bode's stars. I have kept them until now since I considered it handy to be able to distinguish the stars immediately and take its position from the catalogue. But, alas! How bad it is! It can only mislead and spoil the best observations. If he would only not take the omission of his numbers amiss! He will not like my project, as little as it can harm the sales of his atlas.*

Harding to Gauss Lilienthal June 3, 1804

*Until to date observations on Pallas have been impeded by moonlight, clouds and twilight, and close trees still prevent to find her before midnight. I therefore was able only one time, last night, to observe her. She formed together with the two stars of La Lande of 8th magnitude RA 338° 5′ Dec 10° 3′*

338 21 9 52

*an almost equilateral Δ, was very faint and was hard to observe in the already hazy air. Dr. Olbers believes to have seen his Pallas already on March 8. I do not know whether he could assure himself of it because I have not talked to him since then. I have just calculated my observations on Pallas and found to my surprise that the RA is no good. The declination seems to be better. Did I make a mistake reducing the star or did the* Hist. Cel. *give the star erroneously? I will look into both possibilities further.*

| |
|---|
| *On June 1 12h 51′ 38″ m.t. Pallas preceded the star of La Lande* |
| *in time = 9″.5 = 2′ 22″.5 of arc* |
| *and was farther north of the star = 17′ 24″.8* |

*La Lande's star was observed according to* Hist. Cel. Fr, *vol. 1 pag. 40 at Paris on November 23, 1793 22h 33′ 31′ ZD = 38° 56′ 17″ from this I calculated*

| | | |
|---|---|---|
| *1804, Jun 1 appar. RA *=338 29′ 42.″5* | | *app. Decl. = 9° 52′48.″0* |
| *Diff. Pallas =* | *–2 22.5* | *+ 17 24.8* |
| *RA Pallas =* | *338 27′ 20.0* | *Decl. = 10 10 12.8* |

Harding to Gauss Lilienthal June 21, 1804

Harding to Gauss Lilienthal June 21, 1804

*Since Dr. Olbers also found in his observations on Pallas a difference of about 5′ in RA, I start having more trust in my communicated observations of June 1, which need improvement. I made an error calculating the reduction of the star used for comparison. Those observations and the following are after deleting the error:*

| | *Mean time* | *Apparent RA* | *Apparent December S* |
|---|---|---|---|
| *June 1* | *12h 51' 44"* | *338° 27' 10"* | *10° 13' 22"* |
| *June 10* | *12 46 11* | *339 24 2* | *10 39 20* |
| *June 12* | *12 33 40* | *339 34 32* | *10 43 41* |
| *June 13* | *12 28 48* | *339 38 47* | *10 45 35* |
| | *12 57 20* | *339 39 18* | *10 45 37* |
| *June 14* | *12 32 24* | *339 43 38* | *10 48 10* |
| *June 16* | *12 9 40* | *339 51 48* | *10 51 8* |

*For the second observation of the 13^th^ the planet was compared with ξ Pegasi whose I derived from Piazzi. I thus consider it more reliable than the first, which is based on a star of La Lande and am inclined to blame the difference in RA of 31° between both on this star, whose RA seems to need further improvement. Dr. Olbers, whom I met recently in the streets of Bremen, told me that my observations did not accord with his. Because he was too busy I did not dare ask him for a copy of his although I am burning to know to what degree they correspond. I still consider observing Pallas harder than at the beginning of last year because the twilight is cumbersome. I truly did not lack endurance and diligence but if they are a failure I am not to blame.*

*For several days the Moon will impede observations on Pallas. I was unable to find Ceres because she only reaches above the horizon limited by trees when twilight does not permit seeing her anymore.*

## Harding to Gauss Lilienthal July 4, 1804

*I much appreciated your mentioning of my small atlas in your treatise, dearest friend. This is the greatest praise it could get, for what you approve of can hardly be criticized. Since you paved the way for my project I will also make an announcement. Prior to this I would have appreciated Mr. v. Zach's opinion and asked for it already five months ago but have not received an answer yet; and therefore do not know whether he approves or disapproves. But maybe he forgot about it due to his duties, travels and the Duke's decease. I do not know whether he would call it an impertinence if repeated my request. You probably have received my observations on Pallas. How I wish they are good. Since I have not received those of Dr. Olbers, I do not know how they correspond. I saw the planet only once on July 1 and compared it to ξ Pegasi whose position I calculated according to Piazzi in app RA=229° 14' 2."23 Dec=11° 10' 48."50, I obtained for Pallas*

*1804, July 1 11h 50' 50" mean time app RA= 340° 22' 2" app Dec=11° 9' 1"*

*The planet was close to his stationary position then and maybe in this respect you appreciate this observation. I consider it quite accurate because four observations corresponded quite well. Too bad, meridian observations are impossible for the time being. As far as I know, Dr. Olbers will leave for Rehburg and I will eagerly pursue his Pallas in the meantime and as soon as weather permitting look for Ceres. Last night I was able to harvest some observations on Pallas and calculated for*

*July 4, 1804 11h 53' 52" mean time RA=340° 20' 47" Dec=11° 6' 38"*

*This is based on ξ Pegasi as I calculated it for July 1. The aberration and nutation of it will not cause a considerable difference since in three days it would surpass the accuracy gained with a circular micrometer. I consider the RA good; but the differences of the declination deviate by 41", the average of three observations gives the difference in decl. =*

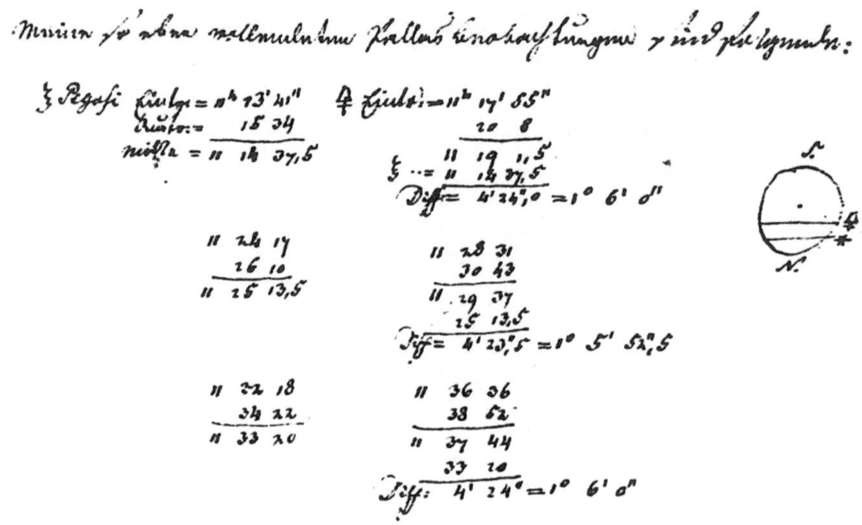

**Fig. 9.6** Excerpt from Harding's letter: postscript July 5, 1804: positions of ξ Pegasi on the *left* and Pallas on the *right*

*4' 10", by what Pallas was further south than ξ Pegasi. The difference in RA was given each time = 4' 27".0 in time = 1° 6' 45" in arc.*

*And again, I could not find Ceres last night. She emerges only in twilight from behind the abhorred trees. But I am hoping to send you also of this planet something new. P.S.* (Fig. 9.6)

*Yesterday the difference in RA was = 4' 27".0 = 1° 6' 45"*

*I assumed the radius of the field = 17' 22" = 138 55*

*The clock was late by 2' 44".*

## Harding to Gauss Lilienthal August 5, 1804

*Finally, on August 2, I had the pleasure of finding and observing Ceres. The skies were clear and I am certain you found the planet, too. I met it right below no. 36 Ceti Flamst, only further south by 2' and slightly more to the East which rendered the observation at the circle difficult. It was of almost the 8$^{th}$ order of magnitude and I recognized it even low on the horizon with a 3f instrument. I did calculate my observations but am enclosing them in the original form in case you would like to calculate for yourself. On August 3, the stars oscillated due to lightning and clouds in the SE that were perhaps close to you, it was muggy and hazy. Ceres had moved southwards and eastwards and followed the star in 11' time, yesterday in 1" so that she had moved within 23 hours about 2' 30" in arc, well, almost as much as she should have according to your ephemeris. From Piazzi's catalogue I calculated for 1804, August 3, RA no. 36 Ceti = 15° 44'*

*45".80 appar. decl. = 7° 49' 6".22 it follows:*

| |
|---|
| 1804 August 2 13h 8′ 40″ mt app. |
| *RA Ceres = 15° 45′ 1″ appar. decl. = 7° 51′ 16″ south* |
| *August 3 12 22 8* |
| *RA Ceres = 15 47 28 appar. Decl. = 7 54 37* |

*According to your ephemeris it should be*

*RA Dec*

| | | | |
|---|---|---|---|
| *Aug 2 RA Ceres = 15° 27′ 40″* | *Dec = 7° 55′ 20″* | *diff +7′ 20″* | *–4′ 4″* |
| *3 15 40 20* | *7 58 40* | *+7 8* | *–4 3* |

*This under the assumption my observations and calculations are correct.*

*The second and third obs. of the inclination of August 3 give a well corresponding product but I consider that of the first better since it also corresponds to the decl. of August 2 very well. I will not tire pursuing Ceres but regarding Pallas I have to take a break since she is now moving into a region where the stars are not well determined. My last observations on this planet are:*

| *appar* | *RA* | *Dec* | |
|---|---|---|---|
| *1804 July 16 10h 52′ 30″* | *339° 51′ 26″* | *10° 44′ 54′* | *compared with ξ Pegasi* |
| *17 11 14 38* | *339 47 8* | *10 41 32* | *compared with ξ Pegasi* |
| *August 2 10 41 28* | *338 3 44* | *9 24 24* | *compared with ζ Pegasi* |
| *3 11 24 49* | *337 53 21* | *9 19 1* | *compared with ζ Pegasi* |

*I would like to thank you for your kind opinion on my last Pallas observations. You can see, that I followed your instructions for circular observations for Ceres. I could hardly apply them on Pallas on August 2 and 3 since ζ Pegasi was away by almost 30′ in decl. All the same, the right ascension does not seem to be bad and do not differ by more than ½″ from each other.*

Harding to Gauss Lilienthal August 8, 1804

*I can only add two observations of August 5 and 6 on Ceres to those recently sent because the other nights were overcast. I was forced to use no. 36 Ceti for comparison again since there was no other star on the planets' parallel. Also this star was not nicely positioned to determine the declination with certainty and omit it entirely regarding the calculation. The right ascension was*

| | |
|---|---|
| *1804 August 5 14h 15′ 0″* | *RA of Ceres = 15° 52′ 16″* |
| *6 12 10 58* | *15 53 48* |

*The difference between your calculated and my observed RA of Ceres remains almost the same as on the 2nd and 3rd. What a pity, the declination could not be further determined. But soon Ceres will be so close to 37 Ceti that I will have both in the field of the telescope. Yesterday, on the 7th, I could only see Ceres low in the horizon. Her light was paler than on the 6th but this might be due to the haziness. Before she could rise any further it became overcast. Before this awful darkness I observed Pallas at the circle, using due to lack of any alternative, no. 49 (σ) Pegasi Flamst. for comparison; it was far from her but had almost the same inclination. I obtained:*

*1804 August 7 11h 5′ 42″ mean time RA of Pallas= 337° 19′ 23.″9 Dec= 8° 48′ 54.″0*

*This gives the difference of the ephemeris = 9′ 24″ in RA just as you calculated.*

*The weather is dismal and I am afraid it might last for weeks. I sent my first observation on Ceres to our dear Dr. Olbers to Rehburg. He was very curious for news on this planet. Without doubt you will bring him further news derived from own observations orally.*

## Harding to Gauss Lilienthal August 19, 1804

*Today I am sending you, dear friend, six further observations on Ceres that I was able to make since my last letter.* [In the *MC* of Nov. 1804 are Harding's positional data on Ceres for Aug. 1804.]

*The observation of the 11th is close to its standstill. On the 9th Ceres was constantly shrouded in mists which impeded the observation incredibly: it took me more than 1 hour to get a good observation. No. 41 Ceti Fl. was the only useful star I could use for comparison. For a determination of the difference in RA it was not sufficient, of course, since the planet was almost 27′ further north. But there were no others and they corresponded well amongst each other so that the RA will hopefully be correct. The declination is very good as are the observations of August 10. On the 11th the air was excellent, Ceres had pure bright light and observing was easy. I consider the one of the 14th best since Ceres and no. 41 had almost the same declination. I let them alternately pass through the middle of the field and very close to the rim and thus obtained the wonderfully corresponding results. The difference of Ceres from the ephemeris is as follows:*

| | | |
|---|---|---|
| *Aug. 9* | *Ceres in RA +7′ 16″* | *in Dec. 3′ 53″* |
| *10* | *6 22* | *4 10* |
| *11* | *7 6* | *4 16* |
| *14* | *7 4* | *3 59* |
| *16* | *6 20* | *4 49* |
| *17* | *7 15* | *2 42* |

*If these differences are correct the observations are less so, since those are so irregular. Have you not observed Ceres yet? I would like to know how you found her position. According to my estimate she does not yet have the 8th order of magnitude and requires a powerful telescope to observe her at the circle. I continue to observe Pallas according to your wish. Here are the latest results:*

| | | |
|---|---|---|
| *Aug 9 11h 0' 45" mean time RA of Pallas=* | *336° 59' 57"* | *Dec = 8° 32' 57"* |
| *11 10 59 26* | *336 39 17* | *8 16 0* |
| *14 10 21 14* | *336 7 47* | *7 49 14* |

*On the 11^th^ and 14^th^ I compared Pallas with 36 Pegasi Fl. because this star was the closest among the well determined stars. Since the moon is now coming close to both asteroids, I have to take a little break from observing. Dr. Olbers wrote me that you arrived in Rehburg on the 12^th^. I am sorry that the weather does not favor your stay, but here it is dismal too.*

*I am very grateful you announced my charts, but in a way that promises more than I may be able to deliver. I regard it as a new obligation to make them as useful and complete as possible. I want to add also tables for aberration and nutation for all stars of Flamsteed. Every day I see the necessity of such tables since one is forced for almost every observation to use different stars for comparison, and it is such a nuisance to always calculate their apparent position anew. La Lande issued such tables again for his latest volume of the* Con. d. Tems *but only for stars of great inclination that can only be used in cometary observations. Also in Zach's tables no stars in the Whale are mentioned and only a few of Pegasus. I do not possess those of Mayer and de Lambre.*

Harding to Lalande Lilienthal November 10, 1804

*I had engaged in an undertaking, in which I had been occupied above a year. Every favourable night I composed a small atlas to represent the zodiac of the two planets recently discovered by Messrs. Piazzi and Olbers. The smallness of these two planets, which, in most of their positions, are only of the eighth or ninth magnitude, requires a perfect knowledge of all the small stars that are in this zodiac. The celestial charts hitherto published are not sufficiently detailed, and cannot convey a knowledge, at the first sight, of those two small planets, because they represent no stars but what are of the seventh or eighth magnitude. This consideration induced me to compose more detailed charts, to contain all the stars down to the ninth and tenth magnitude; an undertaking which would formerly have been immense, but is now greatly facilitated, since you have made astronomers acquainted with fifty thousand stars which are in your* History of the Heavens.

The letter concludes with Harding's account of his discovery of Juno. This section will be printed in the next book in this series.

## References

Bartlett's Book of Anecdotes, (C. Fadiman and A. Bernard, editors). Entry for Augustus, pg. 30. Little Brown & Co., Boston (2000).

Cagnoli, A.: 1802a, *Memoire di matematica e di fisica della Societa Italiana delle scienze* 9. Modena.

Cagnoli, A.: 1802b, Catalogo di stelle boreali. *Memoire di matematica e di fisica della Societa Italiana delle scienze* 10, Modena.

Harari, Y.N.: 2014, *Sapiens: A Brief History of Humankind.* Signal, Toronto.

O'Brian, P.: 1997, *Joseph Banks: A Life*. University of Chicago Press, Chicago.

# Chapter 10
# Herschel's Asteroids

This chapter presents an extensive look at Herschel's first investigation of both Ceres and Pallas, including vitriolic attacks launched against his asteroid research from his fellow Englishmen.

## Herschel's First Paper of 1802: Ceres

William Herschel's (Fig. 10.1) first paper on Ceres was read before the Royal Society on February 18, 1802, but it was never published in the *Philosophical Transactions*, likely because it was quickly superseded by the May 6 paper that covered both Ceres and Pallas, and thus was deemed to be out-of-date information. Here is the text as printed in Dreyer (1912: cix–cx):

> *The discovery of an additional planet of the Solar System by Mr. Piazzi of Palermo, must undoubtedly be highly interesting to all astronomers. Before the elements of its orbit could be well settled, the planet was lost for some time, and when I was upon the lookout for it about the place where by calculation it was likely to be, and where we are now assured it really was, I could perceive no star with any visible disk, whereby I might have distinguished it from the rest. Hence I surmised that it would require fixed instruments to rediscover it; and not being in possession of any, I requested my much esteemed friend Dr. Maskelyne, to give me the earliest notice of its situation, as soon as he should have observed it at Greenwich. Accordingly, the 5$^{th}$ of this month, I received his account of the place where he had seen it early in the morning of the 4$^{th}$, and, by directing my telescopes to the star thus pointed out I obtained the following observations.*
>
> *Feb. 7. 1802. 13h.*
>
> *With a ten-feet reflector and a magnifying power of 600 I viewed the place where I expected the new planet to be, in hopes of distinguishing it from the neighboring stars by its visible disk. Being sufficiently used to direct my telescope to any given part of the heavens, I immediately perceived a star which appeared sufficiently different from another at no great distance, to occasion a surmise that it was the planet. In order to verify my suspicion I put*



C.J. Cunningham, *Studies of Pallas in the Early Nineteenth Century*,
DOI 10.1007/978-3-319-32848-5_10

**Fig. 10.1** A 1785 portrait of William Herschel by Lemuel Francis Abbott (1760-1802). copyright National Portrait Gallery, London

*on a magnifier of 1200, and comparing the supposed planet with the same fixed star, I found a doubt still remaining that there might be a mistake.*

*The 20 feet telescope with a power of 300 and of 600, would not resolve the doubt; but the supposed planet being still too low for very distinct vision, with such high powers, I intended to examine it when in, or near the meridian, as soon as the air should be sufficiently pure.*

*The following days, though cloudy, afforded every now and then an opportunity of ascertaining, by its change of place, that the star I had examined must be the new planet.*

*Feb. 13. 1802 5 o'Clock in the morning.*

*Having long been disappointed by cloudy weather, a favorable change enabled me to see it with great distinctness. When I examined it with a magnifying power of 600, I found, by comparing it alternately many times with the star I had chosen as a standard, that there was a sufficient difference in their appearance, and that a very minute planetary disk might be perceived in the one, which was not to be seen in the other. After having clearly satisfied myself of the planetary nature of the new star, I wished to ascertain its magnitude. The advanced time of the morning, and an apprehension of clouds coming on, would not permit me to apply the lamp and lucid disk micrometers. I therefore had recourse to a comparative view of the Georgian planet, and the newly discovered one, as their situation was such that I could easily change the direction of my telescope from one to the other.*

*When I turned from the new planet to the Georgium Sidus, and compared its diameter with that of the former, I judged it to be apparently from four to six times as large. Immediately after this I directed the telescope again to the new planet, and as the last of luminous objects in succession is apt to make proportionally the strongest impression, its diameter appeared to me now to be nearly one fourth of the diameter of the Georgian planet. On viewing again our known planet, in order to compare it once more with the new one, I estimated its diameter to be not less than 5 or 6 times as large as that of which I was desirous to ascertain the magnitude.*

*Apprehensive of not having soon again so fair an opportunity, I examined the new planet with an attention to its appearance, and found its color is faintly ruddy. Perhaps it appeared rather the more so, on account of my viewing it after the Georgian planet which is of a mild bluish tint.*

*There was no appearance, nor indeed the least suspicion, of any ring surrounding it; its disk, though very minute, being perfectly well defined all round.* [end of Herschel's paper]

## Herschel's Second Paper of 1802: Ceres and Pallas

Table 10.1 summarizes all of Herschel's asteroid observations in 1802. Given here is the complete text of his landmark study of Ceres and Pallas.

## Observations on the Two Lately Discovered Celestial Bodies

*In my early account of the moving star discovered by Mr. Piazzi, I have already shewn that it is of a remarkably small size, deviating much from that of all the primary planets.*

*It was not my intention to rest satisfied with an estimation of the diameter of this curious object, obtained by comparing it with the Georgian planet, and, having now been very successful in the application of the lucid disk micrometer, I shall relate the result of my investigations.*

*But the very interesting discovery of Dr. Olbers having introduced another moving star to our knowledge, I have extended my researches to the magnitude, and physical construction, of that also. Its very particular nature, which, from the observations I shall relate, appears to be rather cometary than planetary, will possibly throw also considerable light upon the circumstances belonging to the other celestial body; and, by that means, enable us to form some judgment of the nature of both the two last-discovered phenomena.*

*As the measures I have taken will oblige me to give a result which must appear extraordinary, it will be highly necessary to be particular in the circumstances of these measures, and to mention the condition and powers of the telescopes that were used to obtain them.*

*Magnitude of the new Stars.*

*April 1, 1802. Having placed a lucid disk at a considerable distance from the eye, but so that I might view it with perfect distinctness, I threw the image of Mr. Piazzi's star, seen in a 7-feet reflector, very near it, in order to have the projected picture of the star and the lucid disk side by side, that I might ascertain their comparative magnitudes. I soon perceived that the length of my garden would not allow me to remove the disk-micrometer, which must be placed at right angles to the telescope, far enough to make it appear no*

**Table 10.1** Herschel's observations of asteroids

| Date | Objects Seen | Instruments & Magnification | Comments |
|---|---|---|---|
| 07 February, 1802 | Ceres | 10 foot: 600 & 1200<br>20 foot: 300 & 600 | Low in sky |
| 13 February | Ceres & Uranus | 10 foot: 600 | Diameters compared and color noted |
| 25 February | Ceres | 20 foot | Satellite search |
| 28 February | Ceres | 20 foot | Satellite search |
| 04 March | Ceres | 20 foot | Satellite search |
| 05 March | Ceres | 20 foot | Satellite search |
| 06 March | Ceres | 20 foot | Satellite search |
| 30 March | Ceres | 20 foot: 300 | Satellite search |
| 31 March | Ceres | 20 foot | Satellite search |
| 01 April | Ceres | 7 foot: 370 | Disk measured |
| 21 April | Ceres & Pallas | 10 foot: 516 | Disk measured; color noted; coma seen around Ceres |
| 22 April | Ceres, Pallas & Jupiter | 10 foot: 516 & 881 | Disks measured; color noted; no coma around Ceres but Pallas appears cometary |
| 28 April | Ceres & Pallas | 10 foot: 516 & 550 | Pallas less than Ceres; coma of Ceres and Pallas noted |
| 30 April | Ceres | | Coma noted |
| 01 May | Ceres & Pallas | 20 foot: 300, 477 | Satellite search: Pallas<br>Coma noted: Ceres & Pallas |
| 02 May | Ceres & Pallas | 10 foot: 516 | Uncommonly calm night for viewing |
| 04 May | Ceres & Pallas | 10 foot: 516<br>20 foot: 477 | Coma of asteroids compared to that of stars |

*larger than the star; and, not having disks of a less diameter prepared, I placed the smallest I had, as far from me as the situation of the star would allow. Then, bringing its image again by the side of the disk, and viewing, at the same time, with one eye the magnified star, while the other eye saw the lucid disk, I perceived that Ceres, which is the name the discoverer has given to the star, was hardly more than one third of the diameter of the disk, and certainly less than one half of it.*

*This being repeated, and always appearing the same, we shall not under-rate the size of the star, by admitting its diameter to have been 45 hundredths of the lucid disk.*

*The power of the telescope, very precisely ascertained, by terrestrial geometrical measures properly reduced to the focus of the mirror on the stars, was 370,42. The distance of the lucid disk from the eye, was 2131 inches; and its diameter 3,4 inches. Hence we compute, that the disk was seen under an angle of 5′ 29″,09; and Ceres, when magnified 370 times, appearing, as we have shewn, 45 hundredths of that magnitude, its real diameter could not exceed 0",40. Had this diameter amounted to as much as was formerly estimated, the power of 370 would have made it appear of 6' 10", which is more than the whole lucid disk.*

*This extraordinary result, raised in me a suspicion, that the power 370 of a 7-feet telescope, and its aperture of 6,3 inches, might not be sufficient to shew the planet's feeble light properly. I therefore adapted my 10-feet instrument to observations with lucid disks; which require a different arrangement of the head of the telescope and finder: I also made some small transparencies, to represent the object I intended to measure.*

*April 21. The night being pretty clear, though perhaps not quite so proper for delicate vision as I could have wished, I directed my 10-feet reflector, with a magnifying power of 516,54, also ascertained by geometrical terrestrial measures reduced to the focus of the instrument on celestial objects, to Mr. Piazzi's star, and compared it with a lucid disk, placed at 1486 inches from the eye, and of 1,4 inch in diameter. I varied the distance of the lucid disk many times; and fixed at last on the above-mentioned one, as the best I could find. There was, however, a haziness about the star, which resembled a faint coma; and this, it may be supposed, must render the measure less satisfactory than it would otherwise have been.*

*From these data we compute, that the disk appeared to the natural eye under an angle of 3' 14",33; while Ceres, when magnified 516 ½ times, was seen by the other eye of an equal magnitude; and that consequently its real diameter, by measurement, was only 0",38.*

*April 22, 11h 38', sidereal time. I used now a more perfect small mirror; the former one having been injured by long continued solar observations. This gave me the apparent diameters of the stars uncommonly well defined; to which, perhaps, the very favorable and undisturbed clearness of the atmosphere might contribute considerably.*

*With a magnifying power of 881,51, properly ascertained, like those which have been mentioned before, I viewed Dr. Olbers's star, and compared it with a lucid disk of 1,4 inch in diameter, placed at 1514 inches from the eye, measured, like the rest of the distances, with long deal rods. The star appeared to me so ill defined, that, ascribing it to the eye-glass, I thought it not advisable to compare the object, as it then appeared, with a well-defined lucid disk. Exchanging the glass for that which gives the telescope a magnifying power of 516 ½, I found Pallas, as the discoverer wishes to have it called, better defined; and saw, when brought together, that it was considerably less in diameter than the lucid disk.*

*In order to produce an equality, I removed the disk to 1942 inches; and still found Pallas considerably less than the disk.*

*Before I changed the distance again, I wished to ascertain whether Ceres or Pallas would appear under the largest angle, especially as the air was now more pure than last night. On comparing the diameter of Ceres with that of the lucid disk, I found it certainly less than the disk. By proper attention, and continued examination, for at least an hour, I judged it to be nearly ¾ of the lucid disk.*

*Then, if we calculate as before, it appears by this observation, in which there is great reason to place confidence, that the angle under which this star appeared, was only 0",22. For, a lucid disk of 1,4 inch diameter, at the distance of 1942 inches, would be seen under an angle of 2' 28",7; three quarters of which are 1' 51",52. This quantity, divided by the power 516,54, gives 0",2159 or, as we have given it abridged, 0",22.*

*13h 7'. I removed the micrometer to the greatest convenient distance, namely, 2136 inches* [178 feet], *and compared Dr. Olbers's star, which, on account of its great altitude, I saw now in high perfection, with the lucid disk. It was, even at this distance, less than the diameter of the disk, in the proportion of 2 to 3.*

*When, by long continued attention, the appearance of Pallas was reduced to its smallest size, I judged it to bear no greater proportion to the diameter of the lucid disk of the micrometer, than as 1 to 2.*

*In consequence of these measures, it appears that the diameter of Pallas, according to the first of them, is 0",17; and according to the last, where the greatest possible distinctness was obtained, only 0",13.*

*If it should appear almost incredible that these curious objects could give so small an image, had they been so much magnified as has been reported, I can say, that curiosity led me to throw the picture of Jupiter, given by the same telescope and magnifying power, on a wall at the distance of 1318 inches* [109.8 feet], *of which it covered a space that measured 12 feet 11 inches. I do not mention this as a measure of Jupiter, for the wall was not perfectly at right angles to the telescope, on which account the projected image would be a little larger than it should have been, nor was I very attentive to other necessary minute circumstances, which would be required for an accurate measure; but we see at once, from the size of this picture, that the power of the telescope exerted itself to the full of what has been stated.*

*As we generally can judge best of comparative magnitudes, when the measures are, as it were, brought home to use; it will not be amiss to reduce them to miles. His, however, cannot be done with great precision, till we are more perfectly acquainted with the elements of the orbits of these stars. But, for our present purpose, it will be sufficiently accurate, if we admit their mean distances from the sun, as the most recent information at present states them; for Ceres 2,6024; and for Pallas 2.8, the geocentric longitudes and north latitudes, at the time of observations, were, for Ceres, about Virgo 20 degrees, 4', 15 degrees 20'; and for Pallas, Virgo 23 degrees 40', 17 degrees 30'. With these data, I have calculated the distances of the stars from the earth at the time of observation, partly by the usual method, and, where the elements were wanting, by a graphical process, which is sufficiently accurate for our purpose. My computed distances were 1,634 for Ceres, and 1,8333 for Pallas; and, by them we find, that the diameter of Ceres, at the mean distance of the earth from the sun, would subtend an angle of 0",35127; and that, consequently, its real diameter is 161,6 miles.*

*It also follows, that Pallas would be seen, at the same distance from the sun, under an angle of 0",3199; and that its real diameter, if the largest measure be taken, is 147 miles; but, if we take the most distinct observation, which gives the smallest measure, the angle under which it would be seen from the sun, will be only 0",2399; and its diameter, no more than 110 1/3 miles.*

Herschel next describes his attempts to find a satellite of Ceres or Pallas. He refers here to Figs. 1, 2, 3 and 4 all of which are given in this book as Fig. 10.4.

*Of Satellites*

*After what has just now been shewn, with regard to the size of these new stars, there can be no great reason to expect that they should have any satellites. The little quantity of matter they contain, would hardly be adequate to the retention of a secondary body; but, as I have made many observations with a view to ascertain this point, it will not be amiss to relate them.*

**Fig. 10.2** Herschel lived in Observatory House in Slough, near Windsor. His telescopes, one of which is shown in Fig. 10.3, were setup outside the house, which was demolished in 1963

**Fig. 10.3** Herschel's 20-foot telescope

*Feb. 25. 20-feet reflector* (Fig. 10.3). *There is no small star near Ceres, that could be supposed to be a satellite.*

*Feb. 28. There is no small star within 3 or 4 min of Ceres, that might be taken for a satellite.*

*March 4. 9h 45', sidereal time. A very small star, south-preceding Ceres, may be a satellite. See Plate V. Fig. 1.* (Figure 10.4) *where C is Ceres, S the supposed satellite, a b c d e f, are delineation stars, c and d are very small. S makes nearly a right angle with them; e is larger then either c or d. There is an extremely faint star f, between e and d.*

**Fig. 10.4** A star chart from Herschel's paper, showing the relative positions of the asteroids and nearby stars

14h 16′ Ceres has left the supposed satellite behind.

*March 5. There are two very small stars, which may be satellites; see Fig. 2.* (Figure 10.4) *where they are marked, 1stS, 2dS. The rest, as before, are delineation stars.*

*March 6. The two supposed satellites of last night remain in their situation, Ceres having left them far behind.*

*10h 16′ There is a very small star, like a satellite, about 75 degrees south-following Ceres. See Fig. 3.* (Figure 10.4) *It is in a line from C to b of last night.*

11h 20′. Ceres has advanced in its orbit; but has left the supposed satellite behind.

*March 30, 9h 35′. A supposed 1st satellite is directly following Ceres: it is extremely faint. A 2d supposed satellite is north-following. See Fig. 4.* (Figure 10.4) *The supposed satellites are so small, that, with a 20-feet telescope, they require a power of 300 to be seen; and the planet should be hidden behind a thick wire, placed a little out of the middle of the field of view, which must be left open to look for the supposed satellites.*

12h 17′. Ceres has changed its place, and left both the supposed satellites behind

*March 31. 9h 20′. There is a very small star, on the north-preceding side of Ceres, which may be a satellite.*

*11h 50′. Ceres has moved forwards in its path; but the supposed satellite remains in its former situation. The nearest star is 20″ of time from Ceres; so that, within a circle of 40″ of time, there certainly is no satellite that can be seen with the space-penetrating power of this instrument.*

*It is evident, that when the motion of a celestial body is so considerable, we need never be long in doubt whether a small star be a satellite belonging to it, since a few hours must decide it.*

*May 1. 12h 51′. I viewed Pallas with the 20-feet reflector, power 300; there was no star within 3′, that could be taken for a satellite.*

*Of the Color of the new Stars*

*Feb. 13. The color of Ceres is ruddy, but not very deep.*

*April 21. Ceres is much more ruddy than Pallas.*

*April 22. Pallas is of a dusky whitish color.*

*Of the Appearances of the new Stars, with regard to a Disk*

*Feb. 7. Ceres, with a magnifying power of 516 ½, shews an ill-defined planetary disk, hardly to be distinguished from the surrounding haziness.*

*Feb. 13. Ceres has a visible disk.*

*April 22. In viewing Pallas, I cannot, with the utmost attention, and under the most favorable present circumstances, perceive any sharp termination which might denote a disk; it is rather what I would call a nucleus.*

*April 28. In the finder, Pallas is less than Ceres. It is also rather less than when I first saw it.*

*Of the Appearances of the new Stars, with regard to an Atmosphere, or Coma*

*April 21. I viewed Ceres for nearly an hour together. There was a haziness about it, resembling a faint coma, which was, however, easily to be distinguished from the body.*

*April 22. I see the disk of Ceres better defined, and smaller, than I did last night. There does not seem to be any coma; and I am inclined to ascribe the appearance of last night to a deception, as I now and then, with long attention, saw it without; at which times, it was always best defined, and smallest.*

*April 28. Ceres is surrounded with a strong haziness. Power 550.*

*With 516 ½, which is a better glass, the breadth of the coma beyond the disk may amount to the extent of a diameter of the disk, which is not very sharply defined. Were the whole coma and star taken together, they would be at least three times as large as my measure of the star. The coma is very dense near the nucleus; but loses itself pretty abruptly on the outside, though a gradual diminution is still very perceptible.*

*April 30. Ceres has a visible, but very small coma about it. This cannot be seen with low powers; as the whole of it together is not large enough, unless much magnified, to make up a visible quantity.*

*May 1. The diameter of the coma of Ceres, is about 5 times as large as the disk, or extends nearly 2 diameters beyond it.*

*13h 19′. 20-feet reflector; power 477. The disk of Ceres is much better defined than that of Pallas. The coma about it is considerable, but not quite so extended as that of Pallas.*

*May 2. 13h 20′. Ceres is better defined than I have generally seen it. Its disk is strongly marked; and, when I see it best, the haziness about it hardly exceeds that of the stars of an equal size.*

*Memorandum. This may be owing to a particular disposition of the atmosphere, which shews all the stars without twinkling, but not quite so bright as they appear at other times. Jupiter likewise has an extremely faint scattered light about it, which extends to nearly 4 or 5 degrees in diameter.*

*April 22. Pallas, with a power of 881 ½, appears to be very ill defined. The glass is not in fault; for, in the day time, I can read with it the smallest letters on a message card, fixed up at a great distance.*

*13h 17′. The appearance of Pallas is cometary; the disk, if it has any, being ill defined. When I see it to the best advantage, it appears like a much compressed, extremely small, but ill defined, planetary nebula.*

*Apr. 28. Pallas is very ill defined; no determined disk can be seen. The coma about it, or rather the coma itself, for no star appears within it, would certainly measure, at first sight, 4 or 5 times as much as it will do after it has been properly kept in view, in order to distinguish between the haziness which surrounds it, and that part which may be called the body.*

*May 1. Pallas has a very ill defined appearance; but the whole coma is compressed into a very small compass.*

*13h 5′. 20-feet reflector; power 477. I see Pallas well, and perceive a very small disk, with a coma of some extent about it, the whole diameter of which may amount to 6 or 7 times that of the disk alone.*

*May 2. 13h 0′. 10-feet reflector. A star of exactly the same size, in the finder, with Pallas, viewed with 516 ½, has a different appearance. In the center of it is a round lucid point, which is not visible in Pallas. The evening is uncommonly calm and beautiful. I see Pallas better defined than I have seen it before. The coma is contracted into a very narrow compass; so that perhaps it is little more than the common aberration of light of every small star. See memorandum to the observation of Ceres, May 2.*

*On the Nature of the new Stars*

*From the account which we have now before us, a very important question will arise, which is, What are these new stars, are they planets, or are they comets? And, before we can enter into a proper examination of the subject, it will be necessary to lay down some definition of the meaning we have hitherto affixed to the term planet. This cannot be difficult, since we have seven patterns to adjust our definition by. I should, for instance, say of planets,*

1. *They are celestial bodies, of a certain very considerable size.*
2. *They move in not very excentric ellipses round the sun.*
3. *The planes of their orbits do not deviate many degrees from the plane of the earth's orbit.*
4. *Their motion is direct.*
5. *They may have satellites, or rings.*
6. *They have an atmosphere of considerable extent, which however bears hardly any sensible proportion to their diameters.*
7. *Their orbits are at certain considerable distances from each other.*

*Now, if we judge of these new stars by our first criterion, which is their size, we certainly cannot class then in the list of planets: for, to conclude from the measures I have taken,*

*Mercury, which is the smallest, if divided, would make up more than 135 thousand such bodies as that of Pallas, in bulk.*

*In the second article, their motion, they agree perhaps sufficiently well.*

*The third, which relates to the situation of their orbits, seems again to point out a considerable difference. The geocentric latitude of Pallas, at present, is not less than between 17 and 18 degrees; and that of Ceres between 15 and 16; whereas, that of the planets does not amount to one half of that quantity. If bodies of this kind were to be admitted into the order of planets, we should be obliged to give up the zodiac; for, by extending it to them, should a few more of these stars be discovered, still farther and farther deviating from the path of the earth, which is not unlikely, we might soon be obliged to convert the whole firmament into zodiac; that is to say, we should have none left.*

*In the fourth article, which points out the direction of the motion, these stars agree with the planets.*

*With regard to the fifth, concerning satellites, it may not be easy to prove a negative; though even that, as far as it can be done, has been shewn. But the retention of a satellite in its orbit, it is well known, requires a proper mass of matter in the central body, which it is evident these stars do not contain.*

*The sixth article seems to exclude these stars from the condition of planets. The small comas which they shew, give them so far the resemblance of comets, that in this respect we should be rather inclined to rank them in that order, did other circumstances permit us to assent to this idea.*

*In the seventh article, they are again unlike planets; for it appears, that their orbits are too near each other to agree with the general harmony that takes place among the rest; perhaps one of them might be brought in, to fill up a seeming vacancy between Mars and Jupiter. There is a certain regularity in the arrangement of planetary orbits, which has been pointed out by a very intelligent astronomer, so long ago as the year 1772; but this, by the admission of the two new stars into the order of planets, would be completely overturned; whereas, if they are of a different species, it may still remain established.*

*As we have now sufficiently shewn that our new stars cannot be called planets, we proceed to compare them also with the other proposed species of celestial bodies, namely, comets. The criteria by which we have hitherto distinguished these from planets, may be enumerated as follows.*

1. *They are celestial bodies, generally of a very small size, though how far this may be limited, is yet unknown.*
2. *They move in very excentric ellipses, or apparently parabolic arches, round the sun.*
3. *The planes of their motion admit of the greatest variety in their situation.*
4. *The direction of their motion also is totally undetermined.*
5. *They have atmospheres of very great extent, which shew themselves in various forms of tails, coma, haziness,* etc.

*On casting our eye over these distinguishing marks, it appears, that in the first point, relating to size, our new stars agree sufficiently well; for the magnitude of comets is not only small, but very unlimited. Mr. Pigott's comet, for instance, of the year 1781, seemed to have some kind of nucleus; though its magnitude was so ill defined, that I probably overrated it much, when, November 22, I guessed it might amount to 3 or 4" in diameter. But, even this, considering its nearness to the earth, proves it to have been very small.*

*That of the year 1783, also discovered by Mr. Pigott, I saw to more advantage, in the meridian, with a 20-feet reflector. It had a small nucleus, which, November 29, was coarsely estimated to be of perhaps 3" diameter. In all my other pretty numerous observations of comets, it is expressly remarked, that they had none that could be seen. Besides, what I have called a nucleus, would still be far from what I now should have measured as a disk; to constitute which, a more determined outline is required.*

*In the second article, their motions differ much from that of comets; for, so far as we have at present an account of the orbits of these new stars, they move in ellipses which are not very excentric.*

*Nor are the situations of the planes of their orbits so much unlike those of the planets, that we should think it necessary to bring them under the third article of comets, which leaves them quite unlimited.*

*In the fourth article, relating to the direction of their motion, these stars agree with planets, rather than with comets.*

*The fifth article, which refers to the atmosphere of comets, seems to point out these stars as belonging to that class; it will, however, on a more particular examination, appear that the difference is far too considerable to allow us to call them comets.*

*The following account of the size of the comas of the smallest comets I have observed, will shew that they are beyond comparison larger than those of our new stars.*

*Nov. 22, 1781. Mr. Pigott's comet had a coma of 5 or 6′ in diameter.*

*Nov. 29, 1783. Another of Mr. Pigott's comets had a coma of 8′ in diameter.*

*Dec. 22, 1788. My sister's comet had a coma of 5 or 6′ is diameter.*

*Jan. 9, 1790. Another of her comets was surrounded by haziness of 5 or 6′ in diameter.*

*Jan. 18, 1790. Mr. Mechain's comet had a coma of 5 or 6′ in diameter.*

*Nov. 7, 1795. My sister's comet had a coma of 5 or 6′ in diameter.*

*Sept. 8, 1799. Mr. Stephen Lee's comet had a coma of not less than 10′ in diameter, and also a small tail of 15′ in length.*

[Note: The comet of 1795 later became known as Encke's comet; see later in this section for a discussion of this comet.]

*From these observations, which give us the dimensions of the comas of the smallest comets that have been observed with good instruments, we conclude, that the comas of these new stars, which at most amount to only a few times the diameter of the bodies to which they belong, bear no resemblance to the comas of comets, which, even when smallest, exceed theirs above a hundred times. Not to mention the extensive atmospheres, and astonishing length of the tails, of some comets that have been observed, to which these new stars have nothing in the least similar.*

*Since, therefore, neither the appellation of planets, nor that of comets, can with any propriety of language be given to these two stars, we ought to distinguish them by a new name, denoting a species of celestial bodies hitherto unknown to us, but which the interesting discoveries of Mr. Piazzi and Dr. Olbers have brought to light.*

*With this intention, therefore, I have endeavored to find out a leading feature in the character of these new stars; and, as planets are distinguished from the fixed stars by their visible change of situation in the zodiac, and comets by their remarkable comas, so the quality in which these objects differ considerably from the two former species, is that they resemble small stars so much as hardly to be distinguished from them, even by very good telescopes. It is owing to this very circumstance, that they have been so long concealed from our view. From this, their asteroidal appearance, if I may use that expression, therefore, I shall take my name, and call them* Asteroids*; reserving to myself, however, the liberty of changing that name, if another, more expressive of their nature, should occur. These bodies will hold a middle rank, between the two species that were known before; so that planets, asteroids, and comets, will in future comprehend all the primary celestial bodies that either remain with, or only occasionally visit, our Solar System.*

*I shall now give a definition of our new astronomical term, which ought to be considerably extensive, that it may not only take in the asteroid Ceres, as well as the asteroid Pallas, but that any other asteroid which may hereafter be discovered, let its motion or situation be whatever it may, shall also be fully delineated by it. This will stand as follows.*

*Asteroids are celestial bodies, which move in orbits either of little or of considerable excentricity round the sun, the plane of which may be inclined to the ecliptic in any angle whatsoever. Their motion may be direct, or retrograde; and they may or may not have considerable atmospheres, very small comas, disks, or nuclei.*

*As I have given a definition which is sufficiently extensive to take in future discoveries, it may be proper to state the reasons we have for expecting that additional asteroids may probably be soon found out. From the appearance of Ceres and Pallas it is evident, that the discovery of asteroids requires a particular method of examining the heavens, which hitherto astronomers have not been in the habit of using. I have already made five reviews of the zodiac, without detecting any of these concealed objects. Had they been less resembling the small stars of the heavens, I must have discovered them. But the method which will now be put in practice, will completely obviate all difficulty arising from the asteroidical appearance of these objects; as their motion, and not their appearance, will in future be the mark to which the attention of observers will be directed.*

*A laudable zeal has induced a set of gentlemen on the Continent, to form an association for the examination of the zodiac. I hope they will extend their attention, by degrees, to every part of the heavens; and that the honorable distinction which is justly due to the successful investigators of nature, will induce many to join in the meritorious pursuit. As the new method of observing the zodiac has already produced such interesting discoveries, we have reason to believe that a number of asteroids may remain concealed; for, how improbable it would be, that if there were but two, they should have been so near together as almost to force themselves to our notice. But a more extended consideration adds to the probability that many of them may soon be discovered. It is well known that the comas and tails of comets gradually increase in their approach to the sun, and contract again when they retire into the distant regions of space. Hence we have reason to expect, that when comets have been a considerable time in retirement, their comas may subside, if not intirely* [sic]*, at least sufficiently to make them assume the resemblance of stars; this is, to become asteroids, in which state we have a good chance to detect them. It is true that comets soon grow so faint, in retiring from their perihelia, that we lose sight of them; but, if their comas, which are generally of great extent, should be compressed into a space so small as the diameters of our two asteroids, we can hardly entertain a doubt but that they would again become visible with good telescopes. Now, should we see a comet in its aphelion, under the conditions here pointed out, and that there are many which may be in such situations, we have the greatest inducements to believe, it would be a favourable circumstance to lead us to a more perfect knowledge of the nature of comets and their orbits; for instance, the comet of the year 1770, which Mr. Lexell has shewn to have moved in an elliptical orbit, such as would make the time of its periodical return only about 5 ½ years; if this should still remain in our system, which is however doubtful, we ought to look for it under the form of an asteroid.*

*If these considerations should be admitted, it might be objected that asteroids were only comets in disguise; but, if we were to allow that comets, asteroids, and even planets, might possibly be the same sort of celestial bodies under different circumstances, the necessary distinction arising from such difference, would fully authorize us to call them by different names.*

*It is to be hoped that time will soon throw a greater light upon this subject; for which reason, it would be premature to add any other remarks, though many extensive views relating to the Solar System might certainly be hinted at.*

*Additional Observations relating to the Appearances of the Asteroids Ceres and Pallas.*

*May 4, 12h 40'. 10-feet reflector; power 516 ½. I compared Ceres with two fixed stars, which, in the finder, appeared to be of very nearly the same magnitude with the asteroid, and found that its coma exceeds their aberration but in a very small degree.*

*12h 50'. 20-feet reflector; power 477. I viewed Ceres, in order to compare its appearance with regard to haziness, aberration, atmosphere, or coma, whatever we may call it, to the same phenomena of the fixed stars; and found that the coma of the asteroid did not much exceed that of the stars.*

*I also found, that even the fixed stars differ considerably in this respect among themselves. The smaller they are, the larger in proportion will the attendant haziness shew itself. A star that is scarcely perceptible, becomes a small nebulosity.*

*10-feet reflector. 13h 10'. I compared the appearance of Pallas with two equal fixed stars; and found that the coma of this asteroid but very little exceeds the aberration of the stars.*

*14h 5', 20-feet reflector. I viewed Pallas; and, with the magnifying power of 477, its disk was visible. The coma of this asteroid is a little stronger than that which fixed stars of the same size generally have.*

[End of Herschel's paper.]

## The Definition of an Asteroid

Table 10.2, which follows, lists the seven criteria Herschel gave in his asteroid papers, and how each applies to planets, comets, Ceres/Pallas, and finally the broader definition of asteroids. Since the concept of criteria is a major element in the critiques of Herschel's paper that follows, a brief digression into what constitutes criteria is in order. Thompson (1971: 30) examines a criteria–symptom distinction that originates with the Austrian philosopher Ludwig Wittgenstein (1889–1951). While some of its ramifications have been questioned by other philosophers, its outline will suffice to address the issue raised by Herschel's use of criteria to distinguish the asteroids. Here the general letter 'A' that stands in for a concept is replaced with 'Asteroid.' The individual *x* can be either Ceres or Pallas.

*Criteria are those tests, operations, features which we ultimately rely on when trying to determine whether a proposition is true or false. In particular we rely on criteria for determining whether a proposition of the form 'The individual x is an Asteroid' is true or not. The criteria of concept Asteroid are types of phenomena whose presence justify a*

**Table 10.2** Herschel's criteria for differentiating asteroids from planets and comets

| HERSCHEL'S CRITERIA | PLANETS | COMETS | CERES/PALLAS | ASTEROID |
|---|---|---|---|---|
| Considerable size (Notes 1, 3, and 5) | yes | no | no | no |
| Small to moderate eccentricity (Notes 3, 4, and 7) | yes | no | yes | not necessarily |
| Low inclination (Notes 2, 3, and 7) | yes | no | no | not necessarily |
| Motion is direct (Note 4) | yes | not necessarily | yes | not necessarily |
| They may possess satellites or rings (Note 8) | yes | no | no | Possibly |
| They have a considerable atmosphere (Note 6) | yes | Yes, in the form of tails, coma and haziness | Yes, in the form of haziness, but no discernible coma or tail | Possibly comas, disks, or nuclei |
| Their orbits are at a considerable distance from each other (Note 2) | yes | yes | no | no |

Note 1: This is the criterion Banks, Olbers, and Regnér believe should be given little weight in distinguishing Ceres and Pallas from planets
Note 2: These are the criteria that Huth says are of no importance in distinguishing Ceres and Pallas from the existing planets
Note 3: These are the criteria that Zach and Regnér believe are of no value in distinguishing Ceres and Pallas from the existing planets
Note 4: These are the criteria Huth says are important in defining a planet
Note 5: This is the criterion Piazzi believes is important in distinguishing Ceres and Pallas from the other planets
Note 6: Bode (1802a) believed that Pallas should be called a comet, even though it did not exhibit a tail or nebulosity. "Maybe the nebulosity around your comet is too thin to be discovered …" he wrote to Olbers on May 4, 1802. "In order not to confuse the astronomical readership I call your moving star a comet because of its path."
Note 7: These are the criteria Laplace believes are insufficient for distinguishing Ceres and Pallas from the planets
Note 8. After 212 years, Herschel's suggestion that asteroids may possess rings was verified when not one but two rings were found orbiting asteroid (10199) Chariklo (Braga-Ribas 2014)

> *judgment that something is an Asteroid. Giving the criteria of Asteroid is thus the same as giving a sufficient condition for being justified in* saying *that an individual is an Asteroid, and is often, though perhaps not invariably, the same as giving necessary conditions for such a judgment.*

As can be seen in the notes that accompany Table 10.2, every criteria Herschel identified was attacked, typically on the grounds of insufficiency. The extension of the above analysis of criteria then comes into play.

*If an indicator of Asteroid is not a criterion, it is a symptom. A symptom of Asteroid may on occasion be used to justify 'An Asteroid is present' but only because it has been empirically established that when the symptom is present, the criterion is also present.*

The critics of Herschel were saying the indicators he listed were not in fact criteria, but merely symptoms or attributes, and they could not be used to justify the presence of an object named 'asteroid' because Herschel had not been able to establish the empirical relationship.

The French physicist and mathematician Pierre Duhem (1861–1916; Fig. 10.5) is also noted for his philosophical work on the indeterminacy of experimental criteria. In related work he views "the set of theoretical statements accepted by scientists *at a given time* as a system that is both theoretical and linguistic". On the theoretical side, it allows the storage of factual information, but it is also "a language in which factual information is stated in neat and compact ways" (Agassi 1973 :187).

In the context of the Solar System, one can think of all the numerical data accumulated by Tycho and others as a database of factual information. Linguistic terms such as planet and comet are then employed to neatly state what the factual information represents. But when too much factual information is stacked, the neatness of the system is disrupted, exactly what happened with the discovery of Ceres and Pallas.

**Fig. 10.5** Pierre Duhem

> *A change of theory then occurs, which is also, of course, a change of the linguistic system (since the theoretical system is the same as the linguistic system). What Duhem claims is that the factual information is now translated from one language into the other: there is no loss of factual information. Only some old verbal formulation of certain information must indeed be rejected, as the result of the rejection of the theories whose terminology is reflected in them.* (Agassi 1973: 187)

This is clearly what began happening in Herschel's 1802 Royal Society paper with his introduction of what was tantamount to a new linguistic system to describe the Solar System. The old verbal formulation was jettisoned without any loss of factual information.

## Critiques of Herschel's Paper on Ceres and Pallas

Here follows a critique of Herschel's paper in the three leading journals of the day, *The Monthly Review, The Monthly Magazine,* and *The Critical Review*. A new magazine of the era, *Annual Review,* or *Register of Literature* (Arthur Aiken, ed.), also published a brief critique that is included here. These are followed by the views of Baron von Zach, and two relatively unknown writers from England who felt constrained to offer their own personal judgments. Brougham's critique in *The Edinburgh Review* was given in an earlier book in this series.

*The Monthly Review*

Herschel's second paper was presented nearly three months after the first (Cunningham 1984). Many journals gave a summary of it, which marked the first full-scale scientific investigation of Ceres and Pallas, but none was more scathing in its commentary than the London publication *The Monthly Review* (1803, vol. 40, 413–414). Like the *Edinburgh Review*, it excoriated Herschel for his choice of the term 'asteroid,' but it went further in criticizing the very nature of the paper itself. With its derisive allusion to a theatrical character, accusation of indolence, and invocation of the world evil, it stands as the most extraordinary attack ever published on Herschel's asteroid studies

> *The surprize of the scientific world on the discovery of two new planets has now subsided: but its curiosity must naturally be awakened by an account of the nature of these stars, from the pen of that astronomer who discovered Georgium Sidus and the Satellites of Saturn. Such remarks are contained in the present paper; which is drawn up in the manner of a journal, in order that every circumstance, however minute, might be noticed in a case so important. Our concern is with results only; and the first result of Dr. H's observations is that the diameter of Ceres (Piazzi's Star) is 161.6 miles, and that of Pallas (Dr. Olbers' Star) 110½ only. After the journal of the investigations by which the diameters of the new planets are determined, follow observations made with the view of ascertaining whether they had comas and satellites; and it appears that the new planets have the former but not the latter. The comas, much less than those which comets have, extend two or three diameters from the disk.*
>
> *Beyond these points, Dr. H's paper seems to have but little interest for the curious inquirer. The remaining part of it is principally occupied with the discussion of this (at best) idle*

*question, 'Are the new stars, planets, or comets?' and, as if the word planet contained an inherent and necessary signification, the Doctor lays down six conditions to which a star must be subject, in order to be intitled [sic] to that appellation. What person, in any degree imbued with astronomical science, having learnt that these new stars agreed with Venus, Mars, Mercury,* etc., *in the great circumstance of revolving round the sun, would have hesitated one moment in calling them planets? The reasons with Dr. H. alleges, to prove that Pallas and Ceres are not planets, are not more philosophical, and they are much less amusing, than those by which Scrub proves the existence of a plot. "Ay, Sir, a plot, a horrid plot – First, it must be a plot because there's a woman in't; secondly, it must be a plot because there's a priest in't; thirdly, it must be a plot, because I don't know what to make on't." (Beaux Strategem.)*

*Having excluded the new stars from the rank of planets, Dr. H. more easily shews that they cannot be called comets: – but, in this age of nomenclatures, are the new stars to be without a name? By no means; Dr. H. gives them the name of Asteroids: and, in order to know what an Asteroid means, we are referred to a definition. Would it not have been more simple to have called them Planets? The peculiar circumstances by which they differ from the other planets might easily have been remembered without a definition; and an evil attached to Dr. H's definition is, that it does not exactly describe the nature of the new stars.*

*We are decidedly adverse to this unnecessary invention of new names. The journal method of Dr. H's paper, also, we do not much approve: we cannot perceive its utility, and the cause of it is at least ambiguous: it might originate from indolence; and it might be suggested by a real and enlightened concern for the interests of science. – It is with reluctance that we animadvert on a philosopher, who has so many and such strong claims on our respect.*

The author of this article mentions Scrub, who is the manservant of Squire Sullen in the 1707 play *The Beaux Stratagem* by George Farquhar (1677–1707). This late Restoration comedy was one of his most popular creations for the stage; the quote comes from Act 4. It cannot be said with a certainty who wrote the *Monthly Review* piece, although it was published at the end of the very long editorship (1749–1803) of Ralph Griffiths.

When the reviewer raises the notion that the word asteroid does not describe the 'new stars' properly, and further attacks it as an evil, his views are consonant with those expressed in the *Encyclopédie*, which used the term logomachies to describe misconceptions that lead to endless technical disputes—essentially "vain arguments about words in which neither party understands the other because their subject is ill defined, or, by its very nature, inexplicable." (Rosenfeld 2001: 18) There was an informed school of thought on such issues in France. How do philosophers deal with complex issues? This was a question was posed by Philippe-Antoine Grouville (1758–1806), secretary of the *Société de Amis de la Révolution de Paris* in 1789, who wrote:

*They begin by defining their words, then carefully, they draw together all their positive conceptions of the given subject in order to compare them with the nature of things, and, finally, they reduce the question to its simple elements only after having considered it in respect to both inviolable principles and necessary proprieties.* (Rosenfeld 2001: 139)

What *The Monthly Review* and other negative responses to the introduction of the word 'asteroid' indicate is that Herschel failed to consider either the principles or the proprieties, or both. In his paper just quoted, one can see he labored to do so, but the widespread impression at the time was that he did not succeed. Since the mid-nineteenth century the word 'asteroid' is used more than any other to describe the small rocky objects of the Solar System, thus transforming all this overwrought angst into interplanetary dust.

*The Monthly Magazine*
In stark contrast to its rival publication just quoted, *The Monthly Magazine* (1802, vol. 14, 117–122) had a cautiously favorable review of Herschel's paper. Its only concern was the diameter measurement of Pallas, compared to that of Schroeter. As part of a long article that included five pages of orbital elements and observational data from across Europe, the magazine had this to say after summarizing Herschel's paper:

> *Observations on the nature of the new planet from the masterly pen of the great Herschel, are intitled (sic) to the most distinguished attention of astronomers; and it is to this accurate and great observer, that we shall most probably be indebted for new and interesting discoveries relative to the nature of those stars. The observations of Dr. Herschel suppose such an extremely small apparent diameter of Pallas, it being, after Professor Seyffer's calculation, only about one-tenth of a second, that it is in itself very wonderful, and does great honor to the acuteness of Dr. Herschel, who seems to be properly calculated for making such extremely minute observations; but, on the other side, we cannot account for a remarkable difference in the measures of Dr. Schroeter, at Lilienthal, from those of Dr. Herschel, the former having found the diameter of Pallas to be four seconds and a half, a difference which is too considerable as to be overlooked.*

It is noteworthy that this English publication mentions Seyffer in connection with the asteroids. In its survey of Ceres and Pallas, the German publication *Goettingische gelehrte Anzeigen* (1802) quoted from the letter Herschel sent to Seyffer on May 22, 1802. To be quite certain its German readership understood Herschel correctly, it inserted some of the original English words such as "asteroidical or star-like appearance."

Brewster (1830: 736) was rather dismissive of the method both astronomers used to determine the asteroid diameters.

> *The results which have been obtained by Dr. Herschel and Mr. Schroeter is measuring the diameters of the new planets with the lucid disc micrometer, are too widely different to inspire us with much confidence in the accuracy of this instrument.*

During the period 1912–2112 the closest approach of Ceres was in 1926 at 1.59 AU while subtending 0.82 arcseconds. Its most distant closest approach of an apparition during that period occurred in 1997 at 2.00 AU and 0.66 arcseconds. During the same period the minimal approach of Pallas will be in 2074 at 1.23 AU while subtending 0.61 arcseconds. Its most distant closest approach of an apparition during that period occurred in 1997 at 2.57 AU and 0.29 arcseconds. Murray (1855: 88) made a calculation in the opposite sense. Suppose you were on Ceres or Pallas. What angle would be subtended by Earth's disc? The answer is 9 seconds.

*The Critical Review*

*We have already announced these two new planets, discovered by MM. Piazzi and Olbers, and have remarked their appropriate situation in a place where there was a considerable chasm in the planetary system, as well as their disproportioned size to the other planets. We, some time since, suggested, that they were perhaps comets brought into less eccentric orbs, by the joint attractions of Jupiter and the sun; and our author's observations seems to confirm the suspicion. From these observations, the diameter of Ceres seems scarcely to exceed 160 miles, and that of Pallas, taking the mean of two measures, about 130 miles. If we compare them with the planets, we shall find that–*

*The Critical Review* here quotes Herschel's paper where he lists the seven criteria and comments on each in turn. The *Review* then begins its disparaging comments on his paper:

*This reasoning is, however, too rigorous. By a similar argument, it might be contended that there should be no more than seven planets, seven colors, &: to which we may add, that the vacant space may be as aptly filled by two smaller bodies as by one larger. Had we found a large planet, three times the united diameter of the two now under our eyes, we should not have contested its title; and we see not, as we shall presently show, that we ought, from any considerations, to combat the claim of either Ceres or Pallas. The other objection is still weaker. If we admit bodies, it is said, of such great geocentric latitudes, we must resign the zodiac. But what power fixed its limits? – the motions of planets, which did not wander beyond it; and now some more eccentric are found, its limits must be, for the same reason, extended. If, however, these bodies be not planets, we may ask, What are they? We know only of three kinds of celestial bodies; planets revolving about the sun, deriving their light from it, with a determined annual parallax, and a diameter subtending a sensible angle; fixed stars shining with a light peculiarly their own, without any parallax, and subtending no sensible angle; and comets, deriving their light from the sun, which they seem to convey in a peculiar form, that of a coma, and a tail projected in a direction opposite to the sun, with a very considerable geocentric latitude – in other words, moving in a plane greatly inclined to that of the earth's orbit. Ceres and Pallas are certainly observed with comae: are they not, therefore, comets?* (p. 17)

The reviewer then quotes Herschel's five distinguishing characteristics of Ceres and Pallas, which gives him a platform to launch his second attack:

*In fact, the smallest coma of a comet exceeds that of Ceres or Pallas above a hundred times; and neither movies in orbs even approaching the eccentricity of a parabola, or is distinguished by a tail. It is also highly probable that the nuclei of comets are very small: they never disturb the planetary motions, though often disturbed by them.*

*Why then are not these bodies planets? We see no reason for any distinction: they revolve round the sun, and are not comets. We must discover another system, before we are allowed to change the appellation. Mr. Herschel would call them asteroids; but he labors for a distinction, which, in the end, will fail him.* (p. 18)

He finally quotes Herschel's definition of the term asteroid, and concludes his critique of the *Philosophical Transaction* paper:

*We shall not extend our article by enlarging on our own original idea, that these bodies may have been comets constrained to revolve within less eccentric orbits; because, in reality, we know little of the nuclei of comets, and have no criterion by which we can measure their density, nor indeed, very correctly, their diameters. The suspicion may remain on record, to be tried by future observations, with little solicitude, in the author, respecting its truth or fallacy.* (p. 19)

The critique has one critical weakness. In the first passage quoted, he believes Ceres and Pallas have comae, and are therefore comets. Unfortunately for his thesis, Ceres and Pallas do not exhibit comae, although Ceres was recently discovered to have localized areas of water vapour (Küppers et al. 2014). His belief that Ceres and Pallas are former comets is not true, although we now know that some asteroids were originally comets, so he glimpsed part of the real situation. If he had not held to the false belief that Ceres and Pallas exhibit comae, he may have been more reticent in criticizing Herschel's idea to discriminate between comets, planets, and asteroids.

*The Annual Review, or Register of Literature*

*The two new planets, Piazzi and Olbers* [The Review uses the names adopted by Lalande in France], *have been examined with glasses of different powers by Dr. Herschel, whence he concludes that the diameter of Piazzi is 161.6 miles, and of Olbers 110 1-3d. He cannot discover any satellite to either, and if there is any it must evidently be very small. The color of Piazzi is ruddy, but not very deep; but still more ruddy than Olbers, which is of a dusky whitish color. The nature of these bodies is examined, and because they move without the zodiac, the fanciful name asteroid is given to them; and they are supposed to be something between a planet and a comet. As it is probable, that there are many bodies of the same kind, that differ only from the other planets in being without the zodiacal limits, (for magnitude is not of any importance) it has been suggested by a gentleman of this country* [Brougham], *that there cannot be any need of a new name, and that the only distinction necessary is to divide the planets into two classes, zodiacal and extra-zodiacal. It is to be remarked that the present zodiacal limits are merely accidental, and formed only on observations previous to the invention of the telescope.* (*The Annual Review* 1803)

*Baron Franz von Zach*

In the July 1802 issue of *The Monthly Correspondence*, Zach (Fig. 10.6) gives a fair and balanced summary of Herschel's paper. He prefaces it by saying:

*Dr. Herschel is of the opinion that these bodies should not be regarded as primary planets if our Solar System. Rather, a new class should be made out of them, in which he not only wants to place Ceres and Pallas, but also all small bodies yet to be discovered which orbit the Sun and which reveal no nebula or tail when viewed with the greatest telescopic magnification.*

After printing a letter from Sir Joseph Banks that disagrees with Herschel (see *Early Investigations of Ceres and the Discovery of Pallas* for this letter), Zach then weighs in with his own negative opinion.

*One sees from this that Dr. Herschel is averse to calling these bodies planets. For him, asteroids constitute a separate class of bodies. It will depend on what astronomers regard as the essential characteristics of planets: a path determined to be near-circular? Or the volume? The latter is a purely relative term, and it would be necessary, for the sake of consistency, to make three classifications out of the nine primary planets, since Mercury, Venus, Mars and Earth are very small compared to Jupiter. If Pallas is a hundred thousand times smaller than Mercury, then again, Mercury is many thousand times smaller than Jupiter.*

*It is possible to raise the objection against Dr. Herschel's above reflections, that if the certain harmonious progression of planetary distances is destroyed through raising the two*

**Fig. 10.6** Baron Franz von Zach. Courtesy of Stiftung Schloss Friedenstein, Gotha

*new bodies to the level of planets, these bodies and asteroids newly classified as special will not be placed here, rather, the same space in the progression will still occur in which the planet belonging in it has not yet been seen. Incidentally, as far as we know, this progression in planetary distances is based on no natural law, it is merely empirical, not proven, drawn from analogous conclusions, and due to this, it is only a possible hypothesis. If we accept Dr. Olbers' hypothesis, that Ceres and Pallas are only rubble of a larger planet belonging in this space, the certain progression will then be saved, and also brings together the rank and magnitude which this planet, in our opinion, should have occupied, although the volumes of these planets, as we all know, develop according to no laws known to us to date; this is why we have no right to allocate size and breadth to it. As remarkable as this well-known harmonious progression indeed is, it is also extremely remarkable that the major axes, and therefore also the sidereal period of both Ceres and Pallas are exactly the same! Inasmuch as it is permitted to philosophize about this harmonious progression and to conclude the existence of a planet from it, it should therefore also be allowed to venture hypotheses of the remarkable paths of these two new bodies.*

*Most astronomers are in agreement that the inclination of the paths cannot be considered here, for the determination of our zodiac is an arbitrary division of ancient astronomy. For us, it is only apparent, because it depends on the greatest apparent geocentric latitude of the planets. There is therefore no available natural law which determines and limits the expanse of our zodiac. Solely the planet with the greatest inclination of its path determines the breadth of the zodiac. The planets which were unknown to us could not indicate the frontiers in the past. For this reason, they were determined though the Moon and Venus, but just as we became acquainted with the smallest inclination of planetary paths through Uranus, Pallas presently reveals itself as having the greatest inclination of these paths.*

> *Incidentally, the dividing up of the zodiac is only arranged, arbitrary and not even necessary, just like Cassini's* Cometen Zodiacus, *which also never occurred.* [See Rost 1718.]
>
> *As far as the smallness of these new planetary bodies is concerned, it is indeed difficult to comprehend how such a small apparent diameter can be measured with certainty. Herschel and Schroeter, both practiced observers, and both equipped with the biggest and best optical tools attempted such measurements. The former found Ceres 18 times, and Pallas 40 times larger than the latter. In order to represent the difficulty of such measurements, Dr. Gauss made the following calculations of the apparent diameters of these bodies from his last elements of the paths. Whether such small diameters can really be measured, let alone see the reflected light of such small surfaces, we leave it to the judgment of others.* [See Chap. 11 in this book for the remainder of the July 1802 paper in the *MC*.]

In a letter of March 2, 1803 to Jan Sniadecki already given in Chap. 9, Zach writes "Mr. Herschel is angry because I fought his asteroids in my journal (and as everybody believes quite successfully)."

*Bartholomew Prescot*

The asteroids feature twice in a book by a Liverpool accountant, Bartholomew Prescot (1803: vii), who refutes the idea of the Copernican system, insisting that Earth is at rest in the center of the Solar System. Writing in Etruria, near Newcastle, the author feels constrained

> *...to offer a few remarks upon certain observations lately presented to the Royal Society, by a celebrated astronomer* [William Herschel], *concerning the planets Ceres and Pallas, lately discovered. I am of opinion, that the philosopher I allude to, does not by any means, consider the discovery as creditable to the Solar System, but rather as making an inroad upon the boasted harmony of it: at any rate, he seems unwilling to allow them to be ranked amongst the Pythagorean world. To all appearance they have all things in common with the rest, and the only degrading circumstance is their diminutive size; but why should that doom them to be thrust out from the rank of planets, and also to be stigmatized by the epithet 'ASTEROID?' Since the astronomers admit that they perform their revolutions round the Sun, the smallness of their appearance can be no solid objection.*
>
> *It is, notwithstanding, a serious charge against the pretended harmony of the Solar System, that primary planets should be discovered belonging to it, whose magnitudes, if repeated five thousand times, would scarcely equal that of our moon, (according to the astronomers) or, if repeated three hundred thousand times, would barely equal the bulk of the earth. The Moon, which the Newtonians term a secondary planet, or a mere satellite, five thousand times larger than one of these primaries! I should not at all be surprised if, (in order to prevent them from bearing witness to the inconsistency of the system, and at the same time to dispose of them to advantage, they were to assert), that such pigmy worlds are mere make-weights to balance the disorders which are said to distract the machine of the universe, by the adventitious effects of planetary attraction.*
>
> *The astronomer I have alluded to, gives the following definition of the new planets; it is so remarkable, that it cannot fail to strike the attention of the man of plain sense, as well as that of the philosopher.* "Asteroids," says he, *"are celestial bodies which move in orbits, either of* little *or of* considerable *excentricity round the sun, the plane of which* may *be inclined to the ecliptic* in any angle whatsoever. *Their motion* may be direct or retrograde*; and they* may, *or* may not, *have considerable atmospheres, very small comas, disks or nuclei." This spectator is determined to allow himself room enough to turn about in; he concludes his observations by remarking, that "many extensive views relating to the Solar System might be hinted at;" yes, "views" not at all creditable to that system. He says "their motion* may be direct *or* retrograde.*" I would observe, that no other motion would at all be suitable to the Solar System. By this observation, however, I would not have it inferred, that*

> *I consider the* direct retrograde, *or* stationary *appearances of planets, any proof of a motion of the earth, because these I believe, have a direct reference or application to the motion of the sun, and I will give my reasons for this belief, confining myself to real appearances.*
>
> [His italics printed here in regular type.]

The author was thus quite derisive of Herschel's criteria, essentially saying it covered nearly every possible contingency and thus had little credibility. He also derides the fact that new worlds are thus disturbing the harmony of the Solar System that had held true for so long.

In a footnote (dated November 19, 1802), we find a truly astounding swipe at the very existence of the newly discovered celestial bodies:

> *Since the commencement of the present century, the discovery of two additional planets has been announced to the public; one of them is named Pallas and the other Ceres. These with the Georgium Sidus, are now incorporated with the Solar System without disturbing in any wise the scales of gravity and attraction which had before been so nicely poised. It may indeed be said, that the two first mentioned are very insignificant worlds; one of them being (according to White's Ephemeris) 140 miles in diameter; and the other about 200; so that upon a rough computation, from these dimensions, the smallest may be said to contain as much surface as Ireland, and the other about the same quantity as Great Britain.*
>
> *The whole quantity of matter contained in both, amounts to, probably, the five hundredth part of the bulk of the moon. Such petty worlds are a disgrace to the Solar System!* (Prescot 1803: 80)

See Chap. 11 in this book for the entry in *White's Ephemeris* he mentions. The best assessment of Prescot is that of Mott (1861: 117). After mentioning Prescot's 2-volume attack on the Newtonian system published in 1822, Mott observes that:

> *He wrote with energy and was a man much respected by his friends; of considerable attainments, and, in some directions, of great ability. His book is one of the curiosities of our local literature. Works of this kind have a peculiar and certainly a melancholy interest. They remind us how easily a man may, in fact, become insane in his relation to any given subject when from any cause, whether ignorance, vanity, natural defect or wilful blindness, he comes to doubt everything rather than himself.*

English writers were still debating the merits of Herschel's criteria into the 1830s. The following example shows that the writer is selectively considering only one of Herschel's criteria—size—to criticize his distinction of the asteroids from planets and comets:

> *Herschel has made repeated attempts to measure the diameters of all, but their extreme smallness seems to render all observations of this sort very uncertain. The same observer considers them to form a class of bodies intermediate between planets and comets; and he has proposed to give them the name of asteroids. But there seems to be no reason why a certain size should be considered as an essential character of planets; it would seem more reasonable to make the difference to consist in the magnitude of the excentricity of the orbit. Of late the discovery of comets revolving within the Solar System, has made it more difficult than ever to distinguish between planets and comets; unless we take for criterion, the degree of ellipticity. This will furnish us with a distinction, at once obvious and easy of application: in the case of the orbit of Juno (the most excentric of the planets) the ratio of the excentricity to the semi-major axis does not exceed 0.26, while for the comet of Encke it is 0.85, and for that of von Biela 0.75.* (Library of Useful Knowledge. Natural Philosophy, 1834: 110–111)

*James Dinwiddie*

James Dinwiddie (1746–1815) visited the Chinese imperial court as official astronomer in Lord Macartney's mission of 1792–1794. When he moved to India in 1795, Dinwiddie was appointed professor of mathematics, natural philosophy, and chemistry at the College of Fort William in Bengal (Proudfoot 1868: 112). He read Herschel's paper about Ceres and Pallas, which prompted Dinwiddie (1805) to write two letters to the *Calcutta Gazette*, a newspaper begun in February 1784 under the avowed patronage of the British government in India. The second letter, given here, represents the only summary and commentary on Herschel's asteroid paper outside of Europe, albeit by a Scotsman. In it, he gives the word 'asteroid' a lukewarm endorsement and pins great hopes on a coordinated European observing program that never got off the ground.

> *SIR, – Be so obliging as to insert in your next* Gazette *the following additional observations on Ceres and Pallas: –*
>
> *Telescopic appearances*
>
> *With a common Achromatic telescope, the magnifying power about thirty, Ceres is just visible. It appears like a fixed star of the eighth or ninth magnitude, with a power even of two hundred; the solid disc is not visible, and the body cannot be distinguished from a very small fixed star. With an excellent Achromatic by Dollond, powers three and four hundred, a disc or nucleus is just discernible, but too small to be measured. The body appears surrounded with a coma or atmosphere, extending, like that of a comet, to a perceptible distance from the solid.*
>
> *It was, however, from Dr. Herschell* [sic] *alone, that the determination of measurements so extremely delicate, could be expected. That distinguished Astronomer, and no less eminent Optician, possesses telescopes, and of his own construction, far superior in power, and still more so in brightness, to those of all other astronomers. From a series of observations with his ten and twenty feet reflectors, and by the assistance of his ingenious lucid disc micrometer, Dr. Herschell was led to the following conclusions, namely, that the diameter of Ceres, at the mean distance from the sun, would subtend an angle of about the third part of a second; and of Pallas, about half that quantity!! From these, and the mean distances, determined more accurately since the Doctor's observations, the diameters in the tables (see my former letter) were calculated. In the measurement of angles so extremely minute, very great accuracy is not to be expected. When future observations, however, shall render some corrections necessary, there is reason to think that those corrections will turn out subtractive, that is, the two bodies will be found to be of less magnitude than they are at present supposed to be.*
>
> *It appears from the tables that the eccentricity of Ceres is less than the tenth part of the mean distance, and consequently that the orbit differs but little from a circle. But the eccentricity of Pallas amounting to nearly a fourth part of the mean distance, the Aphelion will be almost double the Perihelion distance from the sun. Therefore the two orbits intersect each other, that of Pallas sallying between the orbit of Ceres on one side, and falling within it on the other side, to the distance of more than eighty millions of English miles; a very extraordinary phenomenon in astronomy!*
>
> *The inclination of Pallas' orbit is so great, that if this body is to be considered as a planet, the breadth of the Zodiac must be extended to about seventy, instead of sixteen degrees. The breadth, or even the Zodiacal Zone itself is, however, of little use in astronomy.*

*From a careful examination of the phenomena, Dr. Herschell is disposed to consider Ceres and Pallas as belonging to a distinct class of bodies, intermediate between those of planets and comets. And from the great resemblance that bear to small fixed stars, from which indeed they can be distinguished only by the best telescopes, he gives this class the name asteroids. Although this, like most other names, carried no explanation, it may, however, for the sake of distinction, be with propriety adopted.*

*The discovery of these asteroids has given a new interest to astronomy. Observatories are rising rapidly in every part of Europe. Owing to the great improvements lately made in mathematical instrument-making, observatories are now erected on a much smaller scale, and at much less expense than formerly. A coalition has been formed on the Continent. The French, German, and Italian astronomers have entered into an astronomical co-partnership. They have divided the heavens, (at present the Zodiac) in equal shares among them, which they undertake to examine, not, as formerly, by sweeping with an equatorial, but by attending simply to the circumstances of motion, as the heavenly bodies pass the meridian of the Observatory. From this improved mode of observing great expectations are formed. The new celestial firm consists already of upwards of twenty members, and many more will doubtless be added to it; Dr. Herschell speaks of it with rapture, and he will, we may believe, lend his aid to this grand survey. He informs us that he had sweeped or surveyed the zodiac five different times, without seeing the new discovered asteroids, a proof of the imperfection of the old mode of observing, and of how little use an equatorial apparatus, although surmounted with the best telescopes in existence, is in making discoveries, particularly of such small bodies as the asteroids. The instrument may be conveniently used in making continued observations on a body already known, and is also useful in furnishing the young astronomer with materials for his improvement in astronomical calculations; but it is of no use in searching for bodies invisible to the naked eye.*

*When a body is discovered by its motion, and the elements of its orbit ascertained, which may be done by an ordinary transit telescope, application must, at least in the present state of astronomical instruments, be made to Dr. Herschell, to ascertain the magnitude and other particular phenomena of the disc.*

*I am sorry that the want of an Observatory prevents us from contributing to this important survey. Without an Observatory every effort must be fruitless....*

*P.S. – There is at present in this Settlement one of Herschell's ten feet reflectors, by which a satisfactory view might be had of the two asteroids; but the instrument had, unfortunately, lost its two highest astronomical powers before it came into the hands of its present owner, Mr. Hare, Watch-maker* [David Hare, 1775–1842]. *When these shall be replaced, a more delightful view of the heavens will be obtained than has ever been enjoyed in this country. Mr. Hare has also received from London an excellent transit instrument, which I hope soon to put up for him. When this is done, the solution of one very important problem, the true Calcutta time, will be accurately obtained.*

## Two Ladies and a Comet

In a paper by Brewster (1802b) given in full in the *Early Investigations of Ceres and the Discovery of Pallas*, he wrote "Pallas resembles a comet in its motion, in its smallness, in its orbit, and in the inclination of that orbit, we are authorized to rank it among the number of these heavenly bodies." He does not mention eccentricity, but the English natural philosopher Erasmus Darwin (Fig. 10.7) stated a decade

**Fig. 10.7** Erasmus Darwin painting by Joseph Wright. Derby Museum and Art Gallery

after the discovery of Uranus that the prime criteria for distinguishing planets from comets are their eccentricity. While the criteria used by Herschel for distinguishing between the various types of objects is clear, the inherent assumption about the nature of comets is not. Our modern understanding of comets differs markedly from theories current at the time. Here Darwin mentions the comet of 1680, which was also included in a discussion by Baron von Zach in the May 1802 issue of *The Monthly Correspondence* (see Chap. 11 in this book). It was the first comet discovered through a telescope and was used by Newton to verify Kepler's laws. Modern computations suggest it has an orbital period of 9400 years, so a return in 2255 is a bit optimistic.

> *M. de Buffon* [1778] *supposed that the planets were all struck off from the sun's surface by the impact of a large comet, such as approached so near the sun's disk in the year 1680, and is expected to return in 2255. But Mr. Buffon did not recollect that these comets themselves are only planets with more eccentric orbits.* (Darwin 1791: 30)

Darwin himself was inclined to the idea that objects could be ejected from the Sun by sun-quakes or solar volcanoes. Those that came from its equatorial zone became planets, while those ejected from the polar regions would have more eccentric orbits and thus became comets. His informed musings are included in the voluminous notes that accompany his famous scientific poem *The Botanic Garden* which he used "to illumine a scope of subjects with concise couplets supported by imaginative exposition and speculation." (Ryan 2012: 46)

Zach muddied the waters further by *his* speculation that a planet can exhibit a tail. This actually proved prescient, as just such a phenomena was observed in 2013 when the Hubble Space Telescope imaged six tails emanating from a body in the inner asteroid belt designated P/2013 P5 (Jewitt et al. 2013), showing that the distinction between comets and asteroids is not rigid but rather covers a spectrum. Here is Zach (1801):

> *But why can a planet not have a tail as Saturn and Uranus have rings? Do the elements of the comet of 1770 not resemble those of Piazzi's star? I answer this with Lexell's* [Anders Johann Lexell, 1740–1784] *words that Jupiter's influence has totally changed the orbit of this comet. Why was it not seen any earlier and more often? Schröter and Herschel have proven that comets can be visible in a certain period and not in another.*

The comet of 1770 was mentioned several times in *Early Investigations of Ceres and The Discovery of Pallas*. Its role in the asteroid saga continued for decades. In 1826 the hebdomadal English publication *Mechanics' Magazine* printed a letter and a rejoinder regarding the role of the comet of 1770 in imparting an atmosphere to Ceres and Pallas. After recounting orbital work on the comet that had been done by various researchers, the unnamed correspondent to the magazine, then quotes the views of Sir David Brewster. A brief quote from Brewster on this matter was given in *Discovery of the First Asteroid, Ceres*. This more detailed explanation of his hypothesis comes from his entry on comets in the *Edinburgh Encyclopedia*:

> *If the comet of 1770 is lost, this could happen only from its uniting with one of those planets whose orbits it crossed. Now, if such a union took place, two consequences would obviously flow from it. The planet would suffer a sensible derangement in its motions, and its atmosphere would receive a vast accession of that nebulous matter, of which the comets are often wholly composed. Here then, we have two distinct criteria to enable us to ascertain the individual planet by which the comet was attracted. The path of the comet intersects the orbits of Venus, the Earth, Mars, the four new planets, and Jupiter, and therefore it must have united with one of these bodies, or their satellites. Now, since the year 1770, neither Venus, the Earth, Mars, nor Jupiter, have suffered the smallest derangement of this kind, nor have they received any visible addition to their atmospheres. We must, therefore, look to the four new planets for some indication of the presence of a comet, and if they exhibit any phenomena that are unequivocally of this description, we must consider such a coincidence as a strong proof of the theory, or as one of the most wonderful facts in the history of science.*
>
> *Two of the new planets, Ceres and Pallas, exhibit, in the form and position of their orbits, evident marks of some great derangement; but as this may have arisen from that explosive force, by which they seem to have been separated from a larger planet, we are not entitled to regard it as a proof of the present theory. But though we cannot employ our first criterion for or against the theory, the second replies with irresistible force, and we would entreat the particular attention of our readers to this single point. The two planets, Ceres and Pallas, are actually surrounded with atmospheres of an immense size. The atmosphere of Ceres is 675 English miles high, while that of Pallas rises to the height of 468 miles. Now the height of any of these atmospheres is greater than the united heights of the atmospheres of all the other planets, and is above a thousand times higher than it ought to have been, according to the ratio which exists between the globes and the atmospheres of all the other bodies of the system. Astronomers were so forcibly struck with the magnitude of these atmospheres, that a dispute arose whether Ceres and Pallas should be called planets or comets, and the discussion terminated, by giving them the name of asteroids, a class of*

> *bodies which were supposed to partake of the nature both of planets and comets. But to draw this argument still closer upon the subject, let us enquire from what other source these atmospheres could be derived, if they were not imparted by the comet of 1770. If the four new planets are the fragments of a larger body, endowed with an extensive atmosphere, each fragment would obviously carry off a portion of atmosphere proportioned to its magnitude; but two of the fragments, Juno and Vesta, have no atmospheres at all, consequently the atmospheres of Ceres and Pallas could not have been derived from the original planet, but must have been communicated to them at a period posterior to the divergency of the fragments.*
>
> *It would have been a satisfactory addition to the preceding arguments if we had been able to shew, by direct calculation, that Ceres and Pallas were at the same instant with this comet in that part of their orbits which was crossed by its path, and that the position of the planes of the orbits was such as to permit a near approximation. But as we have no data sufficiently correct for such a calculation, we must leave this part of the subject to some future opportunity. There is one fact, however, which in some measure supplies its place, and which is therefore worthy of particular notice. The nodes of the comet of 1770 lie exactly between the nodes of Ceres and Pallas, an arrangement which is absolutely indispensable to the truth of the preceding theory.* (Mechanics' Magazine, 1826a)

A correspondent to the magazine, who signed his letter F. M., took great delight in poking fun at Brewster's ideas. He euphemistically refers to Ceres and Pallas as ladies who had a liaison with a comet, and were thus apparently both expecting a child. Filled with double entendres, it stands as the most risque and amusing text ever written about the asteroids.

> *The comet of 1770 is accused of having paid a visit to the planets Pallas and Ceres, with each of whom he thought proper to leave a portion of his atmosphere, a kind of celestial bouquet, as it were, which these ladies have ever since been seen to wear by way of garland, head dress, or* coma. *Indeed, the tenor of the article, is such as would even induce us to infer that Mr. Comet went so far as actually to divide his very person and substance between them, although it is not actually stated that either of the ladies evinced that increase of personal dimensions, which we might naturally have looked for after such an adventure.*
>
> *The recorder, however, of this bit of scandal assures us, that both Pallas and Ceres, subsequently 'exhibited marks of some great derangement.' This may have been a serious, though not uncommon* accident, *to which all ladies are liable, or we may ascribe the 'derangement' to a storm of jealousy, or even recrimination, for it would really seem, that between the two, they have effectively done for poor Mr. Comet, who has never been heard of since. Such a termination of the affair appears the more remarkable, on referring to the diminutive size both of Pallas and Ceres, neither of them being more than 100 miles broad, and it is said they are as broad as they are long. Mr. Comet must have been the very tiniest of his species, not to have been able to take them both off in his pocket! The old logicians used to say, 'Omne majus continent minus.'* [The greater contains the lesser]. *But, 'nous avons change tout cela,'* [we have changed all that] *and the next discovery will be, that Jonas swallowed the whale! (Mechanics' Magazine,* 1826b)

David Milne (1805–1890; Fig. 10.8) adopted a deadly serious tone to demolish Brewster's contentions. Milne's essay gained the first prize of Dr. Fellowes's prizes, proposed to those who attended the University of Edinburgh "within the last twelve years." Milne (afterwards Milne-Home) was a Fellow of the Royal Society of Edinburgh and subsequently its vice president; he was also personally acquainted

**Fig. 10.8** David Milne

with Brewster in later life, as we learn from a biography written by his daughter (Home 1891). According to a very favorable review in *The London Literary Gazette* (1828), the book was timely.

> *At the present time, when considerable excitation has been wrought in the public mind by the fancied appearance of one comet, and the expectation of another,–when a feverish anxiety and terror has pervaded many classes,– this* Essay on Comets *will be read with interest and curiosity.*

It is important to keep this state of affairs in mind when reading about the relationship between comets and asteroids. Even a full generation after the discovery of the asteroids, and long after the enlightenment, large segments of the populace even in the most advanced countries were awash in superstition about celestial events. Here is an extract from the *Essay* by Milne (1828:117–120):

> *With regard to any derangement in the planetary motions caused by the collision of a comet, I must, in the first place, take notice of a Theory proposed some years ago by Dr Brewster, which attempted to account for two phenomena, that in some respects appeared to form anomalous facts in the planetary system,* viz. *The total disappearance of the Comet of 1770, and, more especially, the prodigious size of the atmospheres of Ceres and Pallas. We have already observed, that, if nothing had occurred to derange the orbit of this Comet, whose period was only 52 years, it ought, since it was last seen, to have reached its*

*perihelion ten times. From this circumstance, Dr Brewster thinks that 'we are therefore entitled to conclude that the Comet of 1770 is lost; which,' says he, 'could happen only from its uniting with one of the planets whose orbit it crossed.' Let us attend to the mode of reasoning which he employs to establish this position.* [Milne then quotes the first portion of Brewster's text just given] *Dr Brewster then comes to apply his second criterion,* viz. *the height of their atmospheres; which are found to be much more considerable than those of the other planets: and out of this single circumstance, he puts forth an inference, which can scarcely be deemed a legitimate one, that there must necessarily have been an addition of nebulous matter to the quantity originally possessed; and he thinks that this addition can have been derived only from the 'lost' Comet of 1770. 'If,' says he, 'the new planets are the fragments of a larger body endowed with an extensive atmosphere, each fragment would obviously carry off a portion of atmosphere proportional to its magnitude.' Let us therefore see what is the bulk and extent of atmosphere which each of these planets has been found to possess. If it appear that the quantity of atmosphere is in exact proportion to the planet's bulk, then the coincidence may undoubtedly be looked on as a fact strongly corroborative of Olbers' theory, that they are the fragments of a larger planetary body, which possessed an extensive atmosphere: but in order to give any plausibility to the notion of Dr Brewster, that two of the fragments since this explosion have received an accession of nebulous matter from the Comet of 1770, it is manifestly necessary, that, in these fragments, as compared with the other two, there shall be a great disproportion between the size of the planet and the extent of atmosphere. What, then, are the measurements of both of these, as deduced by the best observers? According to Herschel, the diameter of Ceres is 163 miles, and the diameter of Pallas, 80 miles (Edinb. Encyclop. (edit. by BREWSTER), Astron. 639.); of the other two, Vesta is the smallest, having a diameter of only 49 miles. Now, what has been ascertained as to their atmospheres? Schroeter makes the atmosphere of Ceres 675 miles in height, and the atmosphere of Pallas 468 miles; each nearly corresponding with the bulk of the fragment. With respect to Juno, which is the next less in size: 'Though there is nebulous appearance around the planet Juno,' says Dr Brewster, 'yet it appears, from the observations made by Schroeter, that it must have an atmosphere more dense than that of any of the old planets of the system.' And with regard to Vesta, which is so very small a body, if it has any atmosphere at all, it is of too limited extent to be yet observable. Thus, then, we perceive, that the only condition which could afford any degree of plausibility to Dr Brewster's theory, is directly contradicted by facts. If it had appeared that Vesta or Juno, the two smallest fragments, were possessed of atmospheres much larger than those belonging to Pallas or Ceres, then there would have been a shadow of presumption thrown upon Dr Brewster's theory, because this disproportion would have indicated some accession of nebulous matter since the period of the explosion; but seeing that the extent of atmosphere belonging to each planet bears a fair proportion to the size of its solid part, the only ground on which his theory can rest is entirely destroyed. Even if the few observations which have yet been made of those telescopic bodies, could warrant any inference less contradictory of the theory, it would be more consistent in those who imagine that they are fragments of a larger planet 'endowed with an extensive atmosphere,' to ascribe the quantity of atmosphere now surrounding each rather to the manner of the rupture itself, than to the subsequent acquirement of nebulous matter by the collision of a comet. But at all events, it is not requisite to have recourse to any such hypothesis, in itself extremely fanciful, to account for the disappearance of the Comet of 1770; that Comet is not lost, as Dr Brewster imagines. The result of the most profound and unimpeachable investigations, has proved, beyond a doubt, that its elements have only undergone such an alteration, through the disturbing influence of Jupiter, as to render the Comet now no longer discernible from the Earth; and this explanation has been deemed so adequate by philosophers, that it is recorded in the annals of human knowledge as one of the noblest efforts which Astronomy has achieved, in unraveling the mysteries of Nature* [Mécan. celeste, Preface 28, & iv 223.].

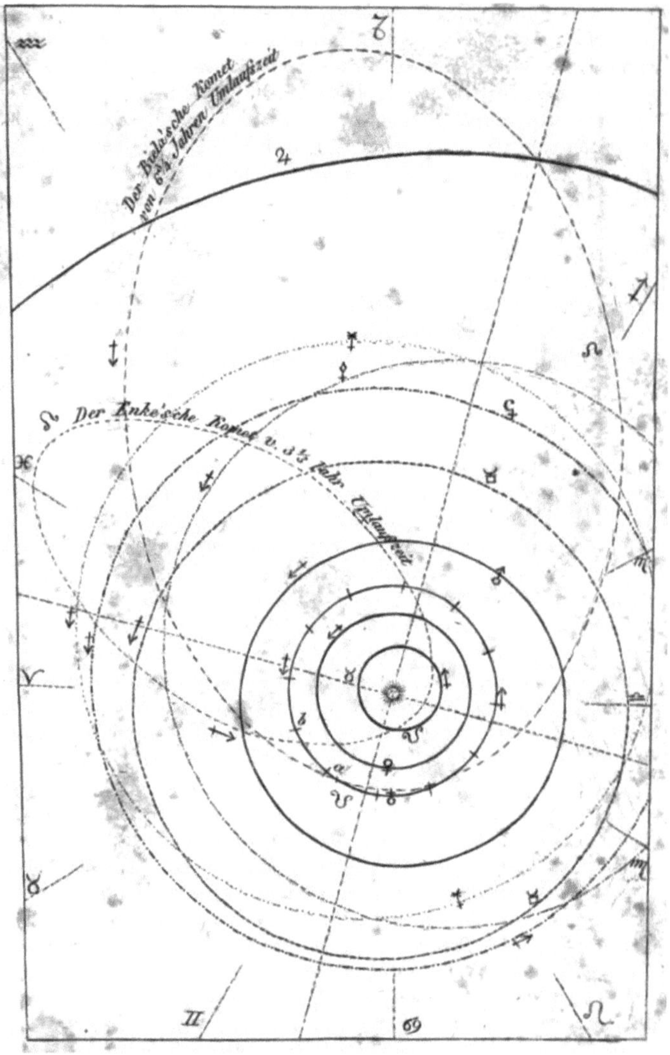


**Fig. 10.9** The path of Encke's comet is overlaid on the orbits of the asteroids, which are denoted by their symbols. From August Gelpke's 1834 book Neue Ansicht über den merkwürdigen Naturbau der Kometen

The fine line between comet and planet was noted in 1835 in a discussion of Encke's comet (Fig. 10.9), which had actually been observed by Caroline Herschel (1750–1848) in 1795.

*The path of this comet is almost in the plane of the ecliptic, and is twice as long as it is broad: one extremity reaches a little beyond the orbit of Pallas…The smallness of its orbit, and the shortness of its period, remove it but little from the nature of a planet.* (*The Quarterly Review*, 1835: 225)

Bode delivered an address to the Berlin Academy about the relationship between planets and comets. The text of this important paper, published in 1806, is in Chap. 11 of this book. The controversial link between asteroids and comets, after all four asteroids had been discovered, will be considered further in a future volume in this series.

## Thomas Kuhn's Paradigm Shift

According to the American philosopher of science, Thomas Kuhn (1922–1996; Fig. 10.10), Herschel's discovery of Uranus in 1781 set the stage for the discovery of the asteroids by enabling a 'shift of vision.'

*A celestial body that had been observed off and on for almost a century was seen differently after 1781 because, like an anomalous playing card (a black four of hearts, for example), it could no longer be fitted to the perceptual categories (star or comet) provided by the*

**Fig. 10.10** Thomas Kuhn, photo by Bill Pierce. Time/Life Pictures; Getty Images

> *paradigm that had previously prevailed...The shift of vision that enabled astronomers to see Uranus, the planet, does, not, however, seem to have affected only the perception of that previously observed object. Its consequences were more far-reaching. Probably, though the evidence is equivocal, the minor paradigm change forced by Herschel helped to prepare astronomers for the rapid discovery, after 1801, of the numerous minor planets or asteroids. Because of their small size, these did not display the anomalous magnification that had alerted Herschel.* (Kuhn 1962:116)

Kuhn alludes to fact that Uranus had been observed at least 22 times between 1690 and 1781 by several astronomers, including Tobias Mayer, none of whom recognized its planetary nature. According to Bourtembourg (2013) it was first seen by Hipparchus in 128 BCE. Gregory Bamford (1989:132) correctly takes exception to the analysis by Kuhn, showing it is based on an unsound premise:

> *No Kuhnian shift of vision was involved in their discovery. Unlike a black four of hearts which one can mistake for a (conventional) four of hearts before noticing the difference between the two, asteroids were virtually indistinguishable from stars for those who discovered them. What such astronomers* saw *was a different pattern of stars, or star-like bodies, at different times, and they* inferred *that this difference was due to the apparent motion of one (or more) of these bodies.*

Kuhn further claims that "astronomers *prepared* to find additional planets were able, *with standard instruments*, to identify twenty [thirteen] of them in the first fifty years of the nineteenth century." Bamford reasons that, in the context of Kuhn's discussion about changes in scientific perception,

> *...this claim is highly misleading. It suggests that a significant, and perhaps the major or only, impediment to their discovery has been merely that earlier astronomers were not primed to notice them. What was principally required, however, was a* practice *that was designed for, or that was at least conducive to, finding such objects. As Herschel himself remarked in 1802, 'From the appearance of Ceres and Pallas it is evident, that the discovery of asteroids requires a particular method of examining the heavens, which hitherto astronomers have not been in the habit of using.'*

Here Bamford is quoting from Herschel's paper about Ceres and Pallas given earlier in this chapter, allowing him to conclusively refute Kuhn's claim.

## References

Agassi, J.: 1973, When Should we Ignore Evidence in Favour of a Hypothesis? *Ratio* 15(2), 183-205.

Bode, J.: 1802a. Letter to Olbers, 4 May. Bremen University Archives.

Bourtembourg, R.: 2013, Was Uranus Observed by Hipparchus? *JHA* 44, 377-387.

Braga-Ribas, F., et al.: 2014, A ring system detected around the Centaur (10199) *Nature,* online edition of 26 March.

Brewster, D.: 1802b, Further notices respecting the two NEW PLANETS, with some remarks tending to shew that they cannot both belong to the Planetary System. *The Edinburgh Magazine* 19, New Series, June, 445-448.

Brewster, D.: 1830, *The Edinburgh Encyclopedia.* William Blackwood, Edinburgh.

Buffon, Comte de: 1778, *Histoire naturelle, générale et particulière contenant les époques de la nature*. De l'Imprimerie, Paris.

Cunningham, C. J.: 1984, William Herschel and the First Two Asteroids. *Minor Planet Bulletin*, 11, 3.

Darwin, E.: 1791, *The Botanic Garden, A Poem in Two Parts*, Part 1. Note 15. J. Johnson, London.

Dinwiddie, J.: 1805, Telescopic Appearances. *Calcutta Gazette*, 23 May issue.

*Goettingische gelehrte Anzeigen*, no. 115, 19 July issue, pg. 1145-1149 (1802).

Home, G.M.: 1891, *Biographical Sketch of David-Milne Home*. David Douglas, Edinburgh.

Jewitt, D., Agarwal, J., Weaver, H., Mutchler, M., and Larson, S., 2013. The extraordinary multi-tailed Main-Belt Comet P/2013 P5. *Astrophysical Journal Letters,* 17 November issue.

Kuhn, T.: 1962, *The Structure of Scientific Revolutions.* University of Chicago Press, Chicago.

Küppers, M., O'Rourke, L., Bockelée-Morvan, D., Zakharov, V., Lee, S., von Allmen, P., Carry, B., Teyssier, D., Marston, A., Müller, T., Crovisier, J., M. Barucci & Moreno, R.: 2014, Localized sources of water vapour on the dwarf planet (1) Ceres. *Nature* 505, 525–527. (23 January 2014). doi:10.1038/nature12918.

Milne, D.: 1828, *Essay on Comets*. Adam Black, Edinburgh.

Mott, A., 1861, On books published in Liverpool. Paper read 10th November 1859. *Transactions of the Historic Society of Lancashire and Cheshire*, new series, 1, 103-120.

Murray, H.: 1855, *The Encyclopaedia of Geography*, vol. 1. Blanchard and Lea, Philadelphia.

Prescot, B.: 1803. *A Defence of the Divine System of the World.* Printed for the author by J. Lang, Liverpool.

Proudfoot, W.J.: 1868, *Biographical Memoir of James Dinwiddie*. E. Howell, Liverpool.

Rosenfeld, S.: 2001, *A Revolution in Language: The Problem of Signs in Late Eighteenth-Century France.* Stanford University Press, Stanford.

Rost, J.L.: 1718, *Astronomisches Handbuch; Das ben Observirung der Cometen*, pg. 502-510. Peter Conrad Monath, Nuernberg.

Ryan, J.C.: 2012, A Poetry of Science or a Science of Poetry?: The Speculative Method of Erasmus Darwin. *The International Journal of Literary Humanities* 10, 45-57.

*The Annual Review*, 1, 879 (1803).

*The London Literary Gazette*, no. 616, November 8 issue, 705-707 (1828).

The Quarterly Review, *Astronomy-The Comet*. December issue, vol. 55, 195-223 (1835).

Thompson, J.: 1971, About Criteria. *Ratio* 13, 30-43.

Zach, F.X., 1801. Letter to Barnaba Oriani, 29 May. Brera Observatory Archives.

# Chapter 11
# Scientific Papers

This chapter begins with the account by Wilhelm Olbers (Fig. 11.1) of his discovery of Pallas, as related to Baron Franz von Zach, and Zach's first observations of the new object. Subsequent observations up to 1834 are covered.

**About a very peculiar new Comet discovered by Dr. Olbers in Bremen**
*MC,* May 1802, pp. 481–501

*On March 31 we got a letter from our esteemed friend Dr. Olbers from Bremen telling us that he had been so lucky as to discover on March 28 a moving star in the northern wing of Virgo, which resembled Ceres in brightness and appearance and cannot be distinguished from a 7$^{th}$ magnitude fixed star, without any nebulosity, retrograde like Ceres but with increasing declination north.*

*Here is the odd story of this discovery. When Dr. Olbers on March 28 was observing Ceres as usual and compared it to Comae Berenices, he mustered the small stars in the northern wing of Virgo with his comet searcher in order to become even better acquainted with them and to find Ceres more easily. Suddenly he looked at 20 Virgo and to his amazement he saw a star of the 7$^{th}$ order of magnitude, which almost formed with 20 Virgo and 191 Virgo of Bode's catalogue an equilateral triangle. Since Dr. Olbers, as our readers know, found Ceres on January 1 at exactly the same place he knew this region quite well and thus knew for certain that in January and February no 7$^{th}$ magnitude star was visible. His first thought was this might be a variable star like Mira Ceti, which has now its greatest luminosity. He immediately compared the stranger to 20 Virgo and continued his observations from 8:45 to 11 o'clock, when it became overcast. The following observations gave the right ascensions smaller and the declinations greater but the differences could not only be blamed on observational errors alone. That very evening he was already almost convinced of the motion of the new star.*

*The following day, March 29, fortunately the weather was fine and Dr. Olbers found that his star had moved considerably. It had moved more than 10 min in right ascension and 20 min in declination. Dr. Olbers compared it once again as accurately as possible with 20 Virgo. On March 30 the weather was favourable again. The new star was already too far north to compare it to 20 Virgo; he determined its location through two stars of La Lande's catalogue* (Conn. d. t. X., *no. 673 and 674*).

*Reassured, Dr. Olbers was so kind as to inform us immediately on March 31 about this peculiar discovery. He sent us his three positions of this new star of March 28, 29 and 30 and wrote: "I would like to ask you, dearest friend, to look for my star as soon as possible, since I need very soon some good observations of it since it will be visible in the meridian*



C.J. Cunningham, *Studies of Pallas in the Early Nineteenth Century*,
DOI 10.1007/978-3-319-32848-5_11

**Fig. 11.1** Wilhelm Olbers, Universitäts-Archiv Rostock

*for only at most four weeks. My observations will not suffice to determine its orbit accurately enough. Of course I am very impatient to hear soon whether you found and observed this new guest."*

*I got this news on Sunday, April 4 at 11 am in the presence of Prof. Buerg, and Prof. Pasquich* [Janos Pasquich, 1754–1829, director of the Buda Observatory]. *Barely twelve hours had passed when the three of us had already at 11 pm paid our respects to this new heavenly body. We did not only find it immediately but even made a very accurate and complete astronomical meridian observation of it.*

*The following day, Monday April 5, they were again favored by the sky; this was the day on which we made sure of this wanderer's own motion. After having finished the observation it showed indeed that this stranger had moved 9′ 43″ eastwards and 18′ 31″ northwards since yesterday. Dr. Olbers' findings were completely verified, only those motions were slightly decreasing.*

*On the 6$^{th}$ the weather was bad but on the 7$^{th}$ we were lucky again and observed the wanderer completely in the meridian; it had moved daily 9′ 23″ in right ascension and 17′ 50″ in declination.*

*On the 8$^{th}$ the sky was overcast. The English sailor leaves port in head wind in order to search the good and the tireless astronomer observes in bad weather in order to find the good. Ceres was invisible this evening but 15 min after its culmination I saw Olbers' star in my passage instrument and observed its RA very accurately through a thin layer of clouds, but it was invisible to Prof. Buerg using a quadrant. And again – in the last 24 hours our new guest had regularly moved 9 min eastwards and 17 min. northwards.*

*Dr. Olbers considers Ceres slightly less bright than his new star. But the latter appeared to have a higher luminosity than the star 191 Virgo of Bode's catalogue. Mr. Schroeter, who found this star on March 30 according to Dr. Olbers' information, and observed it at 288× magnification at the 13-foot telescope found it slightly greater and more clearly bounded than Ceres, the diameter = 4".635 and Ceres' diameter was measured on March 28 only 4".031. This new celestial body appeared to Prof. Pasquich, Buerg and me slightly less bright than Ceres but this estimation became shaky during the next days. But how much do these arbitrary estimations depend on the condition of our atmosphere? I for my part would classify this celestial body to be of the $8^{th}$ order rather than the $7^{th}$ order of magnitude. We could not find any trace of a tail, nebulosity or atmosphere. Should this body be a comet, and it appears so, it is of a new unseen kind. For until today all known comets had a nebulous shape and appearance; this one appears clear and pure and in no way different than a $7^{th}$ or $8^{th}$ magnitude star. Moreover, it appears even brighter and clearer than the neighboring darker planet* Ceres Ferdinandea.

*Dr. Olbers asks: "What is this new star? A strange comet? The entire appearance of this star contradicts this opinion. Is it a planet? What great and paradoxical inclination its orbit must have! I do not dare to judge it yet. What a weird coincidence that I discovered this peculiar celestial body almost at the very place (only 26 min. further north) where I rediscovered Ceres on January 1."* [Lengthy passage here about comets, quoting the definition given by Pingré, with extensive quotes from Hevelius.]

*Dr. Olbers sent us on April 4 three new positions of this star of April 1, 2 and 3. He wrote: "I cannot rest until I know that my small star has been found and observed by you, especially since it is moving so regularly."*

*We had already observed this star on April 4, 5, 7 and 8 when suddenly the fine spring weather turned into bitter frost and a second winter seemed to announce itself by gales and snowstorms. On April 12 and 13 however we observed Ceres between running clouds; but we could see Olbers' star only on the $15^{th}$ and its motion seemed to be already irregularly decreasing.*

*On April 10 we received Dr. Olbers' continued observations of this star, which can be found further below together with ours. He told us that he tried to derive circular elements but learned that those observations cannot be represented by any circle. This body's motion is considerably faster than it is possible to represent in any circle and consequently its orbit comes close to a parabola; but it still can be an ellipse. The strong motion in latitude does already exclude the circle. Dr. Olbers wrote: "This small mysterious heavenly body remains peculiar. As far as I know such a comet has never been seen. It can hardly be distinguished in the telescope from Ceres or a fixed star. Schroeter still states its diameter 4½ sec and with partly clearer edges than Ceres."*

*The question is what shall we think of this extraordinary celestial body, which stands out not only by shape but its uniform motion and which could be named by Dr. Olbers Pallas in order to distinguish it from Ceres.*

*The best thing would be to await a sufficient number of excellent observations and to try then a preliminary calculation of a parabolic orbit which would show whether an ellipse is necessary. A series of* very accurate *observations is even more necessary if we want to find this strange body during this year and its opposition again. But very accurate observation in the meridian field will very soon be impossible due to the twilight. On the sky chart enclosed in this issue, where Ceres' orbit appears in a kind of epicycle we also mapped the apparent orbit of Pallas, which is a straight line as it appears geocentrically.*

*Meanwhile some reveries might be allowed; they were granted in the case of Ceres and behold! they have come true. One of Prof. Wurm's first thoughts regarding Ceres was the following question: "What is to become of this celestial body? A planeto-comet or a cross between both? (i.e., a 'middle-thing')" And indeed, we cannot see why it should not be allowed to consider all comets planets instead of classifying them. There might be genera and species; we have main and minor planets. Even our forefathers classified the comets and Pliny counted twelve* (F. N. by Zach: Natural History, Lib. II n. 2 Cap XXV). *Of course this classification was arbitrary and not based on legitimate grounds but depended on shape, color and direction of the comet, which could be infinite. Based on nothing better were the classifications of Fortunatus Licetus, Hevelius and the* Zodiacus cometicus *by Dom. Cassini. Struyk's classifications seemed to be more physical.* [see F.N. 1 at end of this paper] *The first was based on the perihelion distance of the comet and accordingly he divided the comets into 15 classes. The second was based on the orbits' inclinations and thus there were nine classes. But even Pingré notices that these classifications are based on weak grounds and are arbitrary and justly says that just as well there could be subcategories according to the location of the perihelion or the ascending node. He even suggests the distinction between retrograde and direct comets but discards this idea immediately and remarks:* [French] *"All these divisions are arbitrary, they teach us nothing and are absolutely strange to the nature and properties of comets." How could it be that nobody hit upon the idea to divide them by eccentricities? Why should there not be something – as Wurm calls it – like a planeto-comet? Is the comet of 1531, 1607, 1682, 1759, which will show itself again in 1835 not such a middle-thing? Should the comet of 1770 not be counted among them? Where do the comets of 1680, 1729* [see F.N. 2 at end of this paper], *1742, 1744, 1763, 1773 and 1779 belong to? Should we not distinguish between* elliptical *and* parabolic *and* hyperbolic *comets?* [see Fig. 5.1] *Use the scales* [i.e., weigh the arguments carefully] – *maybe the Olbers' heavenly body brings its shape and motion to bear and becomes a new and thus strangest star in space. What a coincidence, what great luck did not the astronomers have with Ceres' rediscovery? If Olbers' moving star had been discovered by any other astronomer, would he not have considered it the Piazzian star? Would not every astronomer and even Piazzi himself have taken it for the missed one? What confusion would it have caused astronomers if we had started to calculate the orbit, and tried to bring Piazzi's and the new observations together? Gauss' ellipse would have been condemned at once and the search for the true Ceres would have been abandoned. We had worked on calculations and observations of Olbers' Pallas until Ceres would have been lost – and maybe for ever – in the sun's rays. For where should we have searched for it? And who would have given us a hint if we had missed it in 1802? What a triumph for the antagonists of the opinion that there is a planet between Jupiter and Mars? What humiliation for those who believed in Ceres. What annoyance for Piazzi himself, and what a nice field for hypotheses? Certainly, we would have heard enough of those, and as the French say, 'à perte de vuë'* [as far as the eye can see]. *But fortunately exactly the contrary happened. Fortune favors the bold* [tr. from Latin]. *Those who believed boldly in the existence of another main planet had to find it and the new wandering star as well – so no confusion was possible. All hypotheses no longer apply and we are spared.*

*What is to become of Olbers' Pallas we do not know yet. Will Ceres stay with us or disappear in deep space and not return just like other comets? Let us remember this peculiar incidence, which will go down in the annals of astronomy and remain an everlasting warning for us and our grandchildren.!*

*It might become important for us or them to know that there is a star similar to Ceres that roams the universe. We only know one other example in the history of astronomy, which could have led astronomers into a similar maze. But astronomers negotiated every obstacle and escaped by similar luck. It happened with the comet of 1664, which stirred all astrologists and cometo-maniacs, because the famous Hevelius had confused this comet on February 18 with a nebula or any other unusual thing or a little cloud what resulted in an observation incompatible with the motion of this comet. Hevelius and the famous French*

| 1802 | Mittlere Zeit in Bremen | Scheinbare gerade Aufsteig. der Pallas | Scheinbar. nördl. Abweich. der Pallas | Vergleich. mit Sternen |
|---|---|---|---|---|
| | U ′ ″ | ° ′ ″ | ° ′ ″ | |
| März 28 | 9 25 10 | 184 56 49 | 11 33: | Nr. 20 ♍ *v. Zach* |
| — 29 | 8 49 14 | 184 46 36 | 11 52 59 | |
| — 30 | 8 3 17 | 184 36 22 | 12 13 48 | C. d. t. X. 673 674 |
| April 1 | 8 0 4: | 184 15 38 | 12 54 25 | Nr. 225 *Bode*'s Verzeichn. |
| — 2 | 7 56 5: | 184 5 7 | 13 14 28 | 3 ✻ 8 Gr. *LaLande*'s Hist. célest. |
| — 3 | 8 0 37 | 183 54 32 | 13 34 16 | franç. |
| — 4 | 8 1 8 | 183 44 40: | 13 53 0: | Nr. 143 *Bode* |
| — 5 | 8 32 36 | 183 36 38: | 14 11 0: | |
| — 6 | 8 16 0 | 183 25 31 | 14 30 21 | Nr. 109 *Bode* |
| — 7 | 8 33 5 | 183 16 26 | 14 47 25 | |
| — 9 | 8 18 20 | 182 58 27 | 15 20 52 | Nr. 109 und 111 *Bode* |
| — 10 | 8 46 40 | 182 49 34 | 15 37 26 | Nr. 111 *Bode* |
| — 11 | 8 6 28 | 182 41 21 | 15 53 53 | Nr. 111 *Bode*. Starker Sturm |
| — 12 | 8 19 0 | 182 33 28: | 16 9 : | Nr. 111 *Bode* |
| — 13 | 8 33 59 | 182 25 43 | 16 24 35 | Nr. 87 u. Nr. 114. Wolken |
| — 14 | 8 28 20 | 182 18 28 | 16 39 15 | Nr 114 |
| — 17 | 10 11 35 | 181 56 25 | 17 22 5 | Von nun an wurde die *Pallas* |
| — 17 | 13 17 49 | 181 55 40 | 17 23 30 | immer mit Sternen aus un- |
| — 18 | 8 26 21 | 181 50 46 | 17 33 8 | serm im vorigen Hefte S. 386 |
| — 19 | 11 16 7 | 181 43 45 | 17 47 35 | mitgetheilten Verzeichnis- |
| — 20 | 13 25 55 | 181 38 16 | 18 0 5 | se verglichen. Folgende |
| — 21 | 12 18 33 | 181 33 1 | 18 11 29 | Druckfehler müssen noch |
| — 23 | 9 41 2 | 181 23 59 | 18 32 11 | verbes- |

**Fig. 11.2** Observations of Pallas by Olbers from March 28 to April 23, 1802, with comparison stars at right

*astronomer Auzout* [Adrien Auzout, 1622–1691], *who had observed the comet himself, started a quarrel and Hevelius defended and insisted stubbornly on his observation. Luckily several other astronomers had observed this comet in France, Italy, Spain and England and Hevelius' error was fortunately discovered. Auzout uttered a remark, which goes well with our present case of Ceres and Pallas: "That what is most disagreeable in this incident is, that if the sky would have been overcast until Feb. 18 so that no-one could have observed the comet after that day, he* [the comet] *would have embarrassed the present and future astronomers by such an odd observation."*

*If Olbers' Pallas was discovered earlier than Ceres it would have been indeed as Auzout said "she would have embarrassed us." But the below figures prove the contrary and will provide information about this weird celestial body* (Fig. 11.2).

*All of these observations need a more accurate reduction as Dr. Olbers wrote, because of the badly determined small stars, which will be given soon. On April 4 Pallas was close to but below the nebula 143. Then Dr. Olbers found another nebula, which is just as conspicuous but is not to be found in Bode's catalogue. The RA of this nebula is about 184° 3′ and the declination north 14° 4′. Here are our meridian observations of Pallas made at Seeberg Observatory* (Fig. 11.3).

*Of other astronomers only two meridian observations of Pallas have reached us made by Prof. Bode* (Fig. 11.4).

*The president of the Royal Society of Sciences in London, Sir Joseph Banks, wrote that Gilpin, clerk of the Royal Society had found Olbers' Pallas on April 9 upon our message.*

*On April 13 a young promising astronomer named Lee* [Stephen Lee observed from the Grove, Hackney, greater London] *observed Pallas at 11 h 50′ 48″ m.t. RA 182° 24′ and*

| 1802 | | Mittlere Seeberger Zeit | Scheinbare gerade Aufst. der Pallas | Scheinbare nördliche Abweichung der Pallas |
|---|---|---|---|---|
| April | 4 | 11 U 24′ 51,″9 | 183° 44′ 6,″6 | 13° 54′ 52,″0 |
| | 5 | 11 20 17, 3 | 183 34 23, 7 | 14 13 22, 9 |
| | 7 | 11 11 10, 6 | 183 15 38, 5 | 14 49 2, 1 |
| | 8 | 11 6 38, 8 | 183 6 37, 8 | 15 16 10, 0 : |
| | 15 | 10 35 22, 6 | 182 10 16, 5 | 16 53 30, 8 |
| | 18 | 10 22 16, 0 | 181 50 30, 6 | 17 36 7, 4 |
| | 19 | 10 17 55, 8 | 181 44 25, 3 | 17 46 54, 4 : ? |
| | 24 | 9 56 37, 4 | 181 19 37, 8 | 18 42 : Schätzung |
| | 25 | 9 52 25, 2 | 181 15 32, 2 | 18 52 : Schätzung |
| | 26 | 9 48 15, 2 | 181 12 1, 8 | 19 1 49, 4 |
| | 27 | 9 44 6, 3 | 181 8 45, 6 | 19 10 46, 8 |
| | 29 | 9 35 52, 6 | 181 3 16, 6 | 19 27 43, 7 |

**Fig. 11.3** Observations of Pallas by Zach from April 4 to 29, 1802

| 1802 | | Mittl. Berliner Zeit | Scheinbare gerade Aufst. der Pallas | Scheinbare Abweich. der Pallas | |
|---|---|---|---|---|---|
| April | 11 | 10 U 53′ 9″ | 182° 40′ 24,″5 | 15° 55′ 24,″0 | β und o ♌ |
| — | 12 | 10 44 33 | 182 33 23, 5 | . . . . | o und 87 ♌ *Bode* |

**Fig. 11.4** Observations of Pallas by Bode on April 11 and 12, 1802, including the stars he used for comparison

*declination: 16° 27′. The Astronomer Royal Dr. Maskelyne did not mention Olbers' star in a letter from Greenwich dated April 15.*[1,2]

## BAJ, 1805, p. 102 by Olbers

[This contribution by Olbers consisted of five letters. The first two dealt with Ceres, and were dated November 14 and December 10, 1801. They were printed in *Early Investigations of Ceres, and the Discovery of Pallas*. The remaining three letters, regarding Pallas, are presented here.]

***May 7, 1802.*** *When I discovered my moving star, and had noticed its regular retrograde movement and its appearance, which differed from all comets and so closely resembled Ceres, I immediately believed I had found a planet. Soon after the first days, I attempted to calculate its orbital elements, and found, much to my surprise, that the elements resisted representation within a concentric orbit around the Sun. Pallas moved 'faster and faster'*

---

[1]Zach mentions several people who had written about comets: Fortunatus Licetus or Fortunio Liceti (1577–1657; 1623), Italian physician and colleague of Galileo at the University of Padua; Johannes Hevelius (1611–1687; 1668), mayor of Danzig and builder of an observatory in 1641; Domenico Cassini (1625–1712). first director of the Paris Observatory (see Rost 1718: 506); and Nicholaas (Nicolas) Struyck (1687–1769; 1740), who was a Dutch astronomer. The identification and publication dates of their comet work follows. See the references at the end of this chapter for the titles of their works.

[2]It was noted by Humboldt (1845: 37) that "The comet of 1729 is the only one of those hitherto computed which has its perihelion between the orbits of Pallas and Jupiter."

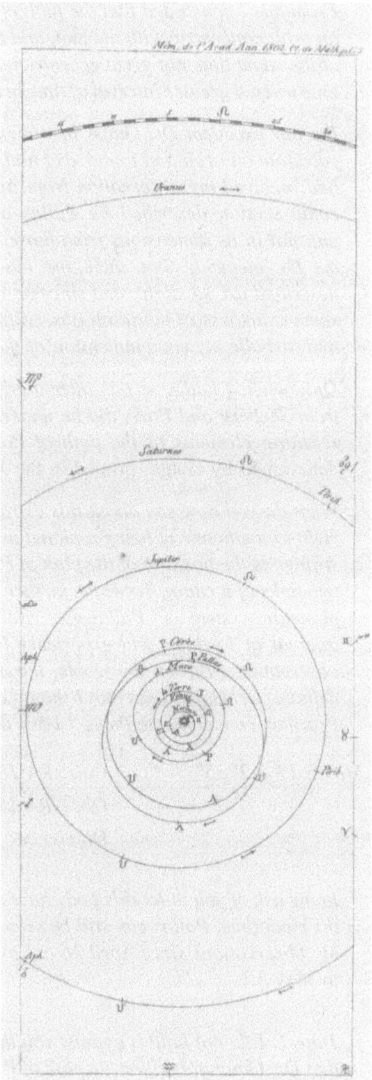

**Fig. 11.5** Map of the Solar System, showing the orbits of Ceres and Pallas. From the Memoirs de l'Academie of 1802, Class de mathematiques

*than was possible within such an orbit. From the position of the visual lines, it was revealed that its distance from the Sun grew. Therefore, its path must differ greatly from an orbit concentric to the Sun, and this confused my suspicions of its planetary nature* (Fig. 11.5).

*Towards the middle of April, I attempted the second extreme—the parabola, and tried to represent the observations of March 28, April 4 and April 10 within a parabola. You know that I have had considerable practice in this type of calculation. But the attempt did not succeed, and it seemed to me that it was impossible to satisfy three complete observations within a parabola; there remained a considerable error either in the longitude or the latitude of the middle observation.*

*From this I concluded that the path of Pallas may be just as different from a parabola than an orbit concentric with the Sun, and that it would form an ellipse of a not insignificant, but at the same time not great eccentricity. I postponed the calculation of this ellipse until that time when a greater interval of time between observations would promise more exact results.*

*But our excellent Dr. Gauss beat me and the rest of us to it. At first, he also attempted to calculate an orbit and found, as I had, that the movement increased in speed. As soon as he had received my observation from April 17, he tried, without any hypothesis, to find the conic section described by Pallas according to his method. He found an ellipse, not unusual in its dimensions from those that I will later share with you. He immediately sent me the elements, but since the interval between observations was so small, and the described arc so slight, and since he had to use my observations which have by far not the same exactness of meridian observations, he regarded these elements with great mistrust, and forbade my communication of them.*

*Only when I had sent this incomparable mathematician several more exact observations from Seeberg and Paris did he undertake the calculation a second time, and here are the resulting elements of the path of Pallas.* [He provides here the first elements of Pallas determined by Gauss, printed in the June *MC*.]

*What do you now say about this extraordinary body? Certainly, Pallas has just as great a right to the honor of being a planet as Ceres. But how great is the inclination! or how very strange is the position of the path of Pallas against that of Ceres? Both are interlocked like two links of a chain. To which further speculations regarding the origins and history of our planetary system will Pallas give rise? Admittedly, the above elements, derived from a time interval of 15 days, will experience fairly considerable corrections as a result of further observations. But on the whole, the dimensions of the path of Pallas can be regarded as definite. Dr. Gauss asks that I should compare the elements with my last observation before I further communicate them. I have done this in all sharpness and have found*

| | | | |
|---|---|---|---|
| *May 5 11 h 2′ 35″* | *Calc. RA 180° 56′ 28″.2* | | *Calc. Decl. 20° 9′ 20″.8* |
| | *Obs. RA 180 56 6* | | *Obs. Decl. 20 8 59* |
| | *Difference* | *+22″.2* | *+21″.8* |

*I now ask of you to let this body have the name Pallas, which seems not improper to me. In the meantime, Pallas can still be seen very well and I hope to be able to track it until June. My observations since April 26 are as follows:* [This table lists observations from April 26 to May 5.]

***June 1.*** *I do not hold it against you that you still doubt the unexpected paradoxical results that Dr. Gauss found for the path of Pallas, since you only had the elements before you that were determined from 15 days. In order to find the conclusions arising from such a small described arc not too risky – conclusions which contradict past analogy and experience – one ought really to have conducted the calculations of this strange body oneself. Meanwhile, you will have received Dr. Gauss's second elements, their comparison with all observations, the ephemeris for June* etc. *from this magnificent scholar himself. If you now consider how strongly the path is marked by the visual lines on such a greatly inclined plane, as the plane of Pallas' path indeed is, then you will be convinced as I am that the main dimensions of the path, as Dr. Gauss now cites them, have a geometrical certainty, and that this path of Pallas has been determined almost as precisely as the path of Ceres had been before its rediscovery.*

*It is impossible to bring the observations in agreement with some parabola. This is not only* a priori *easily recognized (if I may say so): rather, I myself have attempted it.*

*Even now, 29 days after the observations upon which Dr. Gauss based his calculations, these elements, determined with such remarkable exactness, still give the declination approximately 1' too large and the right ascension approximately 40" too small, and this with a regularity which clearly shows that a very moderate change in the elements will be sufficient to completely eliminate this small deviation.*

*I hope to see Pallas until July. I will leave the effort and honor of determining the path of Pallas to Doctor Gauss alone. It would be a futile effort in more than one respect to want to compete with this great genius of measurement.*

*Therefore, the path of Pallas will certainly remain an ellipse of an eccentricity not much greater than the path of Mercury and of an axis almost the same as that of the path of Ceres. The paths of Ceres and Pallas are not interlocked like the links of a chain as I had described earlier; rather, the more elongated and more eccentric path of Pallas lies within the path of Ceres in both nodes. With the descending node of the path of Pallas with that of Ceres, both paths come tremendously close, but we do not yet know the paths well enough to be able to determine whether an actual intersection occurs or had once occurred before the perturbational forces of the remaining planets, especially that of Jupiter, changed the respective positions of both paths toward each other.*

*Where is the analogy here, you ask? Where is the beautiful, regular order that the planets seemed to observe in their distances from each other? It is still too early, I believe, to philosophize about this. We must concern ourselves 'now' just with observing and determining the path in order to have more certain grounds for our conjectures. We will perhaps then be able to decide whether Ceres and Pallas always went through their present paths 'divided' in peaceable proximity or whether both are simply rubble, simply pieces of a formerly great planet, destroyed by some great catastrophe.*

*But, as I say, I don't want to pass this off as anything – not even a conjecture. But peculiar is the small size of these planets, which Herschel finds much smaller in his telescope that of our senior civil servant Schroeter; peculiar is the frequent change in their light intensity, which seems to point to an irregular, not round figure; and it is peculiar to me at least, that in the same area of space, a magnitude 8 star, which had certainly been observed on the 10th of April 1796, according to* Hist. Celest.*, is now missing; this star could have been neither Ceres nor Pallas. Whether we will see Pallas again in the next two years is still uncertain because, nearing its aphelion in these years, it will remain distant from Earth even in opposition. On March 28, 1802, when Pallas was discovered, its distance from Earth was 1.38; now it's 2.10, and in the middle of June, it will amount to 2.31. If, on the other hand, Pallas was in 9Z 7° with 48° north Dec. over Cerberus in opposition at the end of June 1783, its distance from Earth would be 2.62. It would have been very difficult to find it because of our bright twilight. It will be somewhat better in 1804, since Pallas will be in the head of Pegasus, and at the end of August will come to 11Z 3° and 18° north Dec. with the Sun, in opposition. Indeed, its distance will still be 2.40 but the nights will be completely dark.*

*Without doubt we will once again see Pallas in 1805. One will be able to better judge this when we have learned through experience the distances from Earth in which it is still visible.*

*My later observations of Pallas now follow, of which the later ones may call for a somewhat more exact reduction of the star chosen for the purposes of reduction from the* Hist. Cel.

[This table includes observations from May 7-30.]

**June 10.** *I have calculated several parabolas for Pallas; just at this moment, Captain von Wahl has given me one. Here also is one from Dr. Burckhardt:*

| | |
|---|---|
| *Decl. Perihelion* | *1.8432* |
| *Ascending node* | *176° 45′ 34″* |
| *Perih.* | *113 52 3.5* |
| *Incl.* | *54 58 30* |
| *Temp. Perih. 1801 Sept. 29 16 h 48′* | |

*But neither I nor these other gentlemen have been able to bring all observations into line with one parabola. As I hear, all attempts of Burckhardt, Méchain and the great effort of La Place to determine the path of Pallas have been in vain. Burckhardt now maintains an ellipse of 12 years through which he hopes to represent the observations. All the more honor for Dr. Gauss, who has determined the course so accurately through his remarkable method and without any provisional hypotheses regarding the nature of said course.*

*Dr. Gauss' 2nd elements for the path of Pallas from the 27 day long observations from Seeberg until May 1 are as follows:* [Gauss' 2^nd^ elements are printed here; see *MC* June 1802.]

*On May 10, these elements gave the RA as being 10″ too small, and the declination as 18″ to great. Now, in June, the RA is almost 1½ minutes too small, and the declination 2′ too great. I advised Dr. Gauss to devote himself to the further correction of the path until the time that the observations should be fully concluded. As soon as he received the splendid observations of Dr. Maskelyne, who had been so fortunate as to have observed Pallas until May 16 in the meridian, this tireless mathematician determined the elements from 42 day-long observations (in other words, one day longer than had been required in Piazzi's initial observations of Ceres-described arc around the Sun of 11° 24′). These elements now follow:*

[Gauss' 3rd elements are printed here; see MC July 1802]

*You will undoubtedly remark that the major axes and therefore the orbital times for both Ceres and Pallas are very similar.*

[End of Olbers' paper.]

## Continued Reports of a new Highly Remarkable Planet of our Solar System moving between Mars and Jupiter discovered by Dr. Olbers in Bremen

*MC*, June 1802, p. 591

*When we first indicated the remarkable discovery of a new body by Dr. Olbers in the previous issue, the available observations, calculations and accumulated knowledge at that time did not justify us in announcing this curious planet as one belonging to our Solar System, although suspicions did exist, which we did mention, that this body, so very distinguishable by its movement and outer appearance, could not be set in a class of normal so-called comets. Meanwhile, the suspicion of a planetary body was so great upon first observing it, that the discoverer immediately ventured to initially represent its path not in a parabola, as is normally the case, but in a circle concentric with the sun. However, the available observations, as we have cited in the previous issue (p. 486), did not allow representation within an orbit, and as meager as the selection was, the strong latitudinal movement of this body already made it possible to forever disqualify the circular movement in future investigations of the path.*

*Dr. Gauss attempted the same thing and confirmed Olbers' discovery that the movement of Pallas was faster than it should have been within an orbit according to Kepler's laws. As soon as Dr. Gauss had received Olbers' observations of this body from April 17, he undertook the determination of the conic section independent of any hypothesis, according to his own method, the excellence of which Ceres had proven in such a magnificent way. It was foreseeable that the circular micrometer observations were not exact or long enough for these delicate calculations, especially since Pallas was not in a favorable position at the beginning of April, in that the tangent of the geocentric movement went through the perihelion [solar position]. Meanwhile, Dr. Gauss found an ellipse, which, in its main dimensions, was not very different from that which was determined later, and the result generally agreed with the idea that Pallas is a planet located between Mars and Jupiter, the path of which comes very close to that of Ceres at one point and has, with a very strong inclination, a visible eccentricity.*

*The eccentricity was in this connexion 0.3, thus greater than all other planets, but not so great to be able to deny Pallas the designation of a planet. The relation of the major to the minor axis is* $1{:}\sqrt{(1-(0.3)^2)}$*, or, in other words 1:0.95; consequently, the ellipse is not so flat as the disc of Jupiter.*

*Since the observations used for this are naturally by far not as exact as meridian observations, Dr. Gauss expressed a well-founded suspicion and delayed their publication pending future verification. Only when he had received our meridian observations from Seeberg from April 15, 18, 19 and later three Parisian observations from April 10, 12, 13 did he decide, because of the lack of a longer series of exact observations, to improve the previous observations in a way that they followed the six observations from Seeberg of April 4, 5, 7, 15, 18, 19 and the three from Paris, and in this way, he found the following elements of Pallas on the first attempt* (Fig. 11.6):

*With these elements, Dr. Gauss now critically compared the observations he had from Seeberg at that time, as well as the remaining ones, with the observations from Paris and one from Olbers, which gave the following correspondence* (Fig. 11.7):

*The observations give the differences added with reversed signs of the calculations.*

*Dr. Gauss maintains here, as with Ceres, his old custom (which deserves imitation) of deriving the right ascensions and declinations from the elements themselves and of comparing them with the observed ones, and not deriving the observed longitudes and latitudes from the observed RA and declinations. The reason for this method is so plausible that one is forced to wonder why this had not been introduced and followed long ago. In terms of effort, it is exactly the same if one approaches the calculation with the observation in one step or if the calculation leads to the observation. According to the method characteristic of Dr. Gauss, this parallax calculation has become very much easier. The main advantage gained through this is the separation of the good from the somewhat less reliable, otherwise it can easily occur that a good right ascension is ruined by an incorrect declination, leading to not only an incorrect latitude, but rather also a bad longitude. Since the*

| | |
|---|---|
| Epoche 1802 März 31 Mittag in Seeberg . . . . . . . . | 166° 1′ 37″2 |
| Tägliche mittlere tropische Bewegung . . . . . . . . | 800″770 |
| Umlaufszeit . . . . . . . . . . . . . . . . . . . | 1618½ Tage |
| Logarithmus der halben grossen Axe . . . . . . . . | 0.4310494 |
| Sonnenferne } zur Zeit der Epoche siderisch ruhend*) . | 304° 36′ 30″ |
| Knoten } zur Zeit der Epoche siderisch ruhend*) . | 172° 9′ 58″ |
| Excentricität . . . . . . . . . . . . . . . . . . | 0.215708 |
| Neigung . . . . . . . . . . . . . . . . . . . | 33° 39′ 16″6 |

**Fig. 11.6** The First Elements of Pallas

| 1802 | Berechnete AR. der Pallas | Berechnete Abweichung der Pallas | Unterschiede in AR. | Unterschiede in Decl. | Beobachter |
|---|---|---|---|---|---|
| April 4 | 183° 44′ 6″3 | 13° 54′ 51″0 | − 0″3 | − 2″0 | von Zach |
| 5 | 183 34 25.7 | 14 13 17.2 | + 2.0 | − 5.7 | — |
| 7 | 183 15 40.0 | 14 48 55.5 | + 1.5 | − 6.6 | — |
| 8 | 183 6 35.2 | 15 6 9.1 | − 2.6 | − 0.9 | — |
| 10 | 182 48 56.5 | 15 39 40.9 | − 1.5 | − 1.1 | Dr. Burckhardt |
| 12 | 182 32 34.3 | 16 11 4.2 | + 2.8 | + 6.2 | Méchain |
| 13 | 182 24 47.2 | 16 26 6.6 | + 2.8 | + 0.6 | — |
| 15 | 182 10 13.2 | 16 54 33.0 | − 3.3 | + 2.2 | von Zach |
| 18 | 181 50 26.2 | 17 34 26.5 | − 4.4 | . . . | — |
| 19 | 181 44 28.0 | 17 46 51.7 | + 2.7 | − 2.5 | — |
| 24 | 181 19 39.4 | 18 42 27.8 | + 1.6 | . . . | — |
| 25 | 181 15 44.1 | 18 52 18.3 | + 11.9 | . . . | — |
| 26 | 181 12 10.4 | 19 1 43.5 | + 8.6 | − 5.9 | — |
| 27 | 181 8 57.9 | 19 10 44.0 | + 12.2 | − 2.8 | — |
| 29 | 181 3 39.6 | 19 27 31.2 | + 23.0 | − 12.5 | — |
| 30 | 181 1 33.9 | 19 35 18.2 | + 24.0 | − 15.3 | — |
| Mai 1 | 180 59 48.1 | 19 42 42.5 | + 25.6 | + 1.7 | — |
| 1 | 180 59 38.0 | 19 43 38.0 | + 20.0 | + 7.0 | Dr. Olbers |

**Fig. 11.7** Difference between observed and calculated positions, based on the First Elements of Pallas, from April 4 to May 1, 1802

*differences always decrease by the end with the above comparison of observations and the elements of the path, it was foreseeable that given the short duration of the observations, a difference of 25" would still have a great influence on the determination of these elements. A continued investigation which Dr. Gauss immediately carried out following this comparison had shown him that the 15-day observations, on which he based his earlier elements, were not sufficient to deliver an exact orbit determination, but still indicated the nature of the body fairly clearly. He immediately determined new elements from the observations of our Seeberg Observatory which, as well as was possible, followed the observations of April 4 to May 1, and since the arc described on the path already amounted to 7° 23', no doubt as to the planetary nature of Olbers' body could still be harbored. The new ellipse of this planetary body is as follows* (Fig. 11.8):

*Dr. Gauss now compared these elements with the entire series of observations from Seeberg and found their agreement as follows* (Fig. 11.9):

*Dr. Gauss would have easily followed the last right ascensions somewhat more exactly, but because of the less than reliable declinations due to the advancing twilight, he would not have been certain whether he would really have got an 'improvement' of the path from this.*

| | |
|---|---|
| **Epoche 1802 März 31 Mittag in Seeberg** . . . . . . . . | 161° 12′ 43″2 |
| **Tägliche mittl. tropische Bewegung** . . . . . . . . . | 757″166 |
| **Logarithmus der halben grossen Axe** . . . . . . . . . | 0.4472636 |
| **Sonnenferne** } **siderisch ruhend** { . . . . . . . . . | 300° 5′ 4″ |
| **Knoten** . . . . . . . . . . | 172 34 35 |
| **Excentricität** . . . . . . . . . . . . . . . . | 0.2591096 |
| **Neigung** . . . . . . . . . . . . . . . . . . . | 35° 0′ 42″ |

**Fig. 11.8** The Second Elements of Pallas

| 1802 | Berechnete AR. der Pallas | Berechnete Abweichung der Pallas nördlich | Unterschiede in AR. | Unterschiede in Decl. |
|---|---|---|---|---|
| April 4 | 183° 44′ 5″7 | 13° 54′ 55″4 | − 0″9 | + 3″4 |
| 5 | 183 34 24.6 | 14 13 22.6 | + 0.9 | − 0.3 |
| 7 | 183 15 39.6 | 14 49 1.3 | + 1.6 | − 0.7 |
| 8 | 183 6 35.4 | 15 6 15.2 | − 2.4 | + 5.2 |
| 15 | 182 10 17.9 | 16 54 35.7 | + 1.4 | + 4.9 |
| 18 | 181 50 29.4 | 17 34 27.1 | − 1.2 | ... |
| 19 | 181 44 30.7 | 17 46 51.9 | + 5.4 | − 2.5 |
| 24 | 181 19 35.8 | 18 42 27.9 | − 2.0 | ... |
| 25 | 181 15 38.3 | 18 52 18.9 | + 6.1 | ... |
| 26 | 181 12 2.0 | 19 1 44.9 | + 0.2 | − 4.5 |
| 27 | 181 8 46.6 | 19 10 46.3 | + 1.0 | − 0.5 |
| April 29 | 181° 3′ 21″2 | 19° 27′ 36″2 | + 4″6 | − 7″5 |
| 30 | 181 1 11.4 | 19 35 24.9 | + 1.5 | − 8.6 |
| Mai 1 | 180 59 22.9 | 19 42 50.8 | + 0.4 | + 10.0 |
| 2 | 180 57 56.7 | 19 49 53.2 | + 4.8 | + 1.6 |
| 3 | 180 56 52.4 | 19 56 32.9 | − 3.2 | − 14.8 |
| 5 | 180 55 49.6 | 20 8 45.6 | − 5.0 | + 6.9 |
| 6 | 180 55 51.4 | 20 14 19.4 | − 2.7 | ... |
| 7 | 180 56 14.9 | 20 19 32.1 | − 8.9 | − 3.2 |
| 8 | 180 57 0.4 | 20 24 23.7 | − 9.6 | + 3.6 |
| 11 | 181 1 25.5 | 20 36 58.8 | − 16.2 | − 14.4 |

**Fig. 11.9** Difference between observed and calculated positions, based on the Second Elements of Pallas, from April 4 to May 11, 1802

*He therefore wants to await subsequent observations before he will undertake a new correction of these elements.*

*As is presently the case, all previous observations, calculations and comparisons seem to justify the planetary nature of Pallas and safeguards us against all doubt, reservation and uncertain hypotheses. For, as Dr. Gauss remarks very well in his letter: "Nothing can be set against an exact calculation than an equally exact calculation, not empty suspicions and vague reasoning. At any rate, to easily locate Pallas in 1803, my 2*$^{nd}$ *elements will require considerable improvements and hopefully I will receive these if observations succeed. But these improvements will only be considerable in view of the determination of the location of Pallas in 1803, and not in view of the planetary nature or non-planetary nature of Pallas."*

*As was the case with the previously found elements, the path of Pallas, where its ascending node is on the path of Ceres, fell very closely to the path of Ceres. This brought Dr. Olbers to the suspicion of an intersection of the paths of Ceres and Pallas; that both are intertwined, like two links of a chain. According to a rough estimate, which Dr. Gauss made for the distances of Pallas and Ceres in the nodal line of their paths, he found, for the descending node, the distance of Pallas from the sun to be 2.86 and that of Ceres to be 2.93; with the ascending node, he found these distances to be less similar. The series of observations is still too short to deny the possibility that the distances at this place are exactly the same. Dr. Gauss wants the calculations to run their course and mix absolutely nothing hypothetical in with it. It is admittedly still too early to lose oneself in hypotheses since we will, instead of this, soon achieve certain knowledge if the future year sees success in the rediscovery of Pallas.*

*Meanwhile, which grounds for speculation as to the origins and history of our planetary system will this remarkable planet not give? What a completely unexpectedly great inclination for a planet? And how striking the position of its path compared to that of Ceres? This brought Dr. Olbers to the thought that both planets might be but rubble of a single one destroyed by the collision of a comet. Who would have expected something like this in our planetary system, and to which new, important and great information shall and will this planet not lead us, which numbers among the most important astronomical discoveries ever made in more than one respect.*

*Dr. Gauss finds the great agreement of the mean movement of both planets just as remarkable. At present, it is impossible to decide with certainty whether they are perhaps 'exactly' identical. We don't even know the actual daily movement of Ceres to the exact second. But however small the measurements of Ceres and Pallas are, there must be a limit to the difference of their mean movements, within which it cannot fall without having the reciprocal effect brought completely to zero. At present, these mean movements differ only slightly according to the 2nd elements and perhaps they will come much closer together with a more exact knowledge of their paths. In this way, it could perhaps be the case that Ceres and Pallas had the same orbital time, were not always so distant from each other, but without ever having been too close from the very beginning or being able to come closer in future, since the ascending node of Pallas would have been ahead of the descending node of Ceres each time, because of the greatly different equation of a circle with the origin at the center. Then, after a number of years, it will be possible to determine the mass of Pallas from the observations of Ceres and the mass of Ceres from observations of Pallas. According to a rough estimate, Dr. Gauss finds that a 12-second difference in the mean daily movement gives a synodical revolution of almost three hundred years.*

*To which investigations will such a phenomenon, singular in its nature, not give rise? Which new scientific needs will it not awake? For according to all of our knowledge to date, we would not have been prepared for such extraordinary phenomena in the planetary system! How will our methods of calculation, customary until now, be abandoned and inadequate when, for example, we are to calculate the perturbations of this new planetary body. Until now, the eccentricity and inclination of paths have been regarded as infinitely small measures so to speak, according to our usual methods, their development being carried out according to their power, with a disregard of higher values. This can no longer be the case with the two rather strong elements of the path of Pallas, the high powers of which must certainly not be neglected.*

*For this reason, Pallas will perhaps found a necessary new era in physical astronomy. What we know through the most rigorous calculation is the only thing we 'really know' about the constitution of planets [the universe]. But the calculator requires experience and observations; only through these can he transform hypotheses to truths. For this reason, Pallas has been observed with just as much diligence as care. We will communicate to our readers the entire collected series of these precious observations. Our periodical has been the archive of all observations of Ceres, it is to be the same for Pallas and we will spare no effort to make this collection as complete as possible so that all astronomers will find everything together what they would otherwise have to laboriously collect from scattered sheets – and then found only too late and perhaps never at all.*

*One of the most diligent and tireless observers of Pallas was its discoverer himself. We have cited his observations from March 28 to April 23 in the previous issue, p. 499. The continuation follows. The right ascensions are compared according to our communicated positions of the stars and the deviation according to La Lande's* Hist. cèleste française.

*Continued observations of Pallas, conducted by Olbers himself with a circular micrometer*

| *1802* | *Mean Time Bremen* | *Apparent RA* | *Apparent Decl.* |
|---|---|---|---|
| *April 26* | *12 h 37′ 20″* | *181° 11′ 24″* | *19° 2′ 38″* |
| *27* | *12 7 40* | *181 8 19* | *19 12 2* |
| *28* | *12 44 11* | *181 5 35* | *19 19 52* |
| *29* | *12 3 10* | *181 3 15* | *19 27 57* |
| *30* | *12 3 25* | *181 1 10* | *19 35 37* |
| *May 1* | *12 27 15* | *180 59 18* | *19 43 31* |
| *2* | *11 35 20* | *180 58 3* | *19 50 25* |
| *5* | *11 2 35* | *180 56 0* | *20 8 59* |
| *7* | *11 20 27* | *180 56 40* | *20 19 38* |

*Since other astronomers could perhaps have used the same stars for comparison when observing Pallas, the positions of which we have determined very precisely, we will incorporate here, for this use, our 'apparent' determination. If in the declination the seconds occur by tenths, these have also been determined by us, but when the reading is only in minutes, then it is represented only approximately in order to designate the star.* [See pp. 601–602 of the *MC*, June 1802: Apparent position of a few not exactly or still undetermined stars parallel to Pallas, observed from the Seeberg Observatory.]

*We have constantly continued to observe this planet in the meridian from the Seeberg Observatory as often as the weather and twilight allowed. In the previous issue, we cited our observations until April 29; the continuation now follows* (Fig. 11.10):

*In Paris, Burckhardt and Méchain were the first to locate Pallas on April 10 according to the received reports. Méchain observed the planet on that very evening on an achromatic objective equipped with a micrometer and compared it with β Leo and found it at 13 hr 58′ 50″ mean time, RA 182° 45′ 10″, northern declination 15° 41′ 32″.5. Since then, he has observed Pallas in the meridian and here are the observations which he so kindly sent us* (Fig. 11.11).

*As early as April 8, I had sent a report of the discovery of Pallas to Milan. Since the letter was on the way for a very long time and since bad weather had come to Milan, the observations which had been sent were not sufficient to immediately locate the planet. Therefore, in order to guide the finding of this planet somewhat, Oriani calculated an orbit out of Dr. Olbers' observation of March 28 and our observation of April 4. With this, he found the heliocentric longitude for March 28 at the time of the observation to be 184° 46′; the mean tropical movement 1193″.23; distance from the sun 2.10; ascending node 168½°; inclination 24⅔°. With these rough elements, he succeeded in finding the planet on April 25, and here are his observations conducted on the equatorial sector:* [Observations of

| 1802 | Mittl. Seeberger Zeit | Scheinbare AR der Pallas | Scheinbare nördliche Abweichung der Pallas |
|---|---|---|---|
| April 30 | 9u 31′ 48,″3 | 181° 1′ 9,″9 | 19° 35′ 33,″5 |
| May 1 | 9 27 45,1 | 180 59 22,5 | 19 42 40,8 |
| — 2 | 9 23 43,3 | 180 57 51,9 | 19 49 51,6 |
| — 3 | 9 19 43,6 | 180 56 55,6 | 19 56 47,7 |
| — 5 | 9 11 47,7 | 180 55 54,6 | 20 8 38,7 |
| — 6 | 9 7 51,8 | 180 55 54,1 | . . . . . |
| — 7 | 9 3 57,9 | 180 56 23,8 | 20 19 35,3 |
| — 8 | 9 0 5,0 | 180 57 10,0 | 20 24 20,1 |
| — 11 | 8 48 35,4 | 181 1 41,7 | 20 37 13,2 |

**Fig. 11.10** Observations of Pallas by Zach from April 30 to May 11, 1802

| 1802 | Mittl. Pariser Zeit | Scheinbare AR. der Pallas | Scheinbare nördl. Abweichung der Pallas |
|---|---|---|---|
| April 12 | 10u 48′ 33,″8 | 182° 32′ 31,″5 | 16° 10′ 58,″0 |
| — 13 | 10 44 6,8 | 182 24 44,4 | 16 26 5,8 |
| — 15 | 10 35 10,5 | 182 10 2,9 | 16 54 51,4 |
| — 16 | 10 30 52,9 | 182 3 8,0 | 17 8 36,8 |
| — 17 | 10 26 30,5 | 181 56 30,3 | 17 21 49,8 |

**Fig. 11.11** Observations of Pallas by Méchain from April 12 to April 17, 1802

Pallas Olbersiana from the Brera Observatory in Milan, conducted in an equatorial sector by Oriani, from April 25-May 7; For Oriani's complete set of Pallas observations from April 25-June 25, see his Italian paper in this chapter.]

*On the same date, I had, as with Milan, sent Professor Sniadecki in Cracow the report of Olbers' discovery. He received the letter on April 25, immediately searched the sky and found the body on the same evening and assured himself of its movement on the following day. His observations conducted in the meridian now follow. He regards the right ascensions as very precise, since he compared the planet with the well-determined stars in RA – with α, β, η Leonis and η Bootis – according to Dr. Maskelyne's and our star catalogues. He does not regard his observed declinations as so certain, partly because the telescope of his quadrant is not of particularly good quality and tolerates no adequate illumination of the lines, and partly because he regards the determination of the star used in the declination as unreliable. He therefore believes, for instance, that the declination of 33 Bootis was calculated very incorrectly in Bode's star catalogue* (Fig. 11.12).

As we see in the next paper about the discovery of Pallas, what to call it was an immediate problem. Bode refers to it as a "strange moving star" and "Olbers' star." The nomenclature problem posed by the discovery of new Solar System objects was not new. Giuseppe Biancani (1620: Praefatio) wrote this about the Galilean satellites (my emphasis): "It is a remarkable and happy discovery, which has been found by the power of the telescope, that Jupiter is accompanied. And indeed there are always four tiny *stars or planets* about him, running courses round him."

## Observations of Pallas from the Royal Observatory in Berlin

By Bode in 1802

*BAJ* (1805) p. 193 (incorrectly numbered p. 195 in the original)

*Just three months after the rediscovery of Ceres, Dr. Olbers in Bremen has sent the news to me that he has found another strange moving star that seems to move in a highly inclined orbit. Several observations and calculations by Olbers, Gauss and Burckhardt show unequivocally that this unexpected stranger does not move like any other comet, but has a highly inclined orbit. Dr. Olbers' discovery shows us something quite extraordinary, and up to now unheard of. The discoverer gave his reasons and opinions about the true path and the nature of this planetoid that he would like to name Pallas.*

*"If this celestial body will show itself after its return from the Sun next year at exactly that position where it can be expected according to its calculated elliptical motion – we have to*

| 1802 | Mittlere Zeit in Cracau | Scheinbare AR. der Pallas | Scheinbare nördl. Abweichung der Pallas | Sterne womit verglichen worden, Nro. nach *Bode* |
|---|---|---|---|---|
| | U ′ ″ | ° ′ ″ | ° ′ | |
| April 25 | 9 52 32 | 181 15 26,0 | 18 56 | AR η, β ♌ |
| — 26 | 9 48 22 | 181 12 4,3 | . . . . . | AR η, α, β ♌ |
| — 27 | 9 44 13 | 181 8 33,8 | 19 10 10:: | . . . . . . . |
| — 28 | 9 40 6 | 181 5 51,0 | 19 17 57 | 95 Com. Ber. 14, η Bootis |
| — 29 | 9 36 0 | 181 3 17,4 | 19 27 37 | dito und für AR η, α ♌ |
| — 30 | 9 31 55 | 181 0 59,7 | 19 35 14 | dito dito |
| May 1 | 9 27 52 | 180 59 13,0 | 19 41 43 | AR k, θ, ♌ 95 C.B. 14, η Bootis |
| — 2 | 9 23 52 | 180 57 53,0 | 19 48 11 | dito dito |
| — 3 | 9 19 50 | 180 56 52,5 | 19 56 36 | AR α, θ 47 6 ♌ 95 C. B. η, 33 Bootis |

**Fig. 11.12** Meridian observations of Pallas Olbersiana, conducted by Professor Sniadecki at the Cracow Observatory

*recognise it as an extremely extraordinary planet which revolves around the Sun within and close to Ceres' orbit with an exceedingly great inclination on a considerable eccentric orbit – unheard of among the eight major planets until now – and on an ellipse extremely different from that of Ceres but within the same time, if one is not inclined to declare it a planet-like comet, however curious these circumstances might appear. It would then be settled that there, where I expected only one planet between Mars and Jupiter, two of them race around the Sun on equally sized orbits needing the same amount of time, and the beautiful progression of the distances of the planetary orbits would remain intact."*

*On April 5 I received important news from Dr. Olbers about his discovery of his moving star. Since we had a clear sky, I searched and soon found at a certain position south-east a star of 7th magnitude without fog, that I thought and later found out to be the Olbers star. At the wall quadrant they tried to copy my observation without success, due to poor conditions. They tried again on the 8th, but it was too cloudy; on the 9th the air was so unstable that we were not sure of our observation. Later calculations showed the position too large by 15′ 18″.*

| *Mean Time* | *Apparent RA* | *Decl.* |
|---|---|---|
| *April 11 10 h 53′ 9″* | *182° 40′ 24″.5* | *15° 55′ 24″.0* |
| *12 10 48 44* | *182 33 25* | *– – –* |
| *15 10 35 26* | *182 10 31* | *16 54 30* |
| *19 10 17 56* | *181 44 41* | *17 46 40* |
| *May 5 9 11 48* | *180 55 37* | *20 7 47* |

*Even after several observations I was not able to observe it, especially when other stars were in the same parallel as Pallas. I observed Pallas often through the 3½ foot Dollond, since known stars are missing on that parallel, in order to compare them with the circular micrometer.*

| *Mean Time* | *Apparent RA* | *Decl.* |
|---|---|---|
| *May 22 11 h 26′ 41″* | *181 46 34* | *20 59 15* |
| *23 11 8 40* | *181 52 34* | *20 59 25* |
| *24 10 37 30* | *181 58 53* | *20 59 36* |
| *26 10 26 49* | *182 12 18* | *20 59 11* |
| *June 3 11 43 8* | *183 17 26* | *20 48 13* |

*Addendum:*

*The following tables of Ceres & Pallas were published in the* Memoires de l'Academie Royale (Classe de Mathematique, p. 103*) by Bode under the title* Astronomical Observations made at the Royal Observatory in 1802 (Figs. 11.13, 11.14, 11.15).

## **Memoir on the tenth Planet discovered by Dr. Olbers**
## Read in the public sitting of the French National Institute July 5, 1802, by Jerome Lalande

*The Philosophical Magazine (Tilloch's Journal)*, vol. 13, pg. 279.

*When we announced, in the last public sitting, the discovery of a planet by M. Piazzi of Palermo, we were far from thinking that, in three months, we should have to make known a discovery of the same kind. It was also by a fortunate accident that this tenth planet was*

*Obſervation de la nouvelle planète Cérès, au mural, avec des étoiles fixes voiſines.*

| 1802. | Derrière le fil du milieu. T. du prem. mobile. | Hauteur appar. obſervée | Différence entre Cérès & l'étoile dans la culm. | Différence entre Cérès & l'étoile dans la haut. | Calcul de l'apparente ascenſ. dr. | Calcul de l'apparente déclinaiſon. |
|---|---|---|---|---|---|---|
| | H. M. S. | D. M. S. | H. M. S. | D. M. S. | D. M. S. | D. M. S. |
| Mars 1. β ♌ | 11 33 28, 0 | 53.15.24 | —0 47 44, 0 | +0 11 15 | | |
| Cérès | 12 21 12, 0 | 53. 4. 9 | | | 186 40 46 | 15 29 40 B. |
| Mars 5. β ♌ | 11 32 54, 5 | 53.15.19 | —0 45 1, 5 | —0 16 10 | | |
| Cérès | 12 17 56, 0 | 53.21.29 | | | 186 0 0 | 15 57 5 B. |
| Mars 14 β ♌ | 11 31 35, 0 | 53.13.18 | —0 38 4, 5 | —1 12 4 | | |
| Cérès | 12 9 39, 5 | 54.25.22 | | | 184 15 50 | 16 52 54 B. |
| Mars 15. θ ♌ | 10 56 20, 0 | 54. 3.21 | —1 12 20, 0 | —0 27 46 | | |
| β ♌ | 11 31 26, 5 | 53.13.42 | —0 37 13, 5 | —1 17 25 | | |
| ο ♌ | 11 37 59, 0 | 54.17.38 | —0 30 41, 0 | —0 13 29 | | |
| Cérès | 12 8 40, 0 | 54.31. 7 | | | 184 3 7 | 16 58 24 B. |
| Mars 16. β ♌ | 11 31 18, 0 | 53.13.32 | —0 36 23, 0 | —1 23 1 | | |
| ο ♌ | 11 37 50, 5 | 54.17.39 | —0 29 50, 5 | —0 18 54 | | |
| Cérès | 12 7 41, 0 | 54.36.33 | | | 183 50 27 | 17 3 52 B. |
| Mars 19. θ ♌ | 10 55 47, 0 | 54. 2.54 | —1 8 56, 0 | —0 48 22 | | |
| β ♌ | 11 30 54, 0 | 53.13.19 | —0 33 49, 0 | —1 37 57 | | |
| ο ♌ | 11 37 26, 0 | 54.17.13 | —0 27 17, 0 | —0 34 3 | | |
| Cérès | 12 4 43, 0 | 54.51.16 | | | 183 12 8 | 17 18 52 B. |

**Fig. 11.13** Observations of Ceres by Bode, from March 1 to March 19, 1802

*discovered; but accident could favour none but an intelligent and indefatigable astronomer.*

*On the 28th of March, at nine in the evening, Dr. Olbers of Bremen was observing*

*Piazzi's planet, with which astronomers have been engaged for a year. He was examining with his telescope all the small stars in the Virgin's wing, to ascertain their positions, that he might be better able to establish the place of the planet, and had come to the 20th star of the Virgin, near which he had observed the planet in the month of January. He was surprised to see near this star, which is of the 6th magnitude, another smaller of the 7th magnitude. He was very certain that it had not been there at the time of his first observations: he therefore hastened to determine its position; and, having continued to view it for two hours, he perceived that it had changed its place in the course of that interval. The two following nights afforded him the means of being certain of its motion, which was 10 min per day. On the 28th of March, at 9 h 25′ mean time, at Bremen, it had 184° 56′ right ascension, and 11° 33′ north declination.*

*Astronomers have been accustomed to consider as comets, all stars that have motion. This was the case with the planets of Herschel and Piazzi at the time when they were discovered. That of Dr. Olbers had no more resemblance to a comet than the rest. With an achromatic telescope, the magnifying power of which was 180, it could not be distinguished from stars of the 7th magnitude. It was better defined than the planet of Piazzi; and, with a telescope of 13 feet, which magnified 288 times, it seemed to have a diameter of 4 s: but this was an effect of irradiation, or of the dispersion of the rays of light, which always makes the*

| 1802. | Derrière le fil du milieu. T. du prem. mobile. | Hauteur appar. observée | Différence entre Cérès & l'étoile dans la culm. | dans la haut. | Calcul de l'apparente ascens. dr. | déclinaison. |
|---|---|---|---|---|---|---|
| | H. M. S. | D. M. S. | H. M. S. | D. M. S. | D. M. S. | D. M. S. |
| Mars 24. θ ♌ | 10 55 0, 0 | 54. 4.35 | —1 5 16, 5 | —1 8 37 | | |
| β ♌ | 11 30 7, 0 | 53.14.29 | —1 30 9, 5 | —1 58 43 | | |
| o ♌ | 11 36 39, 0 | 54.18.48 | —0 23 37, 5 | —0 54 24 | | |
| Cérès | 12 0 16, 5 | 55.13.12 | | | 182 7 42 | 17 40 16 B. |
| Mars 27. θ ♌ | 10 54 31, 5 | 54. 4.14 | —1 2 4, 0 | —1 20 9 | | |
| β ♌ | 11 29 37, 5 | 53.14.28 | — 26 58, 0 | —2 9 55 | | |
| o ♌ | 11 36 9, 5 | 54.18.37 | — 20 26, 0 | —1 5 46 | | |
| Cérès | 11 56 35, 5 | 55.24.23 | | | 181 29 15 | 17 50 47 B. |
| Avril 1. θ ♌ | 10 53 41, 5 | 54. 4.13 | —0 57 55, 0 | —1 32 6 | | |
| β ♌ | 11 28 47, 5 | 53.14.24 | —0 22 49, 0 | —2 21 55 | | |
| o ♌ | 11 35 20, 0 | 54.18.29 | — 16 16, 5 | —1 17 50 | | |
| Cérès | 11 51 36, 5 | 55.36.19 | | | 180 26 57 | 18 2 48 B. |
| Avril 5. θ ♌ | 10 52 58, 0 | 54. 4.10 | —0 54 47, 5 | —1 37 50 | | |
| β ♌ | 11 28 4, 5 | 53.14.15 | —0 19 41, 0 | —2 27 42 | | |
| o ♌ | 11 34 37, 0 | 54.18.27 | —0 13 8, 5 | —1 23 33 | | |
| Cérès | 11 47 45, 5 | 55.42. 0 | | | 179 39 58 | 18 8 33 B. |
| Avr. 15. β ♌ | 11 26 18, 5 | 53.14.18 | —0 12 59, 5 | —1 24 22 | | |
| o ♌ | 11 32 50, 5 | 54.18.27 | —0 6 27, 5 | —1 20 13 | | |
| Cérès | 11 39 18, 0 | 55.38.40 | | | 177 59 38 | 18 5 12 B. |
| Avr. 19. β ♌ | 11 25 38, 0 | 53.14. 9 | —0 10 53, 0 | —2 16 32 | | |
| o ♌ | 11 32 10, 5 | 54.18.24 | —0 4 20, 5 | —1 12 17 | | |
| Cérès | 11 36 31, 0 | 55.30.41 | | | 177 27 57 | 17 57 29 B. |
| Avr. 29. β ♌ | 11 23 40, 5 | 53.14.22 | —0 7 12, 5 | —1 41 38 | | |
| Cérès | 11 30 53, 0 | 54.56. 0 | | | 176 32 48 | 17 22 19 B. |
| Mai 1. β ♌ | 11 23 13, 0 | 53.14.18 | | | | |
| Cérès | 11 29 59, 5 | 54.46.36 | + 6 46, 5 | +1 32 18 | 176 26 19 | 17 12 59 B. |
| Mai 3. β ♌ | 11 22 47, 0 | 53.14.21 | — 6 25, 5 | —1 21 58 | | |
| Cérès | 11 29 12, 5 | 54.36.19 | | | 176 21 6 | 17 2 39 B. |
| o ♌ | 11 29 19, 5 | | + 0 7, 0 | | | |

**Fig. 11.14** Observations of Ceres by Bode, from March 24 to May 3, 1802

*diameters appears too large; for the satellites of Jupiter appear much larger then the new planets, and yet we know that their apparent diameter is not a second.*

*Dr. Maskelyne, by means of diaphragms placed before the object-glass of his telescope, ascertained that the light of Piazzi's planet is stronger by one half than that of the new planet.*

*Dr. Olbers having observed the new star for four days, he sent notice to different astronomers; and on the 10th of April, C. Burckhardt, when he received his letter, went immediately to the military school to search for it, and next day sent his observation to the Institute.*

*He began to calculate its orbit, trying first a circle, and then a parabola known to be that of comets; but, at the end of three days, his elements were found to err 30 s. He tried also ellipses of different dimensions.*

*Obſervation de Pallas (planète nouvellement découverte, & remarquable par la grande inclinaiſon de ſon orbite, de 35°) au mural, avec des étoiles fixes voiſines.*

| 1802. | Derrière le fil du milieu. T. du prem. mobile. | Hauteur apparente obſervée. | Différence entre Pallas & l'étoile | | Calcul de l'apparente | |
|---|---|---|---|---|---|---|
| | | | dans la cul. | dans la haut. | aſcenſ. dr. | déclinaiſon. |
| | H. M. S. | D. M. S. | H. M. S. | D. M. S. | D. M. S. | D. M. S. |
| Avr. 11. β♌ | 11 26 57, 0 | 53 14 28 | −31 42 5 | −0 14 36 | | |
| o ♌ | 11 33 29, 0 | 54 18 44 | −25 10 5 | −0 49 40 | | |
| Pallas | 11 58 39, 5 | 53 29 4 | | | 182 40 24 | 15 55 24 B. |
| Avr. 15. β♌ | 11 26 18, 5 | 53 14 18 | −29 43 0 | −1 13 40 | | |
| o ♌ | 11 32 50, 5 | 54 18 27 | −23 11 0 | −0 9 31 | | |
| Pallas | 11 56 1, 5 | 54 27 58 | | | 182 10 31 | 16 54 30 B. |
| Avr. 19. β♌ | 11 25 38, 0 | 53 14 9 | −28 0 0 | −2 5 53 | | |
| o ♌ | 11 32 10, 5 | 54 18 24 | −21 27 5 | −1 1 38 | | |
| Pallas | 11 53 38, 0 | 55 20 2 | | | 181 44 41 | 17 46 40 B. |
| Mai 5. Pallas | 11 47 2, 5 | 57 40 42 | | | 180 55 37 environ | 20 7 47 B. |

**Fig. 11.15** Observations of Pallas by Bode, from April 11 to May 5, 1802

*On the 15th of May we were informed, by a letter from baron von Zach, the celebrated astronomer of Gotha, that Dr. Gauss, an astronomer of Brunswick, had found an ellipsis which corresponded to the first observations. On the 22nd we received the details. He found the revolution to be four years seven months, and the inclination 35°. This great inclination seemed to remove it from the order of planets, and some astronomers called it a comet; but its proximity, and continual appearance, will not allow of its being placed among the number of those stars of which we often lose sight for so long a time, and which go to enormous distances.*

*C. Burckhardt, on his part, made similar researches; he made several trials with ellipses very much elongated, which gave him a result very near that of Dr. Gauss.*

*On finding that this planet, like that of Piazzi, was between Mars and Jupiter, and that its motion must be affected by the attraction of Jupiter, C. Burckhardt undertook to calculate these perturbations. The calculation is long and difficult, but it is indispensably necessary to obtain the orbit with more exactness.*

*At last, on the 4th of June, he finished these laborious calculations, and found the following elements:*

*Distance 2.791, or 95,890,000.*
*Revolution, 4 years, 8 months, and 3 days.*
*Eccentricity, 0.2463; equation of the orbit, 28° 25′.*
*Epoch of 1802, 4Z 23° 50′; aphelion 10Z 2° 3′;*
*Node 5Z 22° 28′; inclination 34° 50′ 40″.*

*These elements corresponded to five observations of the 4th, 16th, and 27th of April, and the 7th and 20th of March; the last two made by C. Burckhardt, and Lalande's nephew, who, as well as C. Méchain, Messier and Delambre, continued to observe it as long as it could be seen in the meridian, because such observations are the surest. After the 21st of May, other instruments and other stars were necessary; but it still passed through some included among the 15,000 stars which we have published. On the 15th of June these elements corresponded, within a few seconds, with the observations of Méchain and Messier; which confirms the exactness of the elements found by C. Burckhardt, and assures us, that the motion of the new planet is already known. Baron von Zach has published a great many observations respecting it in his Journal.*

*C. Cabrol de Murol* [see F. N.] *has calculated for us an ephemeris, which gives the situation of this planet to the 21st of October, on which day it will have 227° 7′ of right ascension, and 6° 8′ of declination. It will then set at 7 h 51′: there is therefore reason to think that it may be still observed. It will be above Libra near the Serpent, after passing the legs of the Cow-herd* [Aquila]. *He finds that, in 1806, it will have 33½° of south declination, and that it will then be difficult to see it at Paris; but C. Vidal, who has already observed it this year, will then be better able than we to follow it.*

*Its greatest northern declination will not exceed 26½ degrees, a term at which it will be a year hence. It will be easier to be seen, but its distance will be double, and its light four times less than the present year. In the month of March 1804, it will be at three times the distance; its light will be nine times less, and, in all probability, it will be difficult to observe it.*

*As the orbit of this new planet intersects that of Piazzi, I was curious to know whether the two planets might not meet; but I found that, when they are in the same plane, there will be an interval of about 19 millions of leagues between them.*

*The planet of Dr. Olbers is very small. If we suppose its apparent diameter to be half a second, I find that its real diameter cannot be more than 100 leagues. Dr. Herschel, in a paper which he read before the Royal Society on the 7th of May, makes it be four times less. He says, that on the 22nd of April Piazzi's planet was only 22 hundredths of a second, and that of Olbers 13 hundredths; but it appears to me, that we have no means of determining, with certainty, quantities so small.*

*Dr. Olbers calls his new planet Pallas; but, as I see no sufficient motive for this fabulous denomination, I prefer giving it the name of the person to whom we are indebted for this valuable discovery.*

*Dr. Olbers distinguished himself in 1797 by an excellent treatise on comets, and was worthy of the good fortune with which his labours have been crowned.*

[F.N.: Michel Chabrol de Murol was born in Riom on November 18, 1777, and worked as an astronomer at the Paris observatory. He published a method to calculate eclipses. It is not known where or when he died.]

## Collection of Astronomical Discourses Observations and Reports

*BAJ*, 1802, p. 102, by Olbers

***July 24.*** *I have concluded my observations of Pallas on July 9. This small planet had become too faint, especially since the light of the Moon now combined with the twilight, and I doubted my ability of determining its entry and exit with any certainty.* [Observations from June 2-July 9 printed here.]

*Ellipse No. 3 by Dr. Gauss still coincides splendidly with the observations, but the right ascension seems to be a little too small. Dr. Burckhardt has also found an ellipse for Pallas taking into consideration the perturbations of Jupiter, which coincides fully with the details of Dr. Gauss.*

*Since I saw Pallas relatively well in spite of its low position in the haze of the evening horizon on July 9, I hoped to be able to see it on the occasion of its future opposition in*

*1803. At the end of June 1803, it will be somewhat closer to Earth than it was on July 9, 1802. Admittedly, given its greater distance from the Sun in July 1803, it will be less illuminated by it, but it can still be found high in the sky or on the meridian. And if the dusk should hinder us in northern Germany, it will certainly not escape viewing in southern France or Italy, especially since we will be able to quote its position to within 10′ or 15′.*

*If Burckhardt's two ellipses for Ceres and Pallas are compared, both paths come extremely close in the descending node of Pallas' path and Ceres' path. The distance is only 0.041.*

## Continued Reports regarding the New Primary Planet of our Solar System: Pallas Olbersiana

*MC*, July 1802, p. 71

*Just as Ceres Ferdinandia had been observed with the greatest care and with incessant zeal, Pallas Olbersiana was observed by all astronomers in Europe which just as much (but not more) zeal, and with redoubled attention.*

*The observation of this planet is presently of great importance and is all the more important than that of Ceres, for the latter has already been carefully observed for a year and a half, and its path has already been so closely determined that this new body can never again escape us; but Pallas, on the other hand, has only been observed for the past three months, and since its path could be only less exactly calculated from such a slight described arc, through which its future discovery in 1803, when this planet once again emerges from the rays of the Sun to once again become visible to us, will be made much more difficult, especially if the suspicion should hold true (as our readers will soon discover), that we will be unable to again observe Pallas for the next two years, even with the best telescopes, because of its low light intensity and its great distance from the Earth. But before we present these reasons to our astronomical readers, we will first attempt, with the greatest care, to collect all the observations which have been conducted on Olbers' planet and which have come to our knowledge, as we had done with Ceres.*

*Dr. Maskelyne first began observing this planet on April 23 in the meridian. The right ascensions were observed as usual on the meridian telescope and the declinations on the 8-foot Bird wall-quadrant. The improvement of 3″.8 mentioned in the previous essay about Ceres Ferdinandia has been incorporated in the following observations from Greenwich* (Fig. 11.16).

| 1802 | | Mittl. Zeit in Greenwich | | | Scheinbare gerade Aufst. der Pallas | | | Nördliche Abweich. der Pallas | | |
|---|---|---|---|---|---|---|---|---|---|---|
| April | 23 | 10U | 0′ | 42″ | 181° | 23′ | 42,″0 | 18° | 32′ | 28,″0 |
| | 25 | 9 | 52 | 18 | 181 | 15 | 28, 3 | 18 | 52 | 29, 7 |
| | 26 | 9 | 48 | 7 | 181 | 11 | 50, 2 | 19 | 2 | 2, 6 |
| May | 1 | 9 | 27 | 38 | 180 | 59 | 18, 9 | 19 | 42 | 54, 4 |
| | 2 | 9 | 23 | 36 | 180 | 57 | 57, 3 | 19 | 49 | 56, 7 |
| | 4 | 9 | 15 | 38 | 180 | 56 | 12, 9 | 20 | 2 | 53, 3 |
| | 7 | 9 | 3 | 48 | 180 | 55 | 46, 5 | 20 | 19 | 29, 9 |
| | 9 | 8 | 56 | 7 | 180 | 58 | 21, 0 | 20 | 28 | 42, 5 |
| | 11 | 8 | 48 | 28 | 181 | 1 | 44, 9 | 20 | 36 | 48, 2 |
| | 13 | 8 | 40 | 56 | 181 | 6 | 33, 0 | 20 | 43 | 33, 1 |
| | 14 | 8 | 37 | 11 | 181 | 9 | 29, 4 | 20 | 46 | 26, 7 |
| | 16 | 8 | 29 | 47 | 181 | 16 | 28, 2 | 20 | 51 | 18, 9 |

**Fig. 11.16** Observations of Pallas by Dr. Maskelyne conducted on the Royal Observatory in Greenwich

*In the previous issue, we communicated some observations of Pallas from Milan, which Oriani conducted in an equatorial sector. Here, those observed from the same observatory on a splendid 8-foot Ramsden wall-quadrant by the astronomer Cesaris follow* (Fig. 11.17):

*In Cracow, the skilful and diligent Professor Sniadecki continued to observe Pallas, but unfortunately, he had to give up his meridian observations of it on May 12. Only the following observations have been added to those indicated in the previous issue (p. 605)* (Fig. 11.18):

*In Vilnius, Pallas Olbersiana had still not been found by May 8. It appears that it will never be observed there, which is even more of a shame because the Royal Observatory is equipped with the most splendid Ramsden instruments. The cause of the unsuccessful location lies in the very disorderly postal service, since, as Professor Sniadecki tells us, from time to time it takes up to an entire month to have an answer from Cracow to reach Vilnius. Since the path of Pallas was not well-known at first, the path that this planet would take could therefore not be determined in advance, thus, in the time it took for the news of the first observation in Germany to reach Vilnius, its position had significantly changed, and this is why locating it had become very much more difficult. It was not so difficult with Ceres. Let Gauss's efforts be praised, for it is through these that the approximate position of where Piazzi's planet was to be found again was discovered before it reappeared. As soon as it had been located, it was possible to indicate the error and deviation of Gauss's ephemeris, which had been calculated in advance, and according to this, it was then easy for anyone to immediately find this planet. In the meantime, the honourable veteran*

**Fig. 11.17** Observations of Pallas by Cesaris from May 4 to May 22, 1802

| 1802 | Mittl. Zeit in Mailand | | | Scheinbare ger. Aufsteig. der Pallas | | | Nördliche Abweich. der Pallas | | |
|---|---|---|---|---|---|---|---|---|---|
| May 4 | 9U | 15′ | 43,″1 | 180° | 56′ | 0″ | 20° | 2′ | 50″ |
| 5 | 9 | 11 | 45, 3 | 180 | 55 | 37 | 20 | 8 | 33 |
| 6 | 9 | 7 | 49, 3 | 180 | 55 | 44 | 20 | 14 | 4 |
| 7 | 9 | 3 | 55, 6 | 180 | 56 | 14 | 20 | 19 | 20 |
| 8 | 9 | 0 | 3, 2 | 180 | 57 | 5 | 20 | 24 | 12 |
| 9 | 8 | 56 | 12, 1 | 180 | 58 | 15 | 20 | 28 | 43 |
| 10 | 8 | 52 | 22, 4 | 180 | 59 | 45 | 20 | 32 | 55 |
| 11 | 8 | 48 | 34, 3 | 181 | 1 | 36 | 20 | 36 | 47 |
| 12 | 8 | 44 | 47, 0 | 181 | 3 | 51 | 20 | 40 | 16 |
| 17 | 8 | 26 | 12, 9 | 181 | 20 | 6 | 20 | 53 | 22 |
| 18 | 8 | 22 | 33, 1 | 181 | 24 | 18 | 20 | 55 | 3 |
| 19 | 8 | 18 | 55, 8 | 181 | 28 | 53 | 20 | 56 | 31 |
| 20 | 8 | 15 | 19, 6 | 181 | 33 | 50 | 20 | 57 | 45 |
| 21 | 8 | 11 | 44, 6 | 181 | 39 | 8 | 20 | 58 | 41 |
| 22 | 8 | 8 | 11, 1 | 181 | 44 | 46 | 20 | 59 | 22. |

| 1802 | Mittlere Zeit in Cracau | | | Scheinbare AR. der Pallas | | | Scheinb. nördliche Abweich. | | | Sterne womit verglichen worden |
|---|---|---|---|---|---|---|---|---|---|---|
| May 8 | 9U | 0′ | 11,″5 | 180° | 57′ | 14″ | 20° | 23′ | 52″ | Arcturus und für die AR. α, θ Leonis |
| 9 | 8 | 56 | 20, 0 | 180 | 58 | 5 | 20 | 28 | 57 | |
| 11 | 8 | 48 | 41, 0 | 181 | 1 | 32 | 20 | 38 | 18 | |
| 12 | 8 | 44 | 55, 0 | 181 | 3 | 50 | 20 | . | . | |

**Fig. 11.18** Observations of Pallas by Sniadecki from May 8 to 12, 1802

*Poczobut, who still takes an active part in the newest advances of astronomy, wrote the following verses in honour of Pallas:*

*Sickle be the sign of Ceres; and have with you Pallas*
*who has the Aegida, so that you can protect the*
*sacred works in the skies.*

[The Aegida is a shield used by Pallas, made of goatskin and the head of Medusa; see Sniadecki to Zach: May 24, 1802.]

*After the astronomers are forced to conclude their meridian observations of Pallas because of the approaching dusk in the middle of May, it still remains to observe this planet outside of the meridian in the western sky. If these observations should equal those done in the meridian in quality and sharpness, they could then only be replaced by those conducted on fixed equatorial sectors. We now expect such observations only from Milan and Greenwich. In the meantime, even if this planet cannot be observed with the same sharpness and exactness as on meridian and equatorial instruments, the lack of this will be replaced on the other hand by the longer duration of these observations, through which the arithmetician may receive a greater described arc upon which he may base the determination of this planetary path with greater certainty. Even if the circular micrometrical observations can boast of no extreme precision, especially in declinations, this can be replaced largely by diligence, skilfulness and the repetition of observations. Dr. Olbers is very skilled in this type of observation; his observations always agree relatively well, since he continued to diligently observe Pallas with redoubled attention on the circle micrometer in the present period.*

*Admittedly, these observations depend largely on those stars that must perforce serve in the comparison with the planet. If the positions of these stars are poorly determined (as is largely the case), the observations of the planet can invariably be nothing other than poor. However, these observations do not become useless if the original observations are saved; these stars of uncertain or doubtful position can be precisely determined when they once again culminate at night, and the observations and results be corrected accordingly in future.*

*In this way, in his continued observations of Pallas, Dr. Olbers had to make use of three small magnitude nine stars that appear in no sources other than Lalande's* Histoire cèléste francaise, *p. 68 and 228. Out of this, their positions must now be calculated. The first two, which Dr. Olbers designated as j and ij, have only been observed once; the third star, iij, appears twice. This is why, though their positions may very well be exactly determined, there can however be no confirmation. Dr. Olbers reduced the stars as well as his planetary observations as exactly as possible. He had the goodness of sending me his original observations along with the request to reduce them once again myself. Dr. Olbers continues: "I haven't all the means at my disposal that you have. You certainly know the right ascensions of several stars of this zone very well from your own observations as well as the observations of the declinations of Henry and Barry, and you will therefore be able to conduct a far more exact reduction of these two stars." In order to live up to such flattering confidence to the best of my ability, I calculated, according to the newest aids, the positions of these three required stars as exactly as possible, and made use of very well determined stars in their parallel as a means of comparison. Since these may come in useful to other observers, we will set their determinations here.* [a table of star positions]

*According to these determinations, the observations by Olbers continued from p. 601 now stand thus. So that anyone may repeat and improve the reduction of the three stars j, ij and iij upon future, more precise determination, we have included the original observation as well as the difference between our and Dr. Olbers' own reductions* (Fig. 11.19).

*From May 17 to 28, Dr. Olbers' clock went 8" slower in one stellar day, and from May 30 to June 10, 10" slower than the mean solar time. The above differences in the RA reduced*

| 1802 | Mittlere Zeit in Bremen | Scheinbare gerade Aufsteig. der Pallas nach von Zach's Reduction | Unterschied mit D. Olbers | Scheinbare Abweichung der Pallas nach v. Zach's Reduction | Unterschied mit D. Olbers | Pallas folgt — geht vor + in Zeit } * | Pallas südl. — nördl. + } * |
|---|---|---|---|---|---|---|---|
| May 10 | 11U 35′ 45″ | 180° 59′ 55″ | . . . . | 20° 33′ 26″ | . . . . | . . . . . . | . . . . . |
| 14 | 12 3 43 | 181 9 37 | . . . . | 20 46 29 | . . . . | . . . . . . | . . . . . |
| 17 | 11 1 14 | 181 20 37 | — 1″ | 20 53 42 | 0″ | Nr. j — 3′ 0,″5 | — 0′ 51″ |
| 18 | 10 4 52 | 181 24 52 | — 0 | 20 55 19 | 0 | — 3 17, 5 | + 0 46 |
| 20 | 10 37 34 | 181 34 41 | — 2 | 20 57 59 | 0 | — 3 56, 7 | + 3 25 |
| 21 | 10 21 37 | 181 39 48 | — 3 | 20 58 51 | — 1 | — 4 17, 2 | + 4 17 |
| 22 | 11 17 47 | 181 45 30 | — 13 | 20 59 27 | — 2 | Nr. ij + 1 50, 0 | — 5 51 |
| 23 | 11 13 46 | 181 51 11 | — 13 | 20 59 52 | — 2 | + 1 27, 3 | — 5 26 |
| 24 | 10 50 52 | 181 57 30 | — 13 | 20 59 59 | — 1 | + 1 2, 0 | — 5 18 |
| 25 | 11 7 38 | 182 4 8 | — 13 | 20 59 51 | — 2 | + 0 35, 5 | — 5 27 |
| 26 | 10 45 56 | 182 11 2 | — 11 | 20 59 18 | — 1 | + 0 7, 8 | — 5 59 |
| 28 | 10 50 7 | 182 25 37 | — 9 | 20 58 14 | — 1 | — 0 50, 5 | — 7 3 |
| {30 | 11 40 46 | 182 41 30 | + 1 | 20 56 21 | — 1 | — 1 54, 5 | — 8 56 |
| {30 | 12 14 20 | 182 41 37 | + 6 | 20 56 11 | — 1 | Nr. j — 8 24, 0 | + 1 37 |
| Jun. 2 | 11 13 4 | 183 7 16 | + 2 | 20 51 18 | + 1 | Nr. iij + 0 53, 2 | + 5 48 |
| 3 | 10 46 7 | 183 16 4 | + 9 | 20 49 8 | + 1 | + 0 18, 0 | + 3 38 |
| 6 | 12 53 45 | 183 46 19 | + 14 | 20 41 33 | + 1 | — 1 43, 0 | — 3 57 |
| 8 | 11 9 10 | 184 6 3 | + 18 | 20 35 58 | + 1 | — 3 2, 0 | — 9 32 |
| 10 | 12 33 50 | 184 28 21 | + 22 | 20 29 38 | + 1 | — 4 31, 2 | — 15 52 |

**Fig. 11.19** Pallas observations by Olbers, with differences from Zach's data

*by us and the declinations with the included signs give the locations of the planet as Dr. Olbers had reduced them. In the declination, the difference is completely insignificant; only in the right ascension is it of greater significance where the stars ij and iij have been used. Accordingly, this difference obviously comes from the different reduced right ascensions of both stars. Future, more exact observations of them will completely eliminate this minor difference.*

*Just as tirelessly as the observers of the heavens have followed this new planet with continuously precise observations, so too did Dr. Gauss follow this highly remarkable planet with his strict calculations. In this way, observers and arithmeticians work hand in hand. As the former observed the geocentric course of this planet from day to day and week to week and delivered new material for the calculation of the heliocentric course, the latter processed these and ever more closely approached the true path of this extraordinary stranger in the Solar System.*

*In this way, Dr. Gauss had calculated two paths, which we have communicated to our readers in the previous issue on p. 593 and 596: No. I, which is based on 'seventeen' observations from Seeberg, and No. II, which is based on 'twenty-seven.' Continued observations have made it possible for this accomplished mathematician to calculate the path for the 3rd time, and to bring it into line with 'forty-two' observations from Seeberg and Greenwich. Authorities will recognise with admiration with what tact and exactness this excellent arithmetician came so close to the truth with such poor observations, even if he hitherto limited himself solely to the 'purely elliptical' elements with the exception of any perturbation equations with Ceres as well as with Pallas. It is instructive as well as interesting to hear the reasons which led this astute arithmetician to the method – we do not believe to be able to do it better than by setting his own words down here:* [This passage is printed in Chap. 1 in this book, in the section on the mass of Jupiter.]

*We also hope that the present 3rd elements of the path of Pallas (only if the planet is luminous enough) will suffice to find this planet in 1803, especially if Dr. Gauss were to undertake a 'fourth' improvement of the elements, which he can expect in future from the equatorial observations from Milan and Greenwich. In the meantime, his ellipse, improved for the third time, now follows* (Fig. 11.20):

| | | |
|---|---|---|
| Epoche 1802 März 31 Mittag in Seeberg | | 162° 25′ 45″9 |
| tägliche mittlere tropische Bewegung | | 769″547 |
| Logarithmus der halben grossen Axe | | 0.4425664 |
| Sonnenferne | für die Epoche und siderisch ruhend | 300° 58′ 47″7 |
| Knoten | für die Epoche und siderisch ruhend | 172° 28′ 17″9 |
| Excentricität | | 0.2476402 |
| Neigung | | 34° 39′ 10″7 |

**Fig. 11.20** Third Elements of Pallas Olbersiana

*It is highly remarkable that the mean tropical movement of Pallas always approaches that of Ceres, unintentionally and entirely by itself. According to the 7th elements of the path of Ceres (December issue 1801, p. 647), Dr. Gauss determined the mean movement to be 770".914. This differs by only one and a half seconds from that of Pallas. But since some uncertainty still surrounds both movements, we can't really indicate any difference yet.*

*Dr. Gauss compared the above 3rd elements with the series of observations that follow, and discovered their wonderful agreement, which the following table indicates* (Fig. 11.21).

*From this comparison, it is clear that the elements could only be changed slightly according to this entire series of observations, especially since the last declinations in May from Greenwich (which have a greater influence on the elements than the right ascensions) agree very well with each other. Only the calculated right ascensions must still be decreased by about 4" towards the end. The elements would be only slightly affected by this. In a letter dated June 26, Dr. Gauss reports to us that he had compared the observations of this planet from June 19, 20 and 21 with his 3rd elements and these agreed so well that he could certainly not change anything. We can therefore leave the 3rd elements of the path at that until the observations are finally completed. With that, Dr. Gauss calculated the following* (Fig. 11.22).

*For the beginning of the next year, Dr. Gauss calculated the position of Pallas approximately as follows: on December 31, at midnight in Seeberg, longitude 253° 33', latitude 26° 10' northern declination from Earth 3.779 and from the Sun 3.043. However, it remains very uncertain whether we will be able to see this planet again because of its faintness. Dr. Olbers calculated both following oppositions of Pallas according to the 2nd elements by Gauss. When he discovered it on March 28, its distance from Earth was only = 1.38, by the end of May = 2.00, and on June 15 = 2.31. Next year, in 1803, towards the end of June, Pallas will once again come in opposition with the Sun. It will come to stand over Cerberus in 9Z 8° longitude and 48° northern latitude, but its distance from Earth will be 2.61. Dr. Olbers believes that it ought to be scarcely visible at this distance during the bright northern twilight. It is somewhat better in the opposition of 1804, which will occur towards the end of August at 11Z 3° with 19° N. latitude, therefore in the head of Pegasus. Indeed, Dr. Olbers finds the distance from Earth to be 2.40, but the nights will already be completely dark. But as Dr. Olbers believes, we will most certainly see Pallas again in 1805, since it will have descended to the portion of its path lying closer to the Sun in opposition, and will therefore be much closer to Earth. Meanwhile, these contemplations will hopefully hinder no astronomers in tracking Pallas down as soon as the opportunity presents itself. Large telescopes of great light transmitting capacity may still be sufficient to rediscover this planet, even if it should only appear as a magnitude 12 star. Admittedly, locating it will be associated with somewhat more difficulty. Assuming that all attempts fail in 1803 and 1804, Dr. Gauss still hopes that his elements of the path will still furnish an*

| 1802 | | Berechnete gerade Aufst. der Pallas | Berechnete Abweich. der Pallas | Unterschied in AR. | Unterschied in Declin. | Beobachter |
|---|---|---|---|---|---|---|
| April | 4 | 183° 44′ 5,″6 | 13° 54′ 53,″5 | − 1,″0 | + 1,″5 | v. Zach |
| | 5 | 183 34 24, 4 | 14 13 20, 2 | + 0, 7 | − 2, 7 | v. Zach |
| | 7 | 183 15 38, 5 | 14 48 59, 3 | 0 | − 2, 8 | v. Zach |
| | 8 | 183 6 33, 9 | 15 6 12, 8 | − 3, 9 | + 2, 8 | v. Zach |
| | 10 | 182 48 56, 3 | 15 39 43, 5 | − 1, 7 | + 1, 5 | Dr. Burckhardt |
| | 12 | 182 32 33, 7 | 16 11 6, 4 | + 2, 2 | + 8, 4 | Méchain |
| | 13 | 182 24 46, 8 | 16 26 8, 3 | + 2, 4 | + 2, 3 | Méchain |
| | 15 | 182 10 14, 1 | 16 54 33, 7 | − 2, 4 | + 2, 9 | v. Zach |
| | 15 | 182 10 4, 2 | 16 54 53, 3 | + 1, 3 | + 1, 9 | Méchain |
| | 16 | 182 3 9, 7 | 17 8 30, 3 | + 1, 7 | − 0, 5 | Méchain |
| | 17 | 181 56 33, 7 | 17 21 52, 8 | + 3, 4 | + 3, 0 | Méchain |
| | 18 | 181 50 25, 0 | 17 34 25, 3 | − 5, 6 | . . | v. Zach |
| | 19 | 181 44 26, 4 | 17 46 50, 0 | + 1, 1 | − 4, 4 | v. Zach |
| | 23 | 181 23 41, 9 | 18 32 27, 4 | − 0, 1 | − 0, 6 | Dr. Maskelyne |
| | 24 | 181 19 32, 4 | 18 42 25, 2 | − 5, 4 | . . | v. Zach |
| | 25 | 181 15 35, 3 | 18 52 15, 9 | + 3, 1 | . . | v. Zach |
| | 25 | 181 15 28, 6 | 18 52 33, 1 | + 0, 3 | + 3, 4 | Dr. Maskelyne |
| | 26 | 181 11 59, 4 | 19 1 41, 6 | + 2, 4 | − 7, 8 | v. Zach |
| | 26 | 181 11 53, 3 | 19 1 58, 1 | + 3, 1 | − 4, 5 | Dr Maskelyne |
| | 27 | 181 8 44, 8 | 19 10 42, 7 | + 0, 8 | − 4, 1 | v. Zach |
| | 29 | 181 3 20, 9 | 19 27 31, 5 | + 4, 3 | − 12, 2 | v. Zach |
| | 30 | 181 1 12, 1 | 19 35 19, 6 | + 2, 2 | − 13, 9 | v. Zach. |
| May | 1 | 180 59 24, 6 | 19 42 41, 8 | + 2, 1 | + 4, 0 | v. Zach |
| | 1 | 180 59 21, 7 | 19 42 57, 7 | + 2, 8 | + 3, 3 | Dr. Maskelyne |
| | 2 | 180 57 59, 7 | 19 49 46, 5 | + 7, 8 | − 5, 5 | v. Zach |
| | 2 | 180 57 57, 5 | 19 49 58, 7 | + 0, 2 | + 2, 0 | Dr. Maskelyne |
| | 3 | 180 56 56, 9 | 19 56 25, 9 | + 1, 3 | − 21, 8 | v. Zach |
| | 4 | 180 56 15, 0 | 20 2 53, 0 | + 2, 1 | − 0, 3 | Dr. Maskelyne |
| | 5 | 180 55 56, 9 | 20 8 36, 2 | + 2, 3 | − 2, 5 | v. Zach |
| | 6 | 180 56 0, 3 | 20 14 8, 8 | + 6, 2 | . . | v. Zach |
| | 7 | 180 56 24, 7 | 20 19 20, 4 | + 0, 9 | − 14, 9 | v. Zach |
| | 7 | 180 56 25, 8 | 20 19 29, 4 | + 39, 3 | − 0, 5 | Dr Maskelyne |
| | 8 | 180 57 13, 0 | 20 24 10, 8 | + 3, 0 | − 9, 3 | v. Zach |
| | 9 | 180 58 23, 9 | 20 28 49, 3 | + 2, 3 | + 0, 8 | Dr Maskelyne |
| | 11 | 181 1 44, 6 | 20 36 41, 9 | + 2, 9 | − 31, 3 | v. Zach |
| | 11 | 181 1 48, 3 | 20 36 48, 5 | + 3, 4 | + 0, 3 | Dr. Maskelyne |
| | 13 | 181 6 37, 6 | 20 43 31, 4 | + 4, 6 | − 1, 7 | Dr. Maskelyne |
| | 14 | 181 9 33, 4 | 20 46 25, 2 | + 4, 0 | − 1, 5 | Dr. Maskelyne |
| | 16 | 181 16 26, 5 | 20 51 19, 3 | − 1, 7 | + 0, 5 | Dr. Maskelyne |

**Fig. 11.21** Difference between observed and calculated positions of Pallas from April 4 to May 16, 1802

**Fig. 11.22** Ephemeris for Pallas Olbersiana for midnight, Seeberg

*Ephemeride für die Pallas Olbersiana für Seeberger Mitternacht.*

| 1802 | | AR. der Pallas | | Declin. der Pallas | | Abstand von d. ♁ | Abstand von d. ☉ |
|---|---|---|---|---|---|---|---|
| May | 18 | 181° | 25′ | 20° | 55′ | 1,888 | 2,477 |
| | 21 | 181 | 40 | 20 | 59 | 1,930 | 2,484 |
| | 24 | 181 | 58 | 21 | 0 | 1,973 | 2,492 |
| | 2 | 182 | 19 | 20 | 59 | 2,016 | 2,500 |
| | 30 | 182 | 42 | 20 | 56 | 2,060 | 2,508 |
| Jun | 2 | 183 | 8 | 20 | 51 | 2,104 | 2,516 |
| | 5 | 183 | 36 | 20 | 44 | 2,148 | 2,524 |
| | 8 | 184 | 0 1/2 | 20 | 36 | 2,192 | 2,532 |
| | 11 | 184 | 39 | 20 | 26 1/2 | 2,237 | 2,539 |
| | 14 | 185 | 14 | 20 | 15 | 2,281 | 2,547 |
| | 17 | 185 | 51 | 20 | 3 | 2,326 | 2,555 |
| | 20 | 186 | 30 | 19 | 50 | 2,370 | 2,563 |
| | 23 | 187 | 10 | 19 | 35 | 2,415 | 2,571 |
| | 26 | 187 | 52 1/2 | 19 | 19 1/2 | 2,459 | 2,579 |
| | 20 | 188 | 36 | 19 | 3 | 2,504 | 2,587 |

*adequate connecting thread for 1805 to rediscover Pallas, though admittedly with somewhat more effort than was the case with Ceres.*

[The next section of this paper, starting on p. 87, recounts missives from Olbers about the asteroid explosion hypothesis. This material appears in Chap. 2 of this book. It is followed by Zach's assessment of Herschel's asteroid paper of 1802, in Chap. 10 of this book. The text picks up here with Zach referring to the diameter measurements of Ceres and Pallas.]

*In order to represent the difficulty of such measurements, Dr. Gauss made the following calculations of the apparent diameters of these bodies from his last elements of the paths* (Fig. 11.23).

*Whether such small diameters can really be measured, let alone see the reflected light of such small surfaces, we leave it to the judgment of others.*

*Since the suggested naming of Pallas by Dr. Olbers won the acceptance of all English, French, German and Italian astronomers, and since it has already been generally adopted, it is now time to think of a fitting symbol ⚴ for this new planet. Just as we suggested the sign of a sickle for Ceres, which is now used by all astronomers, we presently suggest for Pallas its lance, as attribute, for its marking. This symbol has already won the favor and approval of its first discoverer, and Dr. Olbers expresses in a letter regarding this as follows: "I very much like the sign for Pallas. The same applies for the sign of Ceres; except, this symbol can easily be mistaken for Venus ♀ from time to time in the case of poor print or over-hasty handwriting. Wouldn't it have been better if you had put a pommel on the handle of the sickle instead of a bar? Then, no confusion would be possible with the sign of Venus, and the sickle with a pommel is just as easy [simple]."*

## Resumed News about the new main Planet of our Solar System, Pallas Olbersiana

*MC*, August 1802, p. 187

*We continue to collect the sparse material and any meridian observation of this planet in order to base this planet's theory on firm ground. Oriani followed Pallas tirelessly with his equatorial sector – just as he had followed Ceres. Here are the Milanese original observations, which are accompanied by our reductions and calculated positions according to the observer's wish. Here are Oriani's own data* [Data from April 25 to July 8, printed later in this chapter in Oriani's Italian-language paper of 1803.]

*It seems that an error has occurred in the observed declination of the star 33 Comae Berenices on July 8; we assume that 8 min should be 10 min and the observation accordingly 18° 10′ 39″, which would match the other observations better. The reading of*

**Durchmesser der Ceres.**

| 1802 | Abstand von der ♁ | scheinbarer Durchmesser, den wahren nach Dr. Herschel = 35 Deutsche Meilen gesetzt. |
|---|---|---|
| März 28 | 1.621 | 0″22 |

**Durchmesser der Pallas.**

| 1802 | Abstand von der ♁ | Scheinb. Durchmesser, wenn man Dr. Herschel's Resultat = 15 Deutsche Meilen als richtig annimmt. |
|---|---|---|
| April 4 | 1.407 | 0″11 |
| — 15 | 1.386 | 0. 10 |
| Juni 23 | 2.415 | 0. 06 |

**Fig. 11.23** A table of distances from Earth, and the corresponding apparent diameters of Ceres and Pallas, assuming the physical diameters given by Herschel

*the clock, necessary for reducing, has already been given with the observations of Ceres on page 183 of this issue. The positions of the employed stars are as follows:* [a table of star positions]

*Our calculation with this data results in the following positions of Pallas; but it has to be said that since no. 24 Comae Berenices is a binary star we did not use the first but the second star, which was following the first in a second. Oriani did not say which of both stars he used for comparing. But the second star provides a better correspondence.*

| *1802* | *Mean Time* | *Apparent RA* | *Apparent Decl.* | *Comp. Star* |
|---|---|---|---|---|
| *Jun 24* | *10 h 10′ 9″* | *187° 23′ 32″.5* | *19° 28′ 41″.7* | *11 Com Ber.* |
| | *10 9 7* | *187 23 29.9* | | |
| | *10 11 11* | *187 23 10.4* | *19 28 41.4* | *24 Com Ber.* |
| *Jun 28* | *10 2 36* | *188 20 48.0* | *19 8 30.7* | *24 Com Ber.* |
| | *10 4 40* | *20 15.3* | | |
| *Jul 1* | *10 18 10* | *189 5 53.8* | *18 52 3.1* | *11 Com. Ber.* |
| | *10 20 14* | *5 40.8* | | |
| | *10 18 10* | *189 5 44.8* | *18 51 45.8* | *23 Com. Ber.* |
| | *10 20 14* | *5 27.7* | | |
| *Jul 2* | *10 18 10* | *189 20 59.8* | *18 46 4.1* | *11 Com. Ber.* |
| | *10 20 17* | *20 38.7* | | |
| | *10 18 14* | *189 20 50.8* | *18 46 7.7* | *25 Com. Ber.* |
| | *10 20 17* | *20 41.8* | | |
| *Jul 3* | *10 6 52* | *189 36 15.8* | *18 39 45.7* | *25 Com. Ber.* |
| | *10 8 55* | *36 11.3* | | |
| *Jul 7* | *9 39 3* | *190 38 58.1* | *18 15 54.1* | *15 Com. Ber.* |
| | *9 41 7* | *39 2.6* | | |
| | *9 39 3* | *190 38 53.9* | *18 15 38.6* | *17 Com. Ber.* |
| | *9 41 7* | *39 13.6* | | |
| *Jul 8* | *9 41 15* | *190 55 32.3* | *18 9 8.1* | *25 Com. Ber.* |
| | *9 43 17* | *55 11.2* | | |
| | *9 41 15* | *190 55 43.3* | *18 9 13.6* | *27 Com. Ber.* |
| | *9 43 17* | *55 19.2* | | |
| | *9 43 17* | *190 55 15.6* | *18 9 18.5* | *33 Com. Ber.* |

*Every astronomer might choose an observation he considers the best. But we would like to inform you that Oriani gave a wrong minute on July 3: it should be 10 h 7′ 54″ instead of 9 minutes. Incessantly and eagerly Dr. Olbers continued to observe this planet at the circular micrometer. Since he again was so kind as to give us all of his original observations we publish those unaltered. His observations are:*

| *Mean time in Bremen* | *RA* | *Decl.* | *Reference star* |
|---|---|---|---|
| *Jun 14 11 h 56 m 14* | *185° 13′ 18″* | *20° 14′ 40″* | *Nor. iij* |
| *19 11 13 55* | *186 16 25* | *19 54 0* | *24 Com B.* |
| *20 11 7 51* | *186 29 34* | *19 49 40* | *— — —* |
| *21 11 8 54* | *186 42 59* | *19 44 58* | *— — —* |

(continued)

(continued)

| | | | |
|---|---|---|---|
| *26 11 18 34* | *187 51 47* | *19 19 23* | *———* |
| *Jul 4 11 28 14* | *189 52 38* | *18 33 44* | *33 Com. B.* |
| *8 11 5 16* | *190 56 9* | *18 8 59* | *———* |
| *8 11 13 20* | *190 56 13* | *18 8 57* | *32 Com. B.* |
| *9 11 53 47* | *191 13 4* | *18 2 34* | *———* |

*The observation of June 14 is not very reliable since it was made in bright moonlight. On the other hand the observation of July 8 is excellent. The planet was almost in the parallel of 32 Comae Berenices, which makes a determination of the difference in declination easy. Regarding the RA Pallas was compared seven times with 33 and six times with 32 Comae Berenices and regarding the declination three times with 33 C.B. and five times with 32 C.B. This positional determination is very accurate as can also be seen from the above double result from both stars. On July 9 the observation was by far not as reliable as that of the day before. Pallas was extremely faint and Dr. Olbers had difficulties noticing when it entered and left. He also believes that this was the last observation, for the moonlight in connexion with twilight makes this inherently faint planet at its lower positions unrecognisable for obtaining reliable observations. At a more southern polar altitude, where twilight is not as hindering, it might be possible to follow Pallas until the end of July. The above Milanese observations are also only until July 8 but Oriani promised in his last letter to be able to send another series weather permitting. But unfortunately at the most southern European observatory Pallas has not been observed. Here is an extract of Prof. Piazzi's letter of June 20 from Palermo:*

*"You guessed right that I would not see Pallas. Right upon reception of your two letters that announced Dr. Olbers' nice and peculiar discovery, which was confirmed some days later by Oriani, I tried to calculate a circular orbit from Dr. Olbers' second observation, your third and one of Oriani. I found a heliocentric longitude for the time of the first observation 183° 46′, the node 170° 12′, inclination 27° 1′. With these elements I searched Pallas on May 20, 23 and 24 – but in vain. It was already too close to the Sun to observe it in the meridian field ... I have published a small brochure on the history of Ceres' discovery. It feeds almost entirely from your letters and* MC. *I wrote it for Palermo only but would like to send you one copy; it contains my observations of this planet from February 21 to May 23 ... I will soon receive an equatorial sector; this and an additional allowance of 100 oz prove me right to call my planet Ceres Ferdinandea."*

*Also from Vilnius, Prague, Kremsmuenster, and Padua we received confirmation that Pallas has not been observed. According to a letter from Dr. Gauss, Dr. Maskelyne had not given up his observation of that planet on July 20 and believes to have observed Pallas as an 11$^{th}$ magnitude star but he was not sure whether it was the planet or a fixed star, which will tell the following observation or a calculation.*

*Dr. Gauss's elements III of Pallas' orbit still match the observed declinations; only the right ascensions differ by a few seconds. Dr. Gauss compared the Milanese observations of Ceres made at an excellent 8-foot wall quadrant by Ramsden, which we published in the last issue on p. 73 and obtained the following correspondence* (Fig. 11.24):

*Dr. Olbers' observations result in the same correspondence. The harmony amongst them show the value of those observations, although made only at a circular micrometer. He, who wants to examine those as thoroughly as we did, will find that they equal those made at equatorial sectors and sometimes even surpass them. This, of course, is rather a proof of an excellent observer than the quality of the instrument, as our readers might see from the following* (Fig. 11.25):

*Vergleichung der Mailänder Meridian-Beobachtungen der Pallas mit den GAUSS'schen III. Elementen.*

| 1802 | Mittl. Zeit in Mailand | Berechnete Gerade Aufsteigung | Berechnete Nördliche Abweichung | Unterschied R—B in AR. | Unterschied R—B in Decl. |
|---|---|---|---|---|---|
| Mai 4 | 9ᵘ 15ᵐ 43ˢ1 | 180° 56′ 15″7 | 20° 2′ 44″1 | + 15″7 | − 5″9 |
| 5 | 9 11 45.3 | 180 55 56.8 | 20 8 38.1 | + 19.8 | + 5.1 |
| 6 | 9 7 49.3 | 180 56 0.3 | 20 14 10.4 | + 16.3 | + 6.4 |
| 7 | 9 3 55.6 | 180 56 24.8 | 20 19 22.1 | + 10.8 | + 2.1 |
| 8 | 9 0 3.2 | 180 57 13.3 | 20 24 12.4 | + 8.3 | + 0.4 |
| 9 | 8 56 12.1 | 180 58 21.9 | 20 28 43.1 | + 6.9 | + 0.1 |
| 10 | 8 52 22.4 | 180 59 52.8 | 20 32 53.0 | + 7.8 | − 2.0 |
| 11 | 8 48 34.3 | 181 1 45.2 | 20 36 43.4 | + 9.2 | − 3.6 |
| 12 | 8 44 47.0 | 181 3 58.8 | 20 40 14,6 | + 7.8 | − 1.4 |
| 17 | 8 26 12.9 | 181 20 17.1 | 20 53 18.0 | + 11.1 | − 4.0 |
| 18 | 8 22 33.1 | 181 24 33.5 | 20 55 3.6 | + 15.5 | + 0.6 |
| 19 | 8 18 55.8 | 181 29 9.4 | 20 56 30.8 | + 16.4 | − 0.2 |
| 20 | 8 15 19.6 | 181 34 4.8 | 20 57 43.0 | + 14.8 | − 2.0 |
| 21 | 8 11 44.6 | 181 39 18.6 | 20 58 39.5 | + 10.6 | − 1.5 |
| 22 | 8 8 11.1 | 181 44 52.9 | 20 59 20.6 | + 6.9 | − 1.4 |

**Fig. 11.24** Comparison of positions from Gauss's Third Elements of Pallas with Oriani's observations in Milan from May 2 to 22, 1802

**Fig. 11.25** Comparison of positions from Gauss's Third Elements of Pallas with Olbers' observations from June 19 to July 9, 1802

| 1802 | Berechnete Gerade Aufſteig. der ⚴ | Berechnete Nördl. Abweich. der ⚴ | Unterſchied in AR. | Unterſchied in Decl. |
|---|---|---|---|---|
| Jun. 19 | 186° 16′ 19″ | 19° 54′ 17″ | − 6″ | + 17″ |
| — 20 | 186 29 12 | 19 49 39 | − 22 | − 1 |
| — 21 | 186 42 39 | 19 44 55 | − 20 | − 3 |
| — 26 | 187 52 3 | 19 19 31 | − 16 | + 8 |
| Jul. 4 | 189 52 12 | 18 33 39 | − 26 | − 5 |
| — 8 | 190 55 44 | 18 8 50 | − 25 | − 8 |
| — 8 | 190 55 50 | 18 8 48 | − 23 | − 10 |
| — 9 | 191 12 36 | 18 2 35 | − 28 | + 1 |

*From all those comparisons and excellent correspondence of the observations with the Gaussian ellipse our hope might be justified that we still possess this planet even if its rediscovery in 1803 and 1804 is impossible due to its faintness. But we are hoping that we will not miss Pallas the next year. Dr. Olbers, who observed this planet on July 9 also believes this and wrote: "It is difficult to see the planet now but I am convinced that you will see it next year with your passage instrument. It is then at its opposition slightly closer to the sun and will be less illuminated but this will be dispensable by its greater altitude above the horizon and the greater distance of the twilight."*

*And also Dr. Gauss believes in a rediscovery of Pallas the next year. Its distance from Earth on January 28, 1803 (around the day of its opposition), will be 2.56. He finds the diameter according to Dr. Herschel's own measurement slightly greater. Dr. Herschel gives on April 22 according to a fairly good observation the diameter = 0″.17; and Dr. Gauss calculated the true diameter 26½ German miles (the distance from earth = 1.562).* [A German mile is 25,000 feet, compared to 5,280 feet in an English mile.] *In his latest letter he expressed his astonishment about Dr. Herschel's and Dr. Schroeter's different results of the diameters because they were made according to one method. "I am very curious to learn what magnifications Dr. Herschel used. A magnification of 500 times would hardly turn an apparent diameter of 0″.17 into a disc, wouldn't it?" I, for my part, could not discern a trace of a disc at 300× magnification of neither Olbers' nor Piazzi's planet.*

*I asked Dr. Olbers about this matter and his answer was: "I never could see a difference between these planets and fixed stars of equal apparent diameter at the greatest magnifications of my Dollond achromat." On page 85 of the last issue we gave an ephemeris of Pallas, which is resumed here. It can no longer be used to facilitate the observations but it still might serve to prevent that mistaken stars are called Pallas* (Fig. 11.26).

## Resumed News about the new main Planet of our Solar System, Pallas Olbersiana

*MC,* Sept. 1802, p. 303

*All studies, efforts and calculations, all opinions, hypotheses and doubts of astronomers about this newly discovered stranger in our Solar System belong to its history. It is very remarkable and instructive to observe and record the course of those studies, caused and influenced by circumstances: how they slowly developed with the objective of truth. Phenomena which can be explained by known, eternal, simple, invariant, universal laws of nature can be fathomed. As soon as you build on those the completion of the building can be expected that neither time, taste, customs, prejudices, superstition, or transcendental beliefs can destroy. When this weird celestial body was first discovered, the busy human mind could not even create any hypothesis but only an opinion and this could only be bold and daring.*

*There comes an unknown heavenly body, moving in space and one dares to think it might be a comet. This does not seem to be very unlikely for uncountable is the armada of comets scattered all over the universe. But the appearance of this body is not comet-like.*

**Ephemeride für die Pallas Olbersiana, für Seeberger Mitternacht.**

| 1802 | | AR. der Pallas | | Declinatio der Pallas | | Abstand von der ♁ |
|---|---|---|---|---|---|---|
| Jul. | 2 | 189° | 22' | 18° | 45' | 2,548 |
| | 5 | 190 | 8 | 18 | 27 | 2,592 |
| | 8 | 190 | 57 | 18 | 9 | 2,635 |
| | 11 | 191 | 46 | 17 | 49 | 2,678 |
| | 14 | 192 | 37 | 17 | 29 | 2,720 |
| | 17 | 193 | 28 | 17 | 9 | 2,762 |
| | 20 | 194 | 21 | 16 | 48 | 2,804 |
| | 23 | 195 | 15 | 16 | 26 | 2,845 |
| | 26 | 196 | 10 | 16 | 5 | 2,886 |
| | 29 | 197 | 6 | 15 | 43 | 2,926 |
| Aug. | 1 | 198 | 2 | 15 | 20 | 2,965 |
| | 4 | 199 | 0 | 14 | 58 | 3,004 |
| | 7 | 199 | 58 | 14 | 35 | 3,042 |
| | 10 | 200 | 57 | 14 | 12 | 3,080 |
| | 13 | 201 | 57 | 13 | 49 | 3,117 |
| | 16 | 202 | 58 | 13 | 27 | 3,153 |
| | 19 | 203 | 59 | 13 | 4 | 3,188 |
| | 22 | 205 | 1 | 12 | 41 | 3,223 |
| | 25 | 206 | 4 | 12 | 18 | 3,257 |
| | 28 | 207 | 7 | 11 | 55 | 3,290 |

**Fig. 11.26** Ephemeris for Pallas Olbersiana, for Seeberg midnight, July 2 to August 28, 1802

*It resembles in shape a recently discovered planet; and one could think it might be a planet and considers this bold. Only continuous observations and calculations will lead us to the path of this celestial body. This is a fact and proven by all knowledge about nature. Archimedes says in his* Mechanics of the Earth: da ponere pedem et terram movebo [give me somewhere to place my foot and I will move the earth] *and Newton in his* Celestial Mechanics: da ponere calculum et astra movebo [let me calculate and I will move the stars]. *All of nature's phenomena are caused by motion in time and space, which can be calculated. The Newton of our times says: "When the laws of chemical attraction will be sufficiently observed so that they can be calculated, then, and only then, chemistry will reach those heights of perfection, to which astronomy has been lifted by the discovery of gravity." If a material communication between the earth and its moon exists, only* calculation *can show its possibility. At least a* calculating *mind uttered this thought first, when a* philosophizing *one could only dream or romanticise.*

*When the first rumour of Olbers' discovery had reached France, the opinion there held that this new body must be a comet. And this view was stubborn and persistent which explains why they stuck longer with parabolic orbits. Dr. Burckhardt, who worked chiefly on this calculation soon found that a parabolic orbit is insufficient and that it must rather be an elliptical one, in fact twice as eccentric as Mercury's. And since he assumed such an eccentricity the error in latitude of the second observation, which is solely responsible for the eccentricity, appeared with the same sign again after passing through zero. From this Dr. Burckhardt believed he had to increase the eccentricity even more in order to serve the latitudes. And in this erroneous opinion he was even more affirmed by the parabolic hypothesis, since this matched those three basic observations.*

*Then he calculated two ellipses according to La Place's method whose eccentricity was 0.8 and 0.6. This confirmed his error even more for this last hypothesis did not represent the three latitudes any longer. Those three orbits were:*

| | | | |
|---|---|---|---|
| *Eccentricity* | *1st order parabola* | *0.785765* | *0.6* |
| *Node* | *176° 25′ 34″* | *175° 52′ 15″* | *174° 45′ 0″* |
| *Inclination* | *54 18 30* | *49 41 59* | *43 28 0* |
| *Position of perihelion* | *113 52 3* | *112 12 57* | *109 59 0* |
| *Anomaly 29 March 1802 8 h* | *73 9 0* | *74 22 23* | *76 10 46* |
| *Distance from the Sun* | *1.8432* | *1.8426* | *1.8139* |
| *Semimajor axis* | *infinite* | *8.6033* | *4.5347* |
| *Period* | *25 years* | *9 yrs. 8 mo* | |
| *Time of transit with the Sun* | *29 Sept. 1802 17 h* | | |

*The eccentricity of the second orbit is that of the comet of 1770. Since Dr. Burckhardt had already calculated a table for this comet in order to find the true anomalies more easily, he rather took this than 0.8, which is not much different from the other.*

*When Dr. B. compared an observation of April 27 with his two elliptical orbits it showed that the one with the smaller eccentricity was to be preferred. He then tested the eccentricities 0.6 and 0.555 but without success since he could never represent the observations better than a minute. This discouraged him and he wanted to wait until all meridian observation of this celestial body were made in order to obtain a greater arc of the orbit.*

*At this time he learned of Gauss's orbital elements, which represent all observations until May 1 so incredibly. Naturally, his failed attempts did not inspire him with confidence in this orbit. But they showed the capability and insights of Dr. Gauss. He compared this orbit with an observation of May 20. An error while calculating led to a result, which he thought he should find in advance, that Gauss's elements would not represent future observations. So Dr. Burckhardt continued his studies unchanged and calculated the following two new orbits with eccentricities 0.555 and 0.65.*

| | | |
|---|---|---|
| *Eccentricity* | *0.555* | *0.65* |
| *Log semimajor axis* | *0.54* | *0.650332* |
| *Node* | *174° 41′ 12″* | *174° 41′ 20″* |
| *Inclination* | *39 0 11* | *41 3 45* |
| *Anomaly* | *85 47 14* | *81 54 20* |
| | *April 4 10 h 51′* | *March 29 8 h 24′* |

*With these elements he found an error of 10 minutes for the position of the node, which was concluded from the second observation. In order to learn whether the eccentricity had to be increased or decreased, he decided to calculate a new parabola:*

| | |
|---|---|
| *Node* | *178° 27′ 7″* |
| *Inclination* | *45 16 22.5* |
| *Log. perihelion distance* | *9.964* |
| *Anomaly* | *95 43 11″.7 April 4 10 h 51′* |
| *Time of transit* | *112 days 10 h 45′ for April 4* |

*Since the error of the second observation amounted to 40 minutes with these elements it was only natural that the eccentricity had to be decreased. Indeed, one of 0.4 reduced the error to six minutes and of 0.2 to two minutes, which resulted in the two following new orbits:*

| | | |
|---|---|---|
| *Eccentricity* | *0.4* | *0.20553* |
| *Log semimajor axis* | *0.475* | *0.440* |
| *Node* | *174° 36′ 45″* | *172° 6′ 0″* |
| *Inclination* | *36 51 35* | *34 1 28* |
| *Anomaly* | *79 43 32.5* | *57 41 0* |
| *on April 4 10 h 51′* | | |

*Only after all those tiring and tedious attempts Dr. B. was led back to the true eccentricity that Dr. Gauss had used; and he recognised the slip of the pen which had misled him. Now Dr. B. calculated new elements and to obtain more accurate results he calculated the perturbations as well that this small heavenly body suffers from Jupiter. But while calculating he took into account only the first powers of the eccentricity, which is not sufficient regarding the latitudes. For the time being such an approximation will suffice because with future observations greater improvements have to be made anyway. Thus Dr. B. found the following influence of Jupiter on Pallas:*

| *1802* | *in longitude* | *in latitude* | *log. Radius Vector* |
|---|---|---|---|
| *April 4* | *−122″.0* | *−174″.4* | *−0.0017641* |
| *27* | *−18.0* | *−184.9* | *−0.0016756* |
| *May 20* | *+77.6* | *−194.6* | *−0.0015837* |

*And at last the following elements of the elliptical orbit were developed:*

| | |
|---|---|
| *Ascending Node* | *172° 28′ 57″* |
| *Proximity to the Sun* | *122 3 2* |
| *Epoch March 31 1802* | *162 51 14.2* |
| *Inclination* | *34 50 40* |
| *Semimajor axis* | *2.791 log 0.4457560* |
| *Eccentricity* | *0.2463 log 9.3914644* |
| *Sidereal period* | *1703.7 days* |
| *Log of diurnal movement* | *2.8813732* |
| *Log of parameters* | *0.4185771* |
| *Log a* $\sqrt{1}$ *− eccen./1 + eccen.* | *9.8907880* |

*These elements represent the observations of April 4, 16, 17 and May 20 perfectly. The first is an observation from Seeberg and the two following are from Paris (National Observatory); the last two were made at the Military School. The error in latitude on April 16 and May 7 is +13″. It could be disposed of at once if the eccentricity was increased by 1 %. Future observations will most likely need major improvements.*

*A common question is: Are Ceres and Pallas really main planets of our Solar System? Since there is no king in the free republic of science, who says* tel est notre plaisir [such is our pleasure], *the astronomers' opinions vary, which is a sign of an unfinished definition of a planet. La Lande has already awarded those two heavenly bodies the planetary diploma – in his letters and even publicly in* Moniteur. *Dr. Herschel on the other hand suggests in his circular letter a new species of celestial bodies named* Asteroids. *In our view both depend on arbitrariness, agreement, or consent of all astronomers. For the definition of planet is rather* nominal *than* real. *If we would like to engage in* didactic *disputes we have to go back to those times, where the knowledge about planets originates and their names were introduced. Democritus did not know the five main planets of our Solar System yet, according to Seneca. Eudoxus brought this knowledge from Egypt to Greece 380 BC. The Greeks called those bodies 'planets', derived from the word* [πλανητηζ], erraticus, *roving/erratic* [German: irrend] *because these celestial bodies* rove *in space in contrast to fixed stars, which were called* Stellas inerrantes *and to comets, called* Stellas comatas. *Now the question is: Would the Greeks have classified Ceres and Pallas as wandering bodies or planets, if they were visible to their eyes? Certainly, for already Artemidorus answered this question by saying that there existed several still invisible planets: "There are not only these five stars running around, but these are the only ones that we have observed. There are countless others besides, moving secretly, unknown either because of the obscurity of their light, or because the position of their orbits is so great that they could only be seen when they came at last to their end." And also Seneca believed in a multitude of planets, by saying: But you think that among the countless stars that distinguish the night with their varied splendour, and which suffer the air to be empty and inactive, there are only five to whom it is permitted to busy themselves, and that the others stand still, a fixed and motionless nation. Just like Seneca then said about the planets "They compel us to be curious" we are captivated by those two new roving bodies (for they wander like the other planets), which arrest our attention. Now is the time of which Seneca once said: "A time will come when the diligence of a distant age will draw those things that lie hidden into the light of day."* [The quotes are from Seneca's *Natural Quest.*] *The day was the first day of the 19*$^{th}$ *century and the diligence that of Piazzi and Olbers.*

*If only Ceres and Pallas had been 50 times brighter in past centuries, where the planets were named: doubtlessly, they would have their place among them and nobody would dispute their right to it. But since this is not the way things are: there might also be room for the following thoughts: why some astronomers were so hard to convince of the new bodies' planetism must have – if not a valid but an apparent reason. And if it is only the following: the believed harmonious distances of the planets – it is well-known that this is a nonproven hypothesis and not based on any law of nature* [he refers here to Bode's Law]. *But that two planets have equal orbital axes and one period of revolution is strange and unheard of in the entire Solar System! Strange indeed. And Dr. Olbers wrote: "I like to hear that Pallas' planetism is hardly believed. For I consider this disbelief a proof of my hypothesis that the actual orbit of Pallas can hardly be its original one and that those two new celestial bodies, Ceres and Pallas, are only debris or pieces of one planet. If we were to find more of those revolving between Mars and Jupiter they might deserve* a separate name *and* in this case *I would agree to a* new species *of celestial bodies."*

*Until then this classification can be deferred, since fortunately neither it nor the names are relevant and will not disturb future observations or our calculations.*

*Maybe the Lilienthal Astronomical Society, which is already active, will succeed in finding some of the fragments; maybe some are among the missing stars. How important it would be to find only one single older observation of Ceres or Pallas! For from time to time here and there we find traces of missing stars. On August 23 Harding wrote from Lilienthal: "Your catalogue of stars, mentioned in the June issue of your* MC *on page 601, with which you compared Pallas surprised me because it contains one star that I have been unable to find since January. I thus considered it missing or enlisted by some mistake. It is no. 96 Virginis of Bode's catalogue, supposedly of the 6th magnitude, and which is mentioned in the* Conn. de tems *An VIII p. 465 no. 28 and An. VII p. 391 no. 19. Consequently, in the first days of Pallas' discovery I was thinking that Pallas might have been observed at that position and informed Dr. Olbers about my suspicion, who answered me on April 8 as follows: 'I am totally convinced that no. 96 has never existed and found its way into the* Conn. d. t. *only by mistake and thus into Bode's catalogue'. It appeared very likely to me too since I could not find it in the* Histoire céléste française, *vol. I. In the sky there was no trace of a star at that position until April 8. But because you observed this star on April 12 I am assuming that it might show a more considerable change of light and becomes invisible during its fainter stages. I would like to ask you to inform me about the order of magnitude of this star on April 12 at the time of your observation..."* (Fig. 11.27).

*That I really had observed on April 12 the star mentioned in the June issue of the MC is certain; but whether the above stars of the two volumes of the* Conn. d. temps *are identical with no. 96 of Bode's catalogue and with mine is not as certain. For apparently the two stars of the* Conn. de temps *VII and VIII are not one and the same star; only that of the* Conn. de temps *VIII is no. 96 of Bode. My April 12 observed star is different by one minute*

**Fig. 11.27** Distances of Pallas from Sun and Earth from February 5 to July 1, 1803

| 1803 | | Mittlere Zeit | | AR. ⚴ | | Decl. ⚴ N. | | Abſtand von der | |
|---|---|---|---|---|---|---|---|---|---|
| | | | | | | | | Sonne | Erde |
| Febr. | 5 | 19 | Uhr | 268° | 1′ | 5° | 49′ | 3, 1172 | 3, 5767 |
| März | 2 | 14 | — | 275 | 13 | 8 | 44 | 3, 1643 | 3, 3734 |
| — | 28 | 4 | — | 280 | 21 | 12 | 40 | 3, 2095 | 3, 1282 |
| April | 10 | 5 | — | 282 | 24 | 14 | 51 | 3, 2312 | 3, 0083 |
| — | 23 | 10 | — | 283 | 22 | 17 | 7 | 3, 2522 | 2, 8860 |
| May | 6 | 20 | — | 283 | 27 | 19 | 19 | 3, 2725 | 2, 7746 |
| — | 20 | 9 | — | 282 | 33 | 21 | 15 | 3, 2920 | 2, 6786 |
| Jun. | 3 | 3 | — | 280 | 42 | 22 | 38 | 3, 3107 | 2, 6093 |
| — | 16 | 23 | — | 278 | 10 | 23 | 19 | 3, 3284 | 2, 5665 |
| Jul. | 1 | 0 | — | 275 | 17 | 23 | 8 | 3, 3452 | 2, 5626 |

*of that. And since I did not observe its declination but only guessed it at the semi-circle of the passage instrument its identity cannot be determined with certainty until this celestial region will emerge from the sun's rays. Meanwhile, since I observed this star at all five threads of the passage instrument and estimated it to be of the 7th order of magnitude and Harding did not see any trace of a star in that region – it does indeed deserve our greatest attention and we will not miss to inform our readers about future successes.*

*We are closing our news about Pallas with an ephemeris for Pallas in order to give an overview of this planet's path in the first part of the coming year. Dr. Olbers calculated them according to Dr. Gauss ellipse no. III.* [Ephemeris from Feb. 8 to July 1, 1803 is given here.]

## Continued Reports regarding the New Primary Planet of our Solar System Pallas Olbersiana

*MC*, Oct. 1802, p. 390

*Just as Ceres had been observed on the Milan equatorial sector until August 5, so too was Pallas observed with this instrument until August 8. The observations reduced by Oriani stand as follows:* [These observations by Oriani can be found in his paper that appears earlier in this chapter.]

*The register of the position and action of the clock has already been cited in the present issue (p. 386) with regard to the observations of Ceres. A more precise reduction of the above observations will only be conducted when some of the stars used for comparison, especially 168, 359, 435, 226, 485, 394, 521 Virginis and 12 Bootis, will be determined more precisely than they appear in Bode's star catalogue, which we expect to make up in future issues.*

*Oriani also ventured to contemplate the perturbations to which Pallas had to be exposed. He finds the calculation of them easy as long as it can be assumed that the two new planets have the same mean distance from the Sun. In this case, the perturbation equations for Ceres are taken and instead of this name, the name of Pallas is used. Thus, if the eccentricity of the path of Pallas is named e, the segments, being A, $_2A$, $_3A$ in this argument, are to be multiplied with 12.284e; $(12.284e)^2$; $(12.284e)^3$. Multiply + 13."70 and – 0.000035 of the argument 4 D – A by $(12.284e)^2$; and likewise – 6."81 and +0.000018 of the argument 5D – A by $(12.284e)^3$. Also multiply the elements with 2H as their argument by 16.910 and insert this argument ± 4° 45′ instead of ±3° 47′. Multiply for the latitude the elements with H in their arguments by 4.112; and insert into the last two elements +1° 54′ instead of –2° 33′. Since the last term has A and H in the argument it must be multiplied by 4.122 × 12.284e.*

*In order to have the perturbations in the latitude still more exact, at least the segments have to be calculated, which depend on the third power [cube] of the inclination and the product of the eccentricities and inclination. It is even essential for the perturbations of longitude and distance to include those differences that depend on the second power [square] and the product of the eccentricities and the inclination. It is perhaps necessary to bring the sharpness to the fifth power of the inclination, but the formulae would then become very long and complicated. Oriani already has all analytical formulae ready; as soon as he finds the time, he will use them on Pallas. In the meantime, he calculated that the tropical movement of the aphelion is 106."1, of the node –7."2, the yearly change of the eccentricity in seconds –1."36, of the inclination +0."81.*

*Dr. Gauss, who had improved his elements for the third time, has ventured a* fourth *correction. These new elements are derived from our first Seeberg observations, from Dr. Maskelyne's last Greenwich meridian observations and Dr. Olbers' observation of July 8.*

| | |
|---|---|
| Epoche 1802 März 31 Seeberger Meridian | 162° 55′ 6″8 |
| Tägliche tropische Bewegung | 769″7263 |
| Logarithmus der halben grossen Axe | 0.4424992 |
| Sonnenferne | 301° 38′ 42″ |
| Excentricität | 0.243888 |
| Knoten ☊ | 172° 26′ 31″ |
| Neigung | 34 36 59 |

**Fig. 11.28** Pallas Elements IV

*Dr. Gauss does not want to pass these off as more certain than the 3rd elements, since he is still expecting a much more exact improvement of the later observations from Milan; meanwhile, these elements are as follows* (Fig. 11.28):

*With these elements, Dr. Gauss calculated the following ephemeride for this planet for the coming year in advance. It is useful to make them known in good time, because the doubtful stars in the area through which Pallas will journey will want to be more exactly determined to help orient and familiarize astronomers with this area, through which locating it will be facilitated. In the last column of the ephemeris, which contains the luminosity of this planet, that which this planet had on July 8 of this year has been taken as a unit; according to this standard, it was as follows:*

| | |
|---|---|
| *April 4 1802 =* | *4.23* |
| *May 4* | *2.75* |
| *July 16* | *0.90* |
| *July 26* | *0.80* |
| *Aug. 7* | *0.70* |
| *Aug. 24* | *0.60* |

*In this, the magnitude* [size] *of the phase, the altitude over the horizon and the distance from the twilight have not been taken into consideration. The first fact is insignificant, the two others depend on the locations of observation and the power of the telescopes. The greatest luminosity attained by Pallas next year will be = 0.656; the same intensity was reached this year of 1802 on August 14. Oriani was still observing this planet on August 8 and saw it with effort on the 17th and 18th of this month, to which the twilight and the low position of this planet apparently contributed much. The pleasant hope still remains that we will see Pallas in the coming year and be able to observe its opposition to the Sun, since the planet will have already been located before this epoch (June 30, 1803); for with the meeting, its visibility or luminosity will once again decrease. We will have little hope of observing this planet in the meridian before the last half of May or the beginning of June, at least in our latitudes.* [An ephemeris for Pallas in 1803, from Feb. 4 to June 28, follows.]

## Resumed News about the new main Planet of our Solar System, Pallas Olbersiana

*MC,* Nov. 1802, p. 499

*We calculated Pallas' observations (October issue, p. 391) by treating them in exactly the same way as we did Oriani's of Ceres. Thus we obtained the following newly reduced positions of this planet* (Fig. 11.29) [Compare this with table in Oriani's paper.]:

*And also Dr. Maskelyne's observation from Greenwich needed a similar reduction since he used the mean position of the compared star of Wollaston's star catalogue, as already mentioned regarding Ceres' observations, and then simply added aberration and nutation. Since this star has a name the original observations can be restored and calculated anew.*

Fig. 11.29 Observations of Pallas by Oriani, from July 10 to August 8, 1802

*Beobachtungen der Pallas von Oriani in Mailand, von dem Herausgeber, berechnet.*

| 1802 | Mittl. Zeit in Mailand | Scheinbare gerade Aufst. der Pallas | Scheinbare nördl. Abweichung der ⚴ | Sterne womit verglichen nach *Bode* |
|---|---|---|---|---|
| Jul. 10 | 9U 40' 42,"7 | 191° 27' 44,"3<br>28 7, 7 | 17° 56' 24,"4<br>56 10, 3 | 109 Com. Beren.<br>117 — — |
| — 18 | 9 36 14, 1 | 193 44 4, 5 | 17 3 7, 3 | 117 — — |
| — 24 | 9 32 16, 2 | 195 31 27, [illegible] | 16 19 47, 0 | 359 Virginis |
| — 25 | 9 [illegible] 11, 6 | 195 49 47, 9 | 16 12 28, 0 | 359 — |
| — 28 | 9 18 6,"4 | 196 44 58, 5<br>44 58, 2 | 15 50 55, 9<br>50 40, 8 | 359 —<br>435 — |
| — 29 | 9 11 48, 3 | 197 3 44, 8<br>3 30, 7<br>3 44, 8 | 15 43 32, 0<br>43 15, 1<br>43 19, 3 | β ♌<br>359 Virginis<br>435 — |
| — 31 | 9 6 9, 5 | 197 41 21,"7 | 15 28 56, 3 | 435 Virginis |
| Aug. 1 | 9 4 13, 8 | 198 0 24, 0<br>0 29, 3<br>0 32, 8 | 15 21 37, 7<br>21 15, 8<br>21 13, 4 | β ♌<br>435 Virginis<br>12 Bootis |
| — 2 | 9 23 37, 2 | 198 19 40,"7<br>19 45, 0 | 15 13 33, 8<br>13 31, 4 | 435 Virginis<br>12 Bootis |
| — 4 | 8 55 37, 7 | 198 57 48, 3<br>57 55, 5 | 14 58 57, 2<br>58 56, 4 | 485 Virginis<br>12 Bootis |
| — 5 | 9 14 56, 7 | 199 17 26, 5<br>17 24, 4 | 14 50 58, 2<br>50 44, 8 | 485 Virginis<br>521 — |
| — 6 | 9 6 26, 7 | 199 36 57, 5<br>36 58, 4 | 14 43 46, 2<br>. . . . | 485 Virginis<br>521 — |
| — 7 | 9 4 17, 7 | 199 56 26, 2<br>56 24, 8 | 14 35 53, 2<br>. . . . | 485 Virginis<br>521 — |
| — 8 | 9 0 14, 5 | 200 16 4, 6 | 14 28 21, 2 | 485 Virginis |

*This star is no. 24 Comae Berenices, whose position we gave already in the August issue on page 189. But this star is a binary star, which Dr. Maskelyne did not notice, and consequently it remains uncertain which star he used for a comparison. Their ascensional difference is 15".2 s. We took the average and then saw that if the first of both stars was meant we have to add 7".6 to our calculated position of the planet and if the second was meant we have to subtract this amount. In Wollaston's catalogue this star is a designated single star in accordance with Flamsteed. The Greenwich observations of Pallas have been calculated* (Fig. 11.30):

*Dr. Maskelyne estimated Pallas' luminosity:*

| | |
|---|---|
| *April 21* | *$9^{th}$ magnitude* |
| *22* | *9* |
| *May 17* | *9* |
| *June 11* | *9* |
| *June 18* | *11* |
| *June 20* | *11 to 12* |
| *June 28* | *10* |

*And the observation of Pallas of May 7 revealed the source of the error mentioned in the July issue (p. 85). Dr. Maskelyne observed Pallas at four threads of the passage instrument only ($1^{st}$, $2^{nd}$, $4^{th}$, $5^{th}$); and the time was wrong by 10" for one thread and hence the sum was so great and the quotient by 10/4 = 2".5 in time = 37".5 in the arc too small. Consequently, the correct RA is 180° 56' 24".0 and the error of Gauss's elements III +1".8.*

*Méchain observed Pallas at the National Observatory with his achromat in July and August as follows* (Fig. 11.31):

| 1802 | Mittl. Zeit Green-wich | Scheinbare gerade Aufsteigung der ⚴ | | Scheinb. Abweichung der ⚴ N. | | Stern womit vergl. |
|---|---|---|---|---|---|---|
| | | nach Dr. *Maskelyne* | nach *v. Zach* | nach D. *Maskelyne* | nach *v. Zach* | |
| | U ′ ″ | ° ′ ″ | ° ′ ″ | ° ′ ″ | ° ′ ″ | |
| Jun. 11 | 11 42 53 | 184 38 57,9 | 184 38 46,7 | 20 26 12 | 20 26 22,1 | 24 Com. Ber. |
| 18 | 11 7 10 | 186 3 30,9 | 186 3 18,3 | 19 58 7 | 19 58 17,5 | — — — |
| 20 | 12 27 37 | 186 30 26,1 | 186 30 13,2 | 19 48 53 | 19 49 4,3 | — — — |
| 28 | 11 53 18 | 188 22 11,8 | 188 21 56,6 | 19 8 11 | 19 8 22,4 | — — — |

**Fig. 11.30** Comparison of Pallas observations by Zach and Maskelyne from June 11 to June 28, 1802

**Fig. 11.31** Observations of Pallas by Méchain from July 12 to August 7, 1802

| 1802 | Mittl. Zeit zu Paris | AR. der ⚴ | Declination der ⚴ nördl. |
|---|---|---|---|
| Jul. 12 | 10U 52′ 47″ | 192° 2′ 10,″4 | 17° 42′ 42,″8 |
| 14 | 10 28 49 | 192 35 45, 7 | 17 29 54, 8 |
| 16 | 9 51 43 | 193 9 33, 7 | 17 16 4, 7 |
| 17 | 10 16 13 | 193 27 11, 2 | 17 9 0, 6 |
| 18 | 10 25 40 | 193 44 56, 2 | 17 1 51, 5 |
| Aug. 4 | 9 36 28 | 198 58 52, 1 | 14 58 28, 8 |
| 5 | 9 33 37 | 199 18 11, 4 | 14 51 6, 1 |
| 7 | 9 22 45 | 199 57 6, 8 | 14 36 13, 2 |

*Astonishingly, La Lande wrote us that Messier was able to observe Pallas late in September. On the 21st of this month at 7 h 28′ he found its right ascension 215° 48′ 46″ and declination 8° 59′ 28″N. Only on September 24 did Messier close his observations of this planet. He estimates the luminosity not less than 10th magnitude. "This is the faintest you can distinguish in an excellent telescope."*

*Dr. Herschel has kindly sent us a special edition of his treatise published in* Philosophical Transactions. *We already informed our readers about the contents by quoting his letters in extracts in the July issue p. 90. Even after a second thorough reading we are not inclined to change our given opinion about it. And furthermore we see from our correspondence with other astronomers, e.g., La Lande, Oriani, Piazzi, and Gauss* etc. *who disapprove of his new classification. Especially the reasons and an accurate remark uttered by Gauss regarding the alleged so-called law of harmony, to which the distances of the planets are supposed to correspond, deserve consideration. This astute master of measurement expressed his thoughts regarding this matter in a letter of October 16 in the following peculiar way:*

*Dr. Herschel still is not willing, as Prof. Huth told me and who visited him in England, to tolerate the new planets, although as far as I know not a single astronomer has approved of his suggestion. Basically, and I agree with you, it depends only on our agreement whether we call Ceres and Pallas planets or not, and people are not saying whether they are planets or not, but whether it is proper and becoming to call these celestial bodies planets that partly resemble the known planets and partly not at all. That the latter is irrelevant you have shown sufficiently in your July issue and that astronomers believed in a circle-like orbit and a dependent perennial character seems to be proven by the circumstance that all astronomers accepted their planetism without hesitation as soon as they learned of the orbit. It even appears to me that, if future finds prove our excellent Olbers' hypothesis right – that Ceres and Pallas are only pieces of a destroyed planet, even then, we do not have to give up calling them planets. I believe it more important to study whether these celestial bodies are entitled to the name planet because of their fundamental characteristics than how they have become it.*

*Strange that the so-called Titius' law is being used as an argument against both planets. This relationship applies to the other planets – against the nature of all truths, which deserve the name law – only incidentally and to Mercury not at all. It appears quite clear to me that the series 4, 4 + 3, 4 + 6, 4 + 12, 4 + 24, 4 + 48, 4 + 96, 4 + 192 with which the distances should correspond, is not a continuous series at all. The element, previous to 4 + 3 needs not be 4, that is 4 + 0, but 4 + 1½. So between 4 and 4 + 3 there should be an endless number, or as Wurm put it (cf. MC, 1801, June issue, p. 594) for Mercury or for* $n = 1$ *the result is* $4 + 2^{n-2}\,3$ *not 4 but 5½.*

*I do not want to disapprove of the fact that one seeks in nature such approximate correspondences. The greatest men subscribed to such* lusus ingenii [games of nature]. *But as proud as Kepler was of his regular bodies reconciled with the distances of planets (as he said he did not want to give his find to the Electorate of Saxony) he certainly would not have used it to challenge Uranus' planetism (if this discovery had been made in his times) simply because it did not match his ideas. Most likely he would have abandoned them immediately. You made observations about Titius' law as well – but you called them dreams, and when they started to contradict the facts, you gave them up.*

*How the planetary castle in the air, built by assumed symmetry, is breaking apart under the destroying eternal force of geometric truth! What will happen to this notorious law of progression, taken as granted by some because of the apparent impossibility to violate it? As much is certain: No geometer has ever accepted it, although we asked La Place repeatedly for his opinion. La Lande and De Lambre called this law a mere* numeric game [German: Zahlen-Spiel]. *Only astronomical adepts trusted in this law more than they should have and now they share the destiny of alchemists. Consequently, Prof. Wurm's apt remark can be parodied as follows: What a shame! Euclid and nature leave no element in this series for Mercury.*

Delambre (1821: 322), speaking of Bode's views, wrote "When we set out to search for the unknown planet, it was not after the ideas of Kepler, it was after the law almost as chimerical, but at least more plausible."

## Resumed News about the new main Planet of our Solar System, Pallas Olbersiana

*MC*, Dec. 1802, p. 579

*Just as we found in last year's August issue an error regarding the observations on Ceres, which we mentioned on p. 186: the observations on Pallas of the same issue on page 188 and 190 showed ambiguities as well, which must be corrected on request as follows.*

*We suspected an obvious slip of the pen regarding the observed declination of the star 33 Comae Berenices on July 8. And really, it occurred by confusion while copying the observations. For Oriani had observed Pallas twice on that day and when he was copying the observations it happened that his piece of writing paper was almost filled but he had only copied the first observation – and unfortunately together with the data of 32 Comae Berenices. Thus both observations of July 8 (of which only the first was mentioned on p. 188 of the August issue) should be* (Fig. 11.32):

*The error regarding the mean time of July 3 was exactly as we mentioned it on p. 190 of the August issue, namely 10 h 7′ 54″.*

*All of Oriani's Milanese observations have been compared by Dr. Gauss with his orbital elements III of this planet and he found the following correspondences* (Fig. 11.33).

Fig. 11.32 Two observations of Pallas by Oriani on July 8, 1802

| Mailand 1802 | Stunden-Winkel | Namen des Geſtirns | Austritt aus der I Stange (Uhr-Zeit) | Eintritt in die II Stange (Uhr-Zeit) | Scheinbare Abweich. mit Inbegriff der Culmination und Refraction |
|---|---|---|---|---|---|
| Jul. 8 | 4U 21/2 | 25 Com. B. | 9U 13′ 1,″0 | 9U 15′ 4,″4 | 18° 10′ 31″ |
| | | 27 — — | 9 22 41, 5 | 9 24 45, 1 | 17 38 55 |
| | | 33 — — | 9 28 :: | 9 30 30, 7 | 18 10 5 |
| | | Pallas | 9 29 36, 5 | 9 31 38, 5 | [illegible] 8 39 |
| Jul. 8 | 4U 12′ | 32 Com. B. | 9 38 8, 7 | 9 40 :. | 18 8 39 |
| | | 33 — — | 9 38 13, 5 | 9 40 21, 0 | 18 10 6 |
| | | Pallas | 9 39 25, 0 | 9 41 28, 7 | 18 8 38 |

Fig. 11.33 Differences between observed and calculated positions of Pallas, based on the Third Elements by Gauss, from July 10 to August 8, 1802

| Mailand 1802 | Berechnete AR der Pallas | Unterſchied | Berechnete nördl. Abweich. der ⚴ | Unterſchied |
|---|---|---|---|---|
| Jul. 10 | 191° 27 33″ | − 11″<br>− 35 | 17° 56′ 20″ | − 4″<br>+ 4 |
| — 18 | 193 43 47 | − 17 | 17 2 29 | − 38 |
| — 24 | 195 39 59 | − 29 | 16 19 59 | + 12 |
| — 25 | 195 49 25 | − 23 | 16 12 40 | + 12 |
| — 28 | 196 44 25 | − 33<br>− 28 | 15 50 56 | + 1<br>+ 16 |
| — 29 | 197 3 9 | − 36<br>− 22<br>− 36 | 15 43 33 | + 1<br>+ 18<br>+ 14 |
| — 31 | 197 40 46 | − 34 | 15 28 55 | − 1 |
| Aug. 1 | 197 59 44 | − 40<br>− 45<br>− 49 | 15 21 18 | − 20<br>+ 2<br>+ 5 |
| — 2 | 198 19 4 | − 36<br>− 41 | 15 13 43 | + 10<br>+ 12 |
| — 4 | 198 57 9 | − 39<br>− 47 | 14 58 52 | − 5<br>− 4 |
| — 5 | 199 16 47 | − 39 | 14 51 13 | + 15<br>+ 28 |
| — 6 | 199 36 7 | − 50<br>− 51 | 14 43 43 | − 3 |
| — 7 | 199 55 38 | − 48<br>− 47 | 14 36 8 | + 15 |
| — 8 | 200 15 9 | − 56 | 14 28 35 | + 14 |

*Dr. Gauss submitted elements improved for the fourth time, but the above Milanese observations have not been compared with those. But he made the following comparison from which we can judge to which degree these elements IV match the observations better than his elements III* (Fig. 11.34).

*The elements III give the RA too small by 1 min and the elements IV too great by approximately 20″. Both declinations correspond, as far as we can see from the accuracy of the observations, quite well.*

*Dr. Gauss improved the orbital elements for a fifth time according to these last observations of Oriani and obtained the following elements* (Fig. 11.35):

*With these elements Dr. Gauss now compared with all the observations of Messier and found:*

**Fig. 11.34** Two observations of Pallas, compared according to the Third and Fourth Elements

**Die IV. Elemente geben**

| 1802 | Die AR. grösser | Die Declination kleiner |
|---|---|---|
| Juli 8 | 24″ | 8″ |
| Aug. 4 | 66 | 0 |

**als die III. Elemente.**

***V. Elemente der Pallas Olbersiana.***

| | | | |
|---|---|---|---|
| **Epoche 1802 März 31 Mittag in Seeberg** | 162° 46′ 58″2 | **Logarithmus der halben grossen Axe** | 0.4425529 |
| **Sonnenferne** . . . . . . . . . . | 301 28 24.0 | **Excentricität** . . . . . . . . . . | 0.244976 |
| **Aufsteigenden Knoten** . . . . . . | 172 27 3.0 | **Neigung der Bahn** . . . . . . . . | 34° 37′ 40″ |
| **Tägliche mittl. tropische Bewegung** . | 769″583 | | |

**Fig. 11.35** The Fifth Elements of Pallas

| *1802* | *Mean Time* | *Calc. RA* | *diff* | *Calc. Decl.* | *Diff.* | *Distance* | *Brightness from Sun* |
|---|---|---|---|---|---|---|---|
| *Paris* | | | | | | | |
| *Sept. 21* | *7 h 28′* | *215° 49′ 14″* | *+28″* | *9° 0′ 16″N* | *+48″* | *3.518* | *0.485* |

*The difference between calculation and observation is still not considerable. Dr. Gauss wrote: "Since the observation of Messier was made at such an unfavourable position (Pallas was only 13 degrees above the horizon) I agree with you that this single observation is too unreliable to use for an improvement of the elements. Furthermore the improvement probably has no influence on Pallas' position during the next year. I have calculated a few positions according to my elements V and found the RA for February 4, March 24, May 11, June 28, 1803, smaller by 3½ min., 5½ min., 7½ min. and 9 min. and all declinations greater by about 1 min. than the ephemeris, calculated according to my elements IV, would indicate."*

*Dr. Gauss has calculated the entire ephemeris of Pallas' orbit in 1803 according to his new elements V and even extended it to one further month* (Fig. 11.36). *Since Pallas will start showing a higher luminosity on April 3, 1803, than it had shown on September 21 this year, and it will consequently be possible to observe it higher above the horizon, it can be taken as granted that we will not miss Pallas with our more powerful telescopes. Here follows:*

*In the third column we assumed the planet's luminosity it would have at a distance 1 from the sun and earth. Accordingly it was in 1802:*

| | |
|---|---|
| *April 4* | *0.08997* |
| *May 16* | *0.04740* |
| *Aug. 10* | *0.01455* |
| *Sept. 21* | *0.01030* |

| Mitternacht in Seeberg | | AR. der ⚴ | | Declin. der ⚴ Nördl. | | Lichtſtärke |
|---|---|---|---|---|---|---|
| Febr. | 4 | 267° | 45″ | 5° | 39′ | 0,00807 |
| | 7 | 268 | 42 | 5 | 56 | 0,00814 |
| | 10 | 269 | 39 | 6 | 14 | 0,00821 |
| | 13 | 270 | 34 | 6 | 33 | 0,00829 |
| | 16 | 271 | 28 | 6 | 53 | 0,00837 |
| | 19 | 272 | 21 | 7 | 15 | 0,00846 |
| | 22 | 273 | 12 | 7 | 36 | 0,00855 |
| | 25 | 274 | 2 | 7 | 59 | 0,00865 |
| | 28 | 274 | 50 | 8 | 23 | 0,00876 |
| März | 3 | 275 | 37 | 8 | 47 | 0,00887 |
| | 6 | 276 | 22 | 9 | 12 | 0,00898 |
| | 9 | 277 | 5 | 9 | 38 | 0,00910 |
| | 12 | 277 | 46 | 10 | 5 | 0,00921 |
| | 15 | 278 | 26 | 10 | 32 | 0,00935 |
| | 18 | 279 | 4 | 11 | 0 | 0,00948 |
| | 21 | 279 | 39 | 11 | 29 | 0,00962 |
| | 24 | 280 | 13 | 11 | 58 | 0,00977 |
| | 27 | 280 | 44 | 12 | 27 | 0,00992 |
| | 30 | 281 | 14 | 12 | 57 | 0,01007 |
| April | 2 | 281 | 41 | 13 | 28 | 0,01023 |
| | 5 | 282 | 5 | 13 | 58 | 0,01039 |
| | 8 | 282 | 27 | 14 | 29 | 0,01056 |
| | 11 | 282 | 47 | 15 | 0 | 0,01073 |
| | 14 | 283 | 4 | 15 | 31 | 0,01090 |
| | 17 | 283 | 19 | 16 | 2 | 0,01108 |
| | 20 | 283 | 30 | 16 | 33 | 0,01125 |
| | 23 | 283 | 39 | 17 | 3 | 0,01143 |
| | 26 | 283 | 45 | 15 | 34 | 0,0116[illegible] |
| | 29 | 283 | 49 | 18 | 4 | 0,01170 |
| May | 2 | 283 | 49 | 18 | 33 | 0,01197 |
| | 5 | 283 | 47 | 19 | 2 | 0,01215 |
| | 8 | 283 | 42 | 19 | 30 | 0,01232 |

| Mitternacht in Seeberg | | AR. der ⚴ | | Declin. der ⚴ Nördl. | | Lichtſtärke |
|---|---|---|---|---|---|---|
| May | 11 | 283° | 34 | 19° | 57′ | 0,01249 |
| | 14 | 283 | 23 | 20 | 22 | 0,01265 |
| | 17 | 283 | 9 | 20 | 47 | 0,01281 |
| | 20 | 282 | 52 | 21 | 10 | 0,01296 |
| | 23 | 282 | 33 | 21 | 32 | 0,01311 |
| | 26 | 282 | 11 | 21 | 52 | 0,01324 |
| | 29 | 281 | 47 | 22 | 11 | 0,01337 |
| Jun. | 1 | 281 | 20 | 22 | 27 | 0,01349 |
| | 4 | 280 | 51 | 22 | 41 | 0,01359 |
| | 7 | 280 | 20 | 22 | 53 | 0,01368 |
| | 10 | 279 | 47 | 23 | 3 | 0,01375 |
| | 13 | 279 | 13 | 23 | 11 | 0,01380 |
| | 16 | 278 | 38 | 23 | 16 | 0,01384 |
| | 19 | 278 | 1 | 23 | 18 | 0,01386 |
| | 22 | 277 | 24 | 23 | 18 | 0,01386 |
| | 25 | 276 | 40 | 23 | 16 | 0,01384 |
| | 28 | 276 | 8 | 23 | 10 | 0,01380 |
| Jul. | 1 | 275 | 31 | 23 | 3 | 0,01375 |
| | 4 | 274 | 53 | 22 | 52 | 0,01368 |
| | 7 | 274 | 17 | 22 | 39 | 0,01359 |
| | 10 | 273 | 41 | 22 | 24 | 0,01347 |
| | 13 | 273 | 7 | 22 | 6 | 0,01333 |
| | 16 | 272 | 34 | 21 | 47 | 0,01313 |
| | 19 | 272 | 3 | 21 | 25 | 0,01303 |
| | 22 | 271 | 34 | 21 | 1 | 0,01287 |
| | 25 | 271 | 7 | 20 | 35 | 0,01269 |
| | 28 | 270 | 43 | 20 | 8 | 0,01249 |
| | 31 | 270 | 20 | 19 | 39 | 0,01228 |
| Aug. | 3 | 270 | 1 | 19 | 9 | 0,01206 |
| | 6 | 269 | 44 | 18 | 37 | 0,01184 |
| | 9 | 269 | 29 | 18 | 5 | 0,01161 |

**Fig. 11.36** An ephemeris for Pallas from February 4 to August 9, 1802, including its luminosity

**Observations of the new Planet Pallas Olbersiana made at the Equatorial Sector**

by Barnaba Oriani.

Ephemerides Astronomicae ad Meridianum Mediolanensem, 1803, p. 22

[This paper was written in 1802, and gives the first full account of the discovery of Pallas for an Italian-language audience.]

*The discovery of the new planet Pallas is due to the very knowledgeable astronomer of Bremen, Dr. Olbers. On the day Mar. 28, 1802, after observing Ceres, Olbers wanted to review with his cometscope the small stars of the boreal side of Virgo to recognise them better and distinguish them from Ceres, which will be in that part of the sky later. He saw right away a planet which looked like a 7th mag. star, and which formed an equilateral triangle with the 20th and the 191st of Virgo, according to Bode's catalogue. He knew perfectly well the position of those two stars and of the ones nearby, because in that part of the sky he had found Ceres on Jan. 1, 1802 for the first time, and he knew that in that position a mag. 7 star did not exist. He at first suspected that it could have been a variable star similar to the one in The Whale* [Cetus], *which was at her maximum splendour* [he is referring to the star Mira Ceti]. *To determine its position exactly he observed it many times from 8:45 pm till 11 pm, comparing it with the 20th of Virgo. He realised immediately that the RA of the new planet was diminishing and its DEC was increasing. Therefore on the same night, in a little more than 2 h., he ascertained that the new planet had a movement and it could not be a fixed star. The next day, Mar. 29, he found it had notably changed position, its RA having diminished 10′ and its DEC increased 20′.*

*The 30th of March, Olbers having studied the motion again, he transmitted the discovery of this new planet to Zach, and named it Pallas. By April 4, Zach found the new planet and after seeing it again on the 5th and 7th, he determined its daily motion in RA to be 9′ 23″ retrograde and the increase in DEC 17′ 50″. He was kind enough to write me about the important discovery on April 6, sending me the first observation of Olbers of March 28, 29 and 30 and his own of April 4 and 5. I received his letter April 23, but a cloudy sky didn't allow me to observe it the very same evening nor the following. I took advantage of this interval to determine the position of Pallas at least probably, supposing it moved in a circle. According to this hypothesis it turned out that its orbit was situated (as is Ceres) between Jupiter and Mars, and that it had an inclination from the ecliptic of many degrees. The evening of April 25 the sky was clear and it allowed me to observe all the small stars, which were around the place supposed for Pallas, and seeing them again the following day April 26 I realized right away that one of them had moved 3½′ in RA with retrograde motion, and that its DEC had increased 9′. With the 3rd observation on April 27 I confirmed even more the finding of the planet and gave the news to some astronomers who hadn't spotted it yet.*

*Using the best observation made at the meridian during one month, Gauss calculated the elliptic elements, which have been published by Zach in his journal of June; and with a larger no. of observations he later corrected those first elements and arrived at such precise results that the calculated position differed only a few seconds from the observed one. These elements are in Zach's journal of the month of July and are the following (these are Elements III, as printed in the* MC*):*

*If we compare these elements with the ones of Ceres we can see immediately that the two new planets' distance from the Sun is the same, so the same avg. motion and the same periodic time pertain. Because of that the harmonic law of analogy between the planetary distances is not perturbed. Furthermore, Olbers derives from this situation the hypothesis that Ceres and Pallas are probably fragments of a big planet which used to be between Mars and Jupiter, and he believes that the frequent light variability which is seen in these two planets comes from different amounts of reflected light by rotation because they are not completely spherical. It is then to be presumed that the number of these fragments is quite big, and perhaps there are some similar ones between Jupiter and Saturn, between Saturn and Uranus, and further than Uranus.*

*Given the small size of Pallas it is very difficult to determine its real diameter. We can only say that its apparent diameter is not quite 3″ due to the fact that in the month of April the planet was totally covered by the hair of the micrometer of which the size is 3″. The astronomers Herschel and Schroeter, equipped with extremely good instruments, and very familiar with measuring small dimensions, disagreed on Pallas' diameter and the one of Ceres* (see Cunningham 2015). *The first astronomer finds these diameters 10 times smaller than the other astronomer. According to Herschel, the apparent diameter of Ceres was, on Mar. 28, only 0″.2 and that of Pallas was, on Apr. 15, only 0″.1. From that it follows that at the beginning of Aug., when it is still possible for me to observe Pallas, its diameter was only 1/20 of a second. Now it seems strange that an opaque body of such a small size can still reflect so much light. Due to the extraordinarily small size of these two planets, and due to the high inclination of their eccentric orbits, Herschel prefers not to classify them as planets, but to create a new class under the name Asteroids. But Baron von Zach objects, saying that if we discriminate planets from asteroids by means of their diameter, we have to put Mercury, Venus, Mars and Earth in the class of asteroids as they are so much smaller than Jupiter.*

*Going back to the Pallas observations, we have to say that having observed this planet at the meridian during the months of April and May, by many observatories, these observations will always be preferred to all the ones with parallactic machines. The following month, it not being possible to perform the observations at the meridian, the one done with the equatorial sector becomes more interesting. I will show the original observations done from June 24 on, and the position of the planet calculated from these observations and the previous ones.*

*The last observation of late August is very dubious; the planet was very small and appeared like a star of mag. 11 or 12; its dim light was weakened by the moonlight and by the atmospheric vapors. According to the elements mentioned above, its distance from Earth was three semidiameters of the terrestrial orbit at the beginning of August. The calculations of Olbers and Gauss show that next year, 1803, this planet will be at 2.5 semidiameters from Earth at opposition with the Sun at the end of June. So we hope we will be able to observe it again at the meridian considering that having a high declination, the atmosphere will not impair the observation.*

*The reduction of the original observations can be done, as mentioned above, speaking about the ones of Ceres. Those who want to use the positions of the stars determined by Zach can use his journal of August and November that lists the RA and the avg. DEC of all the stars I compared to Pallas. There are as well the planet positions deduced separately from each of the following original observations. So from all the various results we can choose the one believed to be the most reliable.*

*In order to convert the clock time in avg. solar time, it will be useful to use the table at the end of the Ceres observations. Apparent positions of Pallas calculated from observations done at the Equatorial Sector:*

| *1802* | *Mean time* | *Apparent RA* | *Apparent Dec* |
|---|---|---|---|
| *April 25* | *10 h 5′ 29″* | *181° 15′ 6″* | *18° 51′ 48″* |
| *26* | *9 52 1* | *181 11 38* | *19 0 53* |
| *27* | *9 19 54* | *181 9 1* | *19 9 53* |
| *29* | *11 52 27* | *181 3 9* | *19 29 12* |
| *30* | *10 0 17* | *181 1 2* | *19 36 17* |
| *May 3* | *9 25 40* | *180 57 8* | *19 56 35* |
| *4* | *9 17 13* | *180 56 9* | *20 3 8* |
| *5* | *10 23 43* | *180 56 13* | *20 9 52* |
| *7* | *9 3 27* | *180 56 34* | *20 19 36* |
| *8* | *9 9 14* | *180 57 15* | *20 25 9* |
| *9* | *9 2 31* | *180 58 35* | *20 29 43* |
| *10* | *9 2 33* | *180 59 46* | *20 33 20* |
| *11* | *8 59 1* | *181 1 45* | *20 36 29* |
| *12* | *9 27 21* | *181 4 5* | *20 39 24* |
| *17* | *9 11 45* | *181 20 27* | *20 53 9* |

(continued)

(continued)

| | | | |
|---|---|---|---|
| *19* | *9 42 47* | *181 29 20* | *20 55 43* |
| *20* | *9 14 13* | *181 34 20* | *20 57 21* |
| *21* | *9 12 51* | *181 39 36* | *20 59 58* |
| *28* | *10 15 2* | *182 25 48* | *20 58 56* |
| *31* | *9 41 36* | *182 49 39* | *20 55 9* |
| *June 1* | *10 27 34* | *182 58 45* | *20 52 41* |
| *10* | *9 44 51* | *184 27 25* | *20 29 33* |
| *11* | *9 55 26* | *184 38 12* | *20 26 48* |
| *12* | *9 44 19* | *184 49 54* | *20 23 0* |
| *14* | *9 43 50* | *185 13 14* | *20 15 19* |
| *15* | *10 3 36* | *185 25 31* | *20 12 20±* |
| *19* | *10 11 4* | *186 15 56* | *19 54 42* |
| *24* | *10 10 8* | *187 23 22* | *19 28 24* |
| *28* | *10 3 40* | *188 20 50* | *19 8 18* |
| *July 1* | *10 19 15* | *189 5 42* | *18 51 45* |
| *2* | *10 19 16* | *189 20 44* | *18 45 46* |
| *3* | *10 7 54* | *189 36 9* | *18 39 24* |
| *7* | *9 40 7* | *190 38 59* | *18 15 31* |
| *8* | *9 42 16* | *190 55 14* | *18 9 5* |
| *10* | *9 40 22* | *191 27 49* | *17 56 2* |
| *18* | *9 36 4* | *193 44 5* | *17 2 52* |
| *24* | *9 32 16* | *195 31 48* | *16 20 5* |
| *25* | *9 44 11* | *195 50 8* | *16 12 46* |
| *28* | *9 8 6* | *196 45 12* | *15 50 59* |
| *29* | *9 11 28* | *197 3 51* | *15 43 32* |
| *31* | *9 6 9* | *197 41 29* | *15 29 11±* |
| *August 1* | *9 4 14* | *198 0 32* | *15 21 35* |
| *2* | *9 23 57* | *198 19 47* | *15 13 51* |
| *4* | *8 55 38* | *198 57 56* | *14 59 17* |
| *5* | *9 14 57* | *199 17 35* | *14 51 17* |
| *6* | *9 6 27* | *199 37 8* | *14 43 53 ±* |
| *7* | *9 4 28* | *199 56 37* | *14 36 13* |
| *8* | *9 0 14* | *200 16 35* | *14 28 40* |

[Oriani then gives his observations of Pallas in 1803: see Fig. 11.37; Fig. 11.38.]

**Astronomical observations and remarks**
by Pastor Fritsch* in Quedlinburg
*BAJ*, 1806, p. 181

*Feb. 22, 1803. In a hurry, I am pleased to bring you the good news that, on Feb. 18 at 14 h 50′ mean time, I succeeded in finding Pallas again. It almost perfectly keeps the orbit calculated by the excellent Dr. Gauss (Yearbook 1805, p. 227), and I owe it to this admirable agreement that I found it in spite of its small size.*

| Giorni 1802 | Angolo Orario | Nome degli Aſtri ſecondo il Catalogo di *Bode* | Sortita dalla I Barra (Tempo dell' Orologio) | Ingreſſo nella II Barra (Tempo dell' Orologio) | Declinazione Boreale apparente compreſo l' errore dello ſtromento e la rifrazione |
|---|---|---|---|---|---|
| | ore | | or ' " | or ' " | or ' " |
| Luglio 8 | 4 2 1/3 | 103 Ch. Ber. | 9 13 1,0 | 9 15 4,4 | 18 10 31 |
| | | 117 Ch Ber. | 9 22 41,5 | 9 24 45,1 | 17 38 55 |
| | | 135 Ch. Ber. | 9 28 :: | 9 30 30,7 | 18 10 5 |
| | | Pallade | 9 29 36,5 | 9 31 38,5 | 18 8 39 |
| 8 | 4 12 | 134 Ch. B r. | 9 38 8,7 | 9 40 :: | 18 8 39 |
| | | 135 Ch Ber. | 9 38 18,5 | 9 40 21,0 | 18 10 6 |
| | | Pallade | 9 39 25,0 | 9 41 28,7 | 18 8 38 |
| 10 | 4 .6 1/3 | 103 Ch. Ber. | 9 8 53,8 | 9 10 57,0 | 18 10 26 |
| | | 117 Ch. Ber. | 9 18 34,0 | 9 20 37,6 | 17 39 4 |
| | | Pallade | 9 27 38,2 | 9 29 41,0 | 17 55 50 |
| | | Anonima 10 | 9 27 52,2 | 9 29 55,5 | 17 55 36 |
| 10 | 4 17 | 135 Ch. Ber | 9 35 9,8 | 9 37 13,3 | 18 10 :: |
| | | Pallade | 9 38 28,0 | 9 40 30,7 | 17 55 50 |
| | | Anon.(jeri) 10 | 9 38 42,0 | 9 40 45,0 | 17 55 35 |
| 18 | 4 24 2/3 | 117 Ch. Ber. | 9 4 50,6 | 9 6 53 9 | 17 38 15 |
| | | Pallade | 9 22 58,7 | 9 25 0,5 | 17 1 52 |
| | | Anonima 10 | 9 23 15,0 | 9 25 17,0 | 17 1 56 |
| 24 | 4 37 | 359 Vergine | 8 59 56,7 | 9 1 58,5 | 16 6 14 |
| | | Pallade | 9 18 52,8 | 9 20 54,4 | 16 19 31 |
| | | Anonima 8 | 9 20 :: | 9 22 12,5 | 16 21 50 |
| | | Anonima 9 | 9 21 :: | 9 23 41,7 | 16 19 51 |
| 25 | 4 52 | 359 Vergine | 9 10 36,4 | 9 12 38,3 | 16 6 23 |
| | | Pallade | 9 30 45,5 | 9 32 47,5 | 16 12 21 |
| | | Anon. (jeri) 8 | 9 30 :: | 9 32 53,0 | 16 22 :: |
| | | Anon. (jeri) 9 | 9 32 :: | 9 34 22,0 | 16 20 :: |
| 28 | 4 24 | 359 Vergine | 8 30 43,3 | 8 32 45,0 | 16 6 15 |
| | | 435 Vergine | 8 43 17,0 | 8 45 18,8 | 15 47 15 |
| | | Pallade | 8 54 32,5 | 8 56 34,3 | 15 50 30 |
| 29 | 4 30 | β Leone | 7 28 49,5 | 7 30 51,0 | 15 40 24 |
| | | 359 Vergine | 8 32 48,5 | 8 34 50,2 | 16 6 27 |
| | | 435 Vergine | 8 45 21,3 | 8 47 23,0 | 15 47 19 |
| | | Pallade | 8 57 51,8 | 8 59 53,3 | 15 43 12 |

**Fig. 11.37** Last obs. of Pallas done at the equatorial sector from July 8 to July 29, 1802

*Johann Heinrich Fritsch (1772–1829) was the first (in 1821) to notice the variability of the star Epsilon Aurigae (*BAJ* for 1824, p. 252). He was superintendent of the church in Quedlinburg.

**British Almanacs**

Several British almanacs gave brief notices about the asteroids beginning with their 1803 editions. *Merlinas Liberatas* by Patridge (1803: 42) told its readers of the discovery of Ceres. "It is about half as large as the *Moon*, and seems to perform its period round the *Sun* in about four years and a half." *Speculum Anni* by Henry Season (1803: 34) gave Ceres a whole paragraph, elevating Maskelyne to a special

(a)

| Giorni 1802 | Angolo Orario | Nome degli Aſtri ſecondo il Catalogo di *Bode* | Sortita dalla I Barra (Tempo dell' Orologio) | Ingreſſo nella II Barra (Tempo dell' Orologio) | Declinazione Boreale apparente compreſo l' errore dello ſtromento e la rifrazione |
|---|---|---|---|---|---|
| | ore ′ | | or ′ ″ | or ′ ″ | ° ′ ″ |
| Luglio 31 | 4 30 | 435 Vergine | 8 37 27,8 | 8 39 29,4 | 15 47 25 |
| | | Pallade | 8 52 28,3 | 8 54 29,8 | 15 28 55± |
| Agoſto 1 | 4 31 | 8 Leone | 7 17 41.8 | 7 19 43,2 | 15 40 24 |
| | | 435 Vergine | 8 34 13,2 | 8 36 14,9 | 15 47 26 |
| | | Anonima 8..9 | 8 50 : : | 8 52 7,5 | 15 12 0 |
| | | Pallade | 8 50 30,0 | 8 52 31,6 | 15 21 18 |
| | | 12 Boote | 9 6 53.4 | 9 8 55,2 | 15 19 4 |
| 2 | 4 53 | 435 Vergine | 8 52 17,1 | 8 54 18.8 | 15 47 34 |
| | | Anon (jeri) 8 .9 | 9 8 11,3 | 9 10 12.7 | 15 12 : : |
| | | Pallade | 9 9 50,5 | 9 11 52,0 | 15 13 44 |
| | | Anonima 8 | 9 19 27.0 | 9 21 : : | 15 24 : : |
| | | Anonima 8. 9 | 9 21 7,5 | 9 23 : : | 15 15 : : |
| | | 12 Boote | 9 24 57,3 | 9 26 59,0 | 15 19 12 |
| 4 | 4 30 | 485 Vergine | 8 33 25,2 | 8 35 26.9 | 14 43 22 |
| | | Anonima 10 | 8 40 32,0 | 8 42 33± | 14 59 : : |
| | | Pallade | 8 41 45,6 | 8 43 47.8 | 14 58 59 |
| | | Anonima 7..8 | 8 50 31,8 | 8 52 33,2 | 15 17 : : |
| | | Anonima 8 | 8 50 33,0 | 8 52 34,6 | 15 8 : : |
| | | 12 Boote | 8 54 20,2 | 8 56 22,3 | 15 19 2 |
| 4 | 4 58 | 485 Vergine | 9 1 30,4 | 9 3 32,6 | 14 43 38 |
| | | Pallade | 9 9 52,8 | 9 11 54.8 | 14 58 58 |
| | | 12 Boote | 9 22 25,8 | 9 24 27,9 | 15 19 16 |
| 5 | 4 52 | 485 Vergine | 8 51 22 3 | 8 53 24.4 | 14 43 28 |
| | | Pallade | 9 1 1 7 | 9 3 3,5 | 14 51 6 |
| | | 521 Vergine | 9 2 37,8 | 9 4 39.7 | 14 50 45 |
| 5 | 5 1 | Pallade | 9 9 51,3 | 9 11 53.4 | 14 51 8 |
| | | 521 Vergine | 9 11 27,2 | 9 13 28 9 | 14 50 48 |
| 6 | 4 46 | 485 Vergine | 8 41 31,9 | 8 43 33,6 | 14 43 31 |
| | | Pallade | 8 52 29,0 | 8 54 30.7 | 14 43 57± |
| | | 521 Vergine | 8 52 47,0 | 8 54 48,9 | 14 51 : : |

(b)

| Giorni 1802 | Angolo Orario | Nome degli Aſtri ſecondo il Catalogo di *Bode* | Sortita dalla I Barra (Tempo del' Orologio) | Ingreſſo nella II Barra (Tempo del' Orologio) | Declinazione Boreale apparente compreſo l' errore dello ſtromento e la rifrazione |
|---|---|---|---|---|---|
| | ore ′ | | or ′ ″ | or ′ ″ | ° ′ ″ |
| Agoſto 6 | 5 4 | 485 Vergine | 8 58 58.7 | 9 1 0.3 | 14 43 46 |
| | | Pallade | 9 9 56,1 | 9 11 57.7 | 14 44 0 |
| | | 521 Vergine | 9 10 13,7 | 9 12 15,6 | 14 51 : : |
| 7 | 4 47 | 485 Vergine | 8 38 2,5 | 8 40 4,1 | 14 43 32 |
| | | 521 Vergine | 8 49 17 8 | 8 51 19,7 | 14 51 : : |
| | | Pallade | 8 50 17 3 | 8 52 18 9 | 14 36 5 |
| 8 | 4 45 | 485 Vergine | 8 32 37,7 | 8 34 39,6 | 14 43 39 |
| | | 521 Vergine | 8 43 52,8 | 8 45 : : | 14 51 : : |
| | | Pallade ✝ | 8 46 10,8 | 8 48 : : | 14 28 31 |

**Fig. 11.38** Last obs. of Pallas done at the equatorial sector from July 31 to August 8, 1802

(undeserved) place. "…early this Year, 1802, it was discovered again, by the Rev. Dr. Maskelyne, his Majesty's Astronomer at Greenwich, where a competent Number of Observations are now making, in order to settle the Elements of its Theory." But only *The Coelestial Atlas* by Robert White (1803: 7; Fig. 11.39) gave its readers of the 1803 edition news about the orbital elements of Ceres and the discovery of Pallas as well:

*Two New Planets*

1. *Young as the present century is, it has produced already the discovery of two new primary planets. The first of these, called Ceres, was discovered on the first day of the present century, or the first of January 1801, by M. Piazzi, astronomer at Palermo in Sicily. It moves in an orbit between Mars and Jupiter; is a very small planet, appearing through the telescope as a star of the 8th magnitude. Its apparent diameter about ¼", and real diameter 200 miles; other elements as follows:*

| | |
|---|---|
| *Place of the ascending node 2Z 20° 58′* | |
| *Inclination of orbit* | *10° 47* |
| *Place of aphelion* | *2 9 0* |
| *Eccentricity of orbit* | *0.0364* |
| *Greater semiaxis* | *23 times the earth's distance* |
| *Periodic time* | *4.13 years* |

2. *The second was discovered March 28, 1802, by Dr. Olbers, of Bremen, who has named it Pallas. It is a very small planet, of about 140 miles in diameter; moves between Mars and Ceres, at a distance about 2½ of the earth's distance, and performs its revolution in about 3 years.*

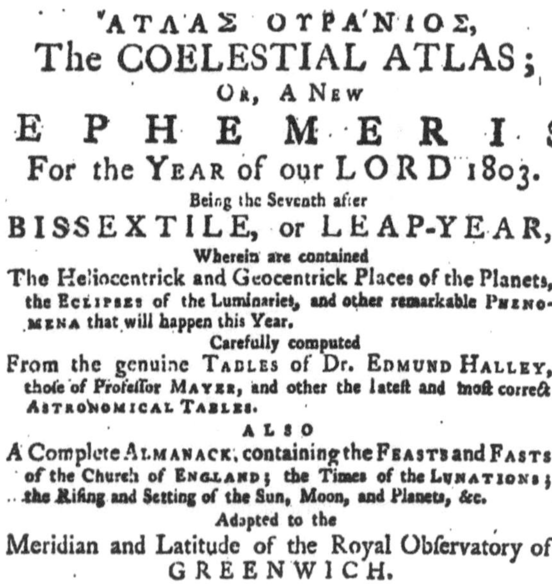
ΑΤΛΑΣ ΟΥΡΑΝΙΟΣ,
The COELESTIAL ATLAS;
OR, A NEW
EPHEMERIS
For the YEAR of our LORD 1803.
Being the Seventh after
BISSEXTILE, or LEAP-YEAR,
Wherein are contained
The Heliocentrick and Geocentrick Places of the Planets, the ECLIPSES of the Luminaries, and other remarkable PHENOMENA that will happen this Year.
Carefully computed
From the genuine TABLES of Dr. EDMUND HALLEY, those of Professor MAYER, and other the latest and most correct ASTRONOMICAL TABLES.
ALSO
A Complete ALMANACK, containing the FEASTS and FASTS of the Church of ENGLAND; the Times of the LUNATIONS; the Rising and Setting of the Sun, Moon, and Planets, &c.
Adapted to the
Meridian and Latitude of the Royal Observatory of GREENWICH.

**Fig. 11.39** The upper half of the title page of the 1803 edition of White's Ephemeris

## Regarding the Search for the Two New Planets Ceres and Pallas in 1803

*MC*, March 1803, p. 276

*Since the time is approaching when the two new planets, Ceres and Pallas, will emerge from the rays of the Sun and will reveal themselves to our keen eyes once again, the attention of all observers of the skies will once again become active and be directed toward these two highly remarkable objects in the Solar System.*

*The path of Ceres has been pretty well cleared up following the most precise observations conducted within a period of time of 19 months and according to the astute calculations of Dr. Gauss, so that, in accordance with the positions calculated in advance from Gauss' elements, one may entertain the hope of being able to find this planet with certainty. But the low light intensity and the low position of this planet over our horizon could be counted among the remaining hindrances to locating this planet again.*

*The situation is different with Pallas. This planet had only been observed four months since its discovery. Dr. Gauss had to base its entire path on these few observations conducted within this short period of time. The ephemeris of its course cannot be given with any measure of reliability for a long time as could be the case with Ceres. Although Pallas will reach a high position this year, its luminosity on the other hand will be so extraordinarily weak that this planet will barely reveal itself as a magnitude twelve star. In the case of this planet, uncertainties are therefore combined with difficulties.*

*In order to counteract this and to make it easier for astronomers to locate this remarkable body, the skilful observer Harding in Lilienthal diligently developed the present chart of the course of Pallas. If it does reach the twelfth stellar magnitude, we hope that this planet will not easily escape us with the help of this chart. Practical astronomers will therefore certainly give the greatest thanks to the observer Harding for this meritorious endeavour.*

*The star chart given with the present issue contains the course of Pallas from February 4 to April 17, 1803. Since the scale had to be made larger to be able to accommodate the registering of even the smallest stars, of which there is a wealth in this area of the heavens, which have to be recognised in order not to miss Pallas, the observer Harding drew the course of this planet on two sheets, since it would have turned out too big for one sheet and too awkward for our paper. This second sheet, which we will therefore include in the April issue (which will be available in March), contains that area of the heavens that the planet will traverse from April 5 until June 28.*

*The most careful of investigations of this area of the heavens yielded the observer, Harding, 1100 stars. Since there was not enough room to register 'all' of these, which would have led to confusion, because they could only be registered by eye, he left them out on the chart, including only those stars which had been observed on instruments of measurement and which appear in the* Hist. cèléste francaise *Vol. 1, in the* Connaissances des tems *and in Professor Bode's newest maps and star catalogues.*

*At the same time, the observer, Harding, sent us an exact catalogue of all stars to be found in his map that also contains the record of where each star was taken. Although this catalogue is not suited for publication since it had not been reduced for this purpose, having been reduced in consideration of 'aberration' and 'nutation', it does however serve to designate each star more closely and to more precisely determine them as required, since the stars appearing in the* Hist. cèléste franc. *are not such that they are exact within a few seconds. On this occasion, a few errors in Professor Bode's catalogue of this area of the heavens have become known to the observer, Harding, which duty dictates be indicated here.*

*For instance, No. 20, 30, 59 and 63 in Poniatowsky's Taurus are missing in the heavens, and must therefore be omitted. No. 69 and 71 are set, in right ascension, some 8 min too far to the east. The observer, Harding, also discovered some errors of printing and representation in the* Hist. cèléste franc. *According to this source, three stars are supposed to be located 279° in RA and 12° in decl., but these are not located 'there.' Their existence is due to an error of printing, and must therefore also disappear with it. They appear on p. 170 of the* Hist. cèl. *and are supposed to have gone through the middle line of the quadrant at 18 h 34′ 31.″5 and 34′ 44″; but if, on the other hand, 18 hr 33′ is used both times, that is, one minute less, then their positions correspond with two magnitude eight stars which appear earlier in the map, which appear on p. 85 of the* Hist. Cel. *at 18 hr 34′ 9.″5 and 34′ 23″. According to this correction, a magnitude eight star at 18 hr 34′ 58.″5, appearing on p. 170, corresponds with a magnitude seven star at 18 h 35′ 38″, appearing on p. 85, which actually does exist. That its magnitude in both positions is different is something that occurs countless times in the* Hist. cèl. *The observer, Harding, himself verified this star and estimated it as being magnitude seven, it is located closely above no. 76 of Poniatowsky's Taurus, according to Bode's catalogue.*

## The rediscovery of Pallas in Febr. of 1803

by Inspector Harding in Lilienthal (from some of his letters)
*BAJ*, 1806, p. 173

*Feb. 22, 1803*

*In a hurry, I am pleased to bring you the good news that, on Febr. 18 at 14 h 50′ mean time, I succeeded in finding Pallas again. It almost perfectly keeps the orbit calculated by the excellent Dr. Gauss (yearbook 1805, p. 227), and I owe it to this admirable agreement that I found it in spite of its small size.*

*On the 18th, I found it at 14 h 50′ between two small telescopic stars west of No. 36 (c) Tauri Poniat of your register, without knowing for sure, however, which star the planet was, but after 2 hours I thought I recognised it by a minor displacement. On the 19th at 16 h 40′, I convinced myself of its presence at first sight. It was just north below No. 36, where no star has been 24 hours before; the star between a and b which at that time has become suspicious was missing, and thus there was no doubt anymore. But the weather was quite unfavorable for micrometric measurements of its position, and the entire sky was already covered with clouds after half an hour. On the 20th, it had advanced to the eastern side of the triangle formed by 3 telescopic stars east of No. 36 and, aided by a pretty clear sky, I could now compare it with No. 36* via *the empty circle. At 15 h 30′ 13″ mean time, it followed this star in 55.″0, and I calculated from this its apparent RA = 272° 38′ 27″. On the 21st, I observed it south above 3 small stars and found its eastern distance from No.36 at 17 h 0′ 11″ mean time = 2′ 7″, hence its linear inclination = 272° 56′ 28″. I could not yet calculate the declination with greatest precision.*

*Pallas appears as a star of 12th or 13th magnitude, in as far as such little stars can be estimated. It cannot be viewed with an excellent 3-foot achromatic telescope that allows me to recognize stars of up to the 11th magnitude.*

***March 6.** Since Feb. 24, the weather here was so miserable that one could not even think of observing Pallas, and now that the Moon is shining all night we shall see it again only after 14 days. Such interruptions, however, are not harmful anymore, because we know that the positions of Pallas, as calculated by Dr. Gauss, fit perfectly with its orbit. On Febr. 21 and 23, when also Dr. Olbers made observations, we compared it with Nos. 36 and 42 Tauri Poniat according to your register; but the position of Pallas opposite these stars was not favorable for observations with the circle micrometer and could not yield the declination with proper precision. On the 21st at 16 h 46′ 37″ mean time, it preceded No. 36 by 2′ 8″*

*and was 19′ 20″ more north, on the 23rd at 15 h 19′ 24″ it preceded No. 42 by 2′15.″5 and was 9′ 46″ more south. The untiring Dr. Gauss has calculated Pallas for these two moments and found that for the 21st the observed RA was smaller by 2′ 2″, the declination larger by 34″; for the 23rd the RA was smaller by 2′ 35″, the declination larger by 57″ than the calculated ones. Accordingly, he has already made a preliminary correction of the elements.*

## Astronomical news and the observations of Pallas

From some of his letters dated Feb. 18, 1803 onward by Dr. Olbers in Bremen
*BAJ,* 1806, p. 175

***February 18, 1803.*** *I was hoping to find again an earlier observation of Pallas, but this hope has been spoiled. On the 1st of March, 1797, according to Burckhardt's elements, Pallas has been in the centre of a zone which, at that time, has been observed in accordance with the* Hist. Cel., *and must have been fairly bright, like a star of 7th magnitude. The elements of Dr. Gauss give Pallas at 14′ south of that zone. I have now searched through this zone; yet, none of those stars of 7th or 8th magnitude observed on March 1, 1797, is missing and, hence, Pallas really must have been south of that zone. I am now expecting a message that the French astronomers have found Pallas again. I myself will not look for it this month. My business does not allow me to sacrifice entire nights. For photometric reasons, I believe to be able to demonstrate that the apparent diameter of Ceres during its opposition in March 1802 could not have been more than 0.″6. The apparent diameter of Pallas must have been significantly smaller. This comes pretty close to Herschel's measurements. I have submitted to Baron von Zach a short article where this result is to be found.*

***February 22.*** *With great pleasure I hurry to announce to you that Mr. Astronomer Harding in Lilienthal was lucky enough to find Pallas again, already on Feb. 18, at 15 h. He found it as a small star of 12th or 13th magnitude, close to No. 36 of Poniatowski's Bull (Taurus) according to your register, almost right at the spot where it should have been according to Dr. Gauss's calculated ephemeris. He recognized it because, in the morning before, he had observed No. 36 and its surroundings very carefully and now found a star at the given position that has not been there the day before. In the morning of the 21st, or on the 20th at 15 h, he saw Pallas advanced as much as required by the theory and he could compare it 5 times with No. 36, which it followed at 15 h 45′ 17″ within a period of 55″. He was so kind as to inform me at once of his important discovery, and in the following night, I too have seen Pallas again and observed it. My observation is not very precise, due to the faint light of the small planet: after a preliminary reduction, it is as follows: Feb. 21, 17 h 0′ RA of Pallas 272° 57′, north declination 7° 32′. With amazement you will notice how accurate Dr. Gauss's ellipse still is. This precise agreement of the calculation is all the more desired as otherwise it would have been almost impossible to find the now so small Pallas within the great host of similar, small telescopic stars.*

*Pallas is so faint in light that Mr. Harding cannot see it at all with a 3-feet achromat. He requires a 7-foot telescope, in which it has a quite lively light. It can be well recognised in my 5-foot Dollond (telescope). As we now know that Dr. Gauss's ephemeris is so reliable, it shall be easy to find Pallas.*

***May 3.*** *For a long time, I have not given you an account of my observations on Pallas. I have the honor to inform you of all of them.* [This data is published in the May & June issues of the *MC*.] *The elements of Dr. Gauss, which have been improved through both of my first observations on February 21 and 23 and which are found in the April issue of* M. C., *still agree so well with the observations up to the middle of April that it is not yet possible to give a certain deviation of them for this oscillating ellipse.*

*Often, it was not easy to find the small planet among the abundance of small stars in the sky where Pallas could be, particularly in the vicinity of the Milky Way. At some occasions, I have had between 23 and 25 stars together with Pallas in the visual field of my telescope. Nevertheless, I have not yet missed Pallas at a single night, nor did I confuse it with a fixed star. Its light has increased, and I now estimate it to be of the 11th magnitude. Yet from one night to the other, it shows also in this year that strange change in light that we have already noticed the year before for Pallas as well as for Ceres.*

*In Paris, it has been observed in March only once (March 12) by Mechain. Oriani in Milan started to observe Pallas on March 22, and David in Prague on March 24.*

*It proves more and more true that neither Ceres nor Pallas are found among the 50,000 stars of the* Hist. Cel. *Once, I had hoped to find Pallas among the stars observed on March 1, 1797: yet at that time it probably was a few minutes south of the zone screened on this day, and there is no star missing in the same area of the sky.*

In this letter, Olbers mentions Oriani, but throughout the study of Pallas the name of Piazzi is curiously absent. He never made any serious contribution to asteroid studies after 1801, but there is a glimmer of interest evident in letters between him and Oriani in 1803. Parts of these letters appear in *Discovery of The First Asteroid, Ceres*:

Letter LXII, Piazzi to Oriani Palermo May 4, 1803

*I have not yet replied* [to Baron von Zach] *because I am waiting first to observe Pallas and Ceres at their meridian passage that I hope to do around the middle of this month.*

*PPS. On May 13 I think I saw Pallas and Ceres. Pallas was very small.*

Letter LXIII Oriani to Piazzi Milan May 19, 1803

*I have seen Pallas again on Mar. 22. The observations are constantly interrupted by clouds.*

**The elements of the orbit of Pallas**

Calculated anew and reported on March 3, 1803, by Dr. Gauss in Brunswick

*BAJ*, 1806, p. 179

*I assume, Reverend Sir, the important news of the early rediscovery of Olbers' planet at Lilienthal has been already received from Mr. Harding himself. Thus, you will most likely be already in the possession of Dr. Olbers' observations. I flatter myself that you will be rather glad about hearing more detailed results. The weather has been very bad here since February 23; in addition, the Moon is now shining. The observations, therefore, shall remain somewhat incomplete.*

*Here is the comparison of Olbers' observation with my 5th elements, according to which the ephemeris had been calculated in the December issue of M.C.*

| *Calculated* | | *Difference* | |
|---|---|---|---|
| *RA* | *Dec* | *RA* | *Dec* |
| *Feb. 21 272 58 47* | *7 30 40* | *+2′ 2″* | *–34″* |
| *23 273 31 14* | *7 45 4* | *+2 35* | *–57* |

*I could not resist the temptation to do a preliminary correction of the elements. I reserve to myself to give better ones as soon as more numerous and more precise observations are available. Reverend Sir, you shall see from the comparison with the 5th *) that the difference is negligible:*

**) See Astronomical Yearbook* 1805, p. 228.

| | |
|---|---|
| *Epoch 1803 in the meridian of Seeberg* | *221° 28′ 54.″0* |
| *Yearly tropical motion (365 days)* | *78° 0′ 36.″9* |
| *Daily* | *769.″416* |
| *Distance from Sun 1803. sid. resting* | *301° 24′ 13″* |
| *Node* | *172° 28′ 8″* |
| *Eccentricity* | *0.245619* |
| *Log. of semimajor axis* | *0.4426160* |
| *Inclination* | *34° 38′ 20″* |

| | *Calculated* | | *Difference* | |
|---|---|---|---|---|
| | *RA* | *Dec* | *RA* | *Dec* |
| *Feb. 21* | *272 56 25.0* | *7 31 29.8* | *−20.″0* | *+15.″8* |
| *23* | *273 28 46.8* | *7 45 52.3* | *+7.8* | *−7.7* |

*Accordingly, one will be able to find the planet this year with all desired ease. Hopefully, also Ceres will soon be found again. This time finding her will probably happen earlier in Italy and France than here due to her low position in the southern sky.*

## **Observations of Ceres and Pallas from the Royal Observatory in Berlin in the year 1803** By Bode
*BAJ*, 1806, p. 242

*Those observations in that year were connected with many difficulties – you could see Ceres only very low in the southerly sky. Pallas, however, was very high on a clear summer night sky but was not easy to find in the middle of the Milky Way, and only after many attempts was I able to observe it. The number of observations is not very great if you don't make it your business to observe only the newly discovered planets. That is why you find very few or no reports of observations of Ceres and Pallas even by very famous astronomers.*

*After some clear nights of observation in May in the area of the Eagle's tail and Cerberus where Pallas should be, I found whole star clusters with my 3½-foot Dollond telescope. I finally discovered on the 16$^{th}$ that a small star observed on the 15$^{th}$ had moved position – that had to be Pallas.*

*I let it go by twice on my circular micrometer at no. 490, and calculated with that the inclination and deviation at 11:30 pm. On the 23$^{rd}$ I marked down a group of stars around Pallas; because of poor conditions I could not find it again the following nights. I tried again on the 8$^{th}$ and 9$^{th}$ without success. Later Mr. von Wisniewsky* [Vincent Wisniewsky] *on July 1 discovered Pallas in his drawn-up star chart; Pallas was positioned southerly under H Cerberus in between a cluster of small stars. On July 2$^{nd}$ I observed Pallas on my circular micrometer and compared it twice to determine the inclination and the deviation with 464 Hercules. With the wall quadrant it was impossible to make observations because of its faint light of a 9$^{th}$ or 10$^{th}$ magnitude star; not the smallest illumination could agree with the threads.*

*I looked for Ceres several nights in May but could not find it because of low clouds at the lower south-easterly sky. Finally I thought I found it on June 24th when I recognised a magnitude 7 star in a westerly triangle with τ Sagittarius and a northerly star. This star (τ Sag) passed my circular micrometer twice and I was able to calculate the inclination and deviation. At 12:15 I wanted to observe with my wall quadrant and passage instrument, but being low on the horizon and having some haze I was unable to see it with the telescope. But I saw τ Sag clearly. On the 25th it was obvious that it was not Ceres I had observed yesterday. I designed a group of stars but haze and clouds in the south made it impossible to relocate. The 26th and 27th were also out because of haze. On the 28th I again searched in the area of Sagittarius, and on the 29th I finally found Ceres by its movement, compared twice with the circular micrometer with no. 180 and τ Sag, equalling 12 h 35' 10" mean time. Its RA was 281° 21' 7" and Dec 28° 10' 58"S. This is as close as I could get it with this method. On July 1st Ceres showed itself clearly in my 3 ½-foot Dollond even though it was very low in the south and in moonlight. I could not recognise it with the finderscope, its position was close to opposition under σ Sag. I observed it with φ and τ Sag with my wall quadrant and calculated the time of culmination 12 h 5' 56" mean time, its RA 280° 45' 17" and Dec 28° 19' 25" South. On July 2nd observations of Ceres with any of my instruments was impossible because of haze in the south. This was also the case on July 11. On July 17 I found Ceres westerly of φ and south-easterly of no. 119 Sag where according to my calculations it should be. At least I found the same three stars in a triangle; of those, I suspect the one most south-easterly is Ceres. I ran this star twice with no. 119 Sag through the field of my circular micrometer and found its inclination and deviation. I will record in my summary of observations for the year 1803 what observations show Ceres and Pallas.*

## Observations of the new planet Pallas

*MC*, April 1803, p. 369

*The proof for the degree of perfection which our actual theoretical and practical astronomy has reached cannot be more apparent and the triumph of this science not greater, even in the eyes of amateurs, than shown by the rediscovery of Pallas. This small planet, hardly observed for four months since its first discovery last year, was, after having approached the rays of the Sun, and being swallowed by these, and it has hardly emerged again from those, after invisibility of six months, rediscovered as a tiny almost invisible spot among myriads of worlds, at exactly that position where theory had placed it. For this rediscovery the greatest possible physical and intellectual powers of humans were combined. On February 18 at 2:50 pm the deft observer Harding at Lilienthal succeeded in finding Pallas at approximately that position where it should be according to Dr. Gauss's calculated ephemeris in the December issue 1802, p. 583. The following night, he had the pleasure of convincing himself of the correctness of its existence. On February 19 at 3 pm he found her close above no. 36 of Poniatowski's Taurus as a small 12th or 13th magnitude star; and he recognised the planet because the previous night he had observed no. 36 and its surroundings very carefully and he now noticed a little star at a position where 24 hours before none had been visible. In the following night this little lucid spot had moved as it was supposed to do according to theory and followed on February 20 at 3 pm 30' 13" mean time 55 time seconds after no. 36. Observer Harding kindly shared his important discovery immediately; he wrote in a letter that he had not been able to discern the planet with a 3ft achromat, which distinctly shows stars down to the 11 order of magnitude. Dr. Olbers, who had already on February 21 the pleasure of seeing his planet again, wrote us that he was able to clearly see Pallas with his 5ft Dollond. But he too, estimated her to be only of the 12th or 13th magnitude and her luminosity equal to that of the fourth satellite of Saturn. Again, just like last year with Ceres, Dr. Gauss deserves the greatest praise and warmest thanks for his incredibly accurate calculation by which only it was possible to find this planet among the hosts of telescopic stars. If those ephemerides of Dr. Gauss had been faulty by 30 to 40 min, I very much doubt that I would have found Pallas, Harding wrote.*

*We estimated Pallas' probable size in the March issue of MC 1803, p. 277, at her first reappearance to be of the 12th order of magnitude; and this assumption was exactly right. We published Harding's map of the probable path of this planet to easily find it again, and this map has become the true orbit; for the differences between the hitherto actual observations and the ephemeris calculated in advance are so small that they cannot be depicted on our map, which is drawn in a very large scale and is almost four times as big as the map of Bode's great stellar atlas. By means of these maps and the ephemeris of the orbit of this planet, published in the December issue of MC 1802, it would be possible, even without the announcement that this calculation is accurate but to a few minutes, to find Pallas where she will be searched after with good telescopes.*

*Mr. Harding's two observations of its right ascension are as follows:*

| *1803* | *Mean time in Lilienthal* | *RA* |
|---|---|---|
| *Feb. 20* | *15 h 30′ 13″* | *272° 38′ 27″* |
| *Feb. 21* | *17 0 11* | *272 56 29* |

*Dr. Olbers, who has a wonderful dexterity in observing with the circle micrometer, has hitherto been able to obtain only two observations. This excellent observer in a letter dated Feb. 23 says, "I find observations of Pallas very difficult, on account of its faint light and therefore they are not very accurate; the declination in particular is somewhat doubtful. In the place where Pallas ought to have stood on the 23rd of February, there were four small telescopic stars, among which I however discovered Pallas, as the brightest, after six comparisons with No. 42 of the Bull; it passed this star on the 23rd of February at 15 h 19′ 24″ and mean time 2′ 15″.5 in time." These two observations are as follows but Dr. Olbers still gives the declinations as very doubtful.*

| *1803* | *Mean time in Bremen* | *Apparent RA of Pallas* | *Apparent Dec of Pallas* | *Star Bode's Catalogue* |
|---|---|---|---|---|
| *Feb. 21* | *17 h 0′ 10″* | *272° 56′ 45″* | *7° 31′ 14″N* | *No. 36 Pon* |
| *23* | *15 24 36* | *273 28 39* | *7 46 1* | *No. 42 Pon* |

*Dr. Olbers in a letter dated March 3 says "the weather has been very unfavourable for observing Pallas; neither Harding nor I have been able to see it again on account of the cloudy state of the atmosphere and as it is now moonlight it will again be lost for sometime. The indefatigable Dr. Gauss could not however withstand the great desire he had to undertake an improvement of the elements intending at a future opportunity to correct them still further from new observations."* [The VI Elements of Pallas are printed here.] *A comparison of these elements with the preceding observations shows that the difference is very small.*

| | *RA tables* | *Dec. Tables* | *RA error* | *Dec error* |
|---|---|---|---|---|
| *1803 Feb. 21* | *272° 56′ 25″* | *7° 31′ 29″.8* | *−20″.0* | *+15.8* |
| *23* | *273 28 46.8* | *7 45 52.3* | *+7.8* | *−7.7* |

*According to our promise we have given in this number two sheets of a chart of the orbit of Pallas constructed by Mr. Harding that represents those districts of the heavens through which the planet will pass from April 5 to June 28 but the places of the planet are borrowed from Dr. Gauss's calculations. The stars marked in these sheets are taken from Bode's catalogue and from Lalande's* Histoire celeste *Vol 1. It is to be regretted the latter does not fully contain several of the zones which fall on this chart and therefore Mr. Harding was obliged to leave some vacancies in it; a vacancy of this kind falls exactly in the orbit of Pallas, which it will pass through in the end of April. Whether all the stars in this chart actually exist in the heavens, Mr. Harding has not yet been able to examine. He doubts the existence of the following stars in Bode's catalogue* viz *No. 436, 463, 471, 472, 485, 488 of Herculis. His doubts however are founded on this circumstance that these stars do not occur in the* Histoire celeste *though in the place where they ought to stand according to Bode, several stars of a less magnitude have been observed by Lalande; No. 471 probably stands about 6 min more to the west than Bode places it; where No. 472 ought to stand the* Histoire celeste *has a star of the eighth magnitude; this is the case also with No. 436. Mr. Harding therefore has omitted both these, and on the other hand introduced Lalande's stars, because they are supported by actual observations. We hope this chart will be of great service, and show with sufficient accuracy to observers the orbit and position of Pallas.*

## Observations of the new planet Pallas

*MC*, May 1803, p. 466

*Continuous bad weather since the beginning of March and moonlight have impeded the observations on Pallas. On the night of March 4 the sky cleared. Pallas was supposed to be on the parallel of no. 58 of Bode's catalogue and close to one small star of the $8^{th}$ order of magnitude mentioned in the* Hist. celeste *on page 87 as follows: RA 18 h 23′ 26″ + 12″ correct. zenith distance 39° 52′ 5″ + 1′ 7″ correction. Dr. Olbers initially compared two stars, one of the $12^{th}$ the other of the $8^{th}$ order of magnitude, with no. 58, but these were too far west and east to be Pallas and the searched for star. Hereupon he picked two other of the same size ratio, and fortunately one of those was Pallas and the other in the* Hist. celeste. *Dr. Olbers reduced the position of these three stars according to his observations as follows:*

| *Magnitude* | *RA 1801* | *N. Dec.* |
|---|---|---|
| *12* | *275° 33′ 47″* | *8° 52′ 51″* |
| *8-9* | *275 35 47* | *8 53 9* |
| *8-9* | *275 59 29* | *8 57 22* |

*For Pallas he obtained:*

*1803 March 14 at 17 h 11′ 41″ mean time*
*RA 275° 52′ 38″ Dec. 8° 58′ 23″ N*

*Only on the $16^{th}$ it was possible again for Dr. Olbers to find Pallas. He found her as a small star that was close to a $6^{th}$ or $7^{th}$ magnitude star, which could not be found in any star catalogue, not even the* Hist. celeste. *He compared both with no. 78 in Poniatowski's Taurus and got at first for the position of this star: RA 278° 37′ 49″ decl. 10° 42′ 14″ and for the planet:*

*1803 March 16 at 14 h 10′ 28″ mean time RA 278° 37′ 16″ Dec. 10° 42′ 21″ N*

*After 15 h Pallas had come close to the $6^{th}$ – $7^{th}$ magnitude star (although still almost 1.5 min. from it) so that she became almost invisible in the 5ft Dollond and Dr. Olbers found it impossible to observe her. Since then he made the following observations.*

| *1803* | *Mean Time* | *RA* | *N. Dec.* |
|---|---|---|---|
| *Mar. 21* | *13 h 40′ 42″* | *279° 37′ 5″* | *11° 30′ 3″ Conn d. Tems* |
| *22* | *13 14 22* | *279 48 10* | *11 38 45 Nr. 70 Bode* |
| *23* | *13 2 12* | *280 10 25* | *11 58 43 Conn d. Tems* |

*The star of the* Conn. d. Tems *is the one La Lande mentions:*

*1790 RA 278° 54′ 20″ Var. +42.″1 Dec. 11° 23′ 45″ Var. +3.″1*

*According to Dr. Olbers' observations Bode's RA of star no. 76 is too small by 42″, which almost matches the* Hist. cel. *On March 27 he observed Pallas very well but had to compare her to a small star of the* Hist. cel. *that is given on page 170 as follows:*

*8.9 mag. RA 18 h 41′ 36.″5 zenith distance 36° 23′ 0″*

*He found*

*March 27 12 h 55′ 23″ Pallas goes forward in time 6.″5*

*and is north 2′ 38″*

*14 h 30 47 Pallas goes forward in time 4.″0*

*and is north 3′ 23″*

*The clock went in one star day only 2.″6 faster than mean time; Dr. Olbers found Pallas this night more luminous than on any other night this year. These elements still correspond excellently with the elements VI of Dr. Gauss's orbit; here is the comparison:*

*March 4 calculated RA 275° 52′ 35.″9 difference −2.″1*

*Calculated Dec 8 58 18. 1 −4.9*

*Since the error of the elements V was so small and well known, the ephemeris given in our December issue will be sufficient to find her during its duration. Its right ascension is now 2.5 min. too great and will be too great on August 9 by 4′ 6″ as far as we can rely on the correspondence of the new elements. The declination is mostly too small by one min. With these improvements, the ephemeris, which is based only on last years' observations, will exactly do what can be expected of the most accurate elements. Observer Harding sketched Pallas' orbit on his wonderful map not according to this ephemeris of the December issue but on the older one of the October issue, which was calculated according to the elements IV, which does not, which could not be expected, to match equally well. Any observer is well advised not to use the position of the map but the improved figures of the ephemeris of the December issue; for around the opposition of the planet the older ephemeris will give its position ¼ deg. too far east. Dr. Gauss has continued the ephemeris of Pallas according to the new elements VI until October 23* (Fig. 11.40). *We will probably not be able to follow her longer since her luminosity will decrease on September 18 and she will be fainter than at the time of her discovery.*

The *Philosophical Magazine* (1803) informed readers of a unique occurrence, an account which did not appear in any other English publication:

*The new planets discovered by Piazzi and Olbers continue to engage the attention of astronomers. Notwithstanding the smallness of the arc which they have passed through in our sight, and notwithstanding the considerable perturbations which they experience from Jupiter, we have already obtained the elements of their orbits with sufficient precision to find again these bodies in the place indicated by calculation, when they become visible, after*

**Fig. 11.40** Ephemeris of Pallas from August 9 to October 23, 1803

| 1803 Mitter-nacht in Seeberg | | Gerade Auf-steigung | | Abweich. nördl. | | Licht-stärke |
|---|---|---|---|---|---|---|
| August | 9 | 269° | 25' | 18° | 6' | 0,01162 |
| | 12 | 269 | 14 | 17 | 33 | 0,01136 |
| | 15 | 269 | 5 | 16 | 59 | 0,01110 |
| | 18 | 268 | 59 | 16 | 25 | 0,01086 |
| | 21 | 268 | 56 | 15 | 50 | 0,01062 |
| | 24 | 268 | 55 | 15 | 15 | 0,01038 |
| | 27 | 268 | 57 | 14 | 39 | 0,01014 |
| | 30 | 269 | 2 | 14 | 4 | 0,00990 |
| Septemb. | 2 | 269 | 10 | 13 | 29 | 0,00967 |
| | 5 | 269 | 20 | 12 | 54 | 0,00944 |
| | 8 | 269 | 33 | 12 | 19 | 0,00921 |
| | 11 | 269 | 48 | 11 | 45 | 0,00898 |
| | 14 | 270 | 5 | 11 | 11 | 0,00876 |
| | 17 | 270 | 25 | 10 | 37 | 0,00854 |
| | 20 | 270 | 47 | 10 | 5 | 0,00834 |
| | 23 | 271 | 12 | 9 | 32 | 0,00813 |
| | 26 | 271 | 38 | 9 | 1 | 0,00794 |
| | 29 | 272 | 7 | 8 | 30 | 0,00775 |
| October | 2 | 272 | 37 | 8 | 0 | 0,00757 |
| | 5 | 273 | 9 | 7 | 31 | 0,00739 |
| | 8 | 273 | 43 | 7 | 3 | 0,00722 |
| | 11 | 274 | 19 | 6 | 36 | 0,00705 |
| | 14 | 274 | 57 | 6 | 9 | 0,00689 |
| | 17 | 275 | 36 | 5 | 44 | 0,00674 |
| | 20 | 276 | 16 | 5 | 19 | 0,00659 |
| | 23 | 276 | 58 | 4 | 56 | 0,00645 |

*having been several months lost in the rays of the sun. The greatest difficulty arises from their extreme smallness, which sometimes causes us to doubt whether we have them in the field of the telescope. This is true in regard to Pallas in particular, which appears sometimes like a star of the 10th or 11th, or even the 12th magnitude, while Ceres appears of the 7th or 8th. But as there is something too arbitrary in this distribution of the stars according to the order of their magnitudes, it will be better to say with Messier, that Pallas is the smallest object that can be distinguished with an excellent telescope. An extraordinary circumstance has given to this imperceptible star for a moment a more sensible diameter and a stronger light. On the 25th of May, the weather being very fine, C. Messier was surprised to find in it a light double to that which it had been before; and yet, according to calculation, the distances of the sun and moon being nearly the same, the brightness of the planet ought not to have changed. The cause of this appearance was soon discovered. The small planet in its course met with a star, to which it appeared to be so close that the least interval could not be observed between them. Forty-two minutes after, a separation took place, and, according to the known motion of the planet, the interval must have been 15". The position of the small star may be determined at leisure; and from the repeated observations which may be made of it, there will result for the moment of the observation of Messier a determination of the place of the planet more exact than any of those which could have been procured in a direct manner. Those observations known under the name of appulses are exceedingly rare. However numerous the small stars may appear, the intervals which they leave between them are sufficiently large for the planets to make the tour of the heavens without concealing one of them, or at least any of those which can be observed.*

A table of geocentric motions of Pallas and Ceres for the month of August was printed after this paragraph.

## Continued Reports regarding the New Primary Planet Pallas

*MC*, June 1803, p. 556

*This planet has finally been observed outside of Germany. Oriani observed it in Milan on March 22 and April 1 on his equatorial sector, but only barely. He writes: "I believe I have seen the bashful Pallas a couple of times – we are in a perpetual state of cloudiness which frustrates the effectiveness of all our instruments. I think that I observed Pallas three times on March 22, and I could only see it between clouds on the 1st of April. The last observation is incomplete, since I could not obtain the declination. Because of the moonlight, I will have to wait eight to ten days before I can once again view this planetary atom and confirm my observation of March 22."*

*Here are these observations* (Fig. 11.41):

*In Prague, Canon David observed Pallas on a circular micrometer, with the following results* (Fig. 11.42):

*Dr. Olbers is continuing to observe his planet on his achromatic Dollond. On March 31, Pallas must have been very close beneath No. 12 in the Eagle, but with the moonlight, he could not even see the small planet through the orbital micrometer with the ocular increased 45 times. He therefore used a magnification of 106, and with this, he saw Pallas clearly. From a mean of several observations at 13 h 24′ 25″ mean time, it followed No. 12 in the Eagle 3.″5 in time, and was only 0′ 21″ to the south. From this, Dr. Olbers calculated:*

| *1803* | *Time of Observation* | *Observed RA* | *Declination* |
|---|---|---|---|
| *March 31* | *13 h 24′ 25″ mean time* | *281° 20′ 18″* | *13° 8′ 50″N* |

*"Is it not unusual," writes Dr. Olbers, "that the planet became visible in the higher magnification of 106 when it remained invisible in the much greater light intensity of a magnification of 45?"*

*With the increasing moonlight, Dr. Olbers had to suspend his observations for some time. On April 10, a close meeting of Pallas with ε of the Eagle was expected. He looked for it for*

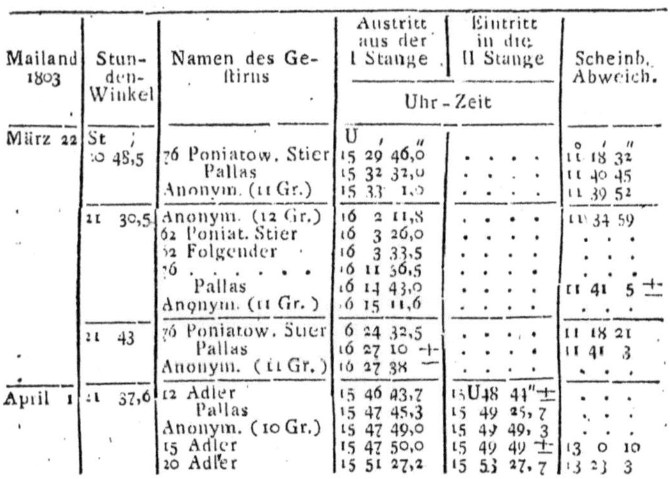

| Mailand 1803 | Stunden-Winkel | Namen des Gestirns | Austritt aus der I Stange (Uhr-Zeit) | Eintritt in die II Stange (Uhr-Zeit) | Scheinb. Abweich. |
|---|---|---|---|---|---|
| März 22 | St ′ | | U ′ ″ | | ° ′ ″ |
| | 20 48,5 | 76 Poniatow. Stier | 15 29 46,0 | . . . . | 11 18 32 |
| | | Pallas | 15 32 32,0 | . . . . | 11 40 45 |
| | | Anonym. (11 Gr.) | 15 33 1,0 | . . . . | 11 39 52 |
| | 21 30,5 | Anonym. (12 Gr.) | 16 2 11,8 | . . . . | 11 34 59 |
| | | 62 Poniat. Stier | 16 3 26,0 | . . . . | . . . |
| | | 52 Folgender | 16 3 33,5 | . . . . | . . . |
| | | 76 . . . . . . | 16 11 36,5 | . . . . | . . . |
| | | Pallas | 16 14 43,0 | . . . . | 11 41 5 ± |
| | | Anonym. (11 Gr.) | 16 15 11,6 | . . . . | . . . |
| | 21 43 | 76 Poniatow. Stier | 16 24 32,5 | . . . . | 11 18 21 |
| | | Pallas | 16 27 10 ± | . . . . | 11 41 3 |
| | | Anonym. (11 Gr.) | 16 27 38 — | . . . . | . . . |
| April 1 | 21 37,6 | 12 Adler | 15 46 43,7 | 15U48 44″± | . . . |
| | | Pallas | 15 47 45,3 | 15 49 25,7 | . . . |
| | | Anonym. (10 Gr.) | 15 47 49,0 | 15 49 49,3 | . . . |
| | | 15 Adler | 15 47 50,0 | 15 49 49 ± | 13 0 10 |
| | | 20 Adler | 15 51 27,2 | 15 53 27,7 | 13 23 3 |

**Fig. 11.41** Oriani's observations of Pallas in Milan on March 22 and April 1, 1803

| Prag 1803 | | Mittlere Zeit | Scheinbare gerade Aufſteigung der ⚴ | Nördliche ſcheinbare Abweich. der ⚴ | Sterne womit Pallas verglichen worden |
|---|---|---|---|---|---|
| März | 24 | 15U 27′ 20,″4 | 280° 11 27″ | 12° 0′ 35″ | mit ι ♌ } v. *Zach*'s Cat. |
| | — | — — — | 28 | 32 | — ε ♍ } |
| | 28 | 15 18 51, 8 | 280 51 58 | 12 36 39 | — α Ophiuch. *Bradley* |
| April | 1 | 15 45 6, 0 | 281 29 40 | 13 18 : | — 20 Adler. *Bode*'s Cat. |
| | — | — — — | 281 29 24 | 13 18 : | — ζ Adler. Conn. d. t. |
| | 2 | 4 50 30, 0 | 281 38 22 | 13 28 : | — 20 Adler |
| | — | — — — | 11 | . . . | — ζ Adler · |
| | 3 | 14 31 45, 7 | 281 46 35 | 13 38 : | — 19 Adler. *Bode*'s Cat. |
| | — | — — — | 42 | 13 37 : | — ζ Adler, Conn. d. t. |

**Fig. 11.42** David's observations in Prague of Pallas from March 24 to April 3, 1803

*half an hour with all magnifications of his Dollond but he could find no trace of it, perhaps because the sky was somewhat hazy. On April 11, Pallas could be seen well. In view of the right ascension, he compared it with ε in the Eagle, but in view of a star in the* Hist. cel., *from the parallel of which it was up to 30" distant, he concluded the following:*

*1803 Apr. 11 12 h 10' 9" obs. RA 282° 43' Decl. 15° 0' 50"N*

*On April 12, it was once again very clear. Pallas was compared with ε in the Eagle. In order to have the declination as exact as possible, Dr. Olbers awaited a magnitude 7 - 8 star in his telescope, which the* Hist. cel., *p. 82, gives as: 18 h 54' 33."5 Z.D. 33° 58' 51" on whose parallel Pallas must be very close. But this star observed on June 13, 1794 in Paris was entirely absent from the heavens. Dr. Olbers believed to have found that 'a simple error of print' had moved this star, for a magnitude 7-8 follows ε in the Eagle 'exactly' in right ascension, as is cited in* Hist. cel, *but just a little more to north than ε. Therefore, the zenith-distance in* Hist. cel. *is probably supposed to read 33° 58' 51" instead of 33° 38' 51".*

*"It would have been desirable," writes Dr. Olbers, "for Lalande to have had the goodness of checking all the stars cited in* Hist. cel., *which are now missing in the heavens, to see whether an error in writing or of printing could have occurred. One could then have reduced the stars – which were certainly observed and are now no longer in existence at the determined location – for the data belonging to their observation, and in this way, one could perhaps happen upon one or several small planets."*

*Dr. Olbers therefore selected another star of the* Hist. cel *for the determination of the declination, and out of this came the following:*

*1803 Apr.14 12 h 12' 18" obs. RA 282° 49' 44" Decl. 15° 11' 17"N*

*Since this time, Dr. Olbers tracked Pallas until April 25, as can be seen from the following* (Fig. 11.43):

| Bremen 1803 | | Mittlere Zeit | Scheinbare gerade Aufſteig. der ⚴ | Nördliche ſcheinbare Abweich. der ⚴ | Sterne womit ⚴ verglichen worden |
|---|---|---|---|---|---|
| April | 13 | 12U 13′ 41″ | 282° 55′ 15″ | 15° 21′ 18″ | mit ε Aquilae |
| | 15 | 12 3 34 | 283. 5 41 | 15 42 11 | — ✻ Hiſtoire céleſte |
| | 20 | 12 55 28 | 283 26 56 | 16 34 51 | — 42 Aquilae nach Bode |
| | 24 | 11 50 56 | 283 37 43 | 17 14 47 | — 35 Aquilae |
| | 25 | 12 16 41 | 283 39 45 | 17 25 31 | — 35 Aquilae u. ✻ Hiſt. cél. |

**Fig. 11.43** Observations of Pallas by Olbers from April 13 to 25, 1803

*Dr. Olbers finds the curious changes of light intensity from one evening to another, which had been observed the previous year in this planet, also very striking this year. In his opinion, Pallas had increased in light in February, and appears as a magnitude 11 star 'from time to time'.*

*Dr. Gauss continues tirelessly to compare Olbers' observations of Pallas with his 6$^{th}$ elements. Here are the results of the observations from March 16 to April 13, through which it will be seen that the elements still continue to correspond very well and that the differences are hardly greater than the uncertainty assumed with the observations themselves, especially in the declinations, which were given as very uncertain by Dr. Olbers himself* (Fig. 11.44):

## Continued News on the new main planet Pallas

*MC*, July 1803, p. 90

*The continuous bad weather since April has interrupted our observations on Pallas. Plus it is extremely difficult to orientate oneself in this region, which the planet is now roaming, because it is so full of stars and of which not enough are in our catalogues and La Lande's* Histoire Celeste. *Dr. Olbers was only able to see this planet again on May 11 since April 25. Here follow the observations, which he was able to obtain* (Fig. 11.45).

*The last observation was not made at a micrometer circle but a micrometer with a thread. On June 10 it was fair weather, but Dr. Olbers compared a small star and not Pallas with no. 477a and 488 according to Bode's catalogue. The cause of this mix-up was the faulty right ascension of no. 488 Bode that is too great by one whole time minute. By the way, no. 488 is not in La Lande's* Histoire Celeste. *Here follow the original observations of Dr. Olbers of May 20, 31 and June 1, because these have been reduced according to Flamsteed's determinations, and consequently a very wrong position might result in the case of no. 112 and 113. His clock was only slower by 2."8 than mean solar time* (Fig. 11.46).

*Foreign astronomers were also struggling with bad weather. Oriani has been able to make only five observations since April 1, which we are communicating here in the well-known way* (Fig. 11.47).

**Fig. 11.44** Differences between observed and calculated positions of Pallas, based on the Sixth Elements by Gauss, from March 16 to April 13, 1803

| Bremen 1803 | Berechnete gerade Aufsteig. der Pallas | Berechnete Abweichung der Pallas nördl. | Unterschied in AR. | Unterschied in Declin. |
|---|---|---|---|---|
| März 16 | 278° 37′ 10,″1 | 10° 43′ 18,″7 | − 5,″9 | + 57,″7 |
| 21 | 279 37 19, 0 | 11 30 22, 1 | + 14, 0 | + 19, 1 |
| 22 | 279 48 30, 2 | 11 ·9 49, 6 | + 20, 2 | + 64, 6 |
| 24 | 280 10 29, 9 | 11 59 11, 0 | + 4, 9 | + 28, 0 |
| 31 | 281 20 15, 1 | 13 8 54, 8 | − 2, 9 | + 4, 8 |
| April 11 | 282 43 35, 5 | 15 0 59, 4 | − 16, 5 | + 9, 4 |
| 12 | 282 49 32, 1 | 15 11 21, 3 | − 11, 9 | + 4, 3 |
| 13 | 282 55 13, 0 | 15 21 42, 8 | − 2, 0 | + 24, 8 |

**Fig. 11.45** Observations of Pallas by Olbers from May 11 to June 1, 1803

| 1803 | Mittlere Zeit | Scheinbare ger. Aufst. der ⚴ | Scheinbare Decl. ⚴ nördl. | Verglichen mit |
|---|---|---|---|---|
| 11 May | 11U 22′ 17″ | 283° 29″ 53″ | 19° 57′ 2″ | 494 Hercul. *Bode* |
| 12 — | 11 50 10 | 283 26 30 | 20 6 4 | 494 — |
| 20 — | 11 38 40 | 282 47 59 | 21 11 21 | 412 — *Flamst.* |
| 31 — | 10 57 47 | 281 24 5[illegible] | 22 22 45 | 113 — — |
| 1 Jun. | { 10 52 5 | 281 15 53 | 22 27 59 | 113 — — |
| | { 11 9 33 | 281 15 39 | 22 28 6. | 113 — — |

| 1803 | Mittl. Zeit | in Zeit | | in Raum | |
|---|---|---|---|---|---|
| May 20 | 11U 38′ 40″ | ⚴ folgte | 7′ 17″ | war 0′ 10″ | südlicher als Nr. 112 |
| 31 | 10 57 47 | ⚴ ging vor | 0 49, 5 | 1 24 | — — Nr. 113 |
| Jun. 1 | 10 52 5 | — — | 1 25, 2 | 3 49 | nördlich. als Nr. 113 |
| | 11 9 33 | — — | 1 26, 1 | 3 56 | — — Nr. 113 |

**Fig. 11.46** The original observations of Olbers of May 20, 31 and June 1

| Mailand 1803 Apr. | Stund. Winkel | Namen des Gestirns | Austritt aus der I Stange (Uhr Zeit) | Eintritt in die II Stang. (Uhr Zeit) | Scheinbare Abweichung Nördl. |
|---|---|---|---|---|---|
| 15 | 20U 4,′5 | 487 Cerberus | 13U 14′ 2,″4 | 13U 16′ 4,″3 | 15° 42′ 57″ |
| | | Pallas | 14 25 32, 7 | 13 27 34, 5 | 15 43 1 |
| | 20 42, 5 | 487 Cerberus | 13 52 9, 3 | 13 54 11, 4 | 15 43 0 |
| | | Pallas | 14 3 40, 0 | 14 5 42, 0 | 15 43 14 |
| 16 | 20 15, 7 | 487 Cerberus | 13 21 19, 0 | 13 23 21, 0 | 15 43 10 |
| | | Pallas | 13 33 7, 8 | 13 35 9, 6 | 15 53 28 |
| | 20 55, 0 | 467 Cerberus | 13 40 23, 8 | 13 42 25, 7 | 15 43 2 |
| | | Anon. (9 Gr.) | 13 50 24, 6 | 13 52 :: | 16 0 :: |
| | | Pallas | 13 52 13, 0 | 13 54 15, 0 | 15 53 32 |
| | 20 53 | 487 Cerberus | 13 59 7, 3 | 14 1 9, 4 | 15 42 59 |
| | | Pallas | 14 10 56, 7 | 14 12 58, 7 | 15 53 37 |
| 17 | 20 48 | 487 Cerberus | 13 49 28, 6 | 13 51 30, 3 | 15 42 53 |
| | | Anon. (9 Gr.) | 13 59 29, 3 | 14 1 31, 0 | 15 59 30 |
| | | Pallas | 14 1 37, 0 | 14 3 38, 5 | 16 3 57 |
| | 21 8 | 457 Cerberus | 14 9 39, 3 | 14 11 41, 0 | 15 42 58 |
| | | Anon. (9 Gr.) | 14 19 40, 7 | 14 21 :: | 15 59 30 |
| | | Pallas | 14 21 47, 5 | 14 23 49, 1 | 16 4 3 |
| 19 | 22 2 | Pallas ? | 15 8 21, 7 | 15 10 23, 5 | 16 25 30 |
| | | Anon. (10 Gr.) | 15 7 :: | 15 9 24, 0 | 16 25 30 |
| | | 47 Adler | 15 13 55, 3 | 15 15 57, 5 | 16 34 25 |
| | | 51 Adler | 15 14 39, 7 | 15 16 41, 9 | 16 34 :: |
| 20 | 20 29 | 457 Cerberus | 13 18 48, 1 | 13 20 50, 0 | 15 43 22 |
| | | Pallas ? | 13 31 40, 8 | 13 33 43, 3 | 16 35 0 |
| | | 47 Adler | 13 37 1, 8 | 13 39 4, 3 | 16 34 25 |
| | | 51 Adler | 13 37 45, 4 | 13 39 48, 1 | 16 34 :: |
| | 20 53, 5 | 18 Adler | 13 51 59, 1 | 13 54 1, 8 | 17 6 33 |
| | | Pallas ? | 13 56 :: | 13 58 17, 5 | 16 35 30 ± |
| | | 47 Adler | 14 1 34, 5 | 14 3 37, 0 | 16 34 38 |
| | | 51 Adler | 14 2 18, 3 | 14 4 21, ± | 16 34 :: |
| | 21 15, 5 | Pallas ? | 14 18 0, 2 | 14 20 3, 3 | 16 35 7 |
| | | 47 Adler | 14 23 20, 5 | 14 25 23, 5 | 16 34 7 |
| | | 51 Adler | 14 24 4, 6 | 14 26 7, 6 | 16 34 :: |

| Voreilung der Uhr vor mittlere Zeit im wahren Mittag | |
|---|---|
| 15 April | + 2′ 33,″0 |
| 16 — | 2 32, 2 |
| 17 — | 2 31, 4 |
| 18 — | 2 30, 2 |
| 19 — | . . . . . . |
| 20 — | 2 28, 3 |

**Fig. 11.47** Observations of Pallas by Oriani, from April 15 to 20, 1803

*This and also that of the last issue of page 557 Milanese observations were used by Professor Buerg for his calculation and he obtained the following positions* (Fig. 11.48):

*Dr. Gauss has resumed the comparison of his elements VI with Olbers' observations and obtained the following correspondence* (Fig. 11.49):

| Mailand 1803 | Mittlere Zeit | Gerade Auffteig. der Pallas | Abweichung der Pallas Nördl. |
|---|---|---|---|
| 22 März * | 15U 29′ 40″ | 279° 48′ 50,″9 | . . . . . |
| | 16 11 51 | 279 49 39, 5 | 11° 40′ 45″ |
| | | 279 48 58, 4 | 11 41 5 ± |
| | 16 24 18 | 279 49 13, 4 | 11 41 3 |
| 1 April * | 15 45 40 | 281 29 52, 8 | . . . . . |
| | | 281 29 49, 3 | . . . . . |
| | | 281 30 17, 1 | . . . . . |
| 15 — | 13 24 1 | 283 5 37, 3 | 15 43 1 |
| | 14 2 9 | 283 5 44, 1 | 15 43 14 |
| 16 — | 13 31 37 | 283 10 14, 8 | 15 53 28 |
| | 13 50 42 | 283 10 23, 1 | 15 53 32 |
| | 14 9 26 | 283 10 24, 6 | 15 53 37 |
| 17 — | 14 0 16 | 283 15 10, 1 | 16 3 57 |
| | 14 20 18 | 283 15 7, 9 | 16 4 3 |
| 19 — | 15 6 53 | 283 22 18, 5 | 16 25 30 |
| | | 283 22 12, 2 | . . . . . |
| 20 — | 13 30 14 | 283 26 22, 5 | 16 35 0 |
| | | 283 26 21, 0 | . . . . . |
| | | 283 26 25, 4 | . . . . . |
| | 13 55 50 | 283 26 45, 5 | 16 35 30 |
| | | 283 26 43, 6 | . . . . . |
| | | 283 26 43, 5 | . . . . . |
| | 14 16 34 | 283 26 32, 1 | 16 35 7 |
| | | 283 26 31, 4 | . . . . . |

**Fig. 11.48** Calculated positions of Pallas by Buerg

| 1803 Bremen | Berechnete AR der Pallas | Berechnete Declination der Pallas | Unterschied in AR. | Unterschied in Decl. |
|---|---|---|---|---|
| April 15 | 283° 5 39,″9 | 15° 42′ 20,″3 | — 1,″1 | + 9,″3 |
| — 20 | 283 26 47, 3 | 16 34 18, 4 | — 8, 7 | — 32, 6 |
| — 24 | 283 38 3, 9 | 17 14 30, 4 | + 20, 9 | — 16, 6 |
| — 25 | 283 40 0, 6 | 17 25 0, 7 | + 15, 6 | — 30, 3 |
| May 11 | 283 29 40, 1 | 19 57 28, 6 | — 12, 9 | + 26, 6 |
| — 12 | 283 26 14, 9 | 20 6 23, 9 | — 15, 1 | + 19, 9 |
| — 20 | 282 48 3, 6 | 21 11 25, 3 | + 4, 6 | + 4, 3 |
| — 31 | 281 24 52, 3 | 22 22 35, 8 | + 2, 3 | — 10, 2 |
| Jun. 1 | 281 15 44, 6 | 22 27 49, 1 | — 8, 4 | — 8, 9 |
| — 1 | 281 15 37, 9 | 22 27 52, 9 | — 1, 3 | — 13, 1 |

**Fig. 11.49** Differences between observed and calculated positions of Pallas, based on the Third Elements by Gauss, from April 15 to June 1, 1803

## Continued News on the new main Planet Pallas

*MC*, August 1803, p. 193

*Not many observations of this planet have been handed in this year. Only one reached us from Paris by Messier of May 28. He saw Pallas united with a small star of the 10$^{th}$ order of magnitude; they parted at 10 40′ true time. At that time the planet had a RA 281° 50′ 34″ and decl. N. 22° 5′ 28″.*

*At Prague Observatory Canon David observed the planet with a 7f. achromat by Dollond equipped with a circular micrometer, as follows* (Fig. 11.50):

| Tage | Mittlere Zeit | Gerade Auffteig. der ⚴ | Nördl. Abweich. der ⚴ | Vergleichungs-Sterne |
|---|---|---|---|---|
| März | U ′ ″ | ° ′ ″ | ° ′ ″ | |
| 24 | 15 27 20,2 | 280 11 28 | 12 0 34 | ι ♌ und ε ♍ *v. Zach* Catal. |
| 28 | 15 18 51,8 | 280 51 58 | 12 36 40 | ϰ Ophiuchi. *Bradley* |
| April | | | | |
| 1 | 15 45 6,c | 281 29 41 | 13 19 12 | mit 20 im Adler. *Bode* Cat. |
| 2 | 14 50 30,c | 281 38 11 | — — — | mit ζ im Adler. *Conn. des tems* |
| 3 | 14 31 46 | 281 46 29 | 13 39 40 | mit 19 im Adler *Bode* Cat. |
| 4 | 15 22 29 | 281 54 52 | 13 51 13 | mit 19 im Adler u. d. Stern 8 Grösse *) |
| 5 | 15 28 33 | 282 2 36 | — — — | Zweifelhaft |

**Fig. 11.50** David's observations of Pallas in Prague from March 24 to April 5, 1803, with comparison stars

| Tage | Mittl. Zeit in Prag | Gerade Aufsteigung der ⚴ | Nördl. Abweich. der ⚴ | Vergleichungs-Sterne |
|---|---|---|---|---|
| May 2 | 14U 6′ 27,″7 | 283° 46′ 42,″5 | 18° 34′ 55″ | mit 111 Herc. Conn. d. tems |
| 3 | 13 0 44, 7 | 283 46 31, 5 | 18 44 32 | Stern 9ter Gröſse |
| 4 | 14 10 5, 0 | 283 45 48, 4 | 18 53 58 | mit 496 Herc. Bode Cat. |
| 6 | 13 22 7, 7 | 283 44 6, 4 | 19 13 7 | mit 496 Herc. — — |

**Fig. 11.51** David's observations in Prague of Pallas from May 2 to 6, 1803

*In the case of star no. 19 of the Eagle David found the declination greater by 40" than indicated in Bode's star catalogue. He observed on April 3 the apparent RA of that star 282° 26′ 5."4, apparent decl. N. 13° 39′ 15". While determining the declination David learnt how difficult it was to get it right when either the star or the planet pass close to the centre of the circular micrometer through the field of the telescope. This was the case on April 2 and 5, and that is why he rather omitted the declination and is not sure about the others. These difficulties induced him to have the Prague watchmaker Simon Mueller build a rhombic micrometer with fine brass lines instead of the threads, which he used instead of the circular micrometer to observe the moments of disappearance of the stars and the planet. He said he could observe better and gave us the following observations on Pallas as proof hereof* (Fig. 11.51).

*The above star of the 9$^{th}$ magnitude with which David compared the planet on May 3, was determined by a comparison with no. 111 Herculis; he found its RA 12′ 32" and the declination 14′ 0" greater than no. 111 Herc. in the* Conn. des tems*; he compared an 11$^{th}$ magnitude star with this 9$^{th}$ magnitude star and with this no. 496 Herc. of Bode's catalogue; but he found the RA of this last star was too great by 2 min of the arc in Bode's catalogue. My continuous absence from Seeberg Observatory during this spring and summer due to other business of which I will inform the readers of the* MC *soon enough is the reason why I was unable to follow the two new planets this year. During my absence Prof. Buerg from Vienna was visiting and he was able to perform the following observations on Pallas at Seeberg Observatory with the Meridian Telescope by Ramsden:*

| *1803* | *Mean time in Seeberg* | *RA of Pallas* |
|---|---|---|
| *July 11* | *10 h 58′ 14."0* | *273° 24′ 57."1* |
| *12* | *10 53 32.9* | *273 13 37.5* |
| *17* | *10 30 16.5* | *272 19 15.9* |
| *18* | *10 25 38.1* | *272 8 37.0* |
| *19* | *10 21 2.3* | *271 58 36.4* |
| *20* | *10 16 26.7* | *271 48 40.2* |
| *23* | *10 2 45.2* | *271 20 8.5* |

## Continued News on the new main Planet Ceres

*MC*, September 1803, p. 288

*In Milan Oriani observed this planet with an 8f mural quadrant by Ramsden and compared it with δφτ in Sagittarius. The positions of these stars were taken from the big star catalogue of Piazzi which has just been published and obtained the following positions of Ceres* (Fig. 11.52):

*Since these observations fall into the time when Ceres was at opposition with the Sun, Oriani calculated from these the following geocentric longitudes and latitudes and compared those with Gauss's elements VIII of this planetary orbit. The employed inclination of the ecliptic was assumed 23° 28′ 3."3. The aberration in geoc. longitude −8."4, in latitude +2."0, the nutation −9."5* (Fig. 11.53).

**Fig. 11.52** Oriani's observations of Ceres in Milan from June 27 to August 2, 1803

| 1803 | Mittlere Zeit in Mailand | Scheinbare AR. der ⚳ | Scheinbare füdl. Abweich. der ⚳ |
|---|---|---|---|
| Jun. 27 | 12° 26′ 54″ | 281° 50′ 15,″8 | 28° 2′ 37,″3 |
| 28 | 12 22 0 | 281 35 46, 0 | 28 6 51, 8 |
| 30 | 12 12 13 | 281 6 41, 3 | 28 15 16, 0 ± |
| Jul. 2 | 12 2 24 | 280 37 32, 5 | 28 23 19, 3 |
| 3 | 11 57 30 | 280 22 54, 2 | 28 27 18, 6 |
| 4 | 11 52 36 | 280 8 20, 5 | 28 31 14, 4 |
| 9 | 11 28 8 | 278 55 57, 5 | 28 49 22, 0 |
| 10 | 11 23 15 | 278 41 49, 3 | 28 52 46, 8 |
| 11 | 11 18 23 | 278 27 42, 8 | 28 55 58, 0 |
| 19 | 10 39 50 | 276 41 4, 6 | 29 19 14, 6 |
| 21 | 10 30 21 | 276 17 0, 2 | 29 24 9, 9 |
| 22 | 10 25 36 | 276 4 44, 6 | 29 26 29, 9 |
| 26 | 10 6 58 | 275 20 49, 3 | 29 34 47, 3 |
| 29 | 9 53 11 | 274 50 39, 3 | 29 40 13, 4 |
| 30 | 9 48 38 | 274 41 39, 4 | 29 42 0, 5 |
| Aug. 2 | 9 35 8 | 274 16 0, ± | 29 46 35, ± |

**Fig. 11.53** Differences between observed and calculated positions of Ceres, based on the Elements VIII by Gauss, from June 28 to July 4, 1803

| Mailand 1803 | Wahre beobachtete geocentrifche. Länge der ⚳ | Wahre beobachtet. füdl. geoc. Breite der ⚳ | Fehler der VIII Gaufs'ifchen Elemente in der Länge | Fehler der VIII Gaufs'ifchen Elemente in der Breite |
|---|---|---|---|---|
| Jun. 28 | 9Z 10° 14′ 51,′3 | 5° 3′ 34,″6 | + 2′ 17,″4 | + 1′ 20,″6 |
| 30 | 9 9 48 32, 1 | 5 9 56, 4 | + 2 21, 8 | + 1 21, 4 |
| Jul. 2 | 9 9 22 14, 2 | 5 16 1, 8 | + 2 23, 1 | + 1 26, 2 |
| 3 | 9 9 9 3, 0 | 5 19 4, 5 | + 2 26, 4 | + 1 25, 1 |
| 4 | 9 8 55 56, 8 | 5 22 4, 5 | + 2 22, 9 | + 1 30, 5 |
| | | Mittlerer Fehler. | + 2 22, 3 | + 1 24, 7 |

*From these Oriani calculated the time of opposition on July 1 at 22 h 39′ 56″ mean time in 9Z 9° 29′ 35.″8 heliocentr. longitude and 3° 24′ 41.″6 south. heliocenter. latitude. The error of the orbital elements VIII mentioned above would be +1′ 32.″5 in longitude and +0′ 55.″1 in latitude.*

*Dr. Gauss studied meanwhile the correction of his latest elements (VIII) resulting from the observations from Bremen and the three of Palermo mentioned in the July issue on page 94. Only a slight change was sufficient to conform these new observations with the old ones. He had to move the node by 3 min compared to the observations of Palermo of 1801. Hopefully this will be justifiable in the future with more complete calculation of the perturbation. Here are the new elements (X):*

| | |
|---|---|
| *Epoch 1803 Seeberg* | *233° 36′ 3.″1* |
| *Aphelion* | *326° 33′ 18″* |
| *Ascending node* | *80 58 22* |
| *Inclination* | *10 37 54* |
| *Eccentricity* | *0.0788941* |
| *Log semimajor axis* | *0.4421516* |
| *Annual tropical movement* | *78 8 7.2* |
| *Daily tropical movement* | *770.″650* |

*These elements correspond to the three mentioned observations by Piazzi as follows:*

| *1803* | *Calculated* | | *Difference* | |
|---|---|---|---|---|
| *Palermo* | *RA* | *Dec* | *in RA* | *in Dec* |
| *May 12* | *288° 19′ 17.″9* | *24° 36′ 22.″0* | *+2.″9* | *+9.″0* |
| *13* | *288 18 57.3* | *24 39 57.2* | *−10.2* | *−4.2* |
| *14* | *288 18 13.1* | *24 43 36.7* | *−1.9* | *+5.6* |

*Because of my continual absence from Seeberg Observatory the news about the new main planet Pallas will be published in the next issue.*

## Continued News on the new main Planet Pallas

*MC*, October 1803, p. 372

*During my absence from Seeberg Observatory Mr. Buerg made the following meridian observations of Pallas:*

| *1803* | *mean time* | *RA of Pallas* |
|---|---|---|
| *July 26* | *9 h 49′ 14.″5* | *9z 0° 54′ 17.″5* |
| *27* | *44 45.9* | *46 6.0* |
| *28* | *40 18.1* | *38 7.0* |
| *29* | *35 52.0* | *30 32.7* |
| *30* | *31 26.9* | *23 13.8* |

*Dr. Olbers started observing Pallas again on August 21; it was very difficult for him to find her again since he could not get his bearings by the neighboring stars. The reason was an error of La Lande's* Hist. cel., *which – of course – had entered Harding's map, which we mentioned in our April issue; with three very close stars on p. 82, which form a distinct stellar group, there is a minute faulty; instead of 58 it has to be 57 min and the observations at the third read 58 instead of 59 min. Consequently, this small stellar triangle is too far east on Harding's map by 15 arc seconds. Here follow Dr. Olbers' observations on Pallas:*

| *1803* | *mean time* | *app RA* | *app Dec comp star* |
|---|---|---|---|
| *Aug. 21* | *11 h 37′ 18″* | *268° 54′ 34″* | *15° 50′ 0″ Hist. Cel.* |
| *24* | *10 55 28* | *268 53 59* | *15 15 6 383 Herc.* |
| *27* | *10 16 58* | *268 56 10* | *14 47 17 383 Herc.* |
| *29* | *11 10 25* | *268 59 13* | *14 16 51 392 Herc.* |

*The first and last of these observations are marked less reliable than those in the middle by Dr. Olbers. On August 29 the moonlight, which causes a small pause for the time being, was obstructive.*

*Dr. Olbers and Dr. Gauss have been tirelessly but in vain looking for these two planets in older stellar catalogues. Dr. Olbers wrote on September 1: "My efforts to find Pallas or Ceres among Flamsteed's observations has been as fruitless as my searching for them in the* Hist. celeste. *Now there are only Mayer and La Caille left. Has Piazzi not yet published a catalogue of the stars missed by him? This would be very important for this study."*

*Piazzi publicized in his new work that was mentioned in the September issue on p. 289 a list of 152 stars observed by different astronomers that he was unable to find again. Due to the importance of this list with regard to the three discovered planets and possible future planets we publish the list here so other astronomers and astronomy afficionados, to whom this foreign precious work might not easily be available can try their luck and discover maybe not the stars but the errors which led to their faulty places in the star catalogues. I told Dr. Gauss, who is honoring me with his visit to Gotha, about the original observations of Mayer for this reason; I have already found a star, missed by Piazzi and suspected by Dr. Olbers of planetism, and correctly placed it. Since these studies will lead to further insights we dedicate an entire article in our journal.*

## Continued News on the new main Planet Pallas

*MC*, December 1803, p. 537

*The last observations that we gave on this planet were by Dr. Olbers, Bremen, of August 29 (October issue, p. 373). Since then this tireless and astute astronomer has made more, even then, when all other astronomers and even the sighter of even the smallest planets, Messier in Paris, had declared her lost (November issue, p. 444). Dr. Olbers followed his foundling four weeks longer and observed her still on October 10. These observations were impeded by the weather but mainly by the huge amount of small stars of that region where the planet passed through. Here the sky is rich in stars and the* Hist. cel. *relatively poor in stars. On September 15 he had together with Pallas 37 stars in the field of his telescope, which is 44 min in diameter and often smaller errors in the star catalogues made it difficult to get one's bearings. On September 8 Pallas had to be compared with no. 405 in Hercules according to Bode. Below no. 405 there is a small 8th magnitude star and on Harding's map (March issue 1803) follows on the same parallel a 7th magnitude star below which is another 9th magnitude star. But Dr. Olbers assures that the two latter are missing; Piazzi did not list those either. The first two are the stars, which are mentioned in the* Hist. cel. *on p. 84 as follows*

| | | |
|---|---|---|
| *8* | *17 h 57′ 59.″5* | *Z.D. 35° 54′ 57″* |
| *6* | *58 7. 5* | *35 45 47* |

*The other (missing) are mentioned on p. 85, observed at the third thread only, as follows:*

| | | |
|---|---|---|
| *8. 9.* | *17 h 59′ 28″* | *Z.D. 35° 55′ 4″* |
| *7* | *59 36* | *35 44 45* |

*Obviously Dr. Olbers shows that these are only a repetition of the first two, where 59 instead of 58 min. were written erroneously. That 58 is the right minute he found by direct comparison of 405 with α Ophiuchi. In the second case the ZD of 405 is smaller by 1 min, probably due to a writing error. – The star on p. 85 18 h 3′ 38.″7 ZD 36° 58′ 30″, which has no size in* Hist. cel., *and which Harding sketched on his map as of the 9th order of magnitude, is 6th to 7th magnitude and by far the brightest among the neighbouring stars. Here are his four observations of Pallas:*

| *1803* | *Mean Time* | *RA* | *Dec. North* | *Comp star* |
|---|---|---|---|---|
| *Sept. 7* | *9 h 8′ 32″* | *269° 26′ 37″* | *12° 32′ 6″* | *no. 405 Bode* |
| *13* | *10 24 34* | *269 57 42* | *11 22 27* | *Hist. Cel.* |
| *14* | *8 46 3* | *270 3 12* | *11 12 5* | *Hist. Cel.* |
| *15* | *9 15 21* | *270 9 39* | *9 15 21* | *Hist. Cel.* |

*Dr. Olbers is satisfied with the last three observations; but he fears that his reduction might not be accurate enough; and therefore gives them here in the original just in case they might deserve more attention. The stars used are on p. 85* Hist. cel. *as follows:*

| | | |
|---|---|---|
| *a* | *18 h 1' 36"* | *Z.D. 37° 19' 14"* |
| *b* | *18 3 8.5* | *37 24 16* |
| *c* | *18 5 34.3* | *37 40 5* |
| *d* | *18 5 51* | *37 46 3* |

*The observations for the indicated times were:*

| | | |
|---|---|---|
| *Sept. 13 Pallas proceeds* | *2' 23."5 in time and is south of b 2' 29"* | |
| *14* | *2 1.5* | *north of c 2 53* |
| *15* | *1 35.7* | *south of d 2 39* |

*To determine the declination he chose stars which were close to Ceres' parallel; Pallas was well visible and not smaller than a $12^{th}$ magnitude star; this confirms the remark that this subtle as well as good observer made in the small photometric treatise (*MC, *October issue 1803, p. 293): that we should not assume, as is generally the case, the distances of the fixed stars are in the ratio of the order of magnitude to which we count them and the $6^{th}$ magnitude stars were only six times farther from us than the $1^{st}$ magnitude stars; at most this assumption fits the stars up to the fourth order of magnitude; Pallas was in April 1802 only of the $7^{th}$ order, now she has according to Dr. Gauss's tables ten times less luminosity, and should thus appear as a star which is 3 1/6 ($=\sqrt{10}$) times distant from us than a $7^{th}$ magnitude star; thus as a $22^{nd}$ magnitude star. Consequently, Dr. Olbers believes that some conclusions in the treatise On the penetrating Power of Telescopes by Dr. Herschel suffer a correction through this remark. Dr. Olbers obtained in October two further observations of his Pallas; the first is very good, the second incomplete. On October 9 the sky cleared after many cloudy days; he found Pallas as an extremely small star of the $14^{th}$ order of magnitude and close to those three neighbouring small stars which Harding mapped on his chart at RA 273° 50' and n. decl. 6° 56'. Only one of these small stars, the most eastern, is mentioned in the* Hist. cel. *p. 88; the other two were added by Harding. These three small stars form a rectangular triangle with its base pointing south and La Lande's star at the eastern acute angle. In the* Hist. cel. *p. 88 this star is mentioned as follows: 6° 14' 43" ZD 41° 53' 12". Dr. Olbers compared Pallas five times with this star and found that Pallas followed this star on October 9 at 7 h 26' 7" m.t. 4."2 in time and was 0' 40" further south. On October 10 she was compared with this very star and was following it now at 7° 2' 10" m.t. in 51" in time; it was impossible to determine the declination this day. He based the reduction of the star on the position of no. 42 of Poniatowski's Taurus as Bode indicates it according to La Lande (Piazzi did not list this star) and he found:*

| *1803* | *Mean time* | *RA of Pallas* | *N. Dec of Pallas* |
|---|---|---|---|
| *Oct. 9* | *7 h 26' 7"* | *273° 51' 48"* | *6° 55' 26"* |
| *10* | *7 2 10* | *274 3 39* | — |

*On October 10 Pallas entered into the tail of the cluster above ν Ophiuchus and it was impossible for him to find her among the myriad of small stars on the $11^{th}$. Therefore he gave up any further observations for this year. Since these were the latest observations, Dr. Gauss compared those to his elements VI and found the following admirable correspondence:*

| *1803* | *Calculated RA diff.* | | *Calculated Dec.diff.* | |
|---|---|---|---|---|
| *Oct. 9* | *273° 52′ 40″* | *+52″* | *6° 55′ 41″* | *+15″* |
| *10* | *274 4 21* | *+42* | *6 46 42* | – |

*This tireless calculator wants to improve his elements VI anew according to all observations from Bremen, Milan and Seeberg, to make it easier for practical astronomers to find her again next year by an accurate ephemeris. And maybe Harding will accompany this with a map of the celestial region where the planet will wander.*

*The royal astronomer Canon David observed at Prague the opposition of this planet with the Sun at his 7f. mural quadrant which is equipped with a rhombus of brass stripes. He compared the planet with π Serpentis whose position was taken from* Conn. des tems; *he found from observing that the RA of this star 446 must be decreased by 16″ and the declination was left untouched. Piazzi does not list this star; he obtained the following results for the positions of Pallas:*

| *1803* | *Mean time in Prague* | *RA* | *Dec.* | *Geocentric* | |
|---|---|---|---|---|---|
| | | | | *Longitude* | *Latitude* |
| *Jan. 27* | *12 h 4′ 27.″0* | *276° 15′ 41″* | *23° 13′ 39″* | *278° 22′ 26.″0* | *46° 30′ 50.″5* |
| *28* | *11 59 40. 0* | *276 8 9* | *23 11 35* | *278 5 33.4* | *46 29 27.3* |
| *29* | *11 54 55. 0* | *275 50 37* | *23 9 6* | *277 48 38.3* | *46 27 41.4* |
| *30* | *11 50 9. 5* | *275 38 16* | *23 6 4* | *277 31 56.0* | *46 25 18.5* |

*In calculating the longitudes and latitudes the inclination of the ecliptic was taken according to the latest determinations of the Parisian astronomers: 23° 28′ 4.″4. It was a very favourable fact that the Sun moved close to the same parallel as Pallas so that the Sun and the planet could be compared at the same time with the star 446 of Hercules. From the time of culmination between the Sun and the star the right ascension of the Sun and from this its apparent longitude were derived; for this time David calculated the same longitude from our solar tables taking account of the corrections mentioned in the MC; the longitude of the tables varied only by 3.″4 from that of the observation and David assigns this fact rather to his observation than to our solar tables. For this reason he uses the solar longitudes as they appear from our solar tables.*

*The Sun's motion within 24 hours from June 29 to the 30$^{th}$ is 57′ 10.″6; that of the planet 16′ 46.″3; the combined motion of both is consequently 1° 13′ 56.″9; from this the opposition of Pallas results on June 30, 1803, at 0 h 54′ 21.″5 m. t. in geo- and heliocentric longitude 277° 39′ 33.″5, geocentr. latitude 46° 26′ 24.″4.*

## Continued News on the two new main Planets Ceres and Pallas
*MC*, March 1804, p. 246

*We published in the October issue of last year on p. 370 an ephemeris of the geocentric path of Ceres for the year 1804 calculated by Dr. Gauss according to his elements improved for the ninth time of this planet's orbit and promised in the December issue p. 535 a map of this path by Harding at Lilienthal. Please find Harding's map in this issue. Dr. Gauss has again corrected the elements of Pallas' orbit according to the latest observations and calculated accordingly the following ephemeris that is again accompanied by a map of the geocentric path of this planet sketched by Inspector Harding.* [Refer here to a table of positions projected for 1804.]

*At some points the difference to the ephemeris given by Prof. Bode in the* Berl. Astr. Yearbook *1806 on p. 91 is considerable. Observing Pallas will be much easier this year than last year. Last year she reached her greatest luminosity on June 20, this year on July 23 and will be even brighter by ¼ on September 7, plus the complete darkness of the nights and the fact that Pegasus and Aquarius are not as rich in stars as the region in which the planet was roaming last year. The luminosity it showed at last years' discovery, was reached this year already on June 5 and on May 18 that of October 10, 1803, when Dr. Olbers observed her for the last time.*

*The latest elements VII according to which Dr. Gauss calculated the above ephemeris of Pallas' orbit are the following*

| | |
|---|---|
| *Epoch 1803 Seeberg* | *221 29 32."0* |
| *Aphelion* | *301° 17' 34.4* |
| *Ascending node* | *172 28 13.7* |
| *Inclination* | *34 38 1.1* |
| *Eccentricity* | *0.2457396* |
| *Log semimajor axis* | *0.4423790* |
| *Annual tropical movement* | *78 4 26.3* |
| *Daily tropical movement* | *770."0446* |

*Last year's December issue contained on p. 536 a small list of those stars which Ceres would come close to on her way taken from Piazzi's great star catalogue. And here follows again a little extract of that catalogue of those stars that Pallas will come close to and can be compared with.* [Table of star positions in Pegasus and Aquarius is printed in the *MC*.]

## Continued News on the new main Planet Pallas
*MC*, April 1804, p. 342

*Here follows the map we promised in the March issue on p. 247 by Inspector Harding of the geocentric path of Pallas* (Fig. 11.54). *In our last issue we published 80 stars of Piazzi's catalogue, which can be compared with the planet for those who want to made differential observations. We have added 25 stars of that catalogue which could be useful for this purpose.* [A table of stars in Pegasus and Aquarius is printed in the *MC*.]

## Continued News on the new main Planet Pallas
*MC*, July 1804, p. 89

*Dr. Olbers was among all astronomers the last who observed his planet last year until October 10 (MC VIII. p. 537) and the first who saw her this year. Thanks to his diligence and thanks to the ephemeris of the accomplished Dr. Gauss (*MC *IX. p. 274) the planet was found so early this year.*

*The night of May 8 was unusually clear and invited Dr. Olbers to search for his planet with his excellent Dollond and he found above 36 Pegasi three extremely small stars of which one must have been Pallas. He was successful and found his planet among these three stars and according to him was able to observe it well. From May 8 till June 1 unclear horizon, moonlight and increasing nocturnal twilight were impeding so he was neither able to*

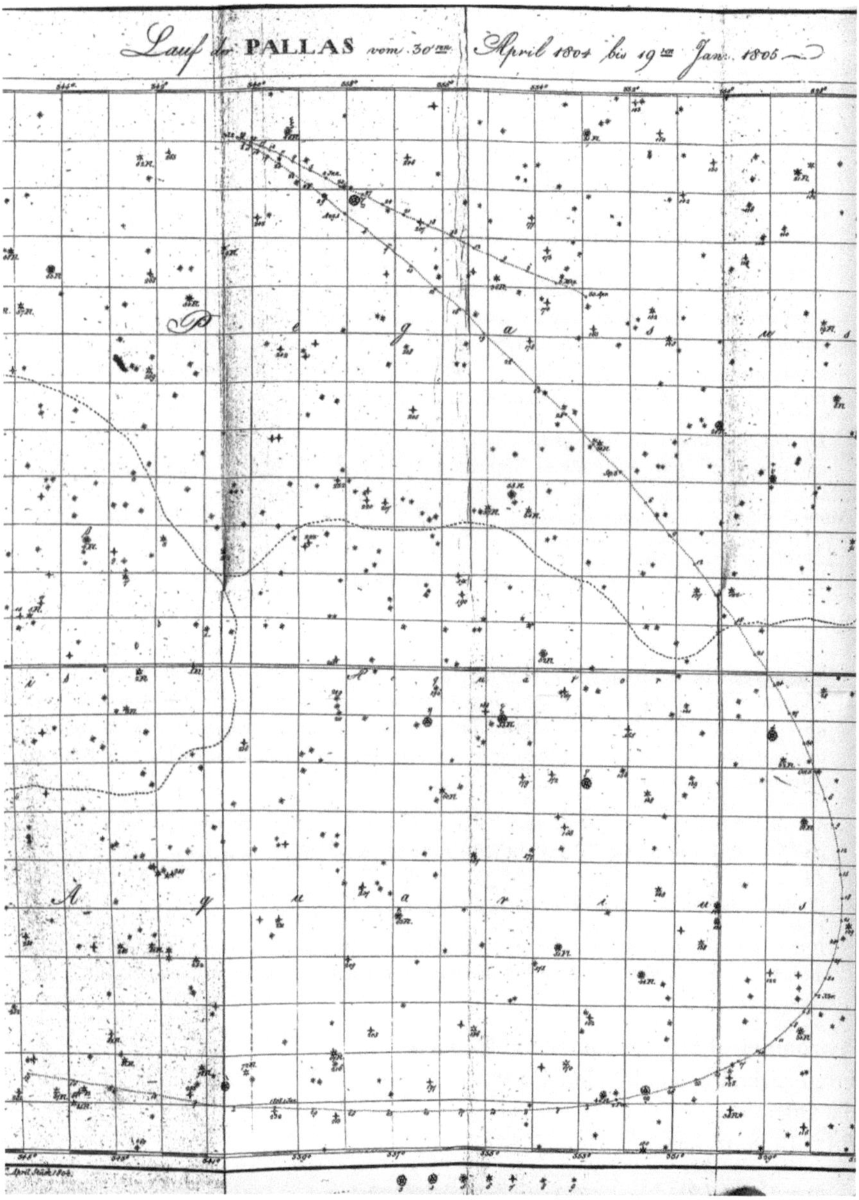


**Fig. 11.54** The path of Pallas from April 30, 1804, to January 19, 1805. (MC April 1804)

*recognise Pallas nor the two small stars with certainty which were so close to the planet on May 8. Finally, June 1 was clear again and he observed Pallas close to ε Pegasi. On June 3 the observation was repeated. Here are his three observations:*

| *1804* | *Mean time* | *RA* | *N. Dec* | *Comp star in Bremen* |
|---|---|---|---|---|
| *May 8* | *13 h 50′ 0″* | *334° 38′ 35″* | *8° 30′ 6″* | *Nr 36 Pegasi* |
| *Jun 1* | *12 4 50* | *338 27 27* | *10 10 32* | *ε Pegasi* |
| *Jun 3* | *11 53 25* | *338 41 25* | — | *ε Pegasi* |

*Dr. Gauss ephemeris is at the beginning of May too small by 4′ in RA and at the beginning of June by 5.5′; the declination in June about 1′ too great, the error in RA will most likely increase, which will not make the finding of the planet more difficult though, so it is not yet necessary to further improve the elements; an error of 5′ is minor taking into account that the planet was observed so little last year and Dr. Gauss was not able to use one single complete meridian observation. These 5′ could be eliminated by small changes in the elliptical elements and would describe all hitherto observations accurately enough without taking into account the perturbations; but Dr. Gauss thinks of taking them into account only in further calculations and more and better meridian observations.*

*Dr. Olbers marks it as strange that around ε Pegasi once again a star of La Lande's* Hist. cel. *is missing; page 40 goes as follows:*

| *Name* | *Middle thread* | *Third thread* | *Zenith Distance* |
|---|---|---|---|
| *ε Pegasi* | *22 h 31′ 15.″6* | ——— | *39° 3′ 5″* |
| *8. 9.* | ——— | *31′ 43.″5* | *38 48 1* |
| *8. 9.* | *31 56.5* | ——— | *38 47 52* |

*The second star is missing. Dr. Olbers wrote: "I am asking you to inform Dr. Burckhardt about the stars of the* Hist. cel *I was unable to find plus the one I indicated in my cometary observations (MC IX p. 504) so this diligent scientist has the opportunity to search the original observations whether these missing stars were truly and accurately observed or an error while writing those observations down occurred."*

*Messier at Paris Observatory found Pallas only on June 2. He compared her with ζ Pegasi and found for this day at 13 h 5′ 36″ true time her RA 248° 36′ and her n. declination 10° 14′ 22″. On the 17th he compared the planet with ξ Pegasi and found at 12 h 41′ 56″ the RA* [Pallas] *239° 56′ 30″, the declination 10° 55′ 23″.*

*It is to be expected that this curious, only little observed planet, now gladly rediscovered planet will be carefully observed in the future of which we in our MC will report in our continued news.*

## Astronomical Observations and Reports

From several letters by Herr Court Councillor and Professor Huth in Frankfurt on the Oder

*BAJ*, 1808, p. 238.

*from Oct 6, 1804*

*My physical observations of Harding's planet (Juno), together with those of Ceres and Pallas revealed, with a magnification of up to 847 times, the following:*

1. *Juno has a dull snow-white light*
2. *Ceres has a blue-white light*
3. *Pallas has a yellow-white light*

*No. 1 and 3 are always very well defined and No. 2 always has a hazy outline with which it is as big as Juno but without which it is only $^2/_3$ to $^3/_4$ in diameter. In diameter, Pallas only appears half the size of Juno.*

*With the Schroeter proj. micrometer, I have measured them several times with different magnifications and different distances from the eye and have obtained the following relationship of apparent diameter.*

| | *Juno* | *Ceres without nebula* | *Pallas* |
|---|---|---|---|
| | *.23* | *15* | *12* |
| *or* | *22* | *16* | *12* |
| *or* | *22* | *18* | *11* |
| *or* | *6* | *4* | *3* |

*The greatest apparent diameter, which I found on the evening of October 5 from 10 to 11 o'clock in the evening amounted to 6.65 seconds, and the smallest from October 1, 9 o'clock p.m. was 4.676 seconds, the mean of 5.66 corresponds fairly exactly with the two other results I obtained on September 28 and October 5, namely, 5.51 and 5.54 seconds.*

*From the above relationships, you can see that I also consider Ceres and Pallas larger in apparent diameter than Herr Herschel.*

***From October 21***

*Measured with the proj. micrometer, I found the apparent diameter of Juno to be 6.1 s yesterday. Pallas was located exactly where it was supposed to be according to Harding's charts, but appeared very small and with pale light. It appeared as though it had become smaller and Juno had become larger.*

*Around 7 o'clock and then from 9 to 10 o'clock, I once again saw and measured Juno through the reflector. A magnification of 200 gave the apparent diameter of 6.34. Nevertheless, with a magnification of 40 times, it appeared only as a magnitude nine star in the reflector, while Pallas appeared as a magnitude 11 to 12 star.*

## Continued Reports Regarding the New Primary Planet Pallas
*MC*, November 1804, p. 476

*Our readers discovered in the June 1804 issue of* Monat. Corr. *that Dr. Olbers had again found and observed this planet on May 8. We have only been able to observe this planet with the intention of making measurements in September due to its perpetual absence from our observatory. Other than that, no observations of this very weak and scarcely observable planet have come to our knowledge* (Fig. 11.55).

*Dr. Gauss's ephemeris (May 1804 issue,* Monat. Corr., *p. 247) gave the right ascension of this planet for the beginning of June 5.5 min too small and the declination one minute too great; now, in the September issue, the former is 8.5 min too small, and the latter again approaches one minute too great.*

*A few observations of this planet which Oriani conducted on the equatorial sector and which he sent to us last year are still to be made up. We have already reported some of these observations in the 7th vol. of the June 1803 issue of* Monat. Corr. *p. 557 and in the 8th vol. of the July 1803 issue, p. 92. The entire sequence of observations as calculated by Oriani himself now follow* (Fig. 11.56).

**Fig. 11.55** Observations of Pallas by Zach in Seeberg from September 6 to October 4, 1804

| 1804 | Mittlere Zeit in Seeberg | Scheinbare gerade Aufsteigung der ⚴ | Scheinbare nördl. Abweichung der ⚴ |
|---|---|---|---|
| Septbr. 6 | 11ᵘ 4' 12,"307 | 331° 50' 35,"97 | . . . . |
| 7 | 10 59 33, 188 | 331 39 46, 05 | . . . . |
| 9 | 10 50 17, 073 | 331 18 38, 16 | . . . . |
| 10 | 10 45 39, 469 | 331 8 11, 06 | . . . . |
| 12 | 10 36 26, 609 | 330 47 52, 11 | . . . . |
| 13 | 10 31 50, 656 | 330 37 49, 80 | . . . . |
| 14 | 10 27 15, 810 | 330 28 4, 19 | 1° 51' 40,"7 |
| 15 | 10 22 4[illegible], 504 | 330 18 38, 69 | . . . . |
| 17 | 10 13 36, 260 | 330 0 2, 30 | . . . . |
| 28 | 9 24 40, 319 | 328 34 34, 29 | 1 4 17, 8 |
| 30 | 9 15 59, 879 | 328 22 22, 98 | 1 28 18, 4 |
| Oct. . | 9 7 28, 699 | 328 12 30, 96 | 1 49 49, 9 |
| 4 | 8 58 53, 177 | 328 1 33, 68 | . . . . |

**Fig. 11.56** Oriani's observations of Pallas in Milan from March 22 to July 28, 1803

| 1803 | Mittl. Zeit in Mailand | Scheinbare gerade Aufst. d. ⚴ | Scheinb. nördl. Abweich. der ⚴ |
|---|---|---|---|
| März 22 | 16ᵘ 12' 7" | 279° 49' 33" | 11° 40' 59" |
| April 1 | 15 44 44 | 281 29 48 | . . . . . |
| 15 | 13 24 1 | 283 6 17 | 15 42 57 |
| 16 | 13 50 42 | 283 11 1 | 15 53 19 |
| 17 | 14 0 7 | 283 15 47 | 16 3 56 |
| 19 | 15 6 54 | 283 22 48 | 16 25 26 |
| 20 | 14 16 34 | 283 26 21 | 16 35 18 |
| May 19 | 13 28 56 | 282 53 7 | 21 4 27 |
| 20 | 12 57 37 | 282 47 28 | 21 11 49 |
| Jun. 12 | 11 15 30 | 279 20 22 | 23 9 54 |
| 15 | 10 42 4 | 278 45 38 | 23 15 38 |
| 16 | 10 58 59 | 278 33 21 | 23 17 20 |
| 17 | 10 47 45 | 278 21 36 | 23 18 11 |
| 18 | 10 45 32 | 278 9 12 | 23 18 57 |
| 19 | 11 5 26 | 277 56 33 | 23 19 36 |
| 23 | 11 47 34 | 277 6 36 | 23 18 50 |
| 26 | 11 11 2 | 276 28 51 | 23 15 45 |
| 27 | 11 10 27 | 276 16 24 | 23 14 8 |
| 28 | 11 51 29 :: | 276 3 16 :: | 23 12 5 :: |
| 29 | 11 32 14 | 275 50 33 | 23 9 55 |
| Julius 1 | 11 7 40 :: | 275 24 52 :: | 23 5 4 :: |
| 9 | 10 2 55 | 273 48 52 | 22 30 54 |
| 11 | 10 9 26 | 273 25 27 | 22 20 4 |
| 12 | 10 3 25 | 273 13 58 | 22 14 36 |
| 19 | 9 44 49 | 271 59 10 | 21 36 35 |
| 21 | 9 47 4[illegible] | 271 39 23 | 21 10 54 |
| 22 | 9 56 13 | 271 29 35 | 21 2 41 |
| 28 | 10 37 49 | 270 37 41 | 20 9 22 |

*From the observation of June 29, being the most certain and closest to the opposition, Oriani attempted to calculate the time and position of the opposition according to Gauss's elements of the path, improved now for the sixth time (*Monat. Corr. *VII, April 1803 issue, p. 374). With this, he found, for 11 h 7° 51' 56."4 Milan mean time, the geocentric longitude of Pallas 9z 7° 51' 56."4, geocentric northern latitude of 46° 27' 4". From his observed right ascension of 275° 50' 33" and from the northern declination of 23° 9' 55", he obtained with regards the inclination of the ecliptic 23° 28' 3", the apparent geocentric longitude 9Z 7° 48' 36" and the apparent northern geocentric latitude 46° 28' 23". If the nutation −9."5 and the aberration −10."1 is set to the longitude, and the aberration −1."2 to the latitude, the true observed geocentric longitude 9Z 7° 48' 16."4 and latitude 46° 28' 22" is obtained. As a result, the error in Gauss's elements in geocentric longitude is +3' 40" and geocentric latitude, −1' 18".*

| | |
|---|---|
| *The calculated long of the Sun* | *3z 7° 8′ 44.″1* |
| *The observed long of Pallas* | *9 7 48 16.4* |
| *Difference* | *6 0 39 32.3* |

*The hourly geocentric movement of*

| | |
|---|---|
| *Pallas in longitude* | *−42.″437* |
| *The hourly movement of the Sun* | *142.923* |
| *Hourly relative movement* | *185.360* |

*If one divides the difference 39′ 32.″3 = 2372.″3 by 185.″36, one finds the interval of time 12 h 47′ 54″, which, adding the time of opposition of Pallas with the Sun to the time of the observation on June 29 of 12 h 32′ 14″, gives the mean time for June 30 of 0 h 20′ 8″. For this moment, one finds:*

| | |
|---|---|
| *Longitude of Pallas* | *9z 7° 39′ 13.″4* |
| *Geocentric N latitude* | *46 27 14.6* |
| *Heliocentric N latitude* | *33 42 18* |

*The error of these elements is accordingly +2′ 19.″5 in geocentric longitude and −49.″5 in heliocentric latitude [Ed: the letter "z" denotes 30 degrees in the sky, so 3z = 90°].*

**Observations of Juno and Pallas**

Comments on the new planets and Lexell's formula by Dr. Olbers in Bremen. From several letters of same.

[Parts of this paper dealing with Juno will be published in a future book in this series.]

*BAJ*, 1808, p. 179

***Dated November 2, 1804*** *Since the major axes of the paths of these asteroids are almost identical but the eccentricities are very different and the aphelion always move toward the nodes, it follows that these paths 'really do intersect' and 'have intersected in the past' at certain times. Now, for example, the path of Pallas comes closer to the path of Ceres from year to year, with the descending node of the former on the latter, and, if I assume Oriani's secular movement of the aphelion and regard the nodes as tropically stationary, they will actually intersect in 282 years. In approximately 8,500 years, I find at least three such intersections.*

***Dated January 1, 1805*** *If it is assumed that the albedo of these three new planets are the same or that the surface of all three reflect the sunlight equally well, then the relationship of their diameters will be very close to the following:*

$$\textit{Dia. of Ceres} = 1.00;\ \textit{Pallas} = 0.74;\ \textit{Juno} = 0.43$$

*At this time, I have been mainly observing Pallas because I fear that astronomers will want to neglect it given its present weak light intensity and difficulty of observation. Here are my observations* (Fig. 11.57):

**Fig. 11.57** Observations of Pallas by Olbers from November 20 to December 31, 1804

| | M. Z. | Scheinb. AR. | Scheinb. Declin. |
|---|---|---|---|
| 1804. Nov. 20 | 8ᵘ 21′ 20″ | 330 7′ 0″ | 8° 22′ 52″ Südlich |
| 25 | 5 49 26 | 330 55 47 | 8 40 26 |
| 27 | 5 54 59 | 331 17 22 | 8 46 27 |
| Dec. 2 | 6 56 36 | 332 15 20 | 8 58 52 |
| 30 | 6 31 30 | 339 1 44 | 9 9 14 |
| 31 | 5 47 24 | 339 17 56 | 9 8 4 |

*In the last two observations, I have compared Pallas with 232 Aquarii according to your catalogue. You attribute the star to Piazzi; but Piazzi did not include it in his great catalogue.*

## Observations of Ceres, Pallas and Juno

By Mr. Wisniewsky in St. Petersburg. Submitted Aug. 21, 1805
*BAJ*, 1808, p. 252

*With the apparent inclination of the ecliptic = 23° 27′ 59.″6 one gets from this the apparent longitudes and latitudes of Ceres; which are reduced to the true ones by aberration and nutation. – For calculating the true geoc. positions of Ceres according to the X elements by Mr. Gauss the longitudes of the Sun were taken from the Sun tables by de Lambre, but with one correction – 8.″63 which follows on average from 5 observations, which the weather of September 17 through 23* (Figs. 11.58 and 11.59).

*From the differences of the observed and calculated places of Ceres from Sept 22 through Oct 5 the arithmetic mean gives at the time of opposition the longitude = 4.″1 and the geocentr. latitude = −9.″67. – Consequently, the opposition of Ceres occurred on September 27 1804 5 h 2′ 46.″7 mean time.*

*Then the true longitude of Ceres was from the mean equin. = 0s 4° 16′ 5.″6, geocentr. latitude 15° 37′ 11″, helioc. latitude 10° 20′ 37.″6 S.*

*I compared the true geocentr. positions of Pallas* (Fig. 11.60) *following from this with those calculated according to the elements VIII by Dr. Gauss; and found the following results* (Fig. 11.61).

[This paper was written by the young Vincent Wisniewski (1785–1855), and shows the asteroids were being observed in Russia. Remainder of the paper, dealing with Juno, will be in the next book in this series.]

**Fig. 11.58** Observations of Ceres by Wisniewsky from September 8 to October 5, 1804

*Beobachtungen der Ceres.*

| 1804. | Mittlere Zeit. | beobacht. AR. der ⚳. | beob. Abw. der ⚳ südlich. |
|---|---|---|---|
| Den 8. Sept. | 13ᵘ 44′ 33″,4 | 13° 42′ 31″,38 | 10° 54′ 56″,44 |
| — 14. — | 13 15 58,8 | 12 42 32,19 | 11 30 11,02 |
| — 16. — | 13 6 39,8 | 12 20 39,88 | 11 41 21,96 |
| — 17. — | 13 1 59,6 | 12 9 34,18 | 11 47 0,27 |
| — 18. — | 12 57 18,3 | 11 58 12,19 | 11 52 38,93 |
| — 22. — | 12 38 25,9 | 11 10 53,17 | 12 13 49,07 |
| — 23. — | 12 33 42,1 | 10 58 52,29 | 12 19 2,17 |
| — 28. — | 12 9 53,8 | 9 56 30,30 | 12 42 56,05 |
| — 30. — | 12 0 20,8 | 9 31 8,74 | 12 51 27,26 |
| — 1. Oct. | 11 55 34,5 | 9 18 31,50 | 12 55 25,19 |
| — 2. — | 11 50 47,8 | 9 5 47,40 | 12 59 19,46 |
| — 5. — | 11 36 27,9 | 8 27 37,93 | 13 10 14,17 |

| 1804. | beob. wahre Länge ⚳ | nach den X Elem. | beob. wahr. Breite südl. | nach den X Elem. |
|---|---|---|---|---|
| Den 8. Sept. | 0ˢ 8° 13′ 41″,3 | + 13″,1 | 15° 26′ 54″,4 | − 4″,6 |
| — 14. — | 0 7 3 5,0 | + 13,3 | 15 35 29,3 | − 5,2 |
| — 16. — | 0 6 37 59,8 | + 10,8 | 15 37 2,9 | + 10,2 |
| — 17. — | 0 6 25 19,4 | + 4,7 | 15 37 50,2 | + 5,0 |
| — 18. — | 0 6 12 23,2 | + 5,2 | 15 38 29,9 | − 0,7 |
| — 22. — | 0 5 19 32,4 | + 6,2 | 15 39 6,0 | + 13,7 |
| — 23. — | 0 5 6 11,9 | + 1,4 | 15 39 6,1 | + 4,5 |
| — 28. — | 0 3 58 20,8 | + 9,1 | 15 36 10,2 | + 3,1 |
| — 30. — | 0 3 31 17,8 | + 6,7 | 15 33 52,3 | + 8,1 |
| — 1. Oct. | 0 3 17 57,1 | + 5,9 | 15 32 28,2 | + 12,3 |
| — 2. — | 0 3 4 32,7 | − 2,6 | 15 30 58,4 | + 14,1 |
| — 5. — | 0 2 24 43,0 | + 2,0 | 15 25 44,1 | + 11,8 |

**Fig. 11.59** Differences of the observed and calculated places of Ceres based on Elements X by Gauss

*Beobachtungen der Pallas.*

| 1804. | Mittlere Zeit. | beobacht. ger. Aufst. der ⚴. | beob. nördl. Abw. der ⚴. |
|---|---|---|---|
| Den 7. Sept. | 10ᵘ 59′ 49″,0 | 331° 40′ 30″,45 | 3° 21′ 10″,50 |
| — 17. — | 10 13 51,0 | 330 0 30,70 | 1 13 56,16 |
| — 18. — | 10 9 19,8 | 329 51 39,31 | 1 1 6,00 |
| — 19. — | 10 4 48,7 | 329 42 50,74 | 0 48 28,92 |
| — 21. — | 9 55 49,8 | 329 26 1,78 | 0 23 2,87 |
| — 22. — | 9 51 21,5 | 329 17 54,97 | 0 10 27,41 |

**Fig. 11.60** Observations of Pallas by Wisniewsky from September 7 to September 22, 1804

| 1804. | beob. wahre Länge der ⚴ | nach den VIII. Elementen. | beob. nördl. Breite d. ⚴ | nach den VIII. Elementen. |
|---|---|---|---|---|
| den 7. Sept. | 334° 55′ 0″,9 | + 18″,1 | 14° 1′ 7″,0 | + 28″,5 |
| — 17. — | 332 32 24,12 | + 8,8 | 12 37 58,5 | + 10,6 |
| — 18. — | 332 19 16,4 | + 1,6 | 12 29 5,5 | + 17,3 |
| — 19. — | 332 6 17,0 | + 1,1 | 12 20 23,1 | + 9,4 |
| — 21. — | 331 41 2,5 | − 0,1 | 12 2 28,6 | + 13,0 |
| — 22. — | 331 28 46,1 | + 1,5 | 11 53 31,5 | + 10,2 |

**Fig. 11.61** Differences of the observed and calculated places of Pallas based on Elements VIII by Gauss

## Continued Reports Regarding the New Primary Planet Pallas

*MC*, January 1805, p. 92

*We still have a few foreign observations of Pallas to make up, which we will now present to our readers. At present, these will have to be the last ones since this planet will not be visible to meridian instruments for a considerable period of time, but rather only on larger equatorial instruments.*

*The last opposition of Pallas, which occurred between August 30 and 31 of last year, was, as far as we know, observed nowhere other than the Brera Observatory in Milan. The observations made there are as follows* (Fig. 11.62):

*and out of this* (Fig. 11.63)

*If the positions of August 29 and 30 are improved through the mean error [deviation] of the elements, one obtains* (Fig. 11.64):

**Fig. 11.62** Oriani's observations of Pallas in Milan from August 27 to 31, 1804

| 1804 August | Mittlere Zeit | | | Scheinbare AR der ⚴ | | | Scheinbare nördl. Abw. | | |
|---|---|---|---|---|---|---|---|---|---|
| 27 | 11 U | 50' | 59" | 333° | 42' | 44,"3 | 5° | 33' | 39,"1 |
| 28 | 11 | 46 | 17 | 333 | 31 | 20, 6 | 5 | 22 | 10, 4 |
| 29 | 11 | 41 | 35 | 333 | 19 | 58, 1 | 5 | 10 | 29, 9 |
| 30 | 11 | 36 | 54 | 333 | 8 | 35, 9 | 4 | 58 | 42, 7 |
| 31 | 11 | 32 | 13 | 332 | 57 | 17, 7 | 4 | 46 | 45, 7 |

**Fig. 11.63** Differences of the observed and calculated places of Pallas based on Elements VII by Gauss

| Wahre beobachtete Länge der Pallas | | | | Beobachtete Breite | | | Fehler der VII Elemente von Gauſs | | | | |
|---|---|---|---|---|---|---|---|---|---|---|---|
| | | | | | | | in der Länge | | | in der Breite | |
| 11 s | 7° | 42' | 29,"3 | 15° | 9' | 50,"0 | — 7' | 24,"2 | | + 2' | 16,"3 |
| 11 | 7 | 27 | 9, 7 | 15 | 13 | 21, 8 | 7 | 27, 7 | | 2 | 11, 8 |
| 11 | 7 | 11 | 47, 6 | 15 | 6 | 41, 3 | 7 | 29, 1 | | 2 | 13, 5 |
| 11 | 6 | 56 | 23, 8 | 14 | 59 | 54, 1 | 7 | 29, 9 | | 2 | 13, 9 |
| 11 | 6 | 40 | 59, 7 | 14 | 52 | 54, 3 | 7 | 30, 1 | | 2 | 18, 1 |
| | | | | Mittlerer Fehler | | | — 7' | 28,"3 | | + 2' | 14,"7 |

**Fig. 11.64** Improved positions of Pallas on August 29 and 30

| August 1804 | Mittlere Zeit | Länge der Pallas | Nördliche Breite | Länge der ☉ |
|---|---|---|---|---|
| | U ' " | s ° ' " | ° ' " | s ° ' " |
| 29 | 11 41 35 | 11 7 11 46, 8 | 15 6 40, 1 | 5 6 18 54, 1 |
| 30 | 11 36 54 | 11 6 56 22, 2 | 14 59 53, 3 | 5 7 16 49, 3 |
| Unterſch. | 23U 55' 19" | — 15' 24"6 | — 6' 46"8 | + 57 55"2 |

*The opposition of Pallas with the Sun therefore occurred on: 30 August, 4 h 56' 36" mean time in Brera. At this time, the longitude of Pallas was 11z 7° 0' 40."0, geocentric latitude 15° 1' 46."8, heliocentric latitude 10° 27' 30."9. The planet was compared with θ Pegasi, the position of which, according to Piazzi, is the following: apparent RA 330° 5' 31."1 apparent north. decl. 5° 14' 46."1.*

*The following elements served in calculating the opposition:*
*Parallax in altitude of Pallas = 2."4*
*Inclination of the ecliptic according to Lalande = 23° 27' 55."5*
*Longitudinal aberration = −12."4*
*Latitudinal aberration = −5."2*
*Nutation = −13."7*

*This opposition had been observed and calculated by the young Carlini who had been occupied at the Brera Observatory with astronomical work for five years while Oriani was occupied with geodetic work. Oriani praises the work of this twenty-year-old man tremendously, and the praise of such a competent judge gives us hope of having found a new asset for science. Following Reggio's death* [Francesco Reggio, 1743–1804] *, the government appointed the young man as a supernumerary astronomer with a salary of 2000 liv[re] (approximately 500 taler), and this honorable distinction will certainly ensure that he will not allow his enthusiasm for astronomy to wane.*

*Canon David diligently observed Pallas from the Royal Observatory in Prague, but was unfortunately hindered in his observations of the opposition by unfavorable weather. His other observations were the following* (Fig. 11.65):

*On May 15, 16, 20 and 21, Canon David compared Pallas with δ Pegasi and a magnitude five star, the right ascension of which was 12° 50' 56."6, and the declination 20'11" greater*

**Fig. 11.65** David's observations of Pallas in Prague from May 15 to September 14, 1804

| 1804 | | Mittl. Zeit in Prag | | | Scheinbare AR. der ⚴ | | | Nördl. Abweich. | | |
|---|---|---|---|---|---|---|---|---|---|---|
| May | 15 | 13 U | 41' | 43" | 335° | 55' | 32" | 9° | 3' | 20" |
| | 16 | 14 | 0 | 35 | 336 | 6 | 10 | 9 | 7 | 21 |
| | 20 | 14 | 4 | 25 | 336 | 46 | 35 | 9 | 25 | 5 |
| | 21 | 14 | 11 | 32 | 336 | 56 | 28 | 9 | 29 | 44 |
| Jun. | 4 | 4 | 6 | 33 | 338 | 48 | 6 | 10 | 21 | 58 |
| Sept. | 6 | 11 | 4 | 7 | 331 | 47 | 50 | 3 | 32 | 48 |
| | 7 | 0 | 59 | 29 | 331 | 37 | 7 | 3 | 20 | 49 |
| | 13 | 10 | 31 | 45 | 330 | 37 | 56 | 2 | 4 | 54 |
| | 14 | 10 | 27 | 11 | 430 | 27 | 45 | 1 | 51 | 24 |

*than that of ε Pegasi. According to the* Connaissance de temps, *from which David borrowed the position of ε, the RA of ε Pegasi on May 15, 1804 = 323° 38′ 37.″3 and the northern declination = 8° 59′ 7.″8. According to this, it would follow for that magnitude 5 star that RA = 336° 29′ 33.″9 and north. decl. = 9° 19′ 18.″8. We have not been able to find this star either in Piazzi's star catalogue or in* Histoire céléste *by Lalande.*

*On September 6, 13 and 14, Canon David compared Pallas with ε Equulei, 88 Aquarii and 58 Pegasi, according to Bode's star catalogue. In the case of the last two stars, he found the difference of 10″ in the RA on the arc and 18″ greater in the declination than cited by Bode.*

*Dr. Olbers delivered the latest observations of Pallas to us, that he observed on October 23 and 24, and November 5 and 6* (Fig. 11.66):

*On the occasion of these observations, Dr. Olbers made the remark that quite a number of magnitude eight stars in the Water Carrier (Aquarius) are missing from the* Histoire céléste.

## Observations of the new Planet Pallas

*MC*, April 1805, p. 376

*Although it has not been possible to observe Pallas during the past months, the diligence of Dr. Gauss enables us to relate some theoretical information about this new planet, which will be of excellent use to facilitate its rediscovery during the next, probably distant, period of visibility. Dr. Gauss, whose tireless diligence and continuous quest to perfect the calculated planetary orbits can only be stopped by a lack of material, has expressed on several occasions his dissatisfaction about the few numbers of available observations of Pallas, which makes it impossible for him the tackle any further improvement of the elements VII of this planet. Certainly the observation of this faint body had many difficulties, and the observations made here by Olbers, Oriani and David are the only ones made the last year. He very much appreciated the opposition of Pallas on August 30, 1804, observed at Brera, when the mean error of the elements VII was in longitude −7′ 28.″3 and latitude +2′ 14.″7 and he based his elements VIII of Pallas on this and all previous observations:*

**Fig. 11.66** Observations of Pallas by Olbers from October 23 to November 6, 1804

| 1804 | | Mittl. Zeit in Bremen | | | Scheinb. AR der Pallas | | | Scheinb. süd. Declin. | | |
|---|---|---|---|---|---|---|---|---|---|---|
| Octbr. | 23 | 7 U | 9' | 36" | 327° | 30' | 49" | 5° | 28' | 57" |
| | 24 | 6 | 39 | 36 | 327 | 32 | 34 | 5 | 32 | 23 |
| Nov. | 5 | 6 | 6 | 21 | 328 | 15 | 20 | 7 | 5 | 17 |
| | 6 | 6 | 13 | 16 | 328 | 20 | 47 | 7 | 11 | 32 |

| | |
|---|---|
| *Epoch Seeberg Meridian 1803 221° 31′ 23.″2* | |
| *1804 299 58 38.1* | |
| *Daily tropical movement* | *771.″6802* |
| *Perihelion 1803* | *301° 1′ 44.″1* |
| *Ascending node 1803* | *172 29 6.8* |
| *Eccentricity* | *0.246101* |
| *Logarithm of the semimajor axis* | *0.4417647* |
| *Inclination of the path* | *34° 37′ 43.″2* |

*According to these, Dr. Gauss calculated the following ephemeris for the geocentric path of Pallas in 1805 and 1806.* [Table follows from July 28, 1805, to April 30, 1806.]

*Dr. Gauss wrote that this ephemeris is based on purely elliptical elements, for he believes it to be not useful at this time, to take the perturbations into account, since partly all hitherto observations can still be quite well described by an ellipsis, and partly the orbit of Pallas does not yet allow a more exact determination due to its short appearance, in order to hope to obtain a degree of accuracy by developing all perturbation equations, which would justify this tedious effort. Every connoisseur will doubtless agree with this man's opinion, who is so well acquainted with all necessary calculations. Most likely further observations of several years will require slight changes in eccentricity, epoch, mean motion* etc. *of Pallas, and since by this also Aequatio centri and radius vector are affected, every calculation of the perturbation carried out now will become faulty and make a repetition necessary, which would be too time consuming to not wait for the epoch when every calculation can be made with certainty with an exactly determined orbit. Were Ceres' mass a little more considerable, this could, partly due to the small distance, partly due to the almost exact mean motion of both planets, cause considerable perturbations of Pallas' orbit. The only observations, the last of Pallas of last year, which we would like to relate to our readers are those made by Dr. Olbers. At no observatory was Pallas so eagerly and continuously observed, as this diligent astronomer did. It was him, who rediscovered Pallas, and it is him, who followed Pallas the longest in the skies. This little planet only showed itself to Dr. Olbers this long, and it is thus only equitable that the father is the last to cast a glance at the child, thus rewarding him for those many nights he sacrificed for her education. The following observations are those Dr. Olbers was able to obtain at the end of last year and he kindly communicated:*

| *1804* | *Mean time in Bremen* | *Apparent RA* | *Apparent Declin.* |
|---|---|---|---|
| *Nov. 20* | *8 h 21′ 20′* | *330° 7′ 0″* | *8° 22′ 52″* |
| *25* | *5 49 26* | *330 55 47* | *8 40 26* |
| *27* | *5 54 59* | *331 17 22* | *8 46 27* |
| *Dec. 2* | *6 56 36* | *332 15 20* | *8 58 52* |
| *30* | *6 31 30* | *339 1 44* | *9 9 14* |
| *31* | *5 47 24* | *339 17 56* | *9 8 4* |

*On this occasion Dr. Olbers made a remark which could have led to a very interesting discovery. Since Pallas was at the beginning of 1751 very luminous with a great southern declination, he hoped to find this planet among the southern stars observed by La Caille. But unfortunately, La Caille started observing on August 6, when Pallas could not have been visible to him and thus the hope to make a similar discovery as in the case of Uranus was thwarted.*

*When Pallas will be visible again in the meridian is not determined exactly, but we are afraid this epoch will occur later than in the case of the other new planets and its rediscovery will be more difficult. In this year's September and October, when Pallas culminates in the morning, it will still be too bright to observe this faint object in the meridian, and in November and December the great southern declination will take away Pallas from our eyes. Consequently, the opposition, happening at the end of November, will not be observed in Germany, since Pallas will have a southern declination of more than 30° and only from observatories like Milan and Palermo can reliable observations be expected. Thus in our region hardly any meridian observations can be made before the end of January or February 1806, when Pallas will culminate with a considerable brightness and in complete darkness. Inspector Harding, who was immediately informed of the above new ephemeris of Pallas, is busying himself with the creditable work to sketch a stellar map for its path. Unfortunately, he will often lack of determinations of stars, since he is not as supported by the* Histoire celeste *in these southern regions as is usually the case by this invaluable collection.*

## Astronomical Observations in Ofen, by Prof. Pasquich

*MC,* April 180, p. 384

*From the letters of Professor Pasquich.*

*January 28, 1805*

[Parts of this paper dealing with Juno will be published in a future book in this series.]

*From October 18 through December 4 no days passed unused; I was only able to observe at the darkened telescope and I saw Pallas several times; but I succeeded only twice to observe her with the greatest strain of my eyes. Since Ceres and Harding's planet often appeared together with several small fixed stars in the telescope, I daresay, that those appeared every time as of the* $9^{th}$ *order of magnitude; but even then Pallas was only visible as a* $12^{th}$ *magnitude star. I could only have thought of observations of good declinations, if I had replaced the very thin silver thread with a thicker one; but this was not tolerated by the weak steel springs of the micrometer.*

## Regarding the True and Apparent Course of Ceres and Pallas and their Interrelation

By Bode (1805) [See also in this chapter Bode's 1806 paper on planetary orbits.]

*BAJ*, 1807, p. 216

*We have seen, since the period of discovery of 1802, these two* new primary planets, *not very different in terms of degrees of longitude, progressing together along the firmament. If the times of their orbits around the Sun of 1681.7 days are and remain constant, and the results of the calculation, derived from the newest observations of Dr. Gauss, indicate their true paths, then both will always remain together and closely united and can be observed appearing in opposition with each other just as little from the Sun as from the earth.*

*In order to represent this new curiosity, which has granted us a hitherto unknown view of things in the Solar System and has offered us extensive material for contemplation, in a new and graphic way, I set the true and apparent course of these two planetary bodies down in a number of sketches of rather grand scale, and used, as an example for an entire orbit, January 1, 1803 to August 10, 1807, and had the newest elliptical elements to form the basis of the calculation of their true paths.*

*The first figure* (Fig. 11.67; this is Bode's Fig. 7) *showed the paths of Ceres and Pallas set down in their form, position and apsis-line on the plane of the ecliptic against the remaining planetary paths of Mercury to Saturn. For the first day of every month, I indicated their true positions on the entire path around the Sun within the aforementioned time period, through which it could already be seen how Pallas trails Ceres while always remaining in its vicinity.*

*The second figure* (Fig. 11.68; this is Bode's Fig. 8) shows *the movement of Ceres and Pallas appearing from the Sun (or heliocentrically) in view of the ecliptic from month to month in the same period. In entering these positions, the calculations according to my tables gave the following data* (Fig. 11.69):

*If these apparent path-lines, constructed from this, are now drawn into simultaneous positions of these planets, this yields the apparent position of them with each other as seen from the Sun. Namely, in the first months of 1803, Pallas appeared north of Ceres, and from then, it will appear continuously westwards toward north or south, and in the year of*

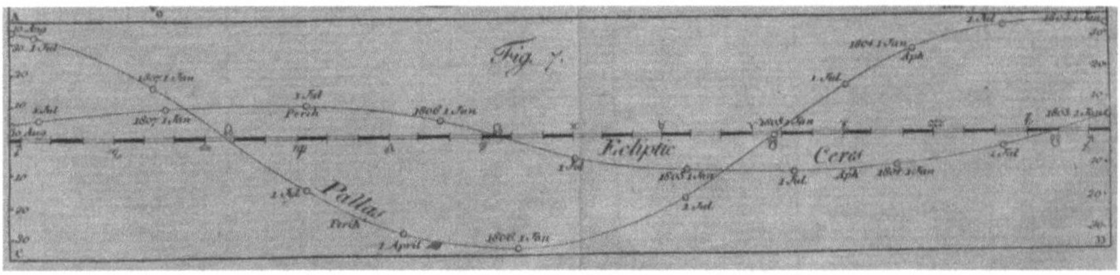


**Fig. 11.67** The paths of Ceres and Pallas set down on the plane of the ecliptic

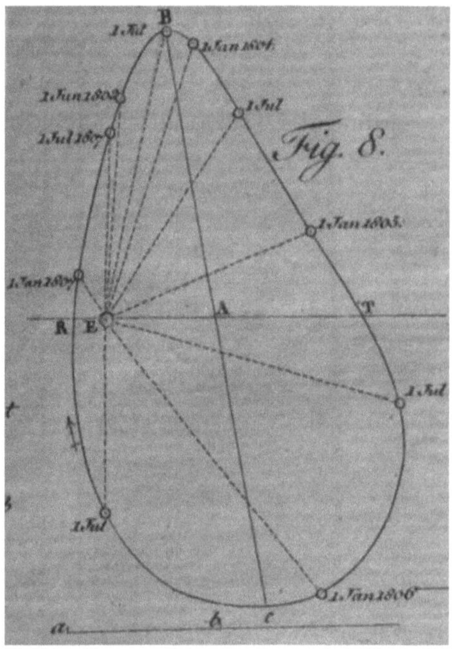


**Fig. 11.68** The paths of Ceres and Pallas appearing from the Sun

Fig. 11.69 Positions of Ceres and Pallas calculated by Bode from Jan 1, 1803, to August 10, 1807

| | Ceres heliocentrische | | Pallas Unterschied | |
|---|---|---|---|---|
| | Länge | Breite | in der Länge | in der Br. |
| 1803 1.Jan. | 8ᶻ 3° 9′ | 3° 17′ N. | 0° 53′ östl. | 29° 58′ N. |
| 1.Apr. | 8 21 28 | 0 5 S. | 0 5 | 34 42 |
| 1.Jul. | 9 9 10 | 3 21 | 1 21 westl. | 37 0 |
| 1.Oct. | 9 26 29 | 6 13 | 3 58 | 37 5 |
| 1804 1.Jan. | 10 13 45 | 8 30 | 7 53 | 35 10 |
| 1.Apr. | 11 0 26 | 9 58 | 12 40 | 31 25 |
| 1.Jul. | 11 17 26 | 10 36 | 18 12 | 25 49 |
| 1.Oct. | 0 5 1 | 10 18 | 24 20 | 18 21 |
| 1805 1.Jan. | 0 22 58 | 9 3 | 29 53 | 8 37 |
| 1.Apr. | 1 11 13 | 6 51 | 34 33 | 2 44 S. |
| 1.Jul. | 2 0 24 | 3 46 | 36 49 | 15 51 |
| 1.Oct. | 2 20 41 | 0 3 | 33 51 | 29 16 |
| 1806 1.Jan. | 3 12 4 | 3 50 N. | 22 52 | 38 24 |
| 1.Apr. | 4 3 59 | 7 18 | 7 43 | 37 11 |
| 1.Jul. | 4 26 54 | 9 43 | 1 36 östl. | 25 25 |
| 1.Oct. | 5 20 18 | 10 38 | 4 2 | 9 22 |
| 1807 1.Jan. | 6 13 11 | 9 51 | 3 10 | 5 42 N. |
| 1.Apr. | 7 4 29 | 7 45 | 1 45 | 17 45 |
| 1.Jul. | 7 24 42 | 4 46 | 0 50 | 26 50 |
| 10.Aug. | 8 3 10 | 3 17 | 0 41 | 29 55 |

*1806 directly to the south and then eastward until April 1807; finally, in July in August, it will appear north of it once again. Both paths twine around each other in such a way so that when Ceres, in its descending and ascending nodes, goes in Sagittarius and Gemini, Pallas attains its greatest northern and southern latitude, and where those of Pallas occur in Pisces and Virgo, Ceres will reach its greatest southern and northern latitude. This is a result of an extraordinarily remarkable condition that the nodal lines of both planets cut through each other at right angles. On the other hand, soon after Pallas traverses its descending node, it then comes next, some 5° Aries, and this occurs in the present orbit in April 1805, but Ceres is moved this time, as always, a considerable arc further to the east. On the other hand, soon after Pallas goes through its ascending node, it will then always (or with every orbit) come together some 5° Libra with Ceres, and the next time this will occur is November 1806.*

*A third figure* (Fig. 11.70; this is Bode's Fig. 9) *developed by myself from the previous one shows the apparent, relative course of Pallas around and against that of Ceres as seen from the Sun. It forms a crooked, pear shaped line which returns to itself, the point of which points to the north. Pallas completes this apparent path once together with Ceres during its orbit around the Sun, and it is from the east to the west, that is: in a backward movement if Ceres, as in its path RAT, is regarded as resting* [stationary]. *I have drawn this heliocentric apparently relative path in miniature in the 8th figure and have noted the positions of Pallas on it from 6 to 6 months. Its major axis BC makes an angle of some 79° with the eastward path of Ceres. Ceres appears far from the middle of this thus constructed path of Pallas towards the east [in E] and also lies on the other side of its plane. The dotted lines also define where Pallas appears vis a vis Ceres and in which approximate distance, according to the scale located below, from = 24260 radius of Earth of the distance of the Sun from Earth. This great distance arises principally from the considerable inclination of the path of Pallas. On November 25, 1806, Pallas passes Ceres [in R.] approximately 4500 Earth radii eastward.*

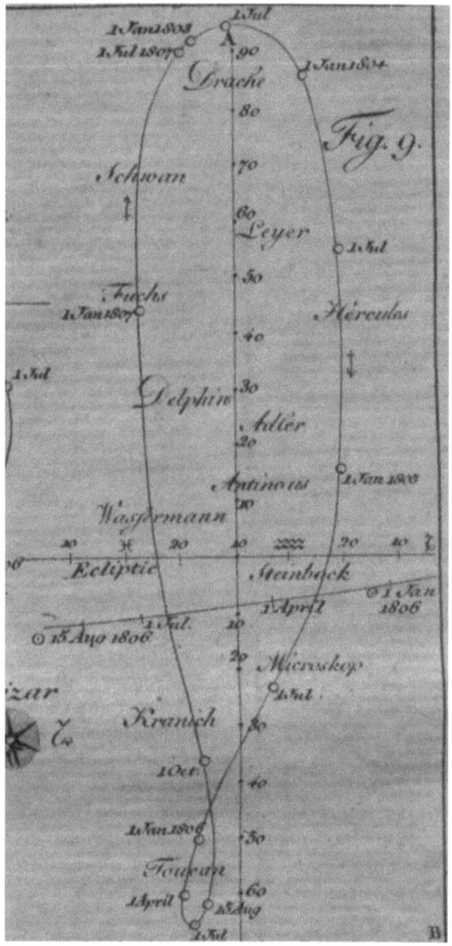


**Fig. 11.70** The course of Pallas around and against that of Ceres

*If one imagines a parallel of the ecliptic drawn through the position of Ceres every time and makes Ceres immobile in it, then the form and position of this apparent relative path comes out only slightly differently.*

*Moreover, in a model, I have physically depicted the size and position of the true paths of Ceres and Pallas against that of Earth and its plane and have drawn from 6 to 6 months lines* [threads] *to the simultaneous positions of Ceres and Pallas to be able to better recognize the true positions distances of them to and from each other. This model also shows how remarkably the true paths of these small planets are intertwined in their nodal lines, which cut through at right angles, and that, especially remarkable, how the path of Pallas lies around in its ascending node, within in its ascending node but outside the path of Ceres: in short, that which has already been noted before regarding the close meeting of both paths.*

*I have also calculated the geocentric positions of both planets during this orbit for the 1st of each month, entered them on both hemispheres of my great celestial charts and have drawn them together with lines, through which the apparent backward and forward course of these in our firmament is produced.*

*Since Pallas and Ceres move in particular paths of identical semi major axes and constantly follow each other in their course, one could suspect that – in spite of their mutually great distance, and although Ceres doesn't lie upon the plane of the path of Pallas and therefore describes no actual path while one describes every orbit around it, still, they have a close relation to each other, and through the general bond with which the all-pervading power positions worlds upon worlds in a reciprocal attraction, are links of a chain.*

*Finally, I have, as an example, calculated the apparent positions of Pallas on the firmament for the entire orbit as seen from Ceres. Since Ceres lies far outside the plane of Pallas' true path, from Ceres, the latter appears on the firmament in no way in the form and position of a greatest orbit of the sphere, but rather contracts itself as a small oval, the major axis of which lies under an angle of approximately 70° with the path of Ceres or the apparent path of the Sun. It constantly includes [encircles] a portion of northern Draco around the pole of the ecliptic, and later the Swan* [Cygnus], *Lyra, the Fox* [Vulpecula], *Cerberus* [an extinct constellation that was located between Hercules, Lacerta and Lyra], *the Eagle* [Aquila], *the Dolphin* [Delphinus], *the Small Horse* [Equuleus, located between Aquarius, Delphinus and Pegasus], *Antinous* [an extinct constellation whose stars are part of Aquila], *Aries, southern Pisces, the Microscope* [Microscopium, located beside Capricornus and Grus], *the Crane* [Grus] *and Toucan* [Tucana], *from 86° northern to 65° southern latitudes.*

*Pallas never comes in opposition with the Sun, as seen from Ceres, but always appears at the side of the Sun – most of the time to the left or east of the same. Less often still does it come in* [upper] *convergence* [meeting] *with the Sun or behind the same, rather, with every orbit, when the Sun appears in Aquarius under the major southern latitude, in the lower position or this side.*

*The peculiar crooked form of the constantly unchanging path of Pallas in the sky as seen from Ceres is indicated in figure 9. The velocity of Pallas in the same seems very irregular, like that represented in the figure for the positions calculated for the current orbit and the details of the following table. A is the north pole of the ecliptic, through which, with Aquarius and Pisces, the apparent path of Pallas goes. Further, the position of the Sun*'s *path and the position of the Sun in this orbit is not noted* (Fig. 11.71).

**Fig. 11.71** Longitude and Latitude figures relating to the diagram in Fig. 11.70

| | Ort der ⊙ in der Ceresbahn. Länge. Z. G. M. | Ort der ⊙ in der Ceresbahn. Breit. G. M. | Abst. d. Pallas von d. ⊙. in d. L. G. M. | Scheinb. Cereri-centrischer Ort der Pallas. Länge. Z. G. M. | Scheinb. Cereri-centrischer Ort der Pallas. Breite. G. M. | |
|---|---|---|---|---|---|---|
| 1803. | | Südl. | Westl. | | Nordl. | |
| 1. Jan. | 2 3 9 | 3 17 | 9 54 | 1 23 15 | 81 24 | Nordl. bey τ Drachen. |
| | | Nrdl. | Oestl. | | | |
| 1. Jul. | 3 9 10 | 3 21 | 28 19 | 4 7 29 | 86 6 | bey ω Drachen. |
| 1804. | | | | | | |
| 1. Jan. | 4 13 45 | 8 30 | 104 10 | 7 27 55 | 77 38 | zwisch. ν u. μ Drachen |
| 1. Jul. | 5 17 26 | 10 36 | 99 1 | 8 26 27 | 54 19 | bey ν ξ Herk. |
| 1805. | | | | | | |
| 1. Jan. | 6 22 58 | 9 3 | 83 15 | 9 16 13 | 15 36 | bey den Füſsen d. Antinous. |
| | | | | | Südl. | |
| 1. Jul. | 8 0 24 | 3 46 | 63 23 | 10 3 47 | 22 57 | im Mikroskop. |
| 1806. | | Südl. | | | | |
| 1. Jan. | 9 12 4 | 3 50 | 37 10 | 10 19 14 | 50 41 | am Hals des Toukan: |
| 1. Apr. | 10 3 58 | 7 18 | 18 13 | 10 22 11 | 60 22 | bey ζ Toukan. |
| | | | Westl. | | | |
| 1. Jul. | 10 26 54 | 9 43 | 6 55 | 10 19 59 | 64 45 | b. Schwanz d. Toukan. |
| 15. Aug | 11 8 22 | 10 23 | 21 13 | 10 17 9 | 61 27 | zwisch. ζ u. π Toukan. |
| 1. Oct. | 11 20 18 | 10 38 | 31 47 | 10 18 31 | 35 37 | bey β Kranich. |
| 1807. | | | | | Nördl. | |
| 1. Jan. | 0 13 11 | 9 51 | 37 22 | 11 5 49 | 45 28 | am südl. Flüg. d. Schw. |
| 1. Jul. | 1 24 42 | 4 46 | 9 25 | 1 15 17 | 80 23 | westl. bey τ Drachen. |
| 10. Aug | 2 3 10 | 3 17 | 7 47 | 1 25 23 | 81 32 ** | Nordl. bey τ Drachen. |

*Bode.*

## Observations of the new planets: Ceres, Pallas and Juno in the years 1803 and 1804, made at the Royal Observatory in Vilnius by the Royal Astronomer Poczobut, and Herr Reschka, Professor of Astronomy at the Royal University

Submitted April 17, 1805.

*Herr Poczobut (77 years old) made the preceding observations with an achromatic telescope of 5½-foot focal length* (Figs. 11.72, 11.73 and 11.74).

[The Juno observations will be published in a future book in this series.]

## Planetary and Cometary Orbits by Bode

Royal Prussian Academy of Science, 1806, p. 147

*BAJ*, 1808, p. 212

Ceres im Jahr 1803.

| Monats-Tage. | Durchgang durch den Meridian. St. M. S. | Scheinb. Abst. v. Scheitelpunct. G. M. S. | Barom. | +α | +β | Scheinbare gerade Aufsteigung. G. M. S. | Scheinbare Abweichung Südlich. G. M. S. | |
|---|---|---|---|---|---|---|---|---|
| 16 Jun. | 13 20 19 | 81 46 35 | 28 0,3 | 59 | 7¾ | 284 19 14 | 27 5 33 | φ ♃ |
| 20 — | 13 1 6,5 | 82 5 17 | 27 9,2 | 60 | 8½ | 283 27 5 | 27 24 15 | |
| 24 — | 12 41 41,5 | 82 23 36 | 28 2,2 | 62 | 12 | 282 31 48 | 27 42 34 | |
| 25 — | 12 36 50 | 82 27 58 | 28 2,4 | 64½ | 16¼ | 282 17 46 | 27 46 58 | |
| 26 — | 12 31 56,5 | 82 32 25 | 28 1,9 | 66 | 14 | 282 3 38 | 27 51 25 | r ♃ |
| 27 — | 12 27 3 | 82 36 43 | 28 1,2 | 65 | 11¼ | 281 49 21 | 27 55 41 | |
| 29 — | 12 17 16,5 | 82 45 0 | 28 0,0 | 63 | 8¾ | 281 20 39 | 28 3 58 | |
| 30 — | 12 12 22 | 82 49 7 | 28 0,8 | 63 | 10¾ | 281 6 0 | 28 8 5 | |
| 1 Jul | 12 7 29 | 82 53 6 | 28 0,7 | 65 | 12¾ | 280 51 17 | 28 12 4 | |
| 7 — | 11 38 2,5 | 83 15 42 | 28 0,8 | 73 | 16 | 279 23 58 | 28 34 40 | |
| 8 — | 11 33 8 | 83 19 24 | 28 2,3 | 73¾ | 16½ | 279 9 26 | 28 38 22 | |
| 18 — | 10 44 39 | 83 49 3 | 27 11,5 | 69 | 14 | 276 52 23 | 29 8 1 | γ ♃ |
| 19 — | 10 39 54 | 83 52 10 | 27 8,7 | 69½ | 15 | 276 40 5 | 29 11 8 | |
| 21 — | 10 30 25 | 83 56 32 | 27 10,6 | 69½ | 13⅔ | 276 15 45 | 29 15 30 | |
| 22 — | 10 25 41 | 83 58 58 | 28 0,0 | 67½ | 12 | 276 3 50 | 29 17 56 | ♌ ♃ |
| 24 — | 10 16 18 | 84 3 12 | 27 11,1 | 67 | 11¾ | 275 41 2 | 29 22 10 | γ ♃ |
| 26 — | 10 7 1 | 84 6 59 | 28 1,3 | 66¼ | 10⅔ | 275 19 48 | 29 25 57 | |
| 29 — | 9 53 12 | 84 12 25 | 28 1,2 | 67½ | 14¼ | 274 49 9 | 29 31 23 | |
| 30 — | 9 43 38,5 | 84 14 5 | 28 0,7 | 68½ | 15¼ | 274 40 22 | 29 33 3 | |
| 1 Aug | 9 39 42 | 84 17 10 | 28 1,2 | 69⅞ | 15 | 274 23 59 | 29 36 8 | ♌ ♃ |
| 3 — | 9 30 41 | 84 20 6 | 28 1,7 | 73 | 19 | 274 6 42 | 29 39 4 | |
| 6 — | 9 17 20,5 | 84 23 42 | 27 11,7 | 73½ | 15⅓ | 273 45 16 | 29 42 40 | |
| 10 — | 9 0 12,5 | 84 27 41 | 28 1,3 | 69 | 12¼ | 273 22 19 | 29 46 39 | ♌ ♃ |

Beobachtung der Ceres an der Rhomboïde der Parallatischen Maschiene.

| | | | | | | | | |
|---|---|---|---|---|---|---|---|---|
| 13 Aug. | 9 22 43 | | 28 1,7 | 71 | 16 | 273 9 33 | 29 57 47 | |
| 14 — | 9 17 40 | | 28 1,9 | 70¾ | 15⅓ | 273 6 2 | 29 58 16 | ♌ ♃ |
| 25 — | 8 44 11 | | 28 1,7 | 67 | 13¾ | 272 54 15 | 30 3 30 | |

Ceres im Jahr 1804.

| | | | | | | | | |
|---|---|---|---|---|---|---|---|---|
| 13 Aug. | 15 34 33 | 63 13 5 | 27 8,0 | 67 | 10¼ | 15 56 8 | 8 32 3 | 173. |
| 12 Spt. | 13 25 6 | 65 57 30 | 28 1,5 | 59¼ | 8 | 13 2 31 | 11 16 28 | Wü |
| 14 — | 13 15 53 | 66 9 7 | 28 1,5 | 61¾ | 13 | 12 42 13 | 11 28 5 | ″ — |
| 15 — | 13 11 12 | 66 15 0 | 28 2,3 | 62 | 13¾ | 12 30 55,5 | 11 33 58 | 1. φ |
| 16 — | 13 6 31 | 66 20 35 | 28 2,8 | 64 | 11¼ | 12 20 19 | 11 39 33 | |
| 17 — | 13 1 53 | 66 26 5 | 28 2,2 | 65 | 13½ | 12 9 47,5 | 11 45 3 | 2. φ |
| 24 — | 12 28 53 | 67 2 51 | 27 8,1 | 53 | 4¼ | 10 46 6,5 | 12 21 49 | 3. φ |
| 27 — | 12 14 29 | 67 17 10 | 27 11,2 | 51 | 6¼ | 10 9 7 | 12 36 8 | 4. φ |

**Fig. 11.72** Observations of Ceres by Poczobut in 1803 and 1804

Ceres im Jahr 1804.

| Monats-Tage. | Durchgang durch den Meridian. St. M. S. | Scheinb. Abst. v. Scheitelpunct. G. M. S. | Barom. | +α | +β | Scheinbare gerade Aufsteigung. G. M. S. | Scheinbare Abweichung Südlich. G. M. S. | |
|---|---|---|---|---|---|---|---|---|
| 6 Oct. | 11 31 33 | 67 52 32 | 28 0,3 | 54½ | 11½ | 8 14 7 | 13 11 30 | |
| 7 — | 11 26 48 | 67 55 18 | 28 2,0 | 52 | 5 | 8 1 49,5 | 13 14 16 | α ♑ |
| 30 — | 9 40 11 | 68 13 0 | 28 1,4 | 35¾ | 4½ | 3 57 39 | 13 31 58 | τ ♒ |
| 31 — | 9 35 44 | 68 11 22 | 28 0,4 | 35½ | 2 | 3 49 29 | 13 30 20 | η W. |
| 1 Nov | 9 31 19 | 68 9 30 | 28 1,4 | 35 | 3½ | 3 42 9 | 13 28 28 | 3 ↓) |
| 19* — | 8 15 32 | 67 6 20 | 28 5,0 | 31½ | 5 | 2 26 39 | 12 25 18 | ♒) |
| **Beobachtungen der Ceres an der Rhomboïde der Paratischen Maschiene.** | | | | | | | | |
| 1 Aug | 14 59 50 | | 28 0,0 | 76 | 15¼ | 15 41 53 | 7 49 18 | 173 |
| 5 — | 14 40 20 | | 28 0,6 | 75 | 16 | 15 51 23 | 8 1 43 | vVIf |
| 10 — | 14 41 10 | | 27 11,8 | 68 | 10 | 15 56 23 | | |
| 11 — | 14 41 10 | | 28 0,5 | 68½ | 12¼ | 15 56 23 | | |
| **Pallas im Jahr 1804.** | | | | | | | **Nördl.** | |
| 24 Jul. | 14 26 30 | 44 25 30 | 27 9,4 | 72 | 15 | 339 8 30 | 10 15 32 | |
| 26 — | 14 17 50 | 44 35 5 | 27 9,5 | 70½ | 12¼ | 338 56 34 | 10 5 57 | |
| 28 — | 14 9 2 | 44 45 35 | 27 10,4 | 73 | 16¾ | 338 42 31 | 9 55 27 | ζ Pe- |
| 29 — | 14 4 35 | 44 51 10 | 28 1,5 | 74 | 16¾ | 338 34 45 | 9 49 52 | gas. |
| 30 — | 14 0 7 | 44 56 55 | 28 1,0 | 76 | 17½ | 338 26 44 | 9 44 7 | ζ et ε |
| 31 — | 13 55 39 | 45 2 58 | 28 1,0 | 77¾ | 17½ | 338 18 42 | 9 38 4 | Peg. |
| 1 Aug | 13 51 11 | 45 9 9 | 28 0,0 | 76 | 15¼ | 338 10 41 | 9 31 53 | |
| 2 — | 13 46 43 | 45 15 27 | 28 1,1 | 76 | 15 | 338 2 40 | 9 25 35 | ζ et ε |
| 3 — | 13 42 13 | 45 22 19 | 28 0,9 | 74 | 13⅔ | 337 54 7 | 9 18 43 | Peg. |
| 4 — | 13 37 42 | 45 29 10 | 28 0,6 | 75 | 16 | 337 45 18 | 9 11 52 | |
| 5 — | 13 33 10 | 45 36 5 | 28 0,6 | 75 | 16 | 337 36 18 | 9 4 57 | |
| 7 — | 13 24 5 | 45 51 30 | 27 9,7 | 73 | 11¾ | 337 17 59 | 8 49 32 | ι Pe- |
| 8 — | 13 19 31 | 45 59 16 | 27 8,4 | 70¼ | 10¼ | 337 8 28 | 8 41 46 | gas. |
| 9 — | 13 14 57 | 46 7 13 | 27 8,2 | 68 | 9⅓ | 336 58 56 | 8 33 49 | |
| 10 — | 13 10 21 | 46 15 14 | 27 11,8 | 68 | 10 | 336 48 21 | 8 25 48 | α Ad- |
| 11 — | 13 5 44 | 46 23 15 | 28 0,4 | 67½ | 10⅓ | 336 38 38 | 8 17 47 | ler. |
| 13 — | 12 56 27 | 46 41 30 | 27 8,0 | 67 | 10¼ | 336 18 8 | 7 59 32 | |
| 15 — | 12 47 15 | 46 59 49 | 27 10,5 | 65 | 9⅓ | 335 55 50 | 7 41 13 | |
| 18 — | 12 33 17,5 | 47 29 0 | 27 8,1 | 63½ | 10⅔ | 335 23 55 | 7 12 2 | |
| 19 — | 12 28 35 | 47 39 3 | 27 8,1 | 61¾ | 7⅔ | 335 11 50 | 7 2 0 | μ Ad |
| 20 — | 12 23 55 | 47 49 15 | 27 7,6 | 63½ | 11½ | 335 0 40 | 6 51 47 | ler. |
| 23 — | 12 9 54 | 48 20 24 | 27 10,3 | 61½ | 9⅔ | 334 27 11 | 6 20 38 | |
| 24 — | 12 5 13 | 48 32 5 | 27 10,6 | 62¼ | 11½ | 334 15 59 | 6 8 57 | |
| 27 — | 11 51 8 | 49 5 45 | 27 10,0 | 60 | 9⅔ | 333 42 23 | 5 35 17 | β E- |
| 29 — | 11 41 47 | 49 28 52 | 28 0,5 | 59¼ | 9⅔ | 333 20 0 | 5 12 30 | quul |
| 1 Spt. | 11 27 43 | 50 4 39 | 27 11,0 | 65 | 15¼ | 332 45 59 | 4 36 23 | ϑ Pe |
| 4 — | 11 13 40 | 50 41 36 | 28 3,6 | 61 | 8⅔ | 332 12 45 | 3 59 26 | gas. |
| 5 — | 11 9 0 | 50 53 37 | 28 3,9 | 60,5 | 8⅔ | 332 1 43 | 3 47 25 | α E- |
| 6 — | 11 4 19 | 51 6 15 | 28 2,9 | 60½ | 9¼ | 331 50 36 | 3 34 47 | qual |

Pallas

**Fig. 11.73** Observations of Ceres and Pallas by Poczobut in 1804

*My former lecture dealt with seven planets and 72 cometary orbits, which were known until then. I was led by an attempt by our unforgettable Lambert in his* kosmologischen Briefen ueber die Einrichtung des Weltbaues, *to a similar venture, to deduce conclusions about the outer appearance, position and shape of the orbits, and their distribution. I started with Halley's tables of 21 comets which appeared from 1337 until 1698, and supplemented this with recent comets resulting in more than three times their number. I took Halley's very incomplete register of comets, with the often found and observed comets from 1698 until 1785, without choosing, since the few number (due to a lack of serious research of those recorded in the first centuries) can change the result only slightly. I studied if something*

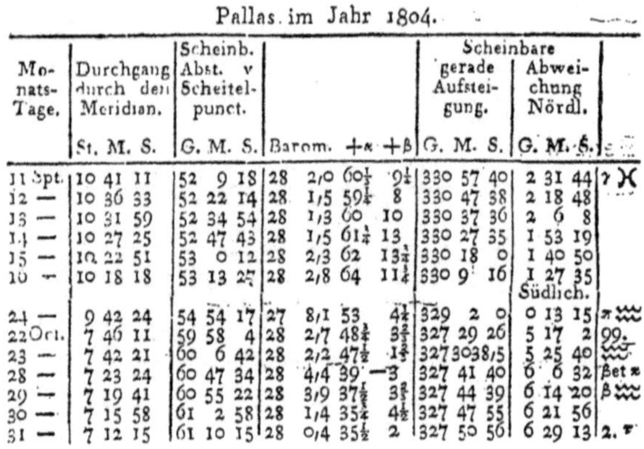

Pallas im Jahr 1804.

| Monats-Tage. | Durchgang durch den Meridian. St. M. S. | Scheinb. Abst. v Scheitelpunct. G. M. S. | Barom. | | +α | +β | Scheinbare gerade Aufsteigung. G. M. S. | Scheinbare Abweichung Nördl. G. M. S. | |
|---|---|---|---|---|---|---|---|---|---|
| 11 Spt. | 10 41 11 | 52 9 18 | 28 | 2,0 | $60\frac{1}{3}$ | $9\frac{1}{4}$ | 330 57 40 | 2 31 44 | ? ♓ |
| 12 — | 10 36 33 | 52 22 14 | 28 | 1,5 | $59\frac{1}{4}$ | 8 | 330 47 38 | 2 18 48 | |
| 13 — | 10 31 59 | 52 34 54 | 28 | 1,3 | 60 | 10 | 330 37 36 | 2 6 8 | |
| 14 — | 10 27 25 | 52 47 43 | 28 | 1,5 | $61\frac{1}{4}$ | 13 | 330 27 35 | 1 53 19 | |
| 15 — | 10 22 51 | 53 0 12 | 28 | 2,3 | 62 | $13\frac{3}{4}$ | 330 18 0 | 1 40 50 | |
| 16 — | 10 18 18 | 53 13 27 | 28 | 2,8 | 64 | $11\frac{1}{4}$ | 330 9 16 | 1 27 35 | |
| | | | | | | | | Südlich. | |
| 24 — | 9 42 24 | 54 54 17 | 27 | 8,1 | 53 | $4\frac{1}{4}$ | 329 2 0 | 0 13 15 | π ♒ |
| 22 Oct. | 7 46 11 | 59 58 4 | 28 | 2,7 | $48\frac{3}{4}$ | $3\frac{2}{3}$ | 327 29 26 | 5 17 2 | 99. |
| 23 — | 7 42 21 | 60 6 42 | 28 | 2,2 | $47\frac{1}{2}$ | $1\frac{2}{3}$ | 327 30 38,5 | 5 25 40 | ♒ |
| 28 — | 7 23 24 | 60 47 34 | 28 | 4,4 | 39 | —3 | 327 41 40 | 6 6 32 | βet ϰ |
| 29 — | 7 19 41 | 60 55 22 | 28 | 3,9 | $37\frac{1}{2}$ | $3\frac{2}{3}$ | 327 44 39 | 6 14 20 | β ♒ |
| 30 — | 7 15 58 | 61 2 58 | 28 | 1,4 | $35\frac{1}{4}$ | $4\frac{1}{2}$ | 327 47 55 | 6 21 56 | |
| 31 — | 7 12 15 | 61 10 15 | 28 | 0,4 | $35\frac{1}{2}$ | 2 | 327 50 56 | 6 29 13 | 2. |

**Fig. 11.74** Observations of Pallas by Poczobut in 1804

*regular regarding the distribution, direction, inclination, position, node* etc. *can be found, how they behave in regards of the position of the solar axis and the solar equator, close to what regions of the Solar System their perihelion and nodes are, and what else could be stated as probable or reliable. I found Lambert's ideas confirmed and also added for a comparison the seven orbits of the planets. On a large sheet of paper I sketched all these 72 cometary orbits on the plane of the Earth, even beyond Mars' orbit, and noted of each the perihelion, nodes, inclination, direction* etc.

*Since that lecture, 19 years ago, 19 new comets have been discovered and the number of all those observed and calculated has increased to 95. Furthermore, during this period, three more major planets of the Solar System between the orbits of Mars and Jupiter were discovered (where I only expected one) bearing the names: Ceres, Pallas and Juno. As it is known, the first was found by Piazzi on January 1, 1801, the second by Dr. Olbers in Bremen on March 28, 1802 and the third by Professor Harding in Lilienthal in September 1, 1804.*

*The number of the known comets and planets amounts at present to 105. On my large sketch I added the 23 cometary orbits, observed and calculated since 1785, plus those parts of the elliptical orbits of the three newly discovered planets, which are within the boundaries of the outer circle. That table of the elements of 72 cometary orbits is already mentioned in the* Memoires *for 1787 and at present those of the 23 comets discovered until 1805 are supplemented; with a sufficient accuracy for a sketch. I added to the same the two newly calculated columns for the longitude of the comets at perihelion reduced to the ecliptic and their heliocentric latitude. At the end of the table the elements follow for all known ten planetary orbits.*

*I would like to start again, as I did in my first lecture, with the perihelion. This point is in all planetary orbits heliocentric, and reduced according to longitude to the plane of the ecliptic, from the 2^nd^ degree of ξ eastward to the 17^th^ degree of ρ, so to an arc of 195°, which would be 21° shorter, so not even be a semi circle, if the perihelion of Uranus was not in the centre of ρ, since the most eastern boundary of it occurs next to Ceres and is in 26° θ. Now, very peculiar, the points of the perihelions, reduced to the ecliptic, of the far greatest part of our hitherto known 95 cometary orbits also fall into that region of the Solar System. For 65 comets reached their perihelion, if I include the last degrees of ω, in the signs ξ μ ν ο π θ and ρ, and only the other 30 came closest to the sun in σ τ ϑ ϖ and in the first part of ω. By the way, through the six northern signs went 58 and the six southern 37.*

*It is, albeit of the very different declinations and mostly much greater inclination of the cometary orbits compared to those of the planets, mainly one region of the solar region where most of the known comets and planets come closest to the sun, a fact that does not seem to be pure coincidence. During the 25 days of the sun's revolution, its axis under the angle of 82.5° keeps its position immobile against the 8° ξ and ρ, to where the solar axis is directed and the poles of the ⊙ fall. Thus the Sun's equator inclines by 7.5° with the plane of the ecliptic and the nodes lie in 8° o and ϑ. So if we see the ⊙ in ξ or ρ, the perihelions of all planets and cometary orbits appear on the east or west side of it, and it thus appears that they have a relation to the position of the rotational axis of the ⊙ and are usually met vertical to this. Around the poles of the ⊙, at 82.5° northern and southern latitude in 8° ξ and ρ on the other hand almost no comet passed through its perihelion.*

*The second to be studied is the distance of the perihelions of all these 95 cometary orbits. If one assumes the mean distance of the earth from the ⊙ = 1000, around the sun went through the region north and south* (Fig. 11.75):

*But if one uses rightly the true distance of the planets from the ⊙, in the region where every comet was at perihelion, the following went through:*

| | |
|---|---|
| *Between the orbits of Sun and Mercury* | *20 comets* |
| *Mercury and Venus* | *36* |
| *Venus and Earth* | *20* |
| *Earth and Mars* | *15* |
| *Mars and Ceres* | *4* |

*The majority of the comets, which reached their perihelion within the earths orbit, shows, that these must be more often visible to us, due to their faster path and greater brightness than those farther away. Since most of the comets came closest to the ⊙ between Mercury's and Venus' orbit, one particular reason for their good visibility can be found in the fact that these reach their perihelion at exactly half the distance between us and the sun, stay for several months in this region and thus can show themselves first at the west or eastern side of the sun in the firmament. Comets, which go down into Mercury's orbit towards the sun, often appear due to their more vivid light and longer tails at a fortunate position of the earth in relation to them and the sun. But those in the vicinity of the earth's orbit, moving inside or outside of it closest to the sun, usually show a fainter illumination, and often describe, due to their closeness in a short period of time, an apparently greater arc and are not visible for long.*

*Since furthermore of our 95 comets 76 pass through their perihelion within the earth's orbit, where the planetary orbits are clustered together, one could deduce that in the larger spaces beyond the earth's orbit, there are even more of these celestial bodies and come closest to the sun. Lambert even increases the number of comets, as the square of the distance of their*

| | | | | | | | | | |
|---|---|---|---|---|---|---|---|---|---|
| zwischen | o u. 100 Theilen | | 4 | Komet. | zwischen | 900 u. 1000 Theilen | | 7 | Komet. |
| — | 100 — 200 | — | 3 | — | — | 1000 — 1100 | — | 9 | — |
| — | 200 — 300 | — | 8 | — | — | 1100 — 1200 | — | 2 | — |
| — | 300 — 400 | — | 7 | — | — | 1200 — 1300 | — | 2 | — |
| — | 400 — 500 | — | 9 | — | — | 1400 — 1500 | — | 1 | — |
| — | 500 — 600 | — | 14 | — | — | 1500 — 1600 | — | 3 | — |
| — | 600 — 700 | — | 8 | — | — | 2100 — 2200 | — | 1 | — |
| — | 700 — 800 | — | 10 | — | — | 3700 — 3800 | — | 1 | — |
| — | 800 — 900 | — | 6 | — | | | | | |

**Fig. 11.75** Distance of the perihelion of 95 comets

*perihelion and finds this assumption in the case of Mercury's and Venus' orbit to be partly in correspondence with the incomplete comet register of Halley. But even the current list almost five times larger, is in fairly good harmony with this. For instance, within Venus' orbit 56 went through their perihelion and within Mercury's orbit 20. Now the circular extent of Mercury's orbit is three times smaller than that of Venus and is in ratio of 56:20, about 3:1.*

*Regarding the seasons the current comet register shows that 39 of the 95 comets went through their perihelion in the 6 summer months from April to September and in the 6 winter months from October to March 56. Since we only see the comets close to their perihelions and by night, their visibility most likely adds to this significant difference or we must discover several during the long winter nights and short summer nights.*

*Third, all ascending nodes of the all ten planetary orbits lay heliocentrically, also around the region of the Solar System where their perihelions are, namely from 16° of ν to 22° ρ, so on an arc of 126 degrees, or slightly above the third part of a circle. Now our list of 95 cometary orbits gives within these degrees of longitude the Ω of 42 comets, about half of the entire number of all observed. If I take the semi circle within which 16° ν and 22° ρ are, so from 19° μ to 19° σ, here occur the Ω of 59 comets, so almost 2/3 of 95. In the 6 northern signs the ascending node of 57 and in the southern that of 38. From this it follows that most of the comets reach together with all other planets in one and the same region of the Solar System their Ω and perihelion.*

*In my* Astronomischen Jahrbuch *[Astronomical Yearbook] for 1789 I published a table, calculated by the late Professor Prosperin in Upsala, regarding the smallest distance from the earth's orbit of all hitherto calculated 72 comets. In Dr. Olbers' treatise about the easiest method for calculating the orbit of a comet, 8. Weimar 1797, this table was resumed until 1795 and contains 84 comets. From this follows that half of it, so 42 can come closest to us at their Ω and 42 at their ☋. In the case of four comets the distance at their perigee before or after the Ω or ☋ was less than 1°, for 11 between 1° and 2°, for 13 this distance was up to 5°. Then the same occurs for the most between 5 and 60°.*

*Assuming the distance of ♁ from ⊙ = 1000, according to this table only three comets came closest to our earth up to 2, 5 and 9 of such parts (each has an absolute value of 20800 miles); 14 up to 50 and 13 up to 100. The rest pass us at a far greater distance. From all this can be seen how negligible the danger is that one of these comets meets the earth at a position when their orbits come close at the same time.*

*Besides that my great chart can serve as a general overview of the true parabolic path of all hitherto calculated 95 comets and the occurring circumstances, like positions and distance of perihelion, node line, direction, inclination* etc., *it also shows that far beyond Mars' orbit, and into the region of Ceres', Pallas' and Juno's orbit, 132 nodes of these 95 comets, around the sun in all regions, namely 74 ascending and 58 descending. But only three of all these intersection points, are approximately (i.e., astronomically) close to several thousand miles, on the earth's orbit; all others are scattered at various distances among the planetary orbits.*

| | |
|---|---|
| *Between the orbits of Sun and Mercury* | *15* |
| *Mercury and Venus* | *25* |
| *Venus and Earth* | *25* |
| *Earth and Mars* | *42* |
| *And beyond the orbit of Mars up to a distance of 2,800* | *25* |

[Bode then prints a table on recent comets; Fig. 11.76]

**Verzeichniſs von 23 seit dem Jahre 1785 erschienenen Kometen, deren Bahnen berechnet worden.**

Fortsetzung der Tafel in den Memoiren 1787 Seite 361.

| No. | Zeit der Sonnennähe<br>Jahre. Mt. T. | ☊<br>Z. G. M. | Neigung der Bahn.<br>G. M. | Sonnennähe in der Bahn.<br>Z. G. M. | Sonnennähe in der Ekliptik.<br>Z. G. M. | Breite in der Sonnennähe.<br>G. M. | Abstand der Sonnennähe.<br>♁ von ☉ = 1000 | Lauf. |
|---|---|---|---|---|---|---|---|---|
| 73 | 1786 Jul. 7 | ♎ 14 23 | 50 54 | ♍ 9 26 | ♍ 20 36 | 26 24 S. | 410 | V. |
| 74 | 1787 Mai 10 | ♋ 16 52 | 48 16 | ♈ 7 44 | ♈ 3 17 | 47 27 N. | 349 | R. |
| 75 | 1788 Nov. 10 | ♍ 7 11 | 12 28 | ♋ 9 8 | ♋ 9 45 | 16 33 N. | 1063 | R. |
| 76 | 1788 Nov. 20 | ♓ 22 24 | 64 30 | ♈ 22 50 | ♈ 6 36 | 27 12 N. | 757 | V. |
| 77 | 1790 Jan. 15 | ♍ 26 12 | 31 54 | ♊ 0 14 | ♉ 26 22 | 28 22 N. | 753 | R. |
| 78 | 1790 Jan. 28 | ♐ 27 9 | 56 58 | ♋ 21 45 | ♋ 11 10 | 20 25 S. | 1064 | V. |
| 79 | 1790 Mai 21 | ♉ 3 11 | 63 52 | ♑ 3 43 | ♐ 11 7 | 51 25 N. | 798 | R. |
| 80 | 1792 Jan. 13 | ♎ 10 46 | 39 47 | ♉ 6 30 | ♐ 1 5 | 16 8 N. | 1293 | R. |
| 81 | 1792 Dec. 27 | ♑ 13 18 | 49 0 | ♌ 16 | ♌ 6 12 | 24 7 N. | 966 | R. |
| 82 | 1793 Nov. 4 | ♋ 18 29 | 60 21 | ♏ 18 42 | ♐ 8 8 | 48 41 S. | 403 | R. |
| 83 | 1793 Nov. 18 | ♈ 2 20 | 51 56 | ♊ 11 9 | ♉ 29 59 | 47 10 N. | 1504 | V. |
| 84 | 1795 Dec. 15 | ♓ 23 14 | 22 10 | ♍ 10 29 | ♍ 11 24 | 4 47 N. | 243 | V. |
| 85 | 1796 April 2 | ♈ 17 2 | 64 55 | ♎ 12 44 | ♎ 15 12 | 3 54 S. | 1578 | R. |
| 86 | 1797 Jul. 9 | ♒ 29 16 | 50 41 | ♉ 19 27 | ♉ 14 0 | 49 40 S. | 527 | R. |
| 87 | 1798 April 4 | ♌ 2 9 | 43 52 | ♋ 14 59 | ♋ 19 36 | 11 48 S. | 485 | V. |
| 88 | 1798 Dec. 31 | ♐ 9 30 | 42 15 | ♉ 3 35 | ♉ 11 18 | 23 14 S. | 775 | R. |
| 89 | 1799 Sept. 7 | ♋ 9 27 | 50 57 | ♈ 3 39 | ♈ 0 17 | 50 35 N. | 840 | R. |
| 90 | 1799 Dec. 25 | ♌ 26 4[illegible] | 77 2 | ♎ 10 20 | ♍ 8 51 | 42 9 S. | 626 | R. |
| 91 | 1801 Aug. 8 | ♉ 14 28 | 21 20 | ♎ 3 49 | ♎ 5 49 | 13 43 S. | 262 | R. |
| 92 | 1802 Sept. 9 | ♒ 10 17 | 57 0 | ♓ 2 8 | ♒ 22 36 | 18 11 N. | 1094 | V. |
| 93 | 1804 Fbr. 13 | ♍ 26 48 | 56 29 | ♌ 28 45 | ♍ 10 24 | 23 5 S. | 1071 | V. |
| 94 | 1805 Nov. 18 | ♓ 14 37 | 15 37 | ♌ 27 51 | ♌ 28 26 | 4 28 N. | 379 | V. |
| 95 | 1805 Dec. 31 | ♐ 10 35 | 16 30 | ♋ 19 22 | ♋ 18 12 | 10 15 S. | 892 | V. |

**Die Planeten.**

| | | | | | | | | |
|---|---|---|---|---|---|---|---|---|
| | ☉ Aequator | ♊ 8 | 7 30 | | | | | |
| I | Merkur ☿ | ♉ 15 52 | 7 0 | ♊ 14 14 | ♊ 14 3 | 3 19 N. | 308 | V. |
| II | Venus ♀ | ♊ 14 48 | 3 24 | ♌ 8 31 | ♌ 8 28 | 2 44 N. | 718 | V. |
| III | Erde ♁ | | | ♋ 9 22 | ♋ 9 22 | 0 0 | 983 | V. |
| IV | Mars ♂ | ♉ 17 59 | 1 51 | ♓ 2 16 | ♓ 2 17 | 1 48 S. | 1382 | V. |
| V | Ceres ⚳ | ♊ 20 59 | 10 38 | ♌ 26 26 | ♌ 26 3 | 9 32 N. | 2689 | V. |
| VI | Pallas ⚴ | ♍ 22 30 | 34 38 | ♌ 1 3 | ♌ 6 35 | 26 23 S. | 2519 | V. |
| VII | Juno ⚵ | ♍ 21 3 | 13 4 | ♉ 23 11 | ♉ 22 33 | 11 32 S. | 2410 | V. |
| VIII | Jupiter ♃ | ♋ 8 20 | 1 19 | ♈ 11 2 | ♈ 11 2 | 1 19 S. | 4953 | V. |
| IX | Saturnus ♄ | ♋ 21 53 | 2 30 | ♊ 28 56 | ♊ 28 57 | 0 58 S. | 9004 | V. |
| X | Uranus ♅ | ♊ 12 49 | 0 46 | ♍ 17 15 | ♍ 17 15 | 0 46 N. | 18174 | V. |

Fig. 11.76 Data on the 25 comets discovered since 1785

## Observations of the new Planets

*MC,* February 1806, p. 185

*Since the discovery of the three new planets we have been relaying any information regarding the three newly discovered planets, their observations, elements, ephemerides, tables and charts of their paths, we continue to do so. Volume XI of our MC contains the newly improved orbital elements by Dr. Gauss according to all observations available and the ephemerides calculated accordingly, on page 285 the elements X of Ceres' orbit, p. 376 the VIII of Pallas' orbit and on p. 475 the elements V of Juno's orbit and the ephemerides. Maps of the geocentric path of these three planets, sketched by Professor Harding, can be found in volume IX and XII of this journal. Since the last enclosed map of the March issue of 1804, describes Ceres' path only until January 19, 1805, we give at present its continuation until May 18, 1806, as long as this small planet would be visible this year.*

*This year's mild and moist winter has been very unfortunate for observing these small planets, which became visible again, even in regions with usually bright skies. The director of Marseilles Observatory, Mr. Thulis, complains about his usually beautiful sky, and the weather has been so unstable in this region that the barometer was always at the limits: either high with unbearable heat, then quickly very low again accompanied with cold, fog and moisture to such an extent that this change has brought upon an epidemic illness.*

*Since we doubted, and expressed this doubt already in the April issue 1805 on page 383, that it would be at all possible to observe Pallas in Germany because of her great southern declination, I asked Mr. Thulis, when visiting him in Marseilles, to take good care of this planet this year because of the fortunate location of his observatory and to follow it. But on January 21 I received a letter from him, telling me that he succeeded only once, on January 14, to observe this planet. He wrote: "I saw Pallas for the first time on the evening of January 14 at the Dollond telescope of our parallactic machine; until then the weather had not permitted to search for this planet; and if I had the satisfaction of observing it, I owe it to your attention to send me its ephemerides as well as those of Ceres and Juno" – I observed it the same evening at the passage instrument; but what difference to the telescope by Dollond! I could hardly see her in the one, whereas I saw it very good in the other even when illuminating much the hairs: I estimated her as a star of $8^{th}$ or $9^{th}$ magnitude in one and of $6^{th}$ magnitude in the other. I volunteer to observe her as often as I can, as well as Ceres and Juno.*

*We received similar news from Viviers, Avignon, Brunswick, Bremen, Goettingen, Buda* etc. *Dr. Gauss wrote: "I looked for Pallas and Juno in the mornings, but in vain, probably due to not quite clear skies." On last year's Dec. 28 Professor Harding wrote from Goettingen: "It hardly seems believable, when I am stating that during my entire stay here it has been impossible to obtain some observations, the sky was almost constantly overcast and hardly ten times was I able to observe the Sun at the mural quadrant. Exactly this constant bad weather rendered it impossible to see Pallas at the meridian or Pallas beyond it and until now I have not been able to obtain confirmation whether a small telescopic star, observed on Nov 10, close to Bode's no. 314 Leonis, really was Juno."*

*No astronomer has observed or sent us an observation; thus we count ourselves lucky that we were able at the Ernestinian Observatory at Seeberg, to make the following meridian observations, namely close to its opposition* (Fig. 11.77).

| 1805. | Mittlere Zeit in Seeberg. | Scheinb. gerade Aufsteigung d. Pallas. | Scheinb. südl. Decl. d. Pallas. |
|---|---|---|---|
| Novbr. 7 | 14U 8′ 40,″389 | 78° 57′ 30,″69 | . . . . |
| 12 | 12 46 54, 382 | 78 25 48, 65 | . . . . |
| 13 | 13 42 29, 003 | 78 18 25, 37 | 29° 24′ 10,″92 |

**Fig. 11.77** Observations of Pallas by Zach in Seeberg from November 7 to 13, 1805

*Thus these observations were extremely welcomed by Dr. Gauss, who still wants to wait further before basing the new elements (IX) on them. According to these observations the error of the VIII elements or the ephemeris given in volume XI of the* MC *on page 376, would be 14 min in RA too small and only a few seconds too great in declination.*

*Juno, the lastly discovered planet, whose elements V by Gauss need the greatest improvement, has not been observed anywhere probably due to its faintness, and we hope to give our readers some observations of it in the next issue, since this planet will com into opposition with the sun on March 10.*

*Dr. Gauss relayed the following observations of Ceres in Brunswick* (Fig. 11.78):

*Accordingly, the right ascension of the ephemeris of October would be too great by five minutes and the declination too small by one minute.*

*The diligent and astute Professor Harding sent us the following observations of Ceres, the first results of his newly established astronomical career in Goettingen and generally of useful and usable observations since the times of Tobias Mayer from this formerly famous observatory, now fallen into disrepair, and which announce a bright future* (Fig. 11.79).

*According to this the RA of Gauss's ephemeris is too great by more than five minutes and the declination too small by about two minutes.*

*Professor Pasquich in Buda was so fortunate to observe Ceres' opposition with the sun, and we owe him the following observations, which we would like to give here in his own words.*

*"I take advantage of this occasion to send you my poor observation of Ceres at her opposition with the sun. The skies here have been so unfriendly that one could not even think of observing. Only on the 3rd, late in the evening the sky cleared and I was stricken by rheumatism that I would have had good reason not to expose myself to the cold and damp*

| 1805. | | Mittlere Zeit in Braunſchweig. | Scheinb. gerade Aufſteig. d. Ceres. | Scheinb. nördl. Abweich. der Ceres. |
|---|---|---|---|---|
| Octobr. | 24 | 11u 22′ 32″ | 109° 28′ 29″ | 23° 27′ 0″ |
| | 28 | 11 31 22 | 110 7 7 | 23 36 59 |
| Novbr. | 14 | 9 22 30 | 111 44 56 | 24 32 1 |
| | 18 | 0 0 37 | 111 50 48 | 24 48 47 |

**Fig. 11.78** Observations of Ceres by Gauss from October 24 to November 18, 1805

| 1805. | | Mittlere Zeit in Göttingen. | Scheinb. gerade Aufſteig. d. Ceres. | Scheinb. nördl. Abweichung d. Ceres. |
|---|---|---|---|---|
| Novbr. | 13 | 11u 4′ 3,″5 | 111° 42′ 34,″4 | 24° 29′ 15,″6 |
| | 14 | 9 22 39, 6 | 111 44 56, 5 | 24 32 28, 2 |
| | 16 | 11 29 19, 6 | 111 49 3, 9 | 24 40 35, 7 |
| | 18 | 10 33 59, 3 | 111 50 59, 4 | 24 58 48, 5 |
| Decbr. | 2 | 10 22 55, 2 | 111 14 6, 5 | 26 1 19, 2 |
| im Meridian | | 14 38 17, 8 | 111 13 11, 6 | 26 2 50, 3 |
| | 10 | 10 7 57, 6 | 110 12 33, 4 | 26 48 53, 6 |
| | 12 | 11 40 42, 4 | 109 53 25, 9 | 27 1 4, 6 |
| | 21 | 10 3 47, 8 | 108 7 27, 8 | 27 57 56, 7 |

**Fig. 11.79** Observations of Ceres by Harding from November 13 to December 21, 1805

*draught, which is omnipresent at our observatory, which is open from all sides. I could not be replaced by our adjunct Huliman, because he was absent; so I walked to the mural quadrant, made my observation as good as possible, and received in return the permission to stay in bed the next day, with such a swollen face I was hardly recognizable. I only mention these circumstances out of duty; they might serve for a proper appreciation of my observation; and at the same time show the reason why the other observation of the fourth was made by Adjunct Huliman. Afterwards it was only possible to observe Ceres on the 22nd which I did; I will connect this observation with future ones.*

*Those two observations depend on the positions of the fixed stars Xi Aurigae and 28 Geminorum: my observation of the third is based on both; that of the fourth is based on the fixed star alone. With a precession of 50."3606 and a parallax of 1."7 the following results were found from the observations made* (Fig. 11.80):

*With this and the inclination of the ecliptic 23° 27′ 54."3, aberration −7."78, nutation −12. "86 in longitude, and nutation +2."86 in latitude, I calculated the true longitude and latitude as follows* (Fig. 11.81):

*The Sun's longitude from the mean aequinoctio +20" for the aberration, plus the hourly motion, I determined for the moment of observation on January 3 according to the new solar tables by Baron von Zach; the hourly motion of Ceres in longitude was derived directly from the observed longitudes of the third and fourth and I based my calculation of the opposition itself on these elements, as follows here* (Fig. 11.82):

*Thus the error of Gauss ephemeris seems to be increasing, here the error at the beginning of January would be too great in RA by more than six and a half minutes and too small by three minutes in declination."*

## Observations of the new Planets

*MC*, March 1806, p. 313

*Since no further observations of Pallas have become known until now than the three, made at Seeberg Observatory, mentioned in the last issue on p. 189, of Juno none at all, we are making haste to give our readers the following news.*

*Dr. Gauss saw Pallas on February 13 again for the first time. But since this observation was only made once and for the purpose to distinguish the planet by its motion on the next*

| 1806. | Mittlere Zeit in Ofen. | Scheinb. gerade Auffteig. d. Ceres. | Scheinb. nördl. Abweich. d. Cer. |
|---|---|---|---|
| Jan. 3 | 12U 8′ 5,″23 | 104°. 53′ 21,″7 | 29° 16′ 7,″2 |
| — 4 | 12 3 7, 21 | 104 37 46, 1 | 29 21 21, 4 |

**Fig. 11.80** Observations of Ceres by Pasquich on January 3 and 4, 1806

**Fig. 11.81** The true latitude and longitude of Ceres

| 1806. | Beobacht. wahre Länge d. Ceres. | Beob. wahre nördl. Breite d. Ceres |
|---|---|---|
| Jan. 3 | 103° 1′ 49,″46 | 6° 28′ 34,″9 |
| — 4 | 102 47 40, 36 | 6 22 24, 2 |

| | | | | | |
|---|---|---|---|---|---|
| 3 Jan. Sonne | 9^z | 12° | 58′ | 58,″059 | |
| Ceres | 9 | 13 | 1 | 49, 460 | |
| Alſo bis zur Oppoſition | | | 2′ | 51,″401 | = 171,″401 |
| Stündl. Beweg. d. Sonne | | | = 2 | 32, 910 | = 152,″910 |
| — — — Ceres | | | = 0 | 35, 502 | |
| Daraus folgt die Zeit bis zur Oppoſition | = | 0^U | 54′ | 34,″97 | |
| Beob. am 3 Januar . . . . | = | 12 | 8 | 5, 23 | |
| Mittlere Zeit der Oppoſition . . | | 13^U | 2′ | 40,″20 | |
| Mit der geo-helioc. Länge . . | | 103° | 1′ | 17,″2 | |
| geocentriſche Breite . . | | 6 | 28 | 43, 6 | |
| heliocentriſche Breite . . | | 4 | 1 | 39, 3 | |

**Fig. 11.82** Elements used by Pasquich in his calculations

*clear evening from the neighbouring stars, which had also to be noticed due to the uncertainty, he did not send it to us because of its lack of accuracy. On the following evening, February 14, it was easy for him to discern Pallas which had moved and appeared as a* $7^{th}$ *or* $8^{th}$ *magnitude star; he compared her that evening with 54 Eridani, and on the following evenings with stars of the* Histoire celeste. *Here are those observations:*

| *1806* | *Mean time in Brunswick* | *Apparent RA* | *Apparent Dec* |
|---|---|---|---|
| *Febr. 14* | *8 h 11′ 16″* | *70° 16′ 31″* | *19° 59′ 13″* |
| *16* | *7 42 28* | *70 42 39* | *19 20 44* |
| *17* | *8 52 38* | *70 56 44* | *19 1 8* |
| *20* | *7 49 35* | *71 39 2* | *18 5 0* |

*According to these observations, Dr. Gauss's ephemeris of* MC *vol. XI, p. 376 gives the RA of this planet too small by 27 min, the declination too great by 5 min: "It is strange," Dr, Gauss said "that the first difference falls to the opposite side, like according to your observations of November last year...; (see last issue, p. 189) the extent of the error does now show, as I believe, the influence of the perturbations."*

*We would like to add to this news, another, certainly just as pleasant to our astronomical readers, that Dr. Gauss informed us that he is currently mainly working on different aspects of his method to determine planetary orbits and soon hopes to work seriously on the completion and publication of his own work dealing with this subject.*

## Observations of the new Planets

*MC*, July 1806, p. 89

[First part of this paper, dealing with Juno, will be in the next book in this series.]

*Only a few observations were made this year of Pallas due to its low position. At Milan the following were made at an 8-foot mural quadrant by Ramsden. The planet was compared with the stars* $\varepsilon$ *in the Hare* [Lepus], $\zeta$ *in the Griffel* [Caelum], *397, v′, 433 and 441 in the Dove* [Columba] *and the positions were taken from Piazzi's star catalogue. With this the following positions of Pallas were obtained* (Fig. 11.83):

*These observations, as well as those of Juno at the equatorial sector, were made and calculated by the young and promising astronomer Carlini, whom we mentioned in volume*

**Fig. 11.83** Observations of Pallas by Carlini from November 5, 1805, to January 15, 1806

| Zeit der Beob. | Mittl. Zeit im Mayland. | Scheinb. beob. ger. Aufsteig. der ⚴. | Scheinb. beob. südl. Abweich. der ⚴. |
|---|---|---|---|
| 1805 Novb. 5 | 14$^h$ 14′ 4″ | 78° 20′ 37,″8 | 27° 16′ 56,″7 |
| — 8 | 14 1 15 | 78 4 46, 7 | 28 5 48, 8 |
| — 11 | 13 48 6 | 77 45 0, 9 | 28 51 45, 5 |
| — 14 | 13 34 46 | 77 21 36, 6 | 29 35 4, 2 |
| — 17 | 13 21 11 | 76 54 48, 2 | 30 14 55, 6 |
| — 21 | 13 2 45 | 76 14 12, 1 | 31 2 21, 0 |
| — 23 | 12 53 25 | 75 52 5, 4 | 31 25 13, 7 |
| — 24 | 12 48 43 | 75 40 35, 1 | 31 33 27, 2 |
| Decb. 3 | 12 5 33 | 73 48 24, 6 | 32 39 57, 5 |
| — 4 | 12 1 4 | 73 35 17, 8 | 32 44 48, 3 |
| — 5 | 11 56 16 | 73 22 5, 2 | 32 49 11, 1 |
| — 6 | 11 51 27 | 73 8 48, 8 | 32 52 44, 3 |
| 1806 Jan. 6 | 9 27 53 | 67 42 51, 1 | 30 55 22, 3 |
| — 7 | 9 25 37 | 67 37 47, 3 | 30 23 34, 3 |
| — 15 | 8 50 36 | 67 14 11, 1 | 28 38 8, 3 |

*XI of the MC on page 94. He calculated the longitudes and latitudes from eight observations close to the opposition and the apparent inclination of the ecliptic 23° 27′ 55.″4, and compared these with the elements VIII of Mr. Gauss (*MC, *vol. XI, p. 378) and also used the latest solar tables by Delambre. With those he obtained the following results* (Fig. 11.84):

*Since here are missing the observations of the time of the opposition, one cannot obtain the heliocentric longitude directly. Thus Mr. Carlini did the following. In order to determine the mean heliocentric error of Gauss's elements VIII, he assumed the distance of the planet from the sun as known. He could do this without noticeable error with Gauss's elements, although the annual parallax is very small. He derived this from the observed elongation and from this the heliocentric longitude, whereby he found the following results* (Fig. 11.85).

*Professor Pasquich at Buda Observatory communicated the following observations of Ceres made there* (Fig. 11.86).

*Professor Pasquich often compared this planet with the stars 62 s and ζ in Gemini and 44X in Auriga, but suspects an erroneous position of one or both. We observed the first very often and found its RA for 1800 in our Tabul. Spec. Aberrat. et Nutat. p. 106 (see also* Conn. des tems, *Année VIII page 397) to be 109° 3′ 19″, only 1″ different from Piazzi's determination. Also the right ascension of X Auriga matches Bradley's determination within 4″ of arc; consequently, positions of both stars can be assumed as accurate.*

**Fig. 11.84** Carlini's comparison of his Pallas data with Elements VIII of Gauss

| 1805. | Beob. Länge der ⚴. | Beob. südl. Breite der ⚴. | Fehler der VIII Elem. in Länge. | Fehler der VIII Elem. in Breite. |
|---|---|---|---|---|
| Novb. 17 | 71° 5′ 27,″0 | 52° 51′ 52,″3 | + 47′ 17,″5 | + 8′ 26,″2 |
| — 21 | 69 55 55, 6 | 53 33 1, 6 | + 48 11, 3 | + 9 8, 4 |
| — 23 | 69 18 22, 3 | 53 50 32, 5 | + 48 30, 1 | + 9 44, 5 |
| — 24 | 68 59 3, 2 | 53 59 0, 2 | + 48 46, 7 | + 9 39, 1 |
| Decb. 3 | 65 57 59, 2 | 54 47 16, 4 | + 49 15, 9 | + 11 42, 8 |
| — 4 | 65 37 54, 4 | 54 49 52, 9 | + 49 8, 9 | + 11 54, 9 |
| — 5 | 65 17 9, 8 | 54 51 59, 4 | + 49 11, 4 | + 12 5, 3 |
| — 6 | 64 56 59, 2 | 54 53 14, 8 | + 49 5, 7 | + 12 29, 8 |

| 1805. | | Elongation der ⚴. | jährliche Parallaxe der ⚴. | helioc. Länge der ⚴. | heliocent. Fehler der VIII Elem. |
|---|---|---|---|---|---|
| Novb. | 17 | 164° 11′ 37,″9 | 8° 7′ 26,″4 | 62° 58′ 0,″6 | +23′ 28,″5 |
| — | 21 | 169 23 24, 6 | 5 30 41, 9 | 64 24 53, 7 | +23 22, 8 |
| — | 23 | 172 1 42, 6 | 4 9 34, 6 | 65 8 47, 7 | +23 19, 7 |
| — | 24 | 173 21 36, 1 | 3 28 19, 6 | 65 30 43, 6 | +23 22, 5 |
| Decb. | 3 | 185 28 30, 1 | 2 55 50, 7 | 68 51 49, 9 | +23 14, 2 |
| — | 4 | 186 49 37, 9 | 3 36 54, 8 | 69 14 29, 2 | +23 12, 8 |
| — | 5 | 188 10 45, 4 | 4 20 0, 0 | 69 57 9, 8 | +23 14, 0 |
| — | 6 | 189 31 40, 9 | 5 2 58, 2 | 69 59 57, 4 | +23 13, 8 |
| Mittlerer Fehler der heliocentrifchen Länge zur Zeit des Gegenfcheins diefes Planeten | | | | | +23′ 18,″3 |

Fig. 11.85 Observed elongation and the heliocentric longitude of Pallas, derived by Carlini

| 1806. | | Mittlere Zeit in Ofen. | Beob. fcheinb. ger. Auffteigung der ⚳. | Beob. nördl. Scheinb. Abweich. der ⚳. |
|---|---|---|---|---|
| Jan. | 22 | 10ᵘ 34′ 39,″7 | 100° 11′ 42,″2 | . . . . |
| — | 29 | 10 1 46, 4 | 98 51 2, 6 | 30° 58′ 2,″7 |
| — | 30 | 9 57 10, 5 | 98 41 2, 6 | 31 0 31, 8 |
| — | 31 | 9 52 37, 2 | 98 31 38, 0 | 31 2 50, 1 |
| März. | 3 | 7 50 23, 9 | 97 27 26, 9 | 31 26 20, 5 |
| — | 11 | 7 19 17, 3 | 98 31 47, 6 | . . . . |
| — | 14 | 7 9 26, 2 | 99 1 0, 9 | 31 20 0, 4 |

Fig. 11.86 Observations of Ceres by Pasquich from January 22 to March 14, 1806

## Observations of the new Planets by Gauss

MC, July issue, p. 187.

Brunswick, July 8, 1806

*With the help of the kindly sent observations from Milan of the planet Pallas, I calculated new elements (IX) of this planetary orbit, which correspond as accurately as possible to Oriani's observations and derived from these the opposition, which had not been possible in such a good way with the elements VIII, which deviate strongly; namely:*

*1805 Nov. 29 11 h 18′ 17″ mean time in Seeberg*

*longitude: 67° 20′ 42.″9*
*south geocentric latitude: 54° 30′ 54.″9*

*I based my new elements for an ephemeris for the next appearance of Pallas on the three oppositions of 1803, 1804 and 1805 and the observations of 1802. It appeared that these different positions could no longer be described with an ellipse, which is doubtless due to the yet neglected perturbations, but this discordance still shows as a trifle, which can hardly surmount one minute. Notably from the observations of 1803 follows an inclination of the orbit greater by 44″ and from those of 1805 smaller by the average of the following elements and the node appears about 45″ further back according to the observations of 1802 and according to those of 1804 as much but further forward than in the assumed average. Here are the new elements (IX) on which the following ephemeris is based* (Fig. 11.87):

*According to these elements the next opposition would occur in 1807 on May 4 at 9 in 223° 53′ longitude and 42° 18′ north. geoc. latitude. Time will tell if this calculation is correct. I sent Mr. Harding my ephemeris at once; he will draw a map accordingly and I hope you*

| | | | | |
|---|---|---|---|---|
| Epoche im Meridian von Seeberg. | 1802 = | 143° | 24' | 1,"1 |
| | 1803 = | 221 | 52 | 56, 4 |
| | 1804 = | 299 | 54 | 42, 5 |
| | 1805 = | 18 | 3 | 37, 8 |
| | 1806 = | 96 | 12 | 33, 2 |
| | 1807 = | 174 | 21 | 18, 5 |

| | |
|---|---|
| Tägliche mittlere tropiſche Bewegung | = 770,"7816 |
| Länge der Sonnenferne 1805 | = 301° 11' 31,"1 |
| Länge des aufſteigend. Knotens 1805 | = 172 30 5, 8 |
| Excentricität . . . . . | 0,2453840 |
| Logar. der mittlern Entfernung | = 0,4421021 |
| Neigung der Bahn . . . | 34° 37' 8,"1 |

**Fig. 11.87** Elements IX of Pallas by Gauss

*will present it in your MC soon.* [Table of the geocentric positions of Pallas for 1806 and 1807 follows.]

## Extract of a letter by Dr. Gauss

*MC* 15, April 1807, p. 577

*Brunswick, March 10, 1807*

*It is my honor to send you my first observations of Pallas of this year. The bad weather has so far hindered their continuation* (Fig. 11.88).

*The first two observations are less reliable, especially regarding the declinations; but the third, made under favourable circumstances, is as accurate as my means permit. For a comparison served on the 5*$^{th}$ *of March two 9*$^{th}$ *magnitude stars of the* Hist. cel., *whose apparent positions were assumed as follows* (Fig. 11.89):

*My ephemeris gives accordingly the declination accurately, the RA's too great by 12 to 13 min; the latter difference seems to increase. Sincere thanks for the observations of Ceres of 1806 from Milan; thus I now have also this opposition with desired accuracy. I only find that the declination on January 4 is too small by 1 min and should be 29° 21' 29."3. This is no typo for also the latitude is too small and Mr. Carlini must have erred also*

| 1807. | Mittl. Zeit. | Scheinb. gér. Aufſteigung ⚴ | Scheinb. nördliche Abw. ⚴ |
|---|---|---|---|
| Februar 28 | 11ᵘ 55' 42" | 236° 48' 6" | 8° 10' 15" |
| März 3 | 12 55 25 | 237 15 5 | 8 54 19 |
| — 5 | 12 22 2 | 237 30 41 | 9 24 52 |

**Fig. 11.88** Observations of Pallas by Gauss from February 28 to March 5, 1807

**Fig. 11.89** Positions of comparison stars used by Gauss

| Gerade Aufſteig. | Nördl. Abweich. |
|---|---|
| 237° 19' 12" | 9° 27' 50" |
| 237 31 13 | 9 28 8 |

*coincidentally by one minute in latitude, calculated from the elements. Please advise this adept astronomer of this at the next opportunity. My last result almost matches that calculated by Mr. Carlini. I find the opposition on January 3 1806 12 h 20′ 12″ mt in Milan in 103° 1′ 9.″4 true longitude and 6° 28′ 32.″4 geoc. northern latitude. I will wait to use this result until this year's opposition has been observed.*

## Continued Reports Regarding the Three Planets – Ceres, Pallas and Juno

*MC,* June 1807, p. 588

*This year, we have supplied our readers reports of these three new planets in the February issue of the* Monat. Corr. *p. 158, in the April issue p. 377 and the May issue p. 509. Since then, we have received further observations of these planets.*

*Professor Harding in Goettingen observed Ceres, as now follows* (Fig. 11.90):

*Dr. Olbers in Bremen, although occupied with his Vesta, did not neglect her modest sister, Pallas, which reveals herself as a magnitude 8 star. He had the goodness of sending us the following observations* (Fig. 11.91):

*Herr Bessel in Lilienthal, as well as Herr Oriani in Milan, promise us further reports of these three planets; the Milanese astronomers find Juno extremely small and find the latest elements by Gauss of this planet very precise, through the help of which they were able to locate this planetary atom again without much trouble. Herr Carlini likewise noted the error of one minute in the declination of Ceres indicated in the April issue of the* Monat. Corr. *p. 378, and has already improved this error in the Milanese Ephemeris which is currently being printed. In Goettingen, Professor Harding hasn't been able to see Pallas;*

| 1807. | | Mittlere Zeit in Göttingen. | Scheinb. gerade Auffteigung ⚳. | Scheinb. füdl. Abweich. ⚳. |
|---|---|---|---|---|
| April | 26 | 12ᵘ 42′ 57,″600 | 224° 33′ 15,″79 | 5° 29′ 27,″8 |
| | 27 | 12 37 11, 208:: | 224 20 0, 15:: | 5 28 17, 7:: |
| | 30 | 12 22 42, 068 | 223 39 32, 02 | 5 23 49, 6 |
| May | 1 | 12 17 52, 270 | 223 26 0, 49 | 5 22 46, 7 |
| | 2 | 12 13 1, 569 | 223 12 17, 65 | 5 21 37, 9 |
| | 4 | 12 3 21, 329:: | 222 45 6, 56:: | 5 19 26, 4:: |
| | 5 | 11 58 30, 381 | 222 31 22, 83 | 5 18 50, 5 |
| | 6 | 11 53 40, 513 | 222 17 47, 09 | 5 17 54, 1 |
| | 7 | 11 48 49, 856 | 222 4 3, 63:: | 5 17 319 |

**Fig. 11.90** Observations of Ceres by Harding from April 26 to May 7, 1807

| 1807. | | Mittl. Zeit in Bremen. | Scheinb. ger. Auffteig. ⚴. | Scheinb. nördl. Abweich. ⚴. |
|---|---|---|---|---|
| März | 12 | 11ᵘ 17′ 18″ | 238° 15′ 10″ | 11° 14′ 16″ |
| | 13 | 11 19 53 | 238 19 52 | 11 31 4 |
| | 15 | 12 17 16 | 238 28 37 | 12 3 43 |
| April | 17 | 10 35 36 | 237 1 32 | 20 56 52 |
| | 18 | 10 22 16 | 236 52 53 | 21 11 1 |
| | 27 | 9 6 58 | 235 1 35 | 23 5 26 |

**Fig. 11.91** Observations of Pallas by Olbers from March 12 to April 27, 1807

*the telescope of the Wall Quadrant there is too weak for this weak spark of light. Several times, he waited for it in vain, or observed a fixed star in its place.*

## Continued Observations of the New Planets

*MC*, December 1807, p. 548

*The observations on Pallas during the past years have been so difficult due to her faintness that only a few astronomers were able to work on this matter and consequently, only a few good observations on this small planet were obtained. The wonderful mural quadrant by Ramsden and the transparent Italian atmosphere made it possible for the diligent astronomers Oriani and Carlini at Milan to observe her successfully and it is probably of interest to our astronomical readers if we publish a short excerpt of the treatise* Observationi di Pallade Presso l'oppositione, fatte negli anni 1803, 1805 e 1807 *da Francesco Carlini which is part of the* Effemeridi astronomiche di Milano per l'anno bisestile *[bisestile = leap year] 1808,* calcolate da Francesco Carlini *which was published lately. Several of the observations and especially the oppositions of 1803 and 1805 are already known to our readers from former volumes and we mention those when Carlini used those for calculating other elements. We will find elements of Pallas which Carlini himself developed but since those were given in an older issue of the* Milan Ephemerides, *which we do not have handy, we cannot judge on what kind of method and observations those elements are based. The elements are the following:*

| *Epoch 1803 Dec 31 mean time in Milan 299° 41′ 39″* | |
|---|---|
| *Mean diurnal tropical* | *12′ 50.″8438* |
| *Log of the mean distance* | *0.4420788* |

| *Aphelion 1804 301° 10′ 53″, annual change +106.″2* | | |
|---|---|---|
| ☊ | *172 28 57* | *−17.2* |
| *Inclination* | *34 37 13* | *+ 0.81* |
| *Eccentricity* | *0.2454266* | *−0.00000661* |

*There was a very good correspondence of these elements with the observations at the time of the opposition of 1803. We would like to stress the following* (Fig. 11.92).

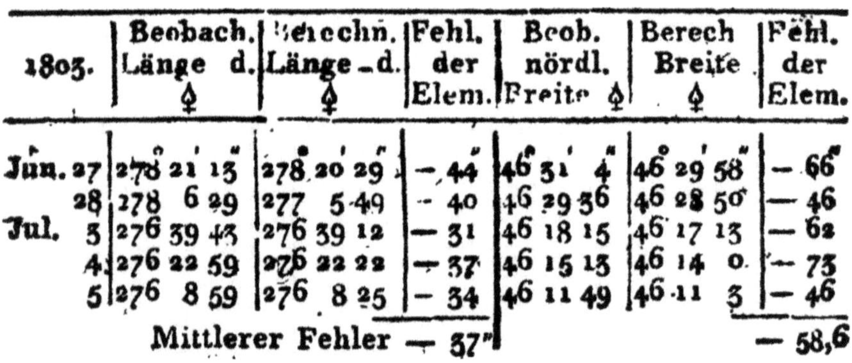

| 1803. | Beobach. Länge d. ⚴ | Berechn. Länge d. ⚴ | Fehl. der Elem. | Beob. nördl. Breite ⚴ | Berech Breite ⚴ | Fehl. der Elem. |
|---|---|---|---|---|---|---|
| Jun. 27 | 278° 21′ 13″ | 278° 20′ 29″ | − 44″ | 46° 31′ 4″ | 46° 29′ 58″ | − 66″ |
| 28 | 278 6 29 | 277 5 49 | − 40 | 46 29 36 | 46 28 50 | − 46 |
| Jul. 3 | 276 39 43 | 276 39 12 | − 31 | 46 18 15 | 46 17 13 | − 62 |
| 4 | 276 22 59 | 276 22 22 | − 37 | 46 15 13 | 46 14 0 | − 73 |
| 5 | 276 8 59 | 276 8 25 | − 34 | 46 11 49 | 46 11 3 | − 46 |
| Mittlerer Fehler | | | − 37″ | | | − 58,6 |

**Fig. 11.92** Difference between observation and the elements of Pallas in 1803

*If we calculate for 0 h June 30 Pallas' longitude and apply the mean error, we obtain 277° 39′ 49″. Longitude of the Sun according to Delambre's latest solar Tables 97° 38′ 23″, relative motion = 1° 14′ 10″ and according to this:*

*Time of opposition, June 1803 0 h 28′ 0″*

*For that time was*

*geocentric longitude of Pallas 277° 39′ 30″*

*— latitude – —   46° 27′ 5″ north*

*For 1804 was found with help of the latest solar tables:*

*Time of*

*⊙ ⚴ 1804 30 Aug 4 h 56′ 0″*

*and for this time*

*geocentric longitude of Pallas 337° 0′ 35.″6*

*— latitude – —   15° 1′ 47.″8 north*

*The comparison of the observations on Pallas from Milan at the opposition of 1805 with Gauss' elements VIII were published in vol. XIV, p. 91, and we consequently give the comparison of those elements with Carlini's elements* (Fig. 11.93):

*These corrections give:*

*Time of*

*⊙ ⚴ 1805 Nov. 29 11 h 14′*

*and for that time was*

*geocentric longitude of ⚴ 67° 20′ 44.″5*

*— latitude – –   54° 31′ 2.″6 south*

| 1805 | | Beobach. Länge d. ⚴ | Berechn. Länge d. ⚴ | Fehler der Elem. | Beobachtete Breite. | Berechnete Breite. | Fehler der Elemente. |
|---|---|---|---|---|---|---|---|
| | | ° ′ ″ | ° ′ ″ | ″ | ° ′ ″ | ° ′ ″ | ′ ″ |
| Novbr. | 17 | 71 5 27,0 | 71 5 19,6 | — 7,4 | 52 51 32,3 | 52 52 50,9 | + 1 18,6 |
| | 21 | 69 55 35,6 | 69 55 29,5 | — 6,1 | 53 33 1,6 | 53 34 22,7 | 1 21,1 |
| | 23 | 69 18 22,3 | 69 18 14,2 | — 8,1 | 53 50 32,5 | 53 52 8,5 | 1 36,0 |
| | 24 | 68 59 3,2 | 68 58 58,8 | — 4,4 | 53 59 0,2 | 54 0 14,6 | 1 14,4 |
| Decbr. | 3 | 65 57 59,2 | 65 57 48,3 | — 10,9 | 54 47 16,4 | 54 48 37,5 | 1 21,1 |
| | 4 | 65 37 34,4 | 65 37 21,7 | — 12,7 | 54 49 52,9 | 54 51 13,2 | 1 20,3 |
| | 5 | 65 17 9,8 | 65 17 1,9 | — 7,9 | 54 51 59,4 | 54 53 16,0 | 1 16,6 |
| | 6 | 64 56 59,2 | 64 56 49,0 | — 10,2 | 54 53 14,8 | 54 54 45,3 | 1 30,5 |
| Mittlerer Fehler | | | | — 8″,5 | | | + 1′ 20″,4 |

**Fig. 11.93** Difference between observation and the elements of Pallas in 1805

*Pallas' opposition in 1807 was again completely observed by Carlini. The planet was compared with $\chi$ Serp. and $\gamma$ Coron* (Figs. 11.94 and 11.95).

*From which the following positions of Pallas resulted* (Fig. 11.96):

*Carlini reduced these positions of Pallas with the apparent obliquity = 23° 27' 50" and with applying aberration and nutation to the ecliptic the following errors of the elements were obtained* (Fig. 11.97):

*From this mean error of Pallas' elements and the new solar tables of Delambre follows the time of the opposition:*

| 1807. | Stern im Parall. | Zeit der Culmination. Stern. | Zeit der Culmination. ⚴ | Scheinb. Zen. Dist. Stern. | Scheinb. Zen. Dist. ⚴ |
|---|---|---|---|---|---|
| | | u ′ ″ | u ′ ″ | ° ′ ″ | ° ′ ″ |
| Apr. 10 | κ Serp. | 15 39 24,10 | 15 50 51,24 | 26 43 2,7 | 26 22 56,0 |
| 19 | | 15 39 18,04 | 15 46 1,12 | 26 42 59,5 | 24 0 40,5 |
| 21 | | 15 39 17,92 | 15 44 43,30 | 26 42 58,7 | 23 33 45,7 |
| 26 | γ Cor. | 15 33 47,18 | 15 41 9,80 | 18 33 4,4 | 22 31 7,7 |
| 27 | | 15 33 46,16 | 15 40 24,28 | 18 33 4,2 | 22 19 26,6 |
| 28 | | 15 33 45,32 | 15 39 37,87 | 18 33 2,6 | 22 8 3,5 |
| 29 | | 15 33 44,50 | 15 38 51,25 | 18 33 4,3 | 21 57 4,7 |
| 30 | | 15 33 43,95 | 15 38 3,88 | 18 33 5,2 | 21 46 20,7 |
| May 1 | | 15 33 43,24 | 15 37 15,44 | 18 33 0,3 | 21 35 53,7 |
| 2 | | 15 33 42,70 | 15 36 26,90 | 18 33 1,0 | 21 25 49,0 |

1807.

**Fig. 11.94** Observations of Pallas by Carlini in Milan from April 10 to May 2, 1807

| 1807. | Stern im Parall. | Zeit der Culmination. Stern. | Zeit der Culmination. ⚴ | Scheinb. Zen. Dist. Stern. | Scheinb. Zen. Dist. ⚴ |
|---|---|---|---|---|---|
| | | u ′ ″ | u ′ ″ | ° ′ ″ | ° ′ ″ |
| May 5 | κ Serp. | 15 39 7,50 | 15 33 38,24 | 26 42 59,6 | 21 16 15,0 |
| 8 | | 15 39 6,08 | 15 31 23,45 | 26 42 56,3 | 20 52 55,0 |
| 10 | | 15 39 5,70 | 15 29 46,80 | 26 42 55,3 | 20 17 59,2 |
| 11 | γ Cor. | 18 33 40,65 | 15 23 56 06 | 26 42 57,0 | 20 11 12,7 |

**Fig. 11.95** Data on Pallas and the comparison stars

| 1807. | Mittl. Zeit in Mayland. | Gerade Aufsteig. ⚴ | Scheinb. Abw. ⚴ |
|---|---|---|---|
| April 29 | 13u 11′ 40″ | 234° 56′ 59,″4 | 23° 30′ 41,″0 |
| 30 | 13 6 57 | 234 45 17, 4 | 23 41 26, 0 |
| May 1 | 13 2 14 | 234 33 21, 5 | 23 51 48, 4 |
| 2 | 12 57 30 | 234 21 21, 5 | 24 1 54, 0 |
| 3 | 12 52 45 | 234 9 13, 7 | 24 11 34, 6 |
| 8 | 12 28 58 | 233 7 8, 6 | 24 54 52, 3 |
| 10 | 12 19 25 | 232 41 49, 1 | 25 9 45, 5 |

**Fig. 11.96** Reduced positions of Pallas from April 29 to May 10, 1807

| 1807. | Beob. Länge ⚴ | Berech. Länge ⚴ | Fehler der Elemente. | Beobacht. Breite. | Berechnete Breite. | Fehler der Elemente. |
|---|---|---|---|---|---|---|
| | ° ′ ″ | ° ′ ″ | ′ ″ | ° ′ ″ | ° ′ ″ | ′ ″ |
| April 29 | 225 9 48,3 | 225 32 4 | + 22 15,7 | 41 40 17,7 | 41 47 44,9 | + 7 27,2 |
| 30 | 224 51 39,4 | 225 14 0,4 | 22 21,0 | 41 47 14,0 | 41 54 42,5 | 7 28,5 |
| May 1 | 224 33 26,4 | 224 55 51,2 | 22 24,8 | 41 53 43,2 | 42 1 18,4 | 7 35,2 |
| 2 | 224 15 12,0 | 224 37 40,4 | 22 28,4 | 41 59 54,0 | 42 17 30,3 | 7 36,3 |
| 3 | 223 56 57,6 | 224 19 23,9 | 22 26,3 | 42 5 37,4 | 42 13 19,0 | 7 41,6 |
| 8 | 222 25 56,7 | 222 48 11,9 | 22 15,2 | 42 28 38,0 | 42 36 37,0 | 7 59,0 |
| 10 | 221 48 55,1 | 222 12 11,0 | 22 15,9 | 42 35 14,0 | 42 43 14,6 | 8 0,6 |
| | | Mittlerer Fehler | + 22′ 21,″0 | | | + 7′ 41,″2 |

**Fig. 11.97** Errors of the elements of Pallas

*⚴ ⊙ 1807 May 4 14° 32′ 16″*

*and for that time was*

*geocentric longitude of Pallas 223° 37′ 22″*

*— latitude – — 42 11 27.″8 north*

*With help of these four oppositions of 1803, 1804, 1805 and 1807 and an observation by Olbers of 1802, Carlini tried to rectify his above elements. On March 29, 1802, 8 h 50′ 48″ mean time at Milan, the apparent RA of ⚴ was 184° 36′ 36″, declination 11° 52′ 59″ north and according to this with 23° 28′ 6″ inclination of the ecliptic and 8° 25′ 44″ follow longitude of Pallas 179° 34′ 14″, latitude 12° 47′ 21″, elongation 171° 8′ 30″. Since the elongation is close to 180° Carlini used for a first approximation the radius vector of the elements and found the annual parallax of the planet 3° 47′ 2″, heliocentric longitude for that time 183° 21′ 16″. With the known inclination and the node the positions of the ecliptic were reduced to the orbital positions and the following results were obtained for those five observations* (Fig. 11.98):

*If one assumes for above elements ε correction of the epoch, a correction of the aphelion (in minutes), e correction of the eccentricity (in thousandths of the mean distance), n correction of the mean motion in tenths of seconds, one obtains the following conditional equations, if the epoch of October 15, 1804, is assumed as the mean moment of the observation* (Fig. 11.99):

| | | | | | |
|---|---|---|---|---|---|
| 3 Jan. Sonne | 9^Z | 12° | 58′ | 58,″059 | |
| Ceres | 9 | 13 | 1 | 49, 460 | |
| Alſo bis zur Oppoſition | | | 2′ | 51,″401 | = 171,″401 |
| Stündl. Beweg. d. Sonne | | = 2 | | 32, 910 | = 152,″910 |
| — — — Ceres | | = 0 | | 55, 502 | |
| Daraus folgt die Zeit bis zur Oppoſition | = | 0^U | 54′ | 34,″97 | |
| Beob. am 3 Januar . . . . . | = | 12 | 8 | 5, 23 | |
| Mittlere Zeit der Oppoſition . . | | 13^U | 2′ | 40,″20 | |
| Mit der geo-helioc. Länge . . | | 103° | 1′ | 17,″2 | |
| geocentriſche Breite . . | | 6 | 28 | 43, 6 | |
| heliocentriſche Breite . . | | 4 | 1 | 39, 3 | |

**Fig. 11.98** Elements of Pallas derived by Carlini

80,"47ε − 124,81n − 20,"47a + 41,"69e = − 3"
40, 02ε − 31,55n + 19, 98a + 17, 20e = + 19
41, 45ε − 3,16n + 18, 55a − 21, 28e = 0
87, 64ε + 59,96n − 27, 64a − 36, 87e = + 3
56, 35ε + 87,50n + 3, 65a + 40, 22e = − 750

**Fig. 11.99** Conditional equations

*From these equations follows by regular elimination:*

$e = -6.9202$
$a = -1.590$
$n = 3.571$
$\varepsilon = -1.635$

*and according to this the improved elements:*

| | | |
|---|---|---|
| *Epoch 1803 Dec. 31 0 h 299° 41′ 44″* | | |
| *Aphelium* | *1804* | *301 9 18* |
| *Daily mean tropical Movement* | | |
| | | *12′ 50.″4867* |
| *Eccentricity* | | *0.2447346* |
| *Log. of mean distance* | | *0.4422130* |

*In order to get the correction of the node and the inclination, Carlini calculated from the observed geocentric latitudes and the radius vector determined by these elements the heliocentric latitudes from which the following errors of the elements resulted* (Fig. 11.100):

*π and i are the correction of the node and the inclination given in seconds, from the above observations the following equations can be derived* (Fig. 11.101):

*But if those equations are treated according to Legendre's* Méthode des moindres quarrés [method of least squares], *it follows:*

$$i = +11'', \pi = +29''$$

| Logar. Dist. ♁ a ☉ | Log. Dist. ⚴ a ☉ | Beob. geoc. Breite. | Beob. helioc. Br. | Berechn hel. Br. | Fehl. d. El. |
|---|---|---|---|---|---|
| | | ° ′ ″ | ° ′ ″ | ° ′ ″ | ″ |
| 9,9997620 | 0,3719288 | 12 47 21,0 B | 7 26 4 | 7 25 0 | − 64 |
| 0,0072116 | 0,5231640 | 46 27 5,0 − | 33 41 15 | 33 40 33 | − 42 |
| 0,0058325 | 0,5154962 | 15 1 47,8 − | 10 27 4 | 10 26 5 | − 59 |
| 9,9937389 | 0,3535541 | 54 31 2,6 A. | 33 41 13 | 33 40 53 | − 20 |
| 0,0039870 | 0,4489700 | 42 11 27,8 B. | 28 14 30 | 28 15 53 | + 83 |

**Fig. 11.100** Errors of the elements

$$+0,''2738\,i - 0,''6667\,\pi = +64''$$
$$+0,\ 9837\,i + 0,\ 1249\,\pi = +42$$
$$+0,\ 3810\,i + 0,\ 6430\,\pi = +59$$
$$+0,\ 9873\,i - 0,\ 1249\,\pi = +20$$
$$+0,\ 8922\,i - 0,\ 3368\,\pi = -83$$

hieraus

$$+3,''5216\,i - 0,''3605\,\pi = +102''$$
$$-0,\ 7850\,i + 1,\ 8963\,\pi = +100$$
$$\pi = 68'' \quad i = +36''$$

**Fig. 11.101** Equations to calculate the correction of the node and the inclination

| Beob. Länge in der Ecliptik. | Berechn Länge in der Bahn. | Länge in der Bahn. | Fehler der Elemente. |
|---|---|---|---|
| 183° 21′ 0″ | 185° 36′ 44″ | 185° 38′ 6″ | + 82″ |
| 277 39 30 | 275 3 59 | 275 1 54 | − 125 |
| 337 0 36 | 333 53 22 | 333 54 31 | + 69 |
| 67 20 44 | 69 56 0 | 69 54 51 | − 69 |
| 223 3 22 | 228 56 11 | 228 57 9 | + 58 |

**Fig. 11.102** Errors of the elements

*and accordingly*

| | |
|---|---|
| *Longitude of the node in 1804 172° 29′ 26″* | |
| *Inclination* | *34 37 24* |

*The reduction of the longitudes of the ecliptic to the orbital longitudes can then be found using the corrected elements; and Carlini found for March 29, 1802, (a typo gives May 29 in the ephemerides) the annual parallaxes 3° 46′ 46″, heliocentric longitude of Pallas = 183° 21′ 0″ and then* (Fig. 11.102):

*Carlini derives from these errors of the elements the following conditional equations* (Fig. 11.103):

*and the following improved elements*

| | |
|---|---|
| *Epoch 1803 Dec. 31 0 h in Milan* | *299° 41′ 47″* |
| *Aphelium* | *301 9 54* |
| ☊ | *172 29 26* |

(continued)

(continued)

| | |
|---|---|
| *Inclination* | *34 37 24* |
| *Daily mean tropical Movement* | *12' 50."4922* |
| *Eccentricity* | *0.2446473* |
| *Log. of mean distance* | *0.4422109* |

*These elements differ only slightly from those given earlier by Dr. Gauss (*MC *vol. XIV, p. 188) and they correspond perfectly to Gauss's given position when Carlini says it was impossible to describe all of the observations by an ellipsis. The greatest error of Carlini*'s *elements is of the opposition of 1803. The differences between the calculated and the observed positions for the oppositions of 1802–1807 are as follows* (Fig. 11.104):

*Dr. Gauss, at present professor for astronomy and mathematics at Goettingen, calculated for 1808 an improved ephemeris that follows here. He sent it shortly before he left for Goettingen and wrote: "Please find enclosed an ephemeris for Pallas for next year; I am unable to find the improved elements, according to which it was calculated, which are probably among already packed papers. As soon as I have found them, I will send them."*

## Excerpt of a Letter

by Dr. Gauss

*MC*, July 1809, p. 78

*Goettingen, August 14, 1809*

*What I can share with you today are a few observations of the new planets that I have made in the last few days. I had an excellent opportunity to look for Pallas at times, because last year, I was not able to improve the elements because of a lack of good, 'complete' observations (I still lack reliable declinations and I almost fear that I will be able to use none other than my own and those from Milan on the equatorial sector). We therefore had to make use of the ephemeris this time, which Professor Bode calculated in the 1811 annual according to the elements determined in 1807. Meanwhile, it didn't require a great deal of*

$$80{,}47\epsilon - 124{,}''81\,n - 20{,}47\,a + 41{,}69\,e = -82''$$
$$40{,}02\epsilon - 31{,}55\,n + 19{,}98\,a + 17{,}20\,e = +125$$
$$41{,}45\epsilon - 3{,}16\,n + 18{,}55\,a - 21{,}28\,e = -69$$
$$87{,}64\epsilon + 59{,}96\,n - 27{,}64\,a + 36{,}87\,e = +69$$
$$56{,}35\epsilon + 87{,}50\,n + 3{,}65\,a + 40{,}22\,e = -58$$

**Fig. 11.103** Conditional equations

| Jahre. | 1802. | 1803. | 1804. | 1805. | 1807. |
|---|---|---|---|---|---|
| Fehler in long. | + 33" | — 126" | + 95" | — 43" | + 34" |
| — in latit. | — 78 | — 27 | — 36 | — 13 | + 83 |

**Fig. 11.104** Differences between the calculated and the observed positions for the oppositions of 1802–1807

*effort to find the planets according to this. On August 8, Professor Harding surveyed the area of Pisces where the planet was supposed to be located and on August 9 we immediately recognised it in the form of a magnitude 9 star that had moved noticeably to the south. On August 10, this rediscovery was confirmed. Here are the results of my observations of these days:*

| *1809* | *mean time in Goettingen* | *Apparent RA* | *Apparent N. Dec.* |
|---|---|---|---|
| *Aug. 9* | *11 h 42′ 48″* | *7° 51′ 44″* | *2° 41′ 6″* |
| *10* | *11 9 58* | *7 50 35* | *2 32 37* |

*From this, it can be seen that the astronomical annuals give the correct declinations, but the right ascensions are some 20′ too small.*

*On August 10, I also located Ceres. A few not yet completely reduced observations show at least that my ephemeris in the May issue of* Monat. Corr. *gives the position correctly to within one minute. Ceres is magnitude 8 to 9.*

## The courses of Pallas, Juno, and Vesta in the year 1809

*BAJ,* 1811, p. 208

*I have calculated the course of Pallas using my tables corrected according to the newest 10th elements of Dr. Gauss. This planet is by the end of February in conjunction with the Sun. In April, it has to be searched for again in the morning in the eastern sky in the southern Fish (Pisces) of the Zodiac. It advances during April, May, and June between the Fishes toward the east. There, in July, it comes to a standstill and becomes retrograde. On Sept. 22, it is opposed to the Sun in the tail of the Whale (Cetus). In November, it comes to a standstill at the waterfall of the Water Bearer (Aquarius) and advances again in December.*

[The remainder of this paper is in a future book in this series.]

## Excerpt from a letter

by Professor Gauss
*MC,* March 1810, p. 276
Goettingen, Feb. 23, 1810

*I have just completed a few calculations of the two last oppositions of Pallas, the results of which I have allowed to be printed in #32 of the Goettingen scholarly journal, and which I will now share with you. In 1808, Pallas was observed from only a few observatories, and then only incompletely – for this reason, no correction of the elements has been possible since 1807. At this observatory, the great dimness of the planet has made all observations on the wall quadrant impossible, and those few observations conducted on a circular micrometer due to special circumstances, can make no claim to great exactness, and your locations of Pallas, precisely observed from the Seeberg observatory on a Passage instrument, were not complete as the declinations were also missing. The determinations made in Milan on an equatorial sector only begin on August 22, and can therefore not be reasonably used for the opposition. Only through the* Astronomical Yearbook *for 1812 did something of the complete observations conducted by Herr David in Prague at the time of the opposition become known to me. From this source, I used the declinations in order to derive the opposition in conjunction with the right ascensions determined at the Seeberg Observatory. Unfortunately, the investigation of those observations from Prague reveal that they do not match very well, since, for instance, the different declinations give results which deviate nearly one minute from each other. Meanwhile, since no better declinations were available, nothing more remained to do than to use what was available as well as possible, and thus arose the fifth result for the 1808 opposition that has been observed to date:*

| | Opposition of Pallas 1808 |
|---|---|
| *26 July 21 h 17′ 32″* | *Meridian of Goettingen* |
| *true longitude* | *304° 2′ 59.″7* |
| *true geocentric latitude* | *37 43 54 north* |

*In 1809, the observations of Pallas made on this observatory appear to have been the earliest; other than these, I used your Seeberg observations and Herr Bouvard's observations conducted in Paris. Since the latter would perhaps otherwise only become known in a number of years, I will cite them here* (Fig. 11.105):

*The result from these observations for the sixth opposition of Pallas is the following:*

*Time of opposition 1809 22 September*

*16 h 10′ 20″ Goettingen meridian*

*True [actual] longitude of Pallas 359° 40′ 4.″4*

*true [actual] geocentric latitude 7 22 10.1 south.*

*I have also based a number of other results on this with regards the elements and am now occupied with a number of calculations regarding this. I have set this aside, together with an analysis of various tricks which I have already used for many years in applying the method (of least squares) developed in the third chapter of the second book of my* Theoria, *for an essay for our society, of which I will send you an excerpt once I have completed the work.*

*I am greatly indebted to Herr Oriani for noting the printing errors in my* Theoria. *He is entirely correct that I had forgotten to add on p. 129 that B, B′, B″ = 0 must be assumed if the equation of condition, in which equation [7] is not applicable, is to have the identical form. Other than that, it is clear that even if B, B′, B″ does not = 0, equation [7] can still not be applicable if the twelve-term-equation developed by Oriani by chance does = 0 or becomes very small. That Euler has already found the theorem from which the theorem ascribed by me to Laplace can very easily be deduced, had already occurred to me, but since the passage on p. 212 had already been printed, I did not wish to set it among the errata since Laplace still was the first to use the above theorem in its initial form. I had already noted most of the printing errors indicated by Oriani. Here are three more overlooked by him:*

**Fig. 11.105** Observations in Paris of Pallas from September 12 to October 13, 1809

| 1809 | | Mittlere Zeit in Paris | Scheinbare AR. der Pallas | Scheinb. südl. Abweich. |
|---|---|---|---|---|
| Septbr. | 12 | 12 51′ 39,″6 | 4° 30 25,″80 | 4° 20′ 10,″4 |
| | 15 | 12 37 41, 7 | 3 57 45, 75 | 5 5 16, 4 |
| | 28 | 11 36 52, 6 | 1 31 50, 25 | 8 20 23, 4 |
| | 29 | 11 32 11, 1 | 1 20 24, 30 | 8 34 57, 4 |
| Octbr. | 2 | 11 18 7, 1 | 0 46 15, 30 | 9 18 22, 5 |
| | 5 | 11 4 5, 9 | 0 12 35, 25 | 10 0 21, 0 |
| | 6 | 10 59 25, 5 | 0 1 37, 50 | 10 14 6, 9 |
| | 7 | 10 54 46, 2 | 359 50 44, 55 | 10 27 36, 8 |
| | 8 | 10 50 7, 1 | 359 39 54, 90 | 10 40 1, 5 |
| | 9 | 10 45 28, 7 | 359 29 16, 50 | 10 54 5, 5 |
| | 13 | 10 27 2, 2 | 358 48 26, 25 | 11 54 59, 5 |

| *Page 1 line 4 v.u. instead of inverse l. composite* | | | |
|---|---|---|---|
| *65* | 2 | – | *cos θ l.cos ζ* |
| *195* | *16* | – | *λ''' l. = λ'''* |

*I can't find the printing error – 380° instead of 180° – on p. 4 of my copy. Instead of the two printing errors on p. 83, one could be indicated; namely ζ + 45 instead of ζ − 45°.*

[The next page contains the table: Printing errors in Dr. Gauss's Theoria motus corporum coelestium, etc., Hamburg 1809 by Senator [Bar.] Oriani]

## Continued Reports Regarding the New Primary Planet Pallas

*MC*, December 1810, p. 588

*A great work regarding the theory of Pallas, newly completed by Professor Gauss, the results of which have been partially made public in the* Goettinger gelehrten Anzeigen *1810 #198 and partly in a private letter to us, now permit us to share the very remarkable results regarding the path of Pallas with our astronomical readers.*

*For three years, Professor Gauss conducted no new observations of the path of Pallas. The observations of 1808 were very scanty and poor given the extremely low light intensity of the planet. These observations were made public only later since it did not seem worth the trouble of improving the elements based on these observations. Only after receiving the observations of 1809, conducted by Bouvard from the Royal observatory in Paris, did he calculate the opposition of 1808 (*Goett. gel. Anz. 1810, *p. 32) along with that of 1809. The further discussion of all six oppositions observed to date gave the result that an elliptical path is no longer sufficient to represent them adequately enough; a result of the great perturbations that Pallas suffers from the remaining planets and especially Jupiter. Professor Gauss indicated this in two ways. First, he calculated three systems of elliptical elements, each consisting of four oppositions, namely, the first of those of 1803, 04, 05, 07; the second from the oppositions of 1804, 05, 07, 08; the third from the oppositions of 1805, 07, 08, 09; among which only minor differences should have been revealed had the movement been purely elliptical. We will now place the three systems side by side* (Fig. 11.106):

*where*

*A is the epoch of the mean longitude 1803 of Goettingen*
*B " " mean daily tropical movement*
*C " " longitude of the perihelion 1803*
*D " " " " " ascending node 1803*

| | I | II | III |
|---|---|---|---|
| A | 221° 39′ 30,″4 | 221° 34′ 56,″7 | 221° 23′ 24,″6 |
| B | 770,″2143 | 770,″4467 | 770,″9265 |
| C | 121° 3′ 11,″4 | 121° 5′ 22,″1 | 120° 58′ 4,″8 |
| D | 172° 28′ 56,″9 | 172° 28′ 46,″8 | 172° 27′ 52,″4 |
| E | 34° 37′ 41,″0 | 34° 37′ 31,″5 | 34° 36′ 49,″4 |
| F | 0,2450198 | 0,2447624 | 0,2446335 |
| G | 0,4423149 | 0,4422276 | 0,4420473 |

**Fig. 11.106** Three systems of elliptical elements, calculated by Gauss

*E " " inclination of the path*
*F " " eccentricity*
*G " " log. of the semi major axis*

*The second way Professor Gauss demonstrated the influence of the perturbations consists in the calculation of purely elliptical elements that agree 'as closely as possible' to all 'six' oppositions and which notwithstanding these, significantly distance themselves from the single observed positions. We will also include this fourth system of elements* (Fig. 11.107):

*The following table represents the error of these elements* (Fig. 11.108):

*Representing all observations through a pure ellipse as well as the great influence of the perturbations of Jupiter on the path of Pallas was irrefutable represented by this development. The need to calculate these perturbations was now at hand and Gauss did not hesitate a moment to undertake this great, arduous work. A period of a few weeks was sufficient for 'his' methods and 'his' abilities of calculation to complete this long and involved calculation. In a letter dated December 2, 1810, which we will quote here, he informed us of the remarkable results of his work: "I hasten to inform you, dear friend", wrote Professor Gauss, "of an important piece of news. I recently wrote to you that I was working on the calculation of perturbations Pallas had incurred from Jupiter in the years 1803–1811. I have shown in the essay, which I have mentioned several times, how poorly pure elliptical elements may represent all six oppositions observed to date. It was therefore greatly satisfying for me, several days after completing the work on these perturbations to find that in view of the perturbations, all six oppositions corresponded far more exactly than I had been led to expect. Here is the table of deviations* (Fig. 11.109):

*As everyone knows, the latitude for 1808 was very poorly determined.*

*"Since the elements had changed 'most significantly' in 1809 with the heliocentric opposition with Jupiter, I was curious to see how much my ephemeris in the October issue*

**Fig. 11.107** The fourth system of elements for Pallas

| | |
|---|---|
| A | 221° 34' 53,"64 |
| B | 770,'5010 |
| C | 121° 8' 8,"54 |
| D | 172° 28' 12,"43 |
| E | 34° 37' 28,"35 |
| F | 0,2447424 |
| G | 0,4422071 |

**Fig. 11.108** Error of the elements

| Opoſition von | Unterſchied der helioc. Länge | der geoc. Breite |
|---|---|---|
| 1803 | — 111,"00 | — 8,"31 |
| 1804 | + 59, 18 | — 36, 67 |
| 1805 | + 19, 92 | + 0, 07 |
| 1807 | + 85, 77 | + 25, 01 |
| 1808 | + 135, 88 | + 28, 72 |
| 1809 | — 216, 54 | + 83, 01 |

**Fig. 11.109** Table of deviations

| | Mittlere Länge | Heliocentr. Breite |
|---|---|---|
| 1803 | + 1″3 | — 1″0 |
| 1804 | — 3. 8 | + 4. 1 |
| 1805 | + 3. 9 | + 6. 0 |
| 1807 | — 3. 3 | + 3. 9 |
| 1808 | + 3. 2 | — 17. 0 |
| 1809 | — 1. 4 | — 3. 9 |

*of* Monat. Corr. *was in error, and calculated a position according to my new results out of consideration for the perturbations. I found the following for noon of January 9, 1811 in Goettingen.*

*RA Pallas 151° 37′ thus 1° 40′ less than in the ephemeris*
*Decl. 22° 10′ south - 11′ more to the south than in the ephemeris.*

*I have just calculated a second position for noon October 26, 1810 in Goettingen.*

*RA Pallas 134° 55′ thus 45′ less than in the ephemeris*
*Decl. 14° 28′ south - 6′ more to the south than the ephemeris.*

*If you wish to calculate single positions yourself, you may use the following elements:*

| | |
|---|---|
| *Epoch 1811, Goettingen*……….. | *126° 32′ 52″* |
| *Daily mean trop. move*………… | *769.″012* |
| *Log. of the semi major axis*…… | *0.44277* |
| *Eccentricity*…………………… | *0.24162 (= sin 13° 58′ 56″)* |
| *Perihel. (for above epoch)*……. | *120° 55′ 5″* |
| *Node*………………………….. | *172° 32′ 30″* |
| *Inclination*………………….. | *34° 35′ 15″* |

*I can still add one more position for noon, December 15 in Goettingen before having to send this letter.*

*RA Pallas 149° 37′ thus 1° 15′ less than in the ephemeris*
*Decl. 21° 37′ thus 12′ more to the south.*

*"I beg of you to employ the first clear night to look for Pallas. To date, it has been impossible here. You can well imagine with what impatience I await confirmation of this remarkable result – a result that has not yet been seen with regards these new planets.*

*"The improvement of the ephemeris in the October issue is the following through interpolation* (Fig. 11.110):

*Such remarkable results and the wish of our most worthy friend was reason enough to undertake the location of Pallas with all possible diligence. Unfortunately, the exceedingly poor weather, unabated for six weeks, thwarted nearly all our efforts. On December 18, it became somewhat clearer and with a parallactic instrument, we saw three stars within the field of the telescope at 2 o'clock a.m., of which one was most likely Pallas; however, a certain determination became impossible since the skies soon clouded over once again.*

| | | | AE. | Decl. |
|---|---|---|---|---|
| 1810 | den 26. October | Mittag | — 45′ | 6′ südl. |
| — | den 7. November | Mitternacht | 49 | 8 |
| — | den 20. November | Mittag | 56 | 10 |
| — | den 2. December | Mitternacht | 65 | 11 |
| — | den 15. December | Mittag | 75 | 12 |
| — | den 27. December | Mitternacht | 87 | 12 |
| 1811 | den 9. Januar | Mittag | 101 | 11 |

**Fig. 11.110** RA and Dec values of Pallas from October 26, 1810, to January 9, 1811. Mittag = noon; Mitternacht = midnight

*Furthermore, we have been able to find no determination of these three stars, either in Piazzi's star catalogue or the* Hist. cèl. *On December 19, we finally succeeded in a meridian observation. At exactly that position supplied us by the improved ephemeris, we observed a magnitude 7–8 star, which was most likely Pallas.*

| *1810* | *Mean time* | *Apparent RA* | *South Decl. in Seeberg* |
|---|---|---|---|
| *Dec. 19* | *16 h 9′ 7.″3* | *150° 20′ 13.″5* | *21° 58.″7* |

*The declination could clearly only be determined on a Passage instrument and is not known to the minute. However, the RA is exact and this corresponds most excellently with the improved determination. In the telescope* [field] *of the quadrant, Pallas was absolutely invisible. A small star in proximity to Pallas, the location of which we found to be*

*RA 149° 28′ 36″, Decl. 21° 59′*

*had maintained its position. We hesitated informing Professor Gauss of this isolated observation in the hope that we would be able to confirm that it was indeed Pallas by means of a second observation. But unfortunately, the skies did not grant us this pleasure and we therefore communicated our first suspected position of Pallas. Professor Gauss, who shares our suspicion, wrote us the following:*

*"I am very pleased that my calculation of the perturbations of Pallas is already being rewarded, because, although the weather here has made it impossible for me to look for this planet, I no longer doubt that the star you observed was indeed Pallas. Harding does not have a star in that position in his chart, but he noted the other one, which was also seen by you on December 18, although it does not appear in the* Histoire cèléste.*"*

*We hope that other astronomers will also want to look for Pallas in order to clear up the small uncertainty that still remains. The first clear morning will resolve this.*

## Continued Reports Regarding the New Primary Planet Pallas

*MC,* October 1811, p. 397

*In the December 1810 issue and the January issue of this year, we began to inform our readers of a few results of the prolonged research by Professor Gauss regarding the theory of this planet, the continuation of which will now follow. Earlier, we communicated our observations of this planet as well as those conducted at the observatory in Mannheim; we will therefore now only cite those locations of Pallas observed in Paris, Berlin and Hamburg by Bouvard, Bode and Schumacher respectively* (Figs. 11.111 and 11.112).

## 1. Pariser Beobachtungen.

| 1811 im Merid. | | AR ⚴ | | | Decl. austr. ⚴ | | |
|---|---|---|---|---|---|---|---|
| Febr. | 17 | 146° | 58′ | 33,″15 | 13° | 29′ | 42,″4 |
| | 18 | 146 | 47 | 55, 95 | 13 | 5 | 59, 4 |
| | 22 | 146 | 6 | 27, 90 | 11 | 28 | 19, 6 |
| | 27 | 145 | 18 | 13, 65 | 9 | 19 | 32, 9 |

**Fig. 11.111** Observations of Pallas in Paris from February 17 to 27, 1811

## 2. Berliner Beobachtungen.

| 1811 | | Mittl. Zeit in Berlin | | | AR ⚴ | | | Declin. austr. | | |
|---|---|---|---|---|---|---|---|---|---|---|
| Febr. | 18 | 11ᵁ | 55′ | 22,″5 | 146° | 48′ | 2 | — | — | — |
| | 19 | 11 | 50 | 34, 0 | 146 | 37 | 32 | 12° | 42′ | 44,″5 |
| | 20 | 11 | 45 | 57, 0 | 146 | 27 | 4 | 12 | 19 | 1 |
| März | 12 | 10 | 16 | 33, 6 | 143 | 45 | 29 | 3 | 32 | 42 |
| | 16 | 9 | 59 | 45, 1 | 143 | 29 | 18 | 1 | 43 | 3 |
| | 17 | 9 | 55 | 35, 1 | 143 | 26 | 13 | 1 | 22 | 11 |

## 3. Hamburger Beobachtungen.

| 1811 im Merid. | | AR ⚴ | | | Decl. austr. ⚴ | | |
|---|---|---|---|---|---|---|---|
| Febr. | 19 | 146° | 37′ | 15,″0 | 12° | 42′ | 45, 3 |
| | 20 | 146 | 26 | 52, 5 | 12 | 18 | 32, 7 |
| | 21 | 146 | 16 | 37, 5 | 11 | 53 | 47, 7 |
| | 22 | 146 | 6 | 36, 0 | 11 | 28 | 53, 5 |
| März | 9 | 144 | 1 | 45, 0 | 4 | 52 | 42, 5 |
| | 15 | 143 | 32 | 39, 0 | 2 | 13 | 59, 0 |
| | 16 | 143 | 29 | 24, 0 | 1 | 47 | 57, 0 |
| | 18 | 143 | 23 | 34, 5 | 0 | 56 | 47, 8 |

**Fig. 11.112** Observations of Pallas in Berlin and Hamburg from February 18 to March 17, and February 19 to March 18, 1811, respectively

*From this, Professor Gauss derived the following result for the opposition:*

| | |
|---|---|
| *1811 Feb. 21* | *19 h 25′ 31″ mean time in Goettingen* |
| *true helio. long.* | *152° 48′ 15.″8* |
| *true geoc. lat. South.* | *23 48 19.2* |

*The correspondence of this opposition with the six previous ones after due consideration of the perturbations of Jupiter, as developed by Professor Gauss, is indeed admirable. After a*

*few exceedingly insignificant corrections had been made on the elements, which had been determined prior to the locating of Pallas, the following differences between observation and calculation were produced, as follows* (Fig. 11.113):

*Hereafter, the calculation of perturbations was continued for another year. Professor Gauss was assisted in this arduous work by Herr Nicolai, who dedicated himself to the mathematical sciences in Goettingen with much enthusiasm and, under the supervision of Professor Gauss, executed the larger portion of these numerical calculations with just as much diligence as skill. Judging by the results of these calculations, the movement of Pallas at the time of next year's opposition can be represented by the following elliptical elements that also include the perturbations* (Fig. 11.114):

*According to these elements, Herr Nicolai calculated ephemerides for the course of Pallas for the years 1812 and 1813.*

[Refer to table, Course of Pallas, 1812 – calculated by Herr Nicolai, pp. 400 to 404.]

*Further, Herr Nicolai calculated the time of the next opposition of Pallas in advance, for which he received the following results from those elements:*

Unterſchied der

| | mittlern Länge | heliocentr. Breite |
|---|---|---|
| 1803 | + 8,"7 | — 1,"1 |
| 1804 | — 7, 0 | — 10, 2 |
| 1805 | + 1, 4 | — 3, 5 |
| 1807 | — 1, 2 | — 2, 4 |
| 1808 | — 3, 1 | — 21, 4 |
| 1809 | + 4, 5 | — 2, 1 |
| 1811 | — 3, 2 | — 12, 1 |

**Fig. 11.113** Differences between observation and calculation for Pallas from 1803 to 1811

*Epoche der mittlern Länge 1812*
*10. Junius Mittags.*

| | | | |
|---|---|---|---|
| Meridian von Göttingen . . . . . | 239° | 4' | 46,"1 |
| Länge des Perihels . . . . . . . . | 121 | 0 | 48, 5 |
| Länge des aufſteigenden Knoten . | 172 | 32 | 44, 5 |
| Neigung der Bahn . . . . . . . | 34 | 34 | 54, 7 |
| Tägliche mittl. tropiſche Beweg. . | 768,"5746 | | |
| Excentricität . . . . . . = Sin | 13° | 59' | 1,"8 |
| Logar. der halben Gr. Axe = | 0,4429521 | | |

**Fig. 11.114** Elliptical elements of Pallas

| *Opposition 1812, June 10 3 hr 31′ mean time in Goettingen* |
|---|
| *Helio. long. of Pallas……………………159° 28′ 53″* |
| *Helio. lat. north………………………… 34 32 37* |
| *Geocentric lat…………………………… 48 16 36* |
| *Log. of the dist. of Pallas from the Sun…..0.504259* |
| *Log. of the dist. of Pallas from Earth……. 0.384915* |

*Later, Professor Gauss imparted us the interesting news that he had completed the latitudinal perturbations in a 'general' theory of the perturbations of Pallas by Jupiter, at least as well as an initial calculation can supply them.*

Recognizing the importance of the treatise by Gauss on the asteroid Pallas, Zach gave a summary of it in his journal. The bulk of that summary is here; the complete treatise is given in Chap. 12.

**Regarding the Elliptical Elements of Pallas**
From a treatise by Professor Gauss
*MC,* November 1811, p. 449

*Some time ago, the author informed us of his treatise –* Disquisito de elementis ellipticis Palladis ex oppositionibus annorum 1803, 1804, 1805, 1807, 1808, 1809 *– appearing in the latest issue of the* Goettinger Commentarien *in a special edition, and we believe that we can provide an important service to our astronomical readers by providing a summary of it, since developments appear here which are to be regarded as additions and elucidations of some of the methods given in the* Theoria motus corpor. coelest. *We referred to the main numerical results of this treatise earlier in this periodical (vol. XXII, p. 588 f.) and will therefore limit ourselves mainly to the theoretical portion of the treatise. Since it would be interesting to juxtapose this with what constitutes the current theory of Pallas, we will now cite the six oppositions of Pallas observed until 1810 as yielded through the reductions of Gauss* (Fig. 11.115):

*The author combined these observed positions in different ways, and obtained three systems of elements which all deviate significantly from each other, and thus irrefutably document the influence of powerful perturbations and the impossibility of representing all observations of this planet in a pure ellipse. Since every one of the three systems is always based upon four observed positions, the author turns to the investigation of how this determination can be most easily obtained. In the* Theoria motus, *the method of determining the*

| Jahr und Tag | Mittl. Zeit in Gött. | | | Zahl d Tag. v. Auf. 1803 | Hel. Länge der Pallas | | | Geoc. Breite der Pallas | | |
|---|---|---|---|---|---|---|---|---|---|---|
| | h | ′ | ″ | | ° | ′ | ″ | ° | ′ | ″ |
| 1803 Jun. 30 | 0 | 27 | 32 | 181.019120 | 277 | 39 | 24,0 | + 46 | 26 | 36,0 |
| 1804 Aug. 30 | 4 | 58 | 27 | 608.207257 | 337 | 0 | 36,1 | + 15 | 1 | 49,8 |
| 1805 Nov. 29 | 11 | 15 | 4 | 1064.468796 | 67 | 20 | 42,9 | − 54 | 30 | 54,9 |
| 1807 May 4 | 14 | 37 | 41 | 1585.609502 | 223 | 37 | 27,7 | + 42 | 11 | 25,6 |
| 1808 Jul. 26 | 21 | 17 | 32 | 2034.887176 | 304 | 2 | 59,7 | + 37 | 43 | 53,7 |
| 1809 Sept. 22 | 16 | 10 | 20 | 2457.673843 | 359 | 40 | 4,4 | − 7 | 22 | 10,1 |

**Fig. 11.115** Latitude and longitude for six oppositions of Pallas from 1803 to 1809

*path from four observations out of which only two are complete is indicated in a most general way; but since this allows for a still more limited treatment in the case that all four observations are oppositions, the solution [resolution] of the question "To find the daily mean and the epoch of the mean longitude with the given four longitudes of the planet in the orbit and the corresponding given times" is a main objective of this treatise. Knowledge of the elements being sought is assumed as being approximated; an assumption that is always permitted since at any time, those data are obtained from an initial calculation. The author's method therefore consists mainly of the following: If the assumed perihelion is drawn from the longitudes observed in the path, the approximated true anomalies u, u', v, v' then follow. If the annual change of the perihelion is known, it can be taken into account. With e = sin φ = the approximated eccentricity, the mean M, M', M'', M''' corresponding to those true anomalies can be calculated by known formulae. If t, t', t'', t''' are the times corresponding to the times of the observations, the differences of t'-t, t''-t', t'''–t'' must be proportional to the differences of M'-M, M''- M', M'''-M'' in the true values of π (perihelion) and "e". If this is not the case, the corrections of φ and π can be determined in the following way: according to p. 15 of the Theoria motus, the relative change of the mean anomaly with regards the true anomaly and eccentricity =* (Fig. 11.116)

*Since the observed longitude in the path will be regarded as free from error, dv = −dπ, and according to this* (Fig. 11.117)

*if* (Fig. 11.118)

*and the same way m', m'', m''', n', n'', n''', for v', v'', v''', the true values of the mean anomaly will be* (Fig. 11.119)

$$= \frac{\cos^3 \varphi}{(1+e.\cos v)^2} . dv - \frac{(2+e.\cos v).\sin v.\cos^2 \varphi}{(1+e.\cos v)^2} . d\varphi.$$

**Fig. 11.116** Equation

$$dM = -\frac{\cos^3 \varphi}{(1+e.\cos v)^2} . d\pi - \frac{\cos^3 \varphi}{(1+e.\cos v)^2} . \frac{(2+e \cos v).\sin v}{\cos \varphi} . d\varphi$$

**Fig. 11.117** Equation

$$m = -\frac{\cos^3 \varphi}{(1+e.\cos v)^2}\,;\quad n = \frac{m(2+e.\cos v).\sin v}{\cos \varphi}$$

**Fig. 11.118** Equation

$$\left\{\begin{array}{ll} M + m\,d\pi + n\,d\varphi; & M' + m'd\pi + n'\,d\varphi; \\ M'' + m''d\pi + n''d\varphi; & M''' + m'''d\pi + n'''\,d\varphi; \end{array}\right.$$

**Fig. 11.119** Equation

*The differences of these mean anomalies divided by the differences of the corresponding times give the following expressions for the mean movement* (Fig. 11.120):

*The time unit which corresponds to 'this' mean movement is that which is expressed in the differences* $t' - t$, $t'' - t'$, $t''' - t''$. *Since these terms must be identical, this then yields the formulae from which the values of* $d\pi$, $d\varphi$ *and then later those of the true mean movement and the epoch are to be determined. If perturbations are to be and can be considered in the process, these must be fixed to observed longitudes earlier. To reduce the observed longitudes to those in the path, the inclination of the path and node must be assumed as being known, as you well know. This can take place with justification, since any errors in these elements have usually only an insignificant influence on any reduction. The way in which the author learned to find the correction of these elements is very peculiar and consists of the following: the heliocentric latitude of the planet can be found first from the observed geocentric and the radius vector calculated from the elliptical elements, and then from the inclination, node and observed heliocentric longitude; the heliocentric latitudes obtained through both methods can only be the same if the correct values have been taken for inclination and node. If this is not the case, the difference of these heliocentric latitudes compared with their corresponding changes for* $d\Omega$, $di$ *will give the value of these corrections. If* $R$ *and* $r$ *are the distance from Earth and that of the planet from the Sun and* $\theta$ *is the observed geocentric latitude* (Fig. 11.121),

*then* $\gamma$, $\gamma'$ *will be the heliocentric latitudes derived from the observed geocentric latitudes. Further, if* $\Omega$, $i$ *and* $\alpha$ *are node, inclination and observed heliocentric longitude* (Fig. 11.122),

*then* $\delta$ *and* $\delta'$ *are the heliocentric latitudes derived from the observed longitudes. If* $\Omega + d\Omega$, $i + di$ *are the true values for node and inclination, then the correct values for* $\gamma$, $\gamma'$, $\delta$, $\delta'$ *will be* (Fig. 11.123)

*The coefficients* $c$, $c'$, $f$, $f'$ *can easily be derived from the formulae (II), it is* (Fig. 11.124):

$$\left\{\begin{array}{l} \dfrac{M'-M}{t'-t} + \dfrac{m'-m}{t'-t}\,.\,d\pi + \dfrac{n'-n}{t'-t}\,.\,d\varphi \\[2ex] \dfrac{M''-M'}{t''-t'} + \dfrac{m''-m'}{t''-t'}\,.\,d\pi + \dfrac{n''-n'}{t''-t'}\,.\,d\varphi \\[2ex] \dfrac{M'''-M''}{t'''-t''} + \dfrac{m'''-m''}{t'''-t''}\,.\,d\pi + \dfrac{n'''-n''}{t'''-t''}\,.\,d\varphi \end{array}\right.$$

**Fig. 11.120** Equation

$$\frac{R.\sin\theta}{r} = \sin(\theta-\gamma);\ \frac{R'.\sin\theta}{r'} = \sin(\theta-\gamma')\ \text{etc.}\quad (I)$$

**Fig. 11.121** Equation

$$\text{tg.}\,i.\sin(\alpha-\Omega) = \text{tg.}\delta;\ \text{tg.}\,i.\sin(\alpha'-\Omega) = \text{tg.}\,\delta';\ldots(II)$$

**Fig. 11.122** Equation

**Fig. 11.123** Equation

$$\gamma + a\,d\,\Omega + b\,di;\quad \delta + c\,d\Omega + f.\,di;$$
$$\gamma' + a'd\,\Omega - b'di;\quad \delta' + c'd\,\Omega + f'.di;\ \text{etc.}$$

**Fig. 11.124** Equation

$$c = -\tfrac{1}{2}.\sin 2\delta.\,\text{cotg.}(\alpha - \Omega);\quad f = \frac{\sin.2\delta}{\sin 2i};$$

$$c' = -\tfrac{1}{2}.\sin 2\delta'.\text{cotg.}(\alpha' - \Omega);\quad f' = \frac{\sin 2\delta'}{\sin 2i};$$

*The relative changes of $\gamma$ in view of $d\Omega$ and di depend, by dint of the means of determining these magnitudes [dimensions], thereon that the radius vector used in their calculation is combined with the function of the node and the inclination and is thus modified through them. According to this, the coefficients a, a', b, b'* etc. *will depend on the relation of the differentials dr, $d\Omega$, di and can be determined just as analytically through appropriate developments. But since the expressions produced in this way would become somewhat lengthy, the author prefers to conduct his calculations with successively changed values for nodes and inclinations and to derive the coefficients a, a', b, b'* etc. *from the difference of the values determined for $\gamma$, $\gamma'$* etc. *in this way. At the same time, this method also has the advantage of immediately revealing the influence of a change in nodes and inclination on the elliptical elements. Since the corrected values for $\gamma$, $\gamma'$, $\delta$, $\delta'$* etc. *must be identical, their connexion gives the following equations of condition for determining $d\Omega$ and di* (Fig. 11.125);

*That these calculations can also take changes of node and inclination into consideration if these are known through the theory, as is the case with the above perihelion, need not be mentioned. If eccentricity and inclination of path are slight, the coefficients a, a', b, b' will become so insignificant that the calculation of them may be omitted, and the values for $d\Omega$ and di can be derived through the simple equation: $\gamma - \delta - cd\ \Omega - f.di$*

*If the values for $d\Omega$ and di are so significant that the reduction can change several seconds in the path, then the first calculation must be repeated and the correction of the elliptical elements be determined anew through the equations. But a new calculation of the coefficients m, m', n, n' will almost always be unnecessary.*

*To demonstrate the application of this method with an example, the author supplies the numerical calculation of the elements for the oppositions of 1805, 1807, 1808, 1809. We will not cite the entire exercise, but will content ourselves with supplying some data to those budding mathematicians who desire practice in these operations that might serve them as an introduction. The following was assumed as known*

*Ω 1803 172° 28′ 46.″8*

*Perihel 1803 308 36 35. 3*

*(sidereally stationary, and must therefore be reduced with application of the precession on other epochs)*

**Fig. 11.125** Equation

$$\gamma - \delta + (a - c).\,d\Omega + (b - f).\,di = 0;$$
$$\gamma' - \delta' + (a' - c').\,d\Omega + (b' - f').\,di = 0;$$
$$\gamma'' - \delta'' + (a'' - c'').\,d\Omega + (b'' - f'').\,di = 0;$$
$$\gamma''' - \delta''' + (a''' - c''').\,d\Omega + (b''' - f''').\,di = 0;$$

*$\varphi = 14° \ 10' \ 4.''08$*

*Inclination = 34 37 31.5*

*For the three equations of condition (452) of the mean daily sidereal movement, the author receives* (Fig. 11.126)

*and for the correction of inclination and node* (Fig. 11.127).

*From which the third system of elements follows, which we indicated in Monat. Corr XXII, p. 589.*

*Since it is impossible, according to what has been said, to be able to represent the course of Pallas over a long period of time with any measure of accuracy without consideration of the perturbations, the author promises to expound his research into the most effective way of conducting these calculations for a path, the inclination and eccentricity of which make all previous methods unusable, in another section. In the present treatise however, he is still concerned with the development of a method to determine the ellipse that satisfies the six oppositions observed to date as well as possible. Indeed, Art. 187 of the* Theoria motus *includes, in general, the developments belonging here; but one part permits a few short cuts to the condition that the observations be oppositions and thus, the author fulfills his promise made on p. 220 of that work to facilitate the practical application of the method of the least squares, with which he also combines a very elegant process of elimination. The method of determining the path to which all observations come the closest can be divided into two parts: first in the development of linear equations which give a difference between the observed and calculated positions through the functions of the sought corrections, and second in determining the most probable value of this correction through the method of the least squares, while here, at almost any time, more equations than unknown* [quantities] *occur, which therefore cannot be done strictly enough. With a view to the first development, the author gives the most comfortable and complete expressions* [terms] *for the relative change between heliocentric longitude, epoch, mean movement, perihelion, eccentricity, node and inclination and then between these elements and the geocentric latitude. We will cite these, since they are of great importance for anyone concerned with similar investigations.*

*Let L, $\zeta$, $\pi$, $e = \sin\varphi$, $\Omega$, i, a, epoch, mean daily sidereal movement in seconds, perihelion, eccentricity, node, inclination and semi major axis; t, number of days from the epoch to the time of the observation; r, v, E, $\mu$, radius vector, true and eccentric anomaly, argument of latitude; $\lambda$, $\gamma$, $\beta$, the heliocentric longitude and latitude calculated from those elements and geocentric latitude; R distance of Earth from the Sun. If the correction of the planetary elements is expressed by dL, $d\pi$* etc.*, then (according to the instructions of art. 15, 16, 52 of the* Theoria motus*)* (Figs. 11.128 and 11.129).

**Fig. 11.126** Equation

$$770,''31398 - 0,0007285.\,d\pi - 0,0061382.\,d\varphi$$
$$770,30718 - 0,\,0011845.\,d\pi + 0,0041933.\,d\varphi$$
$$771,37892 + 0,\,0009780.\,d\pi + 0,0048930.\,d\varphi$$

**Fig. 11.127** Equation

$$+\,62,''48 - 0,''1744.\,d\Omega + 1,''1957.\,di = 0;$$
$$+\,9,\,49 + 0,\,3578.\,d\Omega - 0,\,8172.\,di = 0;$$
$$-\,46,\,79 - 0,\,3292.\,d\Omega - 0,\,8685\,di = 0;$$
$$-\,25,\,55 - 0,\,6915.\,d\Omega + 0,\,2034.\,di = 0;$$

$$
\begin{aligned}
d\lambda = {} & \frac{aa \cos\varphi \cos i}{rr \cos\gamma} \,.\, dL \\
& + \frac{t.\, aa \cos\varphi \cos i}{rr \cos.\, \gamma} \,.\, d\zeta \\
& + \left( \frac{\cos i}{\cos^2\gamma} - \frac{aa \cos\varphi \cos i}{rr \cos^2\gamma} \right) .\, d\pi \\
& + \frac{aa \cos i}{rr \cos^2\gamma} (2 - e \cos E - ee)\, d\varphi \sin E \\
& + \left( 1 - \frac{\cos i}{\cos^2\gamma} \right) .\, d\Omega \\
& - \tan.\, \gamma.\, \cos(\lambda - \Omega)\, di;
\end{aligned}
$$

**Fig. 11.128** Equation

$$
\begin{aligned}
d\beta = {} & - \frac{a \sin\beta \sin(\beta - \gamma).\, \tan.\, \varphi \sin v}{r \sin\gamma} \,.\, dL \\
& + \left( \frac{2 \sin\beta \sin(\beta-\gamma)}{3\zeta.\, \sin\gamma} - \frac{a.t.\sin\beta\sin(\beta-\gamma)\tan.\,\varphi\sin v}{r.\, \sin\gamma} \right) d\zeta \\
& + \frac{a \sin\beta \sin(\beta-\gamma) .\, \tan.\, \varphi \sin v}{r \sin\gamma} \,.\, d\pi \\
& + \frac{a \sin\beta \sin(\beta-\gamma).\, \cos\varphi \cos v}{r \sin\gamma} \,.\, d\varphi \\
& + \frac{2 \sin\beta \cos(\beta-\gamma) .\, \cos\gamma}{\sin 2i} \,.\, di \\
& - \sin\beta \cos(\beta-\gamma) \cos\gamma \cot(\alpha - \Omega) .\, d\Omega.
\end{aligned}
$$

**Fig. 11.129** Equation

*Since there is no question that this method is the best which we have for the case in which more equations than unknown values are available, we are considering printing schemata containing the necessary columns for all operations for the number of three, four, five and six unknown values in order to make their application even more easy; we will distribute them to anyone who is really interested in [arithmetical] astronomy.*

**Astronomical observations and remarks**
by Prof. Gauss in Goettingen. From two of his letters.
*BAJ*, 1814, p. 246

***Feb. 10, 1811.*** *On January 14 this year, I entered a short report in our* Gelehrten Anzeigen [Academic Announcements] *on the perturbations of Pallas by Jupiter. This, however, is not yet the universal theory of those perturbations. This theory shall require much more work, if it is supposed to fulfill all our needs. I have already started with it; but there is no need to rush, because I could continue to deal with these perturbations for many years in the usual way. As soon as the opposition of the current year has been observed, I will continue and, perhaps, deal with Juno and Vesta in a similar way. I request that you carefully observe the opposition of Pallas which will soon occur *) to facilitate this. Here is the improved ephemeris, as the harmony of my new results has been completely confirmed* (Fig. 11.130).

*Prof. Harding has made the following observations of Pallas in Mannheim* (Fig. 11.131):

*The planet is now quite bright; but it occurred to me that Ceres, which is located very close to its ephemeris, is even a little bit brighter.*

**)I have communicated to Prof. Gauss my observations on the meridians of Pallas, from February 18 to March 17, 1811.*

***August 29, 1811.*** *By the way, dear friend, I am pleased to send you the ephemerides of Pallas, Juno, and Vesta for their next appearances *). Three clever students of mine have done this calculation. With respect to Pallas, you will find details listed in Nos. 8 and 130 of the* Goettinger Gelehrten Anzeigen. *Mr. Nicolai, under my supervision, has continued to observe for another year the perturbation of Pallas by Jupiter and is now busy with calculating the ephemeris for 1813* (Fig. 11.132).

**) They follow later.*

*From these observations, which I think are all good, I have deduced the opposition as follows: 1811 May 25. 12 h 44' 30" mean time at Goettingen; 243° 48' 43."9 real*

| | | | | | | | | | | |
|---|---|---|---|---|---|---|---|---|---|---|
| Febr. 11 | AR. 148° | 3' | Decl. 15° | 43' S. | | März 3 | AR. 144° | 43' | Decl. 7° | 32' S. |
| 15 | 147 | 20 | 14 | 16 | | 7 | 144 | 13 | 5 | 44 |
| 19 | 146 | 37 | 12 | 42 | | 11 | 143 | 49 | 3 | 57 |
| 23 | 145 | 55 | 11 | 3 | | 15 | 143 | 31 | 2 | 11 |
| 27 | 145 | 17 | 9 | 19 | | 19 | 143 | 20 | 0 | 29 |

**Fig. 11.130** Ephemeris of Pallas from February 11 to March 19, 1811

| | | | | | | | | | |
|---|---|---|---|---|---|---|---|---|---|
| 1811 Januar 20 | $14^{u}$ | $6^{m}$ | $38^{s}6$ | 151° | 10' | 36"2 | −21° | 7' | 39"7 |
| 21 | 14 | 2 | 18.1 | 151 | 4 | 53.5 | −20 | 57 | 27.7 |
| 22 | 13 | 57 | 58.4 | | | | −20 | 47 | 51.7 |
| 23 | 13 | 53 | 38.6 | 150 | 52 | 48.2 | −20 | 37 | 58.8 |
| 24 | 13 | 49 | 16.4 | 150 | 46 | 0.6 | −20 | 27 | 9.7 |
| 25 | 13 | 44 | 53.0 | 150 | 39 | 9.9 | −20 | 16 | 9.4 |

**Fig. 11.131** Observations of Pallas by Harding from January 20 to 25, 1811

| | | | | | | | | | |
|---|---|---|---|---|---|---|---|---|---|
| 1811 Mai 20 | $12^{u}$ | $27^{m}$ | 45"1 M. Z. | 244° | 42' | 0"9 | −12° | 28' | 10"1 |
| 24 | 12 | 8 | 3.4 | 243 | 42 | 26.1 | −12 | 30 | 1.8 |
| 25 | 12 | 3 | 7.6 | 243 | 27 | 25.3 | −12 | 30 | 47.8 |
| 29 | 11 | 43 | 22.9 | 242 | 27 | 0.7 | −12 | 35 | |
| 30 | 11 | 38 | 27.3 | 242 | 12 | 3.0 | −12 | 35 | 52.8 |

**Fig. 11.132** Observations of Pallas from May 20 to 30, 1811

*longitude, 8° 33′ 59.″5 geocentric northern latitude. With respect to the alteration that Dr. Triesnecker believed had to be made to my results for an opposition of Ceres according to the yearbook of 1813, I would like to remark*

1) *that I did not find any mistakes in the calculation of the 1807 opposition as far as I can tell from the observations at Milan; I rather have to insist on my result of 3 May 3 h 42′ 28″ Paris time and 222° 14′ 9.″1 longitude. Meanwhile, I want to repeat the entire calculation once more, as soon as I find the time to calculate the elements according to the two oppositions in 1809 and 1811 (which have not been used yet).*
2) *The conjecture of Dr. Triesnecker to change the geocentric latitude of the opposition in 1808 from 12° 43′ 44″.8 into 12° 45′ 44.″8 is inadmissible. Just as well, the longitude of the line of the nodes necessarily has to remain 80° 53′ 23″, and by no means can it be assumed to be 5′ larger. Mr. Triesnecker's conjecture has been triggered merely by the less than perfect fit of the latitudes with the elements in my perturbation tables; but I am surprised that he has overlooked what I explicitly have mentioned, namely that rather than applying my old tables, I have used the more recently developed, complete perturbation equations, which have not been published yet and which I, therefore, communicate to you exactly as they have been calculated in the year 1805* (Fig. 11.133).

*The first six equations are the old ones, but corrected; the other nine, however, which may amount to one minute, have been added recently, and it is not surprising that, when considering them all together, the position of the node underwent an alteration by 5′. – These 15 latitude equations are still bearable, but for Pallas I have more than 40 which total 1″.*

The Gauss anagram appears in the following paper:

**Paper about Pallas**
By Gauss
April 25, 1812
Vol.67, p. 657
CGA

*Our journal has already mentioned several times the work of Prof. Gauss on the perturbations of Pallas (issues 8, pg. 322 and 130, pg. 327 of last year). The results of this study indicated there have now been confirmed nicely. Exactly there, where the ephemeris of Mr. Nicolai, calculated according to the perturbation theory, a little star of the 8th magnitude was visible, which had left this place on April 9, and had moved. Some comparisons with 93 Herculi gave the following position, the declination is only estimated.*

(♃ mittl. Länge des ♃, ⚴ mittl. Länge der ⚴)

| | |
|---|---|
| — 9″65 sin (♃ — 74° 13′) | + 8.61 sin (⚴ — ♃ + 95 26) |
| — 14.17 sin (⚴ — 2♃ + 78° 35′) | — 10.31 sin (2⚴ — 2♃ + 94 10) |
| + 27.47 sin (2⚴ — 3♃ + 78 35) | — 6″16 sin (3⚴ — 3♃ + 112° 26′) |
| — 4.62 sin (3⚴ — 4♃ + 78 35) | — 13.69 sin (2♃ — 57° 21′) |
| + 5.51 sin (2⚴ — ♃ — 78 35) | — 6.77 sin (3♃ — ⚴ — 61 25) |
| + 0.99 sin (3⚴ — 2♃ — 78 35) | + 4.30 sin (2⚴ — 4♃ + 72 24) |
| + 1.91 | — 22.51 sin (3⚴ — 5♃ + 70 27) |
| | + 1.65 sin (3♃ — 56° 9′) |

**Fig. 11.133** Perturbation equations for Pallas

*April 9 11 h 32 m 27 s mean time*

*RA of Pallas 269° 6′ 6.″7*

*North Declination 16 38*

*According to this the position of the planets matches the ephemerides completely. Shortly afterwards Prof. Gauss received a message from Dr. Olbers, whom he had asked to search for the planet, that he had already been successful on April 3. The two following observations of Bremen also correspond exactly to the ephemeris of Mr. Nicolai.*

| | *Mean Time* | *RA* | *Dec. N.* |
|---|---|---|---|
| *April 3* | *11 h 26 m 44 s* | *268° 42′ 23.″1* | *15° 25′ 4.″9* |
| *4* | *12 2 33* | *268 47 12.5* | *15 37 41.1* |

*We are hastening to publicise the positions of this planet, whose observation is rendered very difficult this year by its faintness, and invite all astronomers to observe it thoroughly around the time of its opposition (June 10).*

*Prof Gauss has also recently finished the first preliminary calculation of the general theory of the perturbations of Pallas. The number of all periodic equations of the influence of Jupiter, which are yet noticeable, amounts to almost four hundred. Among the very strange results, already showing from these, one is of utmost interest. For some reason, we give it here in a cypher, to which we will give the key in due course: 1111000100101001*

**Observation of Pallas and Juno, calculation of their oppositions, the elements of the latter's orbit, etc.**
By Prof. Gauss in Goettingen
Submitted on Aug. 22, 1812.
[A very similar paper was published in the *MC,* August 1812, p. 199.]
*BAJ,* 1815, p. 245

*If I am not mistaken, I have already reported to you the first discovery of Pallas. The number of local observations are few this time. At the time of the opposition, the position of the planet was too high for our locality to observe it with the usual instrument at the circular micrometer. I therefore had to make use of a more inferior instrument, and even so, poor weather and moonlight permitted only one observation. My total observations are as follows:*

| | *Mean Time* | *App. RA* | *Decl. N.* |
|---|---|---|---|
| *April 9* | *11 h 32 m 27 s* | *269° 5′ 58.″4* | *16° 38′* |
| *May 2* | *10 36 6* | *268 42 5.3* | *21 4 31.″7* |
| *4* | *10 44 44* | *268 31 38.0* | *21 25 4.3* |
| *June 7* | *11 13 51* | *262 50 51.1* | *25 2 56.6* |

*Among nonlocal observations, apart from two earlier observations by Dr. Olbers, only four meridian observations by Dr. Burckhardt from the observatory at the* Military Academy *in Paris have been reported to me, which I had printed in No. 127 of our* Academic Announcements. *From these data Mr. Nicolai whose talent in mathematics you already know, has calculated the opposition:*

| *1812, June 10, 3 h 2' 48" mean time in Goettingen* | |
|---|---|
| *True longitude* | *259° 27' 51."7* |
| *True geocentric latitude* | *48° 16' 26."1 N.* |

*After only minor corrections had been made for the elements, namely*

*−13."29 for the epoch of longitude 1810*
*−0."01200 for the daily motion*
*−17."57 for the perihelial longitude*
*−4."09 for the eccentricity angle*
*+0."80 for the longitude of the node*
*−1."21 for the inclination of the orbit*
*+0.0000045 for the log. of the semimajor axis;*

*the most accurate agreement of these was as follows:*

| | *Difference* | |
|---|---|---|
| *Opposition* | *Mean longitude* | *Heliocentric latitude* |
| *1803* | *−5."06* | *−2."78* |
| *1804* | *−12.95* | *−9.66* |
| *1805* | *−15.20* | *−0.67* |
| *1807* | *+8.28* | *−5.77* |
| *1808* | *+16.04* | *+11.67* |
| *1809* | *+13.00* | *−1.48* |
| *1811* | *+11.73* | *−9.53* |
| *1812* | *−25.92* | *+6.15* |

*I attribute these differences mainly to the perturbations by Saturn and Mars, because until now only the perturbations by Jupiter had been considered. The latter, though, are by far the most significant ones. The first calculation of these gave me approx. 400 equations. A second, more complete calculation which I have started shall yield approx. 1000 which will amount to more than 0."1. The next opposition shall occur according to Mr. Nicolai's calculation*

| *1813, August 18, 8 h 50' 5" mean time in Goettingen.* | |
|---|---|
| *True longitude* | *325° 24' 18"* |
| *True geocentric latitude* | *24° 37' 24" N.* |

[The remainder of this paper, on Juno, will be published in a future book in this series.]

**Continued Reports Regarding Pallas**

*MC,* October 1813, p. 345

[A very similar paper was published in the GGA, vol. 76, p. 1,753 (Nov. 4, 1813).]

*The wonderful confirmation, which the prediction of Pallas' movement, based on the perturbation calculations, produced through the observations of the previous year, and of which we gave an account in* Monat. Corr. *XXVI p. 199, increased the interest with which astronomers are awaiting the reappearance of this planet in the present year. The planet was first observed on the local observatory on September 1, and was found at precisely that location calculated by Herr Nicolai through the ephemeride two years before. Highly unfavourable weather here made it impossible to obtain even one observation in the area of the opposition. Further, the very weak light intensity of the planet, which hardly appeared as a magnitude 10 star, made an observation on the quadrant impossible.*

*Professor Gauss in Goettingen and Herr Burckhardt in Paris were more fortunate. The observations of the former, though conducted on a circular micrometer, turned out excellently precise* (Fig. 11.134):

*Herr Nicolai, the gifted young astronomer, who is now employed as an assistant at the local observatory, took over the calculations of these observations. He compared them most carefully first and foremost with the latest newly improved elements and found the following correspondence* (Fig. 11.135):

*From the observations of August 15, 17, and 19, Herr Nicolai derived the following result for the opposition:*

I. *Meridian-Beobachtungen auf der Seeberger Sternwarte:*

| 1813 | Mittl. Zeit | | | Gerade Aufst. | | |
|---|---|---|---|---|---|---|
| Sept. 1 | 10^U | 25′ | 16,″6 | 316° | 55′ | 15,″9 |
| 3 | 10 | 16 | 9, 6 | 316 | 39 | 25, 5 |
| 4 | 10 | 11 | 35, 4 | 316 | 29 | 48, 7 |

II. *Beobachtungen auf der Göttinger Sternwarte:*

| 1813 | Mittl. Zeit | | | Gerade Aufst. | | | Nördl. Abw. | | |
|---|---|---|---|---|---|---|---|---|---|
| Junius 28 | 11^U | 37′ | 2″ | 327° | 2′ | 13, 3 | 14° | 5′ | 43,″4 |
| Aug. 16 | 10 | 36 | 23 | 319 | 50 | 54, 7 | . | . . | . . |
| 19 | 9 | 31 | 0 | 319 | 17 | 10, 6 | 10 | 0 | 8, 4 |
| 26 | 9 | 23 | 7 | 317 | 59 | 52, 0 | 8 | 44 | 35, 3 |

III. *Meridian-Beobachtungen auf der Sternwarte der Militär-Schule in Paris:*

| 1813 | Sternzeit | | | Gerade Aufst. | | | Nördl. Abw. | | |
|---|---|---|---|---|---|---|---|---|---|
| August 4 | 21^U | 28 | 30,″5 | 322° | 7 | 38, 0 | 12 | 11′ | 58,″5 |
| 9 | 21 | 24 | 42 | . | . . | . | 11 | 32 | 38, 4 |
| 11 | . | . . | . | . | . . | . | 11 | 15 | 28, 1 |
| 15 | 21 | 20 | 7, 2 | 320 | 1 | 48, 0 | 10 | 38 | 48, 2 |
| 17 | 21 | 18 | 35, 2 | 319 | 38 | 48, 5 | 10 | 19 | 29, 3 |
| 19 | 21 | 17 | 4, 3 | 319 | 16 | 4, 5 | 9 | 59 | 18, 3 |
| 26 | 21 | 11 | 56, 1 | 317 | 59 | 1, 8 | 8 | 43 | 23, 4 |

**Fig. 11.134** I. Meridian Observations Conducted on the Seeberg Observatory; II. Observations from the Goettingen Observatory; III. Meridian Observations Conducted on the Observatory of the Military School in Paris

| 1813 | | Ger. Aufſt. | Abweich. | Beobachter |
|---|---|---|---|---|
| Junius | 28 | + 23,"6 | + 22,"4 | *Gauſs* |
| Auguſt | 4 | + 36, 9 | − 5, 9 | *Burckhardt* |
| | 9 | . . . | + 5, 1 | — |
| | 11 | . . . | + 6, 2 | — |
| | 15 | + 26, 9 | + 3, 7 | — |
| | 16 | + 35, 6 | . . . | *Gauſs* |
| | 17 | + 28, 2 | − 6, 6 | *Burckhardt* |
| | 19 | + 28, 5 | + 4, 5 | *Gauſs* |
| | 19 | + 24, 8 | − 7, 4 | *Burckhardt* |
| | 26 | + 22, 4 | − 4, 3 | *Gauſs* |
| | 26 | + 18, 3 | + 11, 1 | *Burkhardt* |
| Septbr. | 1 | + 2[illegible], 8 | . . . | *v. Lindenau* |
| | 3 | + 20, 9 | . . . | — |
| | 4 | + 49, 9 | . . . | — |

**Fig. 11.135** Pallas observations in 1813. Difference of the calculation for RA and Dec; observer at right

| |
|---|
| *Ninth Observed Opposition of Pallas* |
| *Aug. 18, 1813 8 hr 41′ 5″ mean time in Goettingen* |
| *True long. of Pallas.......................325° 23′ 56.″5* |
| *True geoc. Latitude........................ 24 .. 37 36.1 N* |

*Compare this with the prediction in the* Monat. Corr. *XXVI, p. 203.*

*A new improvement of the elements, in order to adapt them to this opposition better still, didn't seem to Professor Gauss be worth the trouble under these conditions.*

*Herr Nicolai himself has added the following supplement to these reports: Since the elements upon which my comparison of the Pallas observations are based in the previous report haven't yet been printed, I will add them here:*

| | |
|---|---|
| *Epoch 1813 Aug. 18 0 hr in Goettingen.......* | *331° 41′ 18.″ 45* |
| *Long. of the perihel. for the same time....* | *121 4 46. 55* |
| *Long. of the ☊.................................* | *172 33 0. 37* |
| *Inclin. of the path..........................* | *34 35 8. 31* |
| *Angle of eccentricity.......................* | *13 58 31. 72* |
| *Daily mean trop. movement..................* | *768.″3934* |
| *Log. of the semi major axis................* | *0.4430003* |

*I have maintained these elements not only for the pre- determination of the next year's opposition but also for the calculation of the geocentric course of Pallas for the occasion of its next reappearance.*

*The forthcoming 'tenth' opposition will occur according to the following:*

| | |
|---|---|
| *1814 Oct. 25 13 h 13′ 13″ mean time in Goettingen* | |
| *True longitude*……………………… | *31° 59′ 53.″3* |
| *Geoc. S. lat*……………………… …. | *37 21 55. 9* |

[Refer to table, Geocentric Course of Pallas from May 26, 1814, to April 3, 1815, pp. 349-351.]

*If the light intensity of Pallas in distance I from the Sun and the Earth is assumed as being a unity, then this light intensity would be in the opposition of 1812 = 0.01666; in 1813 = 0.01475 and in the next announced opposition of 1814, it will be 0.05476; in other words, almost four times greater than that of this year. However, the stronger light might perhaps be somewhat weakened by the relatively low position of this planet.*

[Note on 'equation': An equation in the early nineteenth century signified the angle to be added or subtracted from a mean motion in order to 'correct' it, that is, in order to obtain a theoretical position in agreement with the position observed.]

## Paper about Pallas

*GGA*, part 199, p. 1,985 (Dec. 12, 1814)

*The work by Dr. Gauss, mentioned in our journal already several times, on the complicated theory of Pallas' motion was confirmed nicely by the observations of this year's opposition of October – the tenth since its discovery. We communicate here these observations and the preliminarily deduced results, reserving further interpretation regarding the theory for another occasion. Since Mr. Nicolai's advance calculation, adjunct to Seeberg Observatory, of more than a year ago, that the path for the current visibility was based only on the osculating elements of the $9^{th}$ opposition (what had to be sufficient for refinding her), Prof. Gauss considered it interesting that the improvement of these elements through the constant influence of Jupiter from the $9^{th}$ to the $10^{th}$ opposition would be determined before a reappearance of the planet, and this major task was already at the beginning of period this carried out by Mr. Moebius, who then was here devoted to mathematical and astronomical studies. Of the results we give here only the predetermination of the opposition itself for a comparison with the results of the observations, since one can find it printed in more detail in the* astronomischen Jahrbuch *for 1817.*

| | |
|---|---|
| *1814 October 25. 12 h 39′ 50.″5 mean time in Goettingen* | |
| *True longitude* | *31° 58′ 28.″4* |
| *Heliocentric latitude* | *23 38 32.7 S* |

*Since this proves the correctness of the ephemeris and it was not to be doubted that at other observatories the planet would be observed with fixed instruments, a further observation at the circular micrometer was obsolete. By the way, the above observation was based on several comparisons and was marked in the journal as very good.*

*At Seeberg Observatory the observations were rightly restricted to the meridian telescope, since the planet was too faint to expect reliable declinations from the Dollond quadrant. Messrs.*

*Lindenau and Nicolai observed the following RA's* (Fig. 11.136):

| 1814 | Mittlere Zeit auf Seberg | Scheinbare gerade Aufsteigung |
|---|---|---|
| Sept. 17 | 15ᵘ 19ᵐ 41ˢ5 | 46° 18′ 26″2 |
| 18 | 15 15 52.0 | 46 20 2.8 |
| 19 | 15 12 1.0 | 46 21 17.0 |
| 20 | 15 8 8.0 | 46 22 0.7 |
| Octob. 12 | 13 36 19.2 | 45 2 4.5 |
| 13 | 13 31 52.0 | 44 54 14.6 |
| 15 | 13 22 53.1 | 44 37 25.8 |
| 21 | 12 55 28.0 | 43 39 51.6 |
| 28 | 12 22 45.3 | 42 21 48.5 |

**Fig. 11.136** Observations of Pallas at Seeberg from September 17 to October 28, 1814

*Mr. Schumacher of Mannheim Observatory submitted his observations entirely made at an 8 foot mural quadrant and were reduced here by Mr. Encke carefully, who obtained the following* (Fig. 11.137):

*During the two last observations the planet was so faint, that it was hardly discernible. Prof. Bessel at Koenigsberg Observatory (of whose great works our journal has given already several examples) made the following observations* (Fig. 11.138):

*Also this excellent observer complained about the difficulties observing due to the faintness of the planet.*

*The results of a comparison of all observations with the latest elements yields the following table (it needs to be said that the applied epoch of the mean longitude deviates by a few seconds, due to a method by Dr. Gauss which is explained in more detail elsewhere)* (Fig. 11.139):

*Finally, the result of these observations for the tenth opposition is the following:*

| *1814 Octob. 25. 12 h 33′ 22″ mean time in Goettingen* | |
|---|---|
| *True longitude* | *31° 58′ 11.″3* |
| *Geocentric latitude* | *37 20 53.2 south* |

*A comparison of this result with those found last and the elements, improved by Jupiter's perturbations, gave only such minor differences, that the basic elements do not require any further improvement. So we only give an overview of the correspondence between all hitherto observed oppositions and Dr. Gauss's theory* (Fig. 11.140).

| 1814 | Mittlere Zeit in Mannheim | Scheinbare gerade Aufsteigung | Scheinbare Abweichung |
|---|---|---|---|
| Octob. 12 | 13ᵘ 36ᵐ 17ˢ4 | 45° 2′ 0″3 | 19° 12′ 38″4 S. |
| 15 | 13 22 51.5 | 44 37 23.7 | 20 6 54.6 |
| 21 | 12 55 27.0 | 43 39 57.5 | 21 49 52.1 |
| 22 | 12 50 48.3 | 43 29 15.3 | 22 6 3.8 |

**Fig. 11.137** Observations of Pallas at Mannheim from October 12 to 22, 1814

| 1814 | Mittlere Zeit in Königsberg | Scheinbare gerade Aufsteigung | Scheinbare Abweichung |
|---|---|---|---|
| Octob. 14 | 13h 27m 30s6 | 44° 46′ 15″3 | 19° 48′ 30″8 S. |
| 15 | 13 23 0.4 | 44 37 38.6 | 20 6 20.0 |
| 18 | 13 9 22.9 | 44 10 8.3 | 20 58 58.9 |
| 19 | 13 4 48.3 | 44 0 25.9 | 21 15 52.6 |
| Nov. 1 | 12 3 57.3 | 41 33 59.9 | 24 31 29.8 |
| 2 | 11 59 12.6 | 41 21 46.2 | 24 44 9.7 |
| 3 | 11 54 27.4 | 41 9 24.1 | 24 56 10.6 |
| 4 | 11 49 42.0 | 40 56 59.6 | 25 7 41.8 |

**Fig. 11.138** Observations of Pallas at Koenigsberg from October 14 to November 4, 1814

**Fig. 11.139** Comparison of observed positions with latest elements of Pallas from September 16 to November 4, 1814

| 1814 | Unterschied Ger. Aufst. | Unterschied Abweich. | Beobachter |
|---|---|---|---|
| Sept. 16 | − 9″2 | + 5″9 | Gauss |
| 17 | − 1.6 | | von Lindenau |
| 18 | − 6.2 | | von Lindenau |
| 19 | − 10.7 | | von Lindenau |
| 20 | − 7.5 | | von Lindenau |
| Octob. 12 | − 4.9 | | von Lindenau |
| 12 | − 3.6 | + 2.2 | Schumacher |
| 13 | − 8.8 | | von Lindenau |
| 14 | − 9.4 | + 12.0 | Bessel |
| 15 | − 5.2 | + 6.7 | Bessel |
| 15 | − 6.5 | | von Lindenau |
| 15 | − 7.7 | + 5.7 | Schumacher |
| 18 | − 2.7 | + 13.2 | Bessel |
| 19 | − 4.0 | + 2.4 | Bessel |
| 21 | − 2.2 | | von Lindenau |
| 21 | − 12.0 | + 4.5 | Schumacher |
| 22 | − 0.4 | − 1.3 | Schumacher |
| 28 | + 1.8 | | von Lindenau |
| Nov. 1 | − 0.4 | + 8.5 | Bessel |
| 2 | − 5.1 | + 10.6 | Bessel |
| 3 | − 5.9 | − 2.8 | Bessel |
| 4 | − 7.9 | − 21.9 | Bessel |

## Paper about Pallas by Gauss

*BAJ*, 1818, p. 167

*Goettingen, 1 February 1815*

*The calculations mentioned in the last yearbook for Pallas were again perfectly confirmed. I myself though, was only able to observe the planet once with the circular micrometer*

| *1814* | *Mean time* | *RA* | *Dec.* |
|---|---|---|---|
| *Sept. 16* | *12 h 1′ 1″* | *146°16′ 23.″4* | *11° 4′ 20.″9 S* |

*but through an appreciable series of meridian observations by Messrs. Lindenau, Bessel and Schumacher I was enabled to calculate the opposition well. For the details I refer to the* Goett. gel. Anz. *issue 199 [p. 379] and give here the end result only:*

**Fig. 11.140** Correspondence between observed positions and elements for oppositions of Pallas from 1803 to 1814

| Oppos. | | Unterschied Mittl. Länge | Unterschied Helioc. Breite |
|---|---|---|---|
| I | 1803 | − 1"3 | − 5"3 |
| II | 1804 | − 12.9 | − 11.9 |
| III | 1805 | − 20.4 | − 0.4 |
| IV | 1807 | + 2.5 | − 6.8 |
| V | 1808 | + 23.5 | + 9.0 |
| VI | 1809 | + 21.5 | − 3.1 |
| VII | 1811 | + 14.1 | − 7.2 |
| VIII | 1812 | − 16.1 | + 4.0 |
| IX | 1813 | − 10.9 | − 0.0 |
| X | 1814 | − 14.0 | + 2.4 |

| | |
|---|---|
| *1814 Oct. 25.* | *12 h 33' 22" mean time in Goettingen* |
| *True longitude* | *31° 58' 11."3* |
| *Geocentric latitude* | *37 20 53.2 south* |

*All previous 10 oppositions are described by my theory up to a few seconds.*

*Mr. Encke has taken on the job of doing the calculations for the next in advance. I am giving here his result for the XI. opposition:*

| | |
|---|---|
| *1816 April 9* | *2 h 14' 24" mean time in Goettingen* |
| *True longitude* | *199° 34' 40."5* |
| *Geocentric latitude* | *28 6 16.3 north* |

*At the next opposition Pallas will be brighter than she has been for several years. Here the compared luminosity:*

*1812 0.01666 1813 0.01475 1814 0.05476 1816 0.05997*

*It is my pleasure to enclose a copy of the ephemeris that this young and gifted astronomer calculated.*

## Observations of Ceres in 1816

*Zeitschrift für Astronomie Tübingen.* Vol. 3. 1817, July & August issue, p. 99
Submitted by v. Lindenau July 28, 1816

[Lindenau first prints a table of Ceres data; Fig. 11.141]

*The opposition of Pallas occurs at Geottingen mean time Sept. 8, 1818 18 h 44' 12."5. Pallas longitude 345° 55' 33."8. Geocentric latitude 6° 47' 11."2 north* (Fig. 11.142).

Pál Tittel (1784–1831; Fig. 11.143) was a Hungarian astronomer, and Director of the Eger and Buda observatories. For several years in his youth he was a pupil of Gauss in Göttingen. This paper shows the asteroids were finally being observed from Baron von Zach's homeland.

**a. Beobachtungen, zu Seeberg**

| 1816. | Mz. Zeit Seeberg | Beob. AR. Ceres | Nördl. Decl. |
|---|---|---|---|
| Febr. 24. | 15h 37' 17",6 | 208° 11' 3",3 | - - - |
| Apr. 3. | 12 45 1,7 | 203 26 24,8 | - - - |
| — 4. | 12 40 14,4 | 203 13 32,7 | - - - |
| — 5. | 12 35 27,4 | 203 0 43,5 | 6° 50' 46",4 |
| — 10. | 12 11 28,3 | 201 55 39,9 | 7 6 38,7 |
| — 17. | 11 37 53,9 | 200 24 37,8 | 7 22 16,2 |

**b. Beobachtung in Königsberg.**

| 1816. | Mz. Zeit in Königsberg | Beob. AR. Ceres | Nördl. Decl. |
|---|---|---|---|
| März 26. | 13h 22' 58",7 | 205° 2' 31",8 | 6° 8' 41",3 |
| — 28. | 13 13 35,6 | 204 39 38,6 | 6 17 50,0 |
| Apr. 1. | 12 54 40,4 | 203 51 36,3 | 6 35 1,7 |
| — 2. | 12 49 54,2 | 203 39 9,9 | - - - |
| — 3. | 12 45 8,8 | 203 26 35,7 | 6 43 58,5 |
| — 4. | 12 40 22,2 | 203 13 54,5 | 6 46 44,4 |
| — 9. | 12 16 24,7 | 202 9 13,9 | 7 3 32,3 |
| — 10. | 12 11 36,2 | 201 56 4,1 | 7 6 25,2 |
| — 12. | 12 2 0,3 | 201 29 57,6 | 7 11 44,4 |
| — 14. | 11 52 24,3 | 201 3 50,2 | 7 16 23,2 |
| — 16. | 11 42 49,2 | 200 37 57,0 | 7 20 21,5 |
| — 18. | 11 33 14,9 | 200 12 16,2 | 7 23 34,3 |

**Fig. 11.141** Observations of Ceres in Seeberg and Koenigsberg in 1816

## Paper on Pallas

by Gauss
*BAJ*, 1819, p. 219
*Goettingen, 10 August 1816*

*According to your wish, it is my pleasure, to send you enclosed, dearest friend, ephemerides for the next appearance of Pallas and Juno, to which I added some remarks* (Fig. 11.144). *The ephemeris was calculated by Mr. Westphal from Schwerin, who is devoted to astronomy. The greatest accuracy is to be expected, since he took into account– as is done usually for the last five years – the perturbations according to my theory. The next opposition was calculated by Mr. Westphal in advance, as follows:*

| *1817 July 11. 8 h 45' 5" mean time in Goettingen* | |
|---|---|
| *True longitude* | *289° 3' 22"* |
| *Geocentric latitude* | *43 33 1* |

*Next year Pallas will be very faint and only be of 10th - 11th magnitude. The following is an overview of the last and future opposition:*

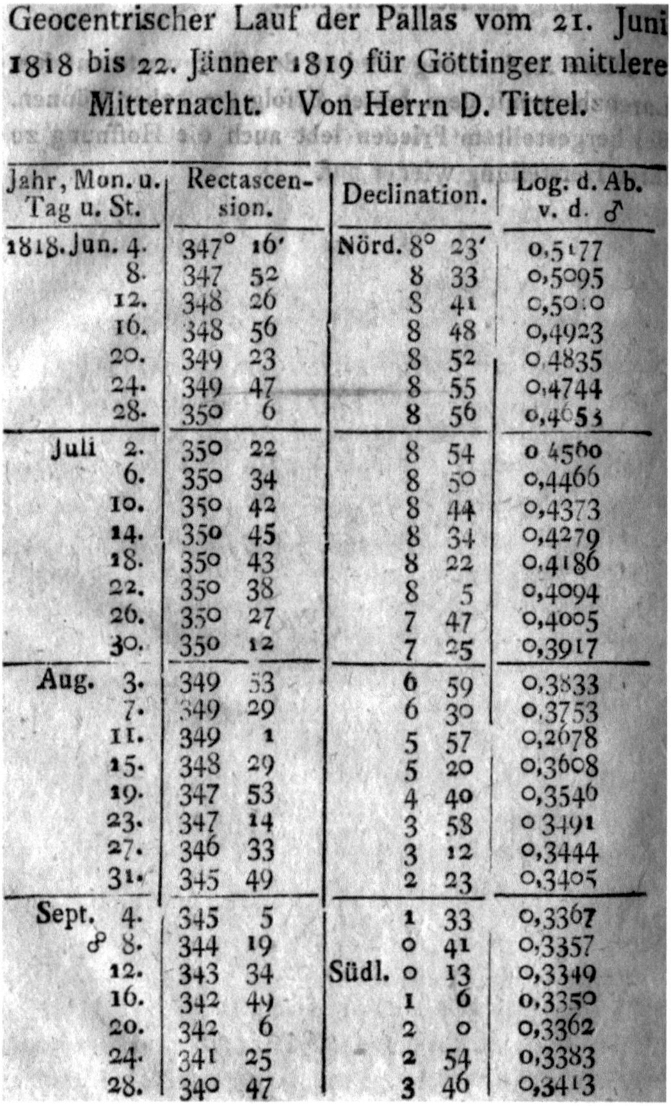

Geocentrischer Lauf der Pallas vom 21. Juni 1818 bis 22. Jänner 1819 für Göttinger mittlere Mitternacht. Von Herrn D. Tittel.

| Jahr, Mon. u. Tag u. St. | Rectascension. | Declination. | Log. d. Ab. v. d. ♂ |
|---|---|---|---|
| 1818. Jun. 4. | 347° 16′ | Nörd. 8° 23′ | 0,5177 |
| 8. | 347 52 | 8 33 | 0,5095 |
| 12. | 348 26 | 8 41 | 0,5010 |
| 16. | 348 56 | 8 48 | 0,4923 |
| 20. | 349 23 | 8 52 | 0,4835 |
| 24. | 349 47 | 8 55 | 0,4744 |
| 28. | 350 6 | 8 56 | 0,4653 |
| Juli 2. | 350 22 | 8 54 | 0,4560 |
| 6. | 350 34 | 8 50 | 0,4466 |
| 10. | 350 42 | 8 44 | 0,4373 |
| 14. | 350 45 | 8 34 | 0,4279 |
| 18. | 350 43 | 8 22 | 0,4186 |
| 22. | 350 38 | 8 5 | 0,4094 |
| 26. | 350 27 | 7 47 | 0,4005 |
| 30. | 350 12 | 7 25 | 0,3917 |
| Aug. 3. | 349 53 | 6 59 | 0,3833 |
| 7. | 349 29 | 6 30 | 0,3753 |
| 11. | 349 1 | 5 57 | 0,2678 |
| 15. | 348 29 | 5 20 | 0,3608 |
| 19. | 347 53 | 4 40 | 0,3546 |
| 23. | 347 14 | 3 58 | 0,3491 |
| 27. | 346 33 | 3 12 | 0,3444 |
| 31. | 345 49 | 2 23 | 0,3405 |
| Sept. 4. | 345 5 | 1 33 | 0,3367 |
| ☍ 8. | 344 19 | 0 41 | 0,3357 |
| 12. | 343 34 | Südl. 0 13 | 0,3349 |
| 16. | 342 49 | 1 6 | 0,3350 |
| 20. | 342 6 | 2 0 | 0,3362 |
| 24. | 341 25 | 2 54 | 0,3383 |
| 28. | 340 47 | 3 46 | 0,3413 |

**Fig. 11.142** First portion of an ephemeris of Pallas from June 21, 1818, to January 22, 1819, by Tittel

| | | | |
|---|---|---|---|
| *1812* | *0.01666* | *1816* | *0.05997* |
| *1813* | *0.01475* | *1817* | *0.01289* |
| *1814* | *0.05476* | *1818* | *0.02122* |

[Section about Juno will be published in a future book in this series.]

**Fig. 11.143** Pál Tittel

## Paper on Pallas
by Gauss
*BAJ,* 1821, p. 212
*Goettingen, 7 September 1818*

*According to your expressed wish, dearest friend, I am sending you enclosed again an ephemeris for Pallas of August 1819 until June 1820. It was calculated by Mr. Dirksen, who is here devoted to mathematics. The very same repeated the calculation of the opposition of this year using the same elements, according to which Dr. Tittel calculated and found a different result, namely:*

| | *Hr. Dirksen* | *Hr. Dr. Tittel* |
|---|---|---|
| *Time in Goettingen* | *Sept. 8 15 h 17′ 5″* | *Sept. 8 18 h 44′ 12″* |
| *True longitude* | *345° 47′ 18″* | *345° 55′ 34″* |
| *Geocentric latitude* | *6 47 25* | *6 47 11* |

*What might be the cause of the calculation error, I cannot say; it is restricted to the calculation of the opposition.*

Enno Heeren Dirksen (1788–1850) was taught by Gauss at Goettingen for three years, graduating in 1817. He studied for his doctorate advised by the mathematician and physicist Johann Mayer (1752–1830), son of Tobias Mayer, and then assisted Gauss in computing the orbits of asteroids and comets. He became professor of mathematics at Berlin University. Gauss continued to observe Ceres and Pallas for many years. Here is a selection as late as 1834 (source: Wikipedia).

| Jahr, Mon. u. Tag u. St. | Rectascension. | Declination. | Log. d. Ab. v. d. ♄ |
|---|---|---|---|
| 1818. Oct. 2. | 340° 12′ | Südl. 4° 37′ | 0,3452 |
| 6. | 339 41 | 5 26 | 0,3499 |
| 10. | 339 14 | 6 12 | 0,3554 |
| 14. | 338 53 | 6 56 | 0,3615 |
| 18. | 338 36 | 7 37 | 0,3681 |
| 22. | 338 24 | 8 15 | 0,3752 |
| 26. | 338 18 | 8 50 | 0,3827 |
| 30. | 338 17 | 9 21 | 0,3905 |
| Nov. 3. | 338 21 | 9 50 | 0,3984 |
| 7. | 338 31 | 10 15 | 0,4066 |
| 11. | 338 45 | 10 37 | 0,4148 |
| 15. | 339 5 | 10 56 | 0,4231 |
| 19. | 339 29 | 11 12 | 0,4314 |
| 23. | 339 57 | 11 26 | 0,4395 |
| 27. | 340 31 | 11 36 | 0,4476 |
| Dec. 1. | 341 8 | 11 44 | 0,4555 |
| 5. | 341 49 | 11 50 | 0,4633 |
| 9. | 342 34 | 11 53 | 0,4708 |
| 13. | 343 23 | 11 54 | 0,4782 |
| 17. | 344 15 | 11 53 | 0,4853 |
| 21. | 345 10 | 11 50 | 0,4921 |
| 25. | 346 8 | 11 45 | 0,4987 |
| 29. | 347 9 | 11 39 | 0,5[illegible]9 |
| 1819. Jan. 2. | 348 12 | 11 30 | 0,5109 |
| 6. | 349 19 | 11 21 | 0,5166 |
| 10. | 350 27 | 11 10 | 0,5220 |
| 14. | 351 37 | 10 57 | 0,5271 |
| 19. | 352 50 | 10 43 | 0,5319 |
| 22. | 334 4 | 10 28 | 0,5364 |

**Fig. 11.144** Ephemerides for the appearance of Pallas from 1818 to 1819

## Observations of Ceres and Pallas in 1825

*Astr. Yearbook*, 1828, p. 171
*Goettingen, 5 August 1825*

> *These observations maybe become more interesting, since these planets were in 1825 for the first time close to that region, where they were discovered and rediscovered in 1802 resp., and maybe Pallas has not been observed elsewhere since the publicised ephemeris deviates so much: I lacked the time to calculate one according to my theory* (Fig. 11.145).

## Observations of Ceres and Pallas in 1826

Paper by Gauss
*Berlin Astr. Yearbook*, 1829, p. 144

*Meridian-Beobachtungen der Pallas auf der Sternwarte in Göttingen.*

| 1825 | Mittlere Zeit in Göttingen | Gerade Aufsteigung | Abweichung |
|---|---|---|---|
| März 3 | $13^u\ 4^m\ 23^s7$ | 177° 34′ 24″3 | 0° 24′ 4″1 Südl. |
| 4 | 12 59 49.6 | 177 24 50.7 | 0 0 18.2 Nordl. |
| 7 | 12 46 4.0 | 176 55 18.1 | 1 14 7.2 |
| 8 | 12 41 27.9 | 176 45 12.3 | 1 38 52.6 |
| 9 | 12 36 51.3 | 176 35 0.3 | 2 3 40.3 |
| 10 | 12 32 14.3 | 176 24 43.2 | 2 28 29.3 |
| 18 | 11 55 14.1 | 175 1 15.0 | 5 44 59.3 |
| 19 | 11 50 37.3 | 174 50 59.1 | 6 8 56.3 |
| 20 | 11 46 0.6 | 174 40 46.2 | 6 32 41.4 |
| 25 | 11 23 6.1 | 173 51 53.7 | 8 27 50.6 |
| 27 | 11 14 1.1 | 173 33 32.8 | 9 11 51.7 |
| 29 | 11 4 59.8 | 173 16 8.2 | 9 54 29.8 |
| April 6 | 10 29 36.6 | 172 16 59.2 | 12 29 56.1 |
| 7 | 10 25 16.7 | 172 10 57.4 | 12 47 32.9 |

*Meridian-Beobachtungen der Ceres auf der Sternwarte in Göttingen.*

| 1825 | Mittlere Zeit in Göttingen | Gerade Aufsteigung | Abweichung |
|---|---|---|---|
| März 9 | $13^u\ 1^m\ 7^s5$ | 182° 40′ 4″6 | 17° 35′ 18″5 Nordl. |
| 10 | 12 56 23.3 | 182 27 57.4 | 17 41 20.9 |
| 18 | 12 18 11.1 | 180 46 26.4 | 18 24 4.4 |
| 19 | 12 13 23.5 | 180 33 28.9 | 18 28 34.0 |
| 20 | 12 8 36.8 | 180 20 32.2 | 18 32 50.0 |
| 29 | 11 25 38.4 | 178 26 37.3 | 19 1 18.0 |
| April 7 | 10 43 31.3 | 176 45 21.7 | 19 10 5.9 |

**Fig. 11.145** Observations of Ceres and Pallas by Gauss in 1825

*My dearest friend, I am enclosing my meridian observations of Pallas and Ceres at the time of this year's opposition for use in your yearbook* (Fig. 11.146).

*The faint light of the planet complicated the observations and decreased their accuracy. On the 27$^{th}$ I got mistaken by a 10$^{th}$ magnitude star which was only 1 min away from Pallas and by this the zenith distance was not adjusted to the considerably fainter planet, and the later observed only at two hairs unreliably* (Fig. 11.147).

*Ceres was of a good 8$^{th}$ magnitude and the observations are all good: only the first RA is based on two hairs only and the least magnification was used.*

## Observations of Ceres in 1829

*AN*, vol. 7, number 161, p. 329

by Gauss

[Gauss gives a table of Ceres data; Fig. 11.148.]

## Observations of Ceres and Pallas in 1830

*AN*, vol. 8, number 185, p. 321

by Gauss

[Gauss gives a table of Ceres and Pallas data; Fig. 11.149.]

*Beobachtungen der Pallas* 1826.

| | M. Z. in Göttingen | Gerade Aufsteig. | Abweich. nördl. |
|---|---|---|---|
| Juni 21 | $12^u$ $11^m$ $35^s4$ | $272^\circ$ $31'$ $4''9$ | $23^\circ$ $52'$ $19''4$ |
| 22 | 12 6 48.7 | 272 18 21.3 | 23 51 24.1 |
| 23 | 12 2 2.9 | 272 5 50.8: | 23 50 1.9 |
| 24 | 11 57 16.2 | 271 53 7.0 | 23 48 32.4 |
| 25 | 11 52 30.0 | 271 40 30.6 | 23 46 39.4 |
| 26 | 11 47 43.6 | 271 27 51.1 | 23 44 24.3 |
| 27 | 11 42 58.0 | 271 15 23.9: | — — — |
| 28 | 11 38 12.4 | 271 2 55.8 | 23 39 10.5 |

**Fig. 11.146** Observations of Pallas by Gauss in 1826

*Beobachtungen der Ceres* 1826.

| | M. Z. in Göttingen | Gerade Aufsteig. | Abweich. südl. |
|---|---|---|---|
| Juni 24 | $12^u$ $30^m$ $16^s6$ | $280^\circ$ $9'$ $33''0$: | $27^\circ$ $40'$ $54''6$ |
| 25 | 12 25 22.6 | 279 54 59.5 | 27 45 11.6 |
| 26 | 12 20 29.0 | 279 40 32.7 | 27 49 20.4 |
| 27 | 12 15 35.2 | 279 26 1.5 | 27 53 25.9 |
| 28 | 12 10 41.1 | 279 11 26.4 | 27 57 26.8 |
| 30 | 12 0 52.3 | 278 42 7.6 | 28 5 20.8 |
| Juli 2 | 11 51 3.6 | 278 12 49.0 | 28 12 57.7 |
| 3 | 11 46 9.3 | 277 58 11.1 | 28 16 37.1 |

**Fig. 11.147** Observations of Ceres by Gauss in 1826

| 1829 | Mittlere Zeit | Gerade Aufsteig. | Abweich. Nordl. |
|---|---|---|---|
| Januar 22 | $10^u$ $21^m$ $46^s2$ | $97^\circ$ $24'$ $25''0$ | $30^\circ$ $22'$ $0''4$ |
| 25 | 10 7 41.4 | 96 50 4.0 | 30 30 8.3 |
| Februar 11 | 8 52 51.7 | 94 49 55.8 | 30 58 55.4 |
| 19 | 8 20 52.8 | 94 41 58.8 | 31 4 46.7 |
| März 5 | 7 29 47.9 | 95 41 35.9 | 31 6 19.2 |
| 14 | 6 59 57.5 | 97 5 4.0 | 31 2 42.0 |
| 16 | 6 53 37.2 | 97 27 56.7 | |
| 18 | 6 47 22.3 | 97 52 14.5 | 30 59 55.9 |

**Fig. 11.148** Observations of Ceres by Gauss in 1829

### Opposition of Pallas in 1834

*AN*, vol. 11, number 262, p. 403
by Gauss
*Gauss in Goettingen made the following meridian observations of Pallas* (Fig. 11.150).

***Ceres.***

| 1830 | | Mittlere Zeit | Gerade Aufsteig. | Abweichung |
|---|---|---|---|---|
| April | 24 | $12^u\ 39^m\ 21^s9$ | $222^\circ\ 20'\ 25''8$ | $4^\circ\ 18'\ 27''6$ S. |
| | 25 | 34 33.0 | 7 9.1 | 16 45.0 |
| | 26 | 29 44.3 | 221 53 54.1 | 15 9.0 |
| | 27 | 24 54.7 | 40 27.3 | 13 38.8 |
| | 28 | 20 5.0 | 26 58.3 | 12 14.6 |
| | 29 | 15 15.3 | 13 28.8 | 10 52.4 |
| | 30 | 10 25.1 | 220 59 52.5 | 9 37.2 |
| Mai | 2 | 0 44.7 | 32 39.1 | 7 29.2 |
| | 5 | 11 46 14.9 | 219 52 0.7 | 5 2.0 |

**Beobachtungen der *Pallas* in Göttingen.**

| 1830 | | Mittlere Zeit | Gerade Aufsteig. | Abweichung |
|---|---|---|---|---|
| April | 24 | $13^u\ 0^m\ 9^s1$ | $227^\circ\ 33'\ 5''2$ | $22^\circ\ 38'\ 1''6$ N. |
| | 25 | 12 55 26.1 | 21 17.4 | 49 48.7 |
| | 26 | 50 42.4 | 9 17.7 | 23 1 16.3 |
| | 27 | 45 58.5 | 226 57 15.4 | 12 19.4 |
| | 28 | 41 14.1 | 45 6.1 | 22 58.4 |
| | 29 | 36 29.3 | 32 51.7 | 33 23.1 |
| | 30 | 81 44.3 | 20 33.1 | 43 21.8 |

**Fig. 11.149** Observations of Ceres and Pallas by Gauss in 1830

| | | | |
|---|---|---|---|
| 1834 Febr. | 6 | $8^u\ 41^m\ 47^s02$ | $-20^\circ\ 44'\ 23''7$ |
| | 9 | 8 39 39.10 | −19 41 52.9 |
| | 10 | 8 38 57.80 | −19 20 0.3 |

**Fig. 11.150** Observations of Pallas by Gauss in 1834

## References


Biancani, G.: 1620, *Sphaera mundi, seu cosmographia demonstrativa, ac facili methodo tradita.* Jul. Cassianus, Modena.

Cunningham, C.: 2015, The Clash Between William Herschel and the Great "Amateur" Johann Schroeter. In *New Insights From Recent Studies in the History of Astronomy: Following in the Footsteps of F. Richard Stephenson*, pg. 205-222. Orchiston, W., Green, D., and Strom, R. (eds.). Springer, New York.

Delambre, J.B.J.: 1821, *Histoire de l'Astronomie Moderne*, vol. 1. Courcier, Paris.

Gauss, C.: 1802, *Monthly Correspondence*, June issue, 597.

*Goettingische gelehrte Anzeigen*, no. 198, 13 Dec. issue, pg. 1969-1973 (1810).

Hevelius, J.: 1668, *Cometographia.* Simon Reiniger, Danzig.

Humboldt, A.: 1845, *Cosmos.* Harper & Bros., New York.

Licetus, F.: 1623, *De novis astris, et cometis libb. sex.* Apud. lo. Guerilium, Venice.

Olbers, W.: 1802. Manuscript archive number III 016, Bremen University, Bremen.

Partridge, J.: 1803, *Merlinus Liberatus. Being an Almanack for the Year of our Redemption*, 1803. W. Thorne, London.

Rost, J.L.: 1718, *Astronomisches Handbuch; Das ben Observirung der Cometen*, pg. 502-510. Peter Conrad Monath, Nuernberg.

Season, H.: 1803, *Speculum Anni: or, Season on the Seasons, for the Year of our Lord 1803.* G. Woodfall, London.

Struyck, N.: 1740, *Inleiding tot de algemeene kennis der cometen, of staartsterren.* Amsterdam.

White, R.: 1803, *The Coeslestial Atlas; or, a new Ephemeris for the Year of our Lord 1803.* M. Brown, London.

# Chapter 12
# Gauss: The Great Asteroid Treatises

*"Don't quote Latin; say what you have to say, and then sit down."*
– Advice given by the Duke of Wellington to new members of Parliament, circa 1820.

Piazzi wrote in Italian, Zach wrote in German, Lalande wrote in French, and Herschel wrote in English. Why did Gauss (Fig. 12.1) feel compelled to write about some of his asteroid work in Latin?

"By the 18th century many people thought that Latin had outlived its usefulness. The less useful it was, the more vigorously its defenders urged its case. Latin grammar was an incomparable intellectual discipline and mental training. Perhaps most important of all, Latin was a sign of class and the emblem of an elite." (Ridley 2001) Latin, then, "became a language of power, bolstering the prestige of those who used it and commanding the respect of those who did not, all the more effectively for being unintelligible." (Jones 2001) For more about the role of Latin in European culture, see Waquet (2001).

Specifically in the case of Gauss, his feelings were personal, as related in Szpiro (2003: 104). "In 1806 Gauss's benefactor, the seventy-year-old duke of Braunschweig-Wolfenbuettel was called up to fight Napoleon. Against his will, the Prussian cabinet had named him supreme commander of the army. The Prussians were badly beaten in the battles at Jena and Auerstaedt. The duke suffered mortal wounds and died a few days later. Gauss was distraught. To him the duke had been one of the noblest representatives of an enlightened monarchy, while Napoleon symbolized the worst dangers of revolution. Gauss's conservative tendencies were reinforced. Henceforth, he refused to write in French and feigned ignorance of the language when circumstances forced him to interact with Frenchmen."

Thus, he wrote much of his important work in Latin to make it as difficult as possible for French researchers to understand it. In France, the move away from Latin had set in more than a century earlier, so that by the time Gauss wrote very few of his French contemporaries could read his works:

> *Most of the books written by our new philosophers are in French. I do not know how it has come about that even the most learned prefer French books to Latin ones. It seems that Latin is being*



C.J. Cunningham, *Studies of Pallas in the Early Nineteenth Century*,
DOI 10.1007/978-3-319-32848-5_12

**Fig. 12.1** Portrait of Carl Gauss painted by Schwarz during a visit by Gauss to Olbers in Bremen, 1803. Used by permission, courtesy of Goettingen Observatory

> *neglected, on the grounds that it is too old, and that this language must suffer the same fate as ordinary philosophy, whose antiquity now seems rebarbative* (La Grange 1675).

In particular, Gauss wrote *Disquisitio de Elementis Ellipticis Palladis* (published in this book for the first time in English); and his 1809 book *Theoria Motus corporum coelestium in sectionibus conicus solem ambientium*. He originally wrote his book in German, and took great pains to rework it into Latin, as we read in an 1807 letter to Olbers:

"I made already a small start, but I do find that the translation costs me more effort than the mere transcription; these subjects have to be thought in Latin all by themselves, if the work should not appear entirely too awkward." He elaborates on this in a letter to Laplace (January 30, 1812), in which he also makes clear his application of least squares to the orbital problems posed by the asteroids:

> *I have used the method of least squares since the year 1795 and I find in my papers, that the month of June 1798 is the time when I reconciled it with the principles of the calculus of probabilities.*
>
> *However my frequent applications of this method only date from the year 1802, since then I use it as you might say every day in my astronomical calculations on the new planets. As I had intended since then to assemble all the methods which I have used in one extensive*

*work (which I began in 1805 and of which the manuscript [his publisher] originally in German, was completed in 1806, but which at the request of Mr. Perthes I afterwards translated into Latin: printing began in 1807 and was finished only in 1809), I am in no hurry to publish an isolated fragment, therefore Mr. Legendre has preceded me* [Full text of this letter is published in Plackett 1972].

A four-page summary of the treatise was given to German readers in the GGA (1810). Here follows the Latin treatise by Gauss, which bears the date of November 25, 1810. The bold text appears in the original

## An Inquiry
## *Concerning the Elliptical Elements of Pallas*
## From the Oppositions of the Years 1803, 1804, 1805, 1807, 1808, 1809

*1.*

*So far in the month of March, of the year 1802, there were actually seven periodic appearances of the primary planet Pallas: And yet, when on the day this planet was first detected it had already gone into opposition with the sun, only six oppositions of it were counted, which for the most part were sufficiently well observed in many astronomical telescopes, and truly they were able to be determined with enough accuracy. We ought to omit one opposition of the year 1808, in which year the planet, seen in the aphelion of its orbit, gave great difficulties for observations due to the weakness of the light, and, to confess the truth, it was somewhat neglected by astronomers. The most distinguished Lindenau even supplied a great enough abundance of right ascensions observed with the usual care at the most excellent Seeberg observatory around the time of opposition: But from a different source I was able to obtain only a few declinations, and far less certain. On account of this it is hardly correct to establish this opposition with the same accuracy as others, and above all the latitude remains very much uncertain.*

*2.*

*The observations of this appearance were determined as well as possible, and I was accustomed to gather them all yearly, to sift them carefully, and to add them to the observations of years past, so that the determination of the elliptical elements elicited from them would be so much more accurate. From the last calculation of this type, (which calculation was established at the end of the year 1807) the elements appeared in Vol. 17 Jan. 1808 of the most renowned Zach's literary Astronomical Correspondence. They were put on public record, uniting all four especially well observed oppositions. In the following year after the fifth opposition a dearth of reliable observations deterred me from new calculations, to which I then finally returned, when observations made in the year 1809 around the sixth opposition allowed me at once to undertake calculations of each opposition, which I have illustrated in our new literature in the year 1810 February 24.*

*Here now is a chart of all six oppositions observed up until that time* (Fig. 12.2).

*3.*

*If Pallas moves in an exact ellipsis, and according to the laws of Kepler, the most recent elements reconciling the four oppositions of the years 1803, 1804, 1805, 1807 especially well should be able to be affected by only the slightest errors to this point, and certainly they ought to be in agreement with the most recent oppositions within one minute or two. But it was far off from so strict a conclusion, as instead those elements demonstrated a discrepancy ascending in the first four minutes already in the fifth opposition, but by the sixth opposition they had deviated in the first twelve minutes. Clearly our planet suffers such great perturbations from the others, and especially from Jupiter, that among the*

| Tempus oppositionis ad meridianum Gottingensem | Dies inde ab initio anni 1803 | Longitudo heliocentrica | Latitudo geocentrica |
|---|---|---|---|
| 1803 Iun. 30. $0^h\ 27^m\ 32^s$ | 181.019120 | 277° 39′ 24″0 | +46° 26′ 36″0 |
| 1804 Aug. 30. 4 58 27 | 608.207257 | 337 0 36.1 | +15 1 49.8 |
| 1805 Nov. 29. 11 15 4 | 1064.468796 | 67 20 42.9 | −54 30 54.9 |
| 1807 Maj. 4. 14 37 41 | 1585.609502 | 223 37 27.7 | +42 11 25.6 |
| 1808 Iul. 26. 21 17 32 | 2034.887176 | 304 2 59.7 | +37 43 53.7 |
| 1809 Sept. 22. 16 10 20 | 2457.673843 | 359 40 4.4 | − 7 22 10.1 |

**Fig. 12.2** Data from six oppositions of Pallas. Columns read from left to right: Time of Opposition at meridian of Goettingen; Days hence from the beginning of 1803; Heliocentric Longitude; Geocentric Longitude

*phenomena an exact and stable consensus and a pure elliptical movement can in no way be obtained. Hence it is immediately made plain,* ***some elements and others are revealed, just as they are built up with some or other quarterly oppositions,*** *which is confirmed through the calculation of two other systems of elements, of which I have recently deduced the one from the oppositions of the years 1804, 1805, 1807 and 1808, and the other from the oppositions of the years 1805, 1807, 1808 and 1809. In this way these diverse systems of elements can be better compared amongst themselves, and I illustrate each here* (Fig. 12.3).

*Longitude of the perihelion 1803*

*Longitude of the ascending node 1803*

*Inclination of the Orbit*

*Eccentricity*

*Logarithm of the Semimajor axis*

*4.*

*In Sect. 2, of Book 2 of Theory of the Movement of the Celestial Bodies, I showed how the elliptical orbit of the planet might be determined from four observations (of which only two are completed). There the solution embraced the handed-down problems with the greatest generality: Nevertheless for this special case, when four observations are oppositions, it is better to call another method into use, the exposition of which I hope will be not unwelcome to astronomers. Certainly this method demands an approximate knowledge of each of the elements, or at least of these four, the inclination of the orbit, the longitude of the node, the longitude of the perihelion and the eccentricity; but nothing should impede us, so let us suppose this knowledge to already be present through an earlier calculation. The essence of the matter hangs on the solution of the following problem:*

*'With four longitudes of a planet in orbit given, with the responding times given, to find the longitude of the perihelion, the eccentricity, the medium daily movement and the epoch of the medium longitude' which I will relate in the following Section.*

*5.*

*Since the longitude of the perihelion is supposed to be approximately known, the true approximate anomalies* $v, v', v'', v'''$ *should arise from this subtraction from the longitudes in the orbit, when that longitude either appears as if it were constant, or its annual variation might be known. It should be accurately conveyed each time. The approximate eccentricity should be* $e = \sin \varphi$. *Using this, the medium anomalies* $M, M', M'', M'''$ *may be*

**I.** ***Elementa elliptica Palladis ex oppositionibus annorum***
**1803, 1804, 1805, 1807.**

| | |
|---|---|
| Epocha longitudinis mediae pro anno 1803 ad meridianum Gottingensem | 221° 39′ 30″4 |
| Motus diurnus medius tropicus | 770″2143 |
| Longitudo perihelii 1803 | 121° 3′ 11″4 |
| Longitudo nodi ascendentis 1803 | 172 28 56. 9 |
| Inclinatio orbitae | 34 37 41. 0 |
| Excentricitas (= sin. 14° 10′ 58″81) | 0.2450198 |
| Logarithmus semiaxis maioris | 0.4423149 |

**II.** ***Elementa elliptica Palladis ex oppositionibus annorum***
**1804, 1805, 1807, 1808.**

| | |
|---|---|
| Epocha longitudinis mediae 1803 | 221° 34′ 56″7 |
| Motus medius tropicus diurnus | 770″4467 |
| Longitudo perihelii 1803 | 121° 5′ 22″1 |
| Longitudo nodi ascendentis 1803 | 172 28 46. 8 |
| Inclinatio orbitae | 34 37 31. 5 |
| Excentricitas (= sin. 14° 10′ 4″08) | 0.2447624 |
| Logarithmus semiaxis maioris | 0.4422276 |

**III.** ***Elementa elliptica Palladis ex oppositionibus annorum***
**1805, 1807, 1808, 1809.**

| | |
|---|---|
| Epocha longitudinis mediae 1803 | 221° 23′ 24″6 |
| Motus medius tropicus diurnus | 770″9265 |
| Longitudo perihelii 1803 | 120° 58′ 4″8 |
| Longitudo nodi ascendentis 1803 | 172 27 52. 4 |
| Inclinatio orbitae | 34 36 49. 4 |
| Excentricitas (= sin. 14° 9′ 36″63) | 0.2446335 |
| Logarithmus semiaxis maioris | 0.4420473 |

**Fig. 12.3** Elliptical elements of Pallas from the oppositions of the years 1803, 1804, 1805 and 1807. Entries on left read Epoch of the mean longitude for the year 1803 at the meridian of Goettingen; and mean tropical daily movement

*computed through known methods responding to the true anomalies* $v$*,* $v'$*,* $v''$*,* $v'''$*. By designating the time as* $t$*,* $t'$*,* $t''$*,* $t'''$*, the differences of these* $t' - t$*,* $t'' - t'$*,* $t''' - t''$ *ought to be proportional to the differences* $M' - M$*,* $M'' - M'$*,* $M''' - M''$*, if we have progressed from valid truths of the perihelion and of the angle* $\varphi$*: if it should turn out otherwise, the corrections to be applied to these quantities are elicited in the following manner. By determining* (Fig. 12.4) *from example 15 of the Theory of the Movement of the Celestial Bodies, it is clear that, if the perihelion should be corrected by the slight increment of* $d\Pi$*, and the angle* $\varphi$ *by the increment* $d\varphi$*, the correct values of the medium anomalies become* (Fig. 12.5)

$$m = -\frac{\cos\varphi^3}{(1+e\cos v)^2}, \qquad n = \frac{m(2+e\cos v)\sin v}{\cos\varphi}$$

$$m' = -\frac{\cos\varphi^3}{(1+e\cos v')^2}. \qquad n' = \frac{m'(2+e\cos v')\sin v'}{\cos\varphi}$$

$$m'' = -\frac{\cos\varphi^3}{(1+e\cos v'')^2}, \qquad n'' = \frac{m''(2+e\cos v'')\sin v''}{\cos\varphi}$$

$$m''' = -\frac{\cos\varphi^3}{(1+e\cos v''')^2}, \qquad n''' = \frac{m'''(2+e\cos v''')\sin v'}{\cos\varphi}$$

**Fig. 12.4** Equation

*If indeed the powers and products of the corrections dΠ, it is permissible to neglect dφ. And so three expressions of the mean daily movement with respect to the perihelion will be considered, consider* (Fig. 12.6) *from the symmetry of which dΠ are unknown, dφ is elicited. With these substituted in one expression, the correct medium movement will appear, from whence using a known method even the semimajor axis will be determined. At last with the values of the corrections dΠ substituted, dφ in the expression for another mean anomaly, the correct value of this, and thence the epochs of the mean longitude willingly transferred at the appropriate time will result.*

*The rest is easily known, how, if there is need, the calculation firstly of the periodic perturbations, and then of the cycles, can be made. Evidently it is necessary to remove from those only the longitudes given in the orbit, before they are brought into the calculation; moreover, through those following it is necessary to transfer both the longitude of the perihelion and the eccentricity placed before the arbitrary epochs at each time t, t', t'', t'''.*

*6.*

*Now let us see, how these may be utilised for the determination of the orbit from the four observed oppositions. Let the longitude of the planet in the first ecliptic opposition be* ϐ α, *the geocentric latitude is, the distance of the earth from the sun* ***R****, the approximate longitude of the ascending node Ω, and the approximate inclination of the plane of the orbit at the ecliptic i. I mark the quantities looking at the remaining oppositions through similar*

**Fig. 12.5** Equation

$$M + m\,d\Pi + n\,d\varphi$$
$$M' + m'\,d\Pi + n'\,d\varphi$$
$$M'' + m''\,d\Pi + n''\,d\varphi$$
$$M''' + m'''\,d\Pi + n'''\,d\varphi$$

**Fig. 12.6** Equation

$$\frac{M'-M}{t'-t} + \frac{m'-m}{t'-t}\,d\Pi + \frac{n'-n}{t'-t}\,d\varphi$$
$$\frac{M''-M'}{t''-t'} + \frac{m''-m'}{t''-t'}\,d\Pi + \frac{n''-n'}{t''-t'}\,d\varphi$$
$$\frac{M'''-M''}{t'''-t''} + \frac{m'''-m''}{t'''-t''}\,d\Pi + \frac{n'''-n''}{t'''-t''}\,d\varphi$$

*characters with one index, the distinct ones with two or three indices. From* $\alpha$, $\Omega$, *i, the longitude in orbit will be deduced using the known formula, and a similar calculation will be constructed with three longitudes remaining in the orbit. From these four longitudes in the orbit through distinguished learned skills the elliptical elements will be derived, and hence through known formulae the four vectors of the radius* $r$, $r'$, $r''$, $r'''$. *Now, by determining* (Fig. 12.7) *the four heliocentric longitudes will clearly be* $\gamma$, $\gamma'$, $\gamma''$, $\gamma'''$, *deduced from the geocentric latitudes. Let us derive these same from the longitudes, by establishing* (Fig. 12.8)

*And it is clear, that it ought to be* $\gamma = \delta$, $\gamma' = \delta'$, $\gamma'' = \delta''$, $\gamma''' = \delta'''$, *if true values are received for* $\Omega$ *and* **i**, *and if indeed it were possible to satisfy through the same ellipsis with all four oppositions (which bring back problems more than can be determined). If it should turn out otherwise, the corrections d* $\Omega$, ***di*** *with quantities* $\Omega$, *to be applied to* ***i*** *will be elicited in the following way. How far these corrections, as if scant quantities of the first order, appear (the powers and products of which it is permissible to set aside), evidently with these values used,* $\Omega + d\Omega$, ***i*** + ***di***, *with the place of the longitudes* $\gamma$, $\delta$, $\gamma'$, $\delta'$ etc. *other forms of this will appear* (Fig. 12.9)

*Where the coefficients c, f, c', f'* etc. *will be found through these formulae* (Fig. 12.10)

*We are also able to determine the coefficients* ***a, b***, etc. *through analytic operations, but I prefer the following method. Again, I seek only the indicated calculation, using in the place of the node of the longitude* $\Omega$ *another, chosen however you please, differing only slightly in quantity from that longitude, from whence the numerical values of the coefficients* ***a, a'***, etc. *are easily elicited; a new repetition of the same calculation, with the accepted inclination slightly changed will supply the values of the coefficients* ***b, b'*** etc. *in a similar manner. In this way it will immediately be clear what great corrections the elliptical elements have suffered, from the slight corrections of the quantities* $\Omega$, ***i***. *With these things so done, for (or instead of) determining the corrections d*$\Omega$, ***di*** *there are four equations* (Fig. 12.11)

*It is scarcely possible to be exactly satisfied by these, owing to the inevitable errors of observations; therefore the values of the unknowns d*$\Omega$, ***di*** *will be determined particularly capably through the beginning explications in Sect. III. of the second Book of the Theory of the Motion of the Heavenly Bodies. Henceforth, the corrections of the remaining elements will at once be easily derived, through these things, which we have now discussed.*

$$\frac{R\sin 6}{r} = \sin(6 - \gamma), \qquad \frac{R'\sin 6'}{r'} = \sin(6' - \gamma') \quad \text{etc.}$$

**Fig. 12.7** Equation

$$\operatorname{tang} i \sin(\alpha - \Omega) = \operatorname{tang}\delta, \qquad \operatorname{tang} i \sin(\alpha' - \Omega) = \operatorname{tang}\delta' \text{ etc.}$$

**Fig. 12.8** Equation

**Fig. 12.9** Equation

$$\begin{array}{ll} \gamma + a\,\mathrm{d}\Omega + b\,\mathrm{d}i, & \delta + c\,\mathrm{d}\Omega + f\,\mathrm{d}i \\ \gamma' + a'\mathrm{d}\Omega + b'\mathrm{d}i, & \delta' + c'\mathrm{d}\Omega + f'\mathrm{d}i \quad \text{etc.} \end{array}$$

$$c = -\tfrac{1}{2}\sin 2\delta \operatorname{cotang}(\alpha - \Omega), \qquad f = \frac{\sin 2\delta}{\sin 2i}$$
$$c' = -\tfrac{1}{2}\sin 2\delta' \operatorname{cotang}(\alpha' - \Omega), \qquad f' = \frac{\sin 2\delta'}{\sin 2i} \quad \text{etc.}$$

**Fig. 12.10** Equation

$$\begin{aligned}
\gamma - \delta + (a - c)\,d\Omega + (b - f)\,di &= 0\\
\gamma' - \delta' + (a' - c')\,d\Omega + (b' - f')\,di &= 0\\
\gamma'' - \delta'' + (a'' - c'')\,d\Omega + (b'' - f'')\,di &= 0\\
\gamma''' - \delta''' + (a''' - c''')\,d\Omega + (b''' - f''')\,di &= 0
\end{aligned}$$

**Fig. 12.11** Equation

*7.*

*It should be permitted at this point to add certain observations to these rules.*

*I. There will scarcely be need for advice, for our formulae to be able also to pay heed to this chance, when the nodes do not appear to be at rest, if only the supposed value of the longitude of the node might be reduced correctly from its own epoch at the four times, and these four diverse values should be used in those formulae for the four separate places. Likewise we are certain concerning the inclination, if indeed the variation of its cycle should become known, and it would seem valuable to make a calculation of it. In addition, if it is decided to include the perturbations of the heliocentric latitude in the calculation, in the preceding equations these should clearly either be transferred from* $\gamma$, $\gamma'$, $\gamma''$, $\gamma'''$, *or added to* $\delta$, $\delta'$, $\delta''$, $\delta'''$.

*III. As often as the inclination of the orbit and the eccentricity are moderate, the coefficients* **a, b, a', b'** etc. *should be avoided so small, so as to neglect them, and truly in the second and third calculation it should be permissible to refrain from the above rule. Then, when the corrections have been discovered as* $d\Omega$, *di (unless by chance they are too small and are avoided), to create a new calculation of the longitudes in the orbit with the correct values of the longitude of the node and the inclination consider* $\Omega + d\Omega$, *i +* **di**, *and it is necessary to seek again the calculation of the elliptical elements (as it should normally be done, as in chap. 5), if indeed the longitudes, corrected from the earlier incorrect values, will appear diverse by many minutes and seconds. As for the rest, there will scarcely be a need to calculate the coefficients* **m, n, m'** etc. *again, since obviously the values of these are already found to be precise enough through the first hypothesis.*

*8.*

*For a greater illustration of the preceding rules, I place here an entire calculation, through which the third system of the elements from the oppositions of the years 1805, 1807, 1808, 1809 is determined. I have supposed the longitude of the ascending node according to the beginning of the year 1803, as 172° 28' 46"8 =* $\Omega$, *and the inclination of the orbit as 34° 37' 31"5; I have reduced it to the times of the individual oppositions, by adding the precession, 2' 26" 01.3' 37" 50.4' 39" 12.5' 37"11. Hence I have derived the heliocentric latitudes from the longitudes.* (Fig. 12.12)

**Fig. 12.12** Equation

$$\delta = -33^0\ 40'\ 50''63$$
$$\delta' = +28\ \ 14\ \ 51.24$$
$$\delta'' = +27\ \ 20\ \ 55.86$$
$$\delta''' = -\ \ 4\ \ 52\ \ 28.44$$

$$c = +0.1252, \qquad f = -0.9870$$
$$c' = -0.3366. \qquad f' = +0.8917$$
$$c'' = +0.3609. \qquad f'' = +0.8727$$
$$c''' = +0.6803, \qquad f''' = -0{,}1811$$

*Further on the conclusions of the latitude appear*

257° 25′ 6″73
56 24 5.96
126 2 55.07
188 36 2.69

*From which it is necessary to elicit the elliptical elements. To this end I have established the angle φ = 14° 10′ 4″08 (as was found in the second system of the elements), the longitude of the perihelion, according to the beginning of the year 1803, as = 121° 5′ 22″.1. The perihelion (in the same way as the node) is supposed from the immobile backdrop of the fixed stars, and truly its distance is constant from the ascending node = 308° 36′ 35″.3. Hence we have the following through our known formulae:*

| | |
|---|---|
| *v = 308° 48′ 31″43* | *M = 328° 15′ 45″08* |
| *v′ = 107 47 30.66* | *M′ = 79 46 27.05* |
| *v″ = 177 26 19.77* | *M″ = 175 54 28.87* |
| *v‴ = 239 59 27.39* | *M‴ = 266 29 57.59* |
| *m = −0.68517* | *n = + 1.18569* |
| *m′ = −1.06482* | *n = −2.01318* |
| *m″ = −1.59701* | *n = −0.12921* |
| *m‴ = −1.18352* | *n = + 1.93947* |

*Therefore, by denoting the correct longitude of the perihelion as dΠ, and the correction of the angle φ as dφ, we will have the medium movement relating to the stars.*

*From the first opposition to the second within days 521.140706*

$$111^\circ\, 30'\, 41''97 - 0.37965 d\Pi - 3.19887 d\varphi$$

*From the second opposition to the third within days 449.277674*

$$96^\circ\, 8'\, 1''82 - 0.53219 d\Pi + 1.88397 d\varphi$$

*From the third opposition to the fourth within days 422.786667*

$$90^\circ 35'\, 28''72 + 0.41349 d\Pi + 2.06868 d\varphi$$

*Hence three expressions will be elicited for the medium daily value relating to the stars*

$$\begin{aligned} &770''31398 - 0.0007285 d\Pi - 0.0061382 d\varphi \\ &770.30718 - 0.0011845 d\Pi + 0.0041933 d\varphi \\ &771.37892 + 0.0009780 d\Pi + 0.0048930 d\varphi \end{aligned}$$

*From the comparison of which these equations result:*

$$\begin{aligned} 0 &= 0''00680 + 0.0004560 d\Pi - 0.0103315 d\varphi \\ 0 &= -1.07174 - 0.0021625 d\Pi - 0.0006997 d\varphi \end{aligned}$$

*And hence* $d\Pi = -488''82$, $d\varphi = -20''92$, *and so then the correct longitude of the perihelion from the beginning of the year 1803, is*

$$120^\circ\, 57'\, 13''.28,$$

*and the correct value of the angle is* $\varphi = 14^\circ\, 9'\, 43''16$.

*Further on the mean daily movement sidereal becomes 770"7985*

*the mean correct anomaly for the first opposition 328° 20' 55"20*

*the mean longitude for the same epoch 89° 20' 34"49*

*At last the logarithm of the semimajor axis, 0.4420439, is derived from the mean movement related to the stars, and indeed the logarithm of the semi parameter, 0.4152361. Now, since we have the correct values of the true anomalies*

$$\begin{aligned} v &= 308^\circ\, 56'\, 40''25 \\ v' &= 107\ 55\ 39.48 \\ v'' &= 177\ 34\ 28.59 \\ v''' &= 240\ 7\ 36.21 \end{aligned}$$

*the logarithms of the radii of the vectors (radius vectors)*

$$logr = 0.3531088$$
$$logr' = 0.4492406$$
$$logr'' = 0.5369700$$
$$logr''' = 0.4716739$$

*Hereafter from the Solar tables we have*

$$log\mathbf{R} = 9.9937332$$
$$log\mathbf{R}' = 0.0039862$$
$$log\mathbf{R}'' = 0.0065917$$
$$log\mathbf{R}''' = 0.0011160$$

*from whence we at last deduce*

$$\gamma = -33^\circ\, 39'\, 48''15$$
$$\gamma' = +28\, 15\, 0.73$$
$$\gamma'' = +27\, 20\, 9.07$$
$$\gamma''' = -\, 4\, 52\, 53.99$$

*In order to elicit the coefficients a, b* etc. *I make a second hypothesis, by retaining the inclination, but by increasing the longitude of the node by one minute in the first, so that at the beginning of the year 1803, it is 172° 29' 46".8. Then, as in the first hypothesis, by establishing the longitude of the perihelion for the same epoch as = 121° 5' 22."1, the true anomalies are found*

$$v = 308^\circ\, 48'\, 40''93$$
$$v' = 107\, 47\, 34.06$$
$$v'' = 177\, 26\, 22.25$$
$$v''' = 239\, 59\, 15.00$$

*And hence (by also establishing that* $\varphi$ *= 14° 10' 4".08) the mean anomalies*

$$M = 328^\circ\, 15'\, 51''59$$
$$M' = 79\, 46\, 30.67$$
$$M'' = 175\, 54\, 32.83$$
$$M''' = 266\, 29\, 42.93$$

*For the rest of this calculation there is no need to use the common method, but it suffices, with the values of the mean anomalies found in the first hypothesis, to add the products from the positively obtained coefficients* **m, m', m", m'"** *into the differences between the respective values of the true anomalies in each hypothesis, consider*

$$\begin{aligned} +9''50\,x\,0.68517 &= +6''51 \\ +3.40\,x\,1.06482 &= +3.62 \\ +2.48\,x\,1.59701 &= +3.96 \\ -12.39\,x\,1.18352 &= -14.66 \end{aligned}$$

*Hence by retaining the values of the coefficients* **m, n** etc. *we elicit*

*d$\Pi$ = −468" 21, the longitude of the perihelion from the beginning of the year 1803, 120° 57' 33" 89*

*d$\varphi$ = − 20" 62, $\varphi$ = 14° 9' 43"46*

*the mean sidereal daily movement = 770"7761*

*the logarithm of the semimajor axis = 0.4420523*

*the mean longitude in the first opposition = 89° 20' 47"85*

$$\begin{aligned} \gamma &= -33^\circ\,39'\,51''10 \\ \gamma' &= +28\;1\;51.27 \\ \gamma'' &= +27\;20\;10.97 \\ \gamma''' &= -4\quad 52\;\;54.66 \end{aligned}$$

*At length I make a third hypothesis by establishing Ω = 172° 28' 46"6,* ***i*** *= 34° 38' 31"5, from which, just as was deduced before*

*the longitude of the perihelion from the beginning of the year 1803, 120° 55' 34"46*

$$\varphi = 14^\circ\,9'\,52''63$$

*the mean daily sidereal movement = 770"8398*

*the logarithm of the semimajor axis = 0.4420283*

*the mean longitude in the first opposition = 89° 20' 20"65*

$$\begin{aligned} \gamma &= -33^\circ\,39'\,35''63 \\ \gamma' &= +28\;1\;55.20 \\ \gamma'' &= +27\;20\;9.32 \\ \gamma''' &= -4\;52\;52.65 \end{aligned}$$

*A comparison of the values of the heliocentric latitudes* $\gamma$, $\gamma'$, $\gamma''$, $\gamma'''$ *found in the three hypotheses produces*

$$\begin{aligned} a &= -0.0492, & b &= +0.2087 \\ a' &= +0.0212, & b' &= +0.0745 \\ a'' &= +0.0317, & b'' &= +0.0042 \\ a''' &= -0.0112, & b''' &= +0.0223 \end{aligned}$$

*And so the four equations are considered*

$$\begin{aligned} &+62''48 - 0.1744 d\Omega + 1.1957 di = 0 \\ &+9.49 + 0.3578 d\Omega - 0.8172 di = 0 \\ &-46.79 - 0.3292 d\Omega - 0.8685 di = 0 \\ &-25.55 - 0.6915 d\Omega + 0.2034 di = 0 \end{aligned}$$

*from which we deduce the following through the first of the four lesser*

$$\begin{aligned} d\Omega &= -54''41 \\ di &= -42.06 \end{aligned}$$

*so that this might be considered*

*the longitude of the ascending node from the beginning of the year 1803, 172° 27′ 52″39*

*the inclination of the orbit 34 36 49.44*

*The remaining elements I have either elicited through a comparison of them, which appeared in each of the three hypotheses, or, which is more accurate, they were able to be determined through a new calculation of the longitudes in orbit and a repetition of the operations explained in* **art. 5**. *The first method supplies*

| | |
|---|---|
| *the longitude of the perihelion 1803* | *120° 58′ 3″86* |
| *the angle* $\varphi$ | *14 9 36.25* |
| *the mean longitude in the first opposition* | *89 20 32.08* |
| *the mean daily sidereal movement* | *770″7899* |
| *the logarithm of the semimajor axis* | *0.4420471* |

*Through another method we find*

| | |
|---|---|
| *the longitude of the perihelion 1803* | *120° 58′ 4″81* |
| *the angle* $\varphi$ | *14 9 36.63* |
| *the mean longitude in the first opposition* | *89 20 31.81* |
| *the mean daily sidereal movement* | *770″7893* |
| *the logarithm of the semimajor axis* | *0.4420473* |

*These things come together with these elements, which I have related above (art. 3).*

*9.*

*However great these perturbations may be, which Pallas suffers from the remaining planets, nevertheless, as experience shows, the elliptical elements in these four oppositions adapted to the movement of the planet within this whole time satisfy us very well: moreover, they differ truly little from the* **preceding and following movement**, *unless an excessive interval of time might be assumed, so that, for example, the second element related in* ***art 3***, *differs in the three first minutes from the observed heliocentric longitude in the opposition of the year 1803, and in the first five minutes in the opposition of the year 1809. Since this is so, it seems to me that it will always be of the greatest advantage to use purely elliptical elements to create a journal of the future movement of the planet, elements derived from the most recently preceding oppositions, even if the multitude of equations arising from the perturbations is so truly great that even the calculation of only one heliocentric location of the planet should progress with far less work than the calculation of the elliptical elements through the rules related above. Thus it will always be permissible to predict the geocentric location of the planet without risk, and with a safe margin of a few of the first minutes, with precision enough to find the planet.*

*10.*

*And yet the dignity of science demands that it should be examined with a more steadfast consensus, which clearly cannot be obtained before the perturbations are introduced into the calculation. Indeed, in my opinion, it would be a premature calculation, so abundant and full of tedium, as long as the abundance of observations spanned an exceedingly scant time, and the perturbations arising from the remaining planets were scarcely manifesting themselves. But now, when the elliptical movement no longer suffices for reconciling all of the observed locations among themselves, it seems the time is at hand when we should examine a more accurate theory. How I intend that the calculation of these perturbations, which Pallas suffers (especially from Jupiter), should be solved (since it is scarcely possible to use the methods utilised with the other planets on account of the excessive eccentricity and inclination), I will soon explain in a more expanded fashion: however, in the following passages I will still speak concerning the elliptical elements, which seem especially suitable, so that the calculation of the perturbations can be built upon them. Certainly I propose that I should elicit the elliptical elements, which approximately satisfy not these or those oppositions precisely, but all the oppositions that have thus far been observed. Indeed, I have already described the method by which we might be permitted to pursue this business quite succinctly, in the Theory of the Movement of the Heavenly Bodies* ***art. 187***. *However, since it is not only true that certain compendia should receive those things that I have discussed there generally, in a special instance when the observed locations are oppositions, but also that certain practical skills, through which I have been accustomed for a long time now to make the application of the method of the lesser quadruples more easy, ought to be desired in this work, I hope it will not be unwelcome to astronomers if I discuss these calculations a little more fully here. Since the whole business is hinged upon the determination of the corrections to be added to the approximate elements, which are supposed to err very little from all of the observed locations, the entire labour will depend on two important factors: clearly the first that should be made are the linear equations, which each of the observed locations supply, and then from these equations the values of the unknowns should be particularly capably elicited.*

*11.*

*Let the following be the approximate elements*

**L** *the mean longitude of the planet for the arbitrary epoch*
**t** *the number of days thence elapsed from the epoch to the moment of observation*

*7* *the mean daily sidereal movement in minutes and seconds*
*Π* *the longitude of the perihelion*
***e*** *= sin φ eccentricity*
***a*** *the semimajor axis*
***r*** *the radius vector*
**v** *The true anomalies*
***E*** *the eccentric anomalies*
*Ω* *the longitude of the ascending node*
***i*** *the inclination of the orbit*
***u*** *the argument of the latitude*
*λ* *the heliocentric longitude*
*γ* *the heliocentric latitude*
*β* *the geocentric latitude*
***R*** *the distance of the Earth from the Sun*
*Moreover, from observation I suppose these*

*α geocentric latitude*

*ϭ heliocentric longitude*

*At last through* ***dL****, d7, dΠ* etc. *I denote the corrections of the quantities* ***L****, 7, Π* etc.

*And so* ***dL*** *+ td7 becomes the correction of the medium longitude,*

*dL +* ***t*** *d7 − dΠ the correction of the mean anomaly,*

*and thus through* ***art.*** *15, 16 of the Theory of the Movement of the Heavenly Bodies* (Fig. 12.13)

*Further on we have the correction of the argument of the latitude,* ***du*** *= dv + dΠ − dΩ, and through* ***art.*** *52 of the Theory of the Movement of the Heavenly Bodies the correction of the heliocentric longitude:* (Fig. 12.14)

*Hence it is combined* (Fig. 12.15)

*Further on when this is considered* (Fig. 12.16)

*through differentiation it becomes* (Fig. 12.17)

$$\mathrm{d}v = \frac{aa\cos\varphi}{rr}(\mathrm{d}L + t\mathrm{d}7 - \mathrm{d}\Pi) + \frac{aa}{rr}(2 - e\cos E - ee)\sin E\,\mathrm{d}\varphi$$

$$\mathrm{d}r = \frac{r}{a}\mathrm{d}a + a\,\mathrm{tang}\,\varphi\sin v\,(\mathrm{d}L + t\mathrm{d}7 - \mathrm{d}\Pi) - a\cos\varphi\cos v\,\mathrm{d}\varphi$$

**Fig. 12.13** Equation

$$\mathrm{d}\lambda = \mathrm{d}\Omega - \mathrm{tang}\,\gamma\cos(\lambda - \Omega)\,\mathrm{d}i + \frac{\cos i}{\cos\gamma^2}\mathrm{d}u$$

**Fig. 12.14** Equation

**Fig. 12.15** Equation

$$\begin{aligned} d\lambda = {} & \frac{aa\cos\varphi\cos i}{rr\cos\gamma^2}\,dL \\ & + \frac{taa\cos\varphi\cos i}{rr\cos\gamma^2}\,d7 \\ & + \left(\frac{\cos i}{\cos\gamma^2} - \frac{aa\cos\varphi\cos i}{rr\cos\gamma^2}\right) d\Pi \\ & + \frac{aa\cos i}{rr\cos\gamma^2}(2 - e\cos E - ee)\sin E\,d\varphi \\ & + \left(1 - \frac{\cos i}{\cos\gamma^2}\right) d\Omega \\ & - \operatorname{tang}\gamma\cos(\lambda - \Omega)\,di \end{aligned}$$

**Fig. 12.16** Equation

$$\begin{aligned} a^{\frac{3}{2}}7 &= \text{Const.} \\ r\sin(\beta - \gamma) &= R\sin\beta \\ \operatorname{tang}\gamma &= \operatorname{tang} i\sin(\alpha - \Omega) \end{aligned}$$

**Fig. 12.17** Equation

$$\frac{da}{a} = -\frac{2d7}{37}$$

$$\frac{dr}{r} + \operatorname{cotang}(\beta - \gamma).(d\beta - d\gamma) = \operatorname{cotg}\beta\,d\beta, \text{ sive}$$

$$d\beta = \frac{\sin\beta\cos(\beta-\gamma)}{\sin\gamma}\,d\gamma - \frac{\sin\beta\sin(\beta-\gamma)}{r\sin\gamma}\,dr$$

$$d\gamma = \frac{\sin 2\gamma}{\sin 2i}\,di - \tfrac{1}{2}\sin 2\gamma\operatorname{cotang}(\alpha - \Omega)\,d\Omega$$

*From whence with the value utilised of dr it is combined with that explained above* (Fig. 12.18)

*Hence the values of the heliocentric longitude and of the geocentric latitude from the corrected values of the elements $\lambda + d\lambda$, $\beta + d\beta$, and indeed any opposition will supply the two equations* (Fig. 12.19)

*12.*

*By applying these rules to the six oppositions of Pallas related in* ***art.*** *2, if we make our calculation with then following system of elements outlined in* ***art.*** *3, following the two equations we obtain:*

*From the first opposition, where the*

*the calculated longitude is = 277° 36′ 20″07,*

*the geocentric latitude is = + 46° 26′ 29″19:* (Fig. 12.20)

*From the second opposition, where the*

*the calculated longitude is = 337° 0′ 36″04,*

$$\begin{aligned} d\beta = &-\frac{a \sin\beta \sin(\beta-\gamma) \operatorname{tang}\varphi \sin v}{r \sin\gamma} dL \\ &+\left\{\frac{2\sin\beta\sin(\beta-\gamma)}{37\sin\gamma} - \frac{at\sin\beta\sin(\beta-\gamma)\operatorname{tang}\varphi\sin v}{r\sin\gamma}\right\} d7 \\ &+\frac{a\sin\beta\sin(\beta-\gamma)\operatorname{tang}\varphi\sin v}{r\sin\gamma} d\Pi \\ &+\frac{a\sin\beta\sin(\beta-\gamma)\cos\varphi\cos v}{r\sin\gamma} d\varphi \\ &+\frac{2\sin\beta\cos(\beta-\gamma)\cos\gamma}{\sin 2i} di \\ &-\sin\beta\cos(\beta-\gamma)\cos\gamma \operatorname{cotang}(\alpha-\Omega)\, d\Omega \end{aligned}$$

**Fig. 12.18** Equation

$$\begin{aligned} \alpha &= \lambda + d\lambda \\ \delta &= \beta + d\beta \end{aligned}$$

**Fig. 12.19** Equation

$$\begin{aligned} 0 = &-183''93 + 0.79363\,dL + 143.66\,d7 + 0.39493\,d\Pi + 0.95920\,d\varphi \\ &-0.18856\,d\Omega + 0.17387\,di \\ 0 = &-\ \ 6''81 - 0.02658\,dL + 46.71\,d7 + 0.02658\,d\Pi - 0.20858\,d\varphi \\ &+0.15946\,d\Omega + 1.25782\,di \end{aligned}$$

**Fig. 12.20** Equation

*the geocentric latitude is = + 15° 1' 46"71:* (Fig. 12.21)

*From the third opposition, where the*

*the calculated longitude is = 67° 20' 42"88,*

*the geocentric latitude is = − 54° 31' 3"88:* (Fig. 12.22)

*From the fourth opposition, where the*

*the calculated longitude is = 223° 37' 25"39,*

*the geocentric latitude is = + 42° 11' 28"07:* (Fig. 12.23)

*From the fifth opposition, where the*

$$\begin{aligned} 0 = &-\ \ 0''06 + 0.58880\,dL + 358.12\,d7 + 0.26208\,d\Pi - 0.85234\,d\varphi \\ &+0.14912\,d\Omega + 0.17775\,di \\ 0 = &-\ \ 3''09 + 0.01318\,dL + 28.39\,d7 - 0.01318\,d\Pi - 0.07861\,d\varphi \\ &+0.91704\,d\Omega + 0.54365\,di \end{aligned}$$

**Fig. 12.21** Equation

$$0 = -\ 0''02 + 1.73436\,dL + 1846.17\,d7 - 0.54603\,d\Pi - 2.05662\,d\varphi - 0.18833\,d\Omega - 0.17445\,di$$

$$0 = -\ 8''98 - 0.12606\,dL - 227.42\,d7 + 0.12606\,d\Pi - 0.38939\,d\varphi + 0.17176\,d\Omega - 1.35441\,di$$

**Fig. 12.22** Equation

$$0 = -\ 2''31 + 0.99584\,dL + 1579.03\,d7 + 0.06456\,d\Pi + 1.99545\,d\varphi - 0.06040\,d\Omega - 0.33750\,di$$

$$0 = +\ 2''47 - 0.08089\,dL - 67.22\,d7 + 0.08089\,d\Pi - 0.09970\,d\varphi - 0.46359\,d\Omega + 1.22803\,di$$

**Fig. 12.23** Equation

*the calculated longitude is = 304° 2′ 59″71,*

*the geocentric latitude is = + 37° 44′ 31″82:* (Fig. 12.24)

*From the sixth opposition, where the*

*the calculated longitude is = 359° 34′ 46″67,*

*the geocentric latitude is = − 7° 20′ 12″13:* (Fig. 12.25)

*But out of these twelve equations we reject ten entirely, since the observed geocentric latitude is too uncertain.*

*13.*

*Since we are thus unable to determine the six unknowns* ***dL****, d7* etc. *in order to satisfy exactly all 11 equations, i.e. so that the individual functions of the unknowns which (are on the right side of the equations at once) become = 0, we will elicit those values,* ***through***

$$0 = +\ 0''01 + 0.65311\,dL + 1329.09\,d7 + 0.38994\,d\Pi - 0.08439\,d\varphi - 0.04305\,d\Omega + 0.34268\,di$$

$$0 = +\ 38''12 - 0.00218\,dL + 38.47\,d7 + 0.00218\,d\Pi - 0.18710\,d\varphi + 0.47301\,d\Omega - 1.14371\,di$$

**Fig. 12.24** Equation

$$0 = -317''73 + 0.69957\,dL + 1719.32\,d7 + 0.12913\,d\Pi - 1.38787\,d\varphi + 0.17130\,d\Omega - 0.08360\,di$$

$$0 = +117''97 - 0.01315\,dL - 43.84\,d7 + 0.01315\,d\Pi + 0.02929\,d\varphi + 1.02138\,d\Omega - 0.27187\,di$$

**Fig. 12.25** Equation

***which the squares of these functions should make the total as low as possible.*** *It is indeed easily observed, if in general the linear functions of the unknowns* **p, q, r, s** etc. *should be put forward* (Fig. 12.26)

*and the conditional equations, so that* **ww + w'w' + w''w''** + etc. = Ω *might become the least, to be these* (Fig. 12.27)

*by designating for the sake of brevity* (Fig. 12.28)

**p, q, r, s** etc. *ought to be determined through elimination from the equations* (Fig. 12.29)

*But still, as often as the multitude of the unknowns* **p, q, r, s** etc. *is somewhat greater, the elimination requires the most tedious and exceedingly drawn out labour, which can be diminished in the following remarkable manner. Besides the coefficients [an], [aa] [ab]* etc. *(of which the majority becomes ½(**ii** + 3i), if the majority of the unknowns =* **i***), I even suppose this calculation* **nn + n'n' + n''n'' + n'''n'''** + etc. = *[**nn**], and it is easily observed to become* (Fig. 12.30)

*And so by designating*

$$[an] + [aa]p + [ab]q + [ac]r + [ad]s + etc.\, per\, A$$

**Fig. 12.26** Equation

$$\begin{aligned} n + ap + bq + cr + ds + \text{etc.} &= w \\ n' + a'p + b'q + c'r + d's + \text{etc.} &= w' \\ n'' + a''p + b''q + c''r + d''s + \text{etc.} &= w'' \\ n''' + a'''p + b'''q + c'''r + d'''s + \text{etc.} &= w''' \end{aligned}$$

**Fig. 12.27** Equation

$$\begin{aligned} aw + a'w' + a''w'' + a'''w''' + \text{etc.} &= 0 \\ bw + b'w' + b''w'' + b'''w''' + \text{etc.} &= 0 \\ cw + c'w' + c''w'' + c'''w''' + \text{etc.} &= 0 \\ dw + d'w' + d''w'' + d'''w''' + \text{etc.} &= 0 \\ \text{etc.} \quad \text{sive} & \end{aligned}$$

**Fig. 12.28** Equation

$$\begin{aligned} &an + a'n' + a''n'' + a'''n''' + \text{etc.} \quad \text{per } [an] \\ &aa + a'a' + a''a'' + a'''a''' + \text{etc.} \quad \text{per } [aa] \\ &ab + a'b' + a''b'' + a'''b''' + \text{etc.} \quad \text{per } [ab] \\ &\text{etc.} \\ &bb + b'b' + b''b'' + b'''b''' + \text{etc.} \quad \text{per } [bb] \\ &bc + b'c' + b''c'' + b'''c''' + \text{etc.} \quad \text{per } [bc] \\ &\text{etc. etc.} \end{aligned}$$

$$\begin{aligned}
&[an]+[aa]p+[ab]q+[ac]r+[ad]s+\text{ etc.} &= 0\\
&[bn]+[ab]p+[bb]q+[bc]r+[bd]s+\text{ etc.} &= 0\\
&[cn]+[ac]p+[bc]q+[cc]r+[cd]s+\text{ etc.} &= 0\\
&[dn]+[ad]p+[bd]q+[cd]r+[dd]s+\text{ etc.} &= 0\\
&\text{etc.}
\end{aligned}$$

**Fig. 12.29** Equation

$$\begin{aligned}
\Omega = [nn] &+ 2[an]p + 2[bn]q + 2[cn]r + 2[dn]s + \text{ etc.}\\
&+ [aa]pp + 2[ab]pq + 2[ac]pr + 2[ad]ps + \text{ etc.}\\
&+ [bb]qq + 2[bc]qr + 2[bd]qs + \text{ etc.}\\
&+ [cc]rr + 2[cd]rs + \text{ etc.}\\
&+ [dd]ss + \text{ etc.}\\
&\text{etc.}
\end{aligned}$$

**Fig. 12.30** Equation

*It reveals its individual parts $A^2/[aa]$, which involve the factor p, to be contained in Ω, and thus $\Omega - A^2/[aa]$ is the function freed from p. Therefore by establishing* (Fig. 12.31)

*it will be* (Fig. 12.32)

*Which function we designate with Ω ′.*

*In a similar manner by setting*

$$[bn,1]+[bb,1]q+[bc,1]r+[bd,1]s+etc. = B$$

*$\Omega' - B^2/[bb,1]$ will be the function freed from* ***q****, which we establish as Ω ″. In the same manner we will make* (Fig. 12.33)

**Fig. 12.31** Equation

$$\begin{aligned}
&[nn]-\frac{[an]^2}{[aa]} = [nn,1]\\
&[bn]-\frac{[an].[ab]}{[aa]} = [bn,1]\\
&[cn]-\frac{[an].[ac]}{[aa]} = [cn,1]\\
&[dn]-\frac{[an].[ad]}{[aa]} = [dn,1] \quad \text{etc.}\\
&[bb]-\frac{[ab]^2}{[aa]} = [bb,1]\\
&[bc]-\frac{[ab].[ac]}{[aa]} = [bc,1]\\
&[bd]-\frac{[ab].[ad]}{[aa]} = [bd,1] \quad \text{etc. etc.}:
\end{aligned}$$

$$\Omega - \frac{A^2}{[aa]} = [nn,1] + 2[bn,1]q + 2[cn,1]r + 2[dn,1]s + \text{etc.}$$
$$+ [bb,1]qq + 2[bc,1]qr + 2[bd,1]qs + \text{etc.}$$
$$+ [cc,1]rr + 2[cd,1]rs + \text{etc.}$$
$$+ [dd,1]ss + \text{etc.}$$
$$\text{etc.}$$

**Fig. 12.32** Equation

$$\Omega = \frac{A^2}{[aa]} + \frac{B^2}{[bb,1]} + \frac{C^2}{[cc,2]} + \frac{D^2}{[dd,3]} + \text{etc.} + [nn,\mu]$$

**Fig. 12.33** Equation

*and*

$$[cn,2] + [cc,2]r + [cd,2]s + etc. = C$$

*From which* $\Omega'' - C^2/[cc,2]$ *will be the function also freed from* **r**. *We will progress in the same way until in the progression* $\Omega$, $\Omega'$, $\Omega''$ etc. *we shall have come to the end, freed from all of the unknowns, which will be* $[nn, \mu]$, *if the majority of the unknowns p, q, r, s* etc. *is denoted with* $\mu$. *And so we have* (Fig. 12.34)

*Now since* $\Omega = \omega\omega + \omega'\omega' + \omega''\omega''$ etc. *is by its very nature not able to obtain a negative value, it is easily demonstrated that the divisors [aa], [bb,1], [cc,2], [dd,3]* etc. *necessarily must avoid positives (nevertheless, for the sake of brevity I will not pursue this demonstration more fully here). Hence on its own account it follows that the lowest value of* $\Omega$ *appears, if* $A = 0$, $B = 0$, $C = 0$, $D = 0$ etc. *And so from these* ($\mu$) *equations the unknowns p, q, r, s* etc. *ought to be determined, which, with the order inversed, shall be extremely simple to bring about, since clearly the final equation implies a unique unknown, and the second from final implies two, and so on. At once this method recommends for the reason that the lowest value of the attached* $\Omega$ *is spontaneously made known from it, which is indeed clearly* $[n\,n, \mu]$.

***14.***

*Let us now apply these rules to our example, where p, q, r, s* etc. *are* **dL**, *d7,* $d\Pi$, $d\varphi$, $d\Omega$, *di. When the calculation is accurately made, I find these numerical values:* (Fig. 12.35)

[The treatise then concludes as follows:]

*From whence we deduce*

$$di = -3''15$$
$$d\Omega = -34''37$$
$$d\varphi = -4''29$$
$$d\Pi = +166''44$$
$$d\mathbf{7} = +0''054335$$
$$d\mathbf{L} = -3''06$$

$$
\begin{aligned}
[nn] &= 148848\\
[an] &= -371.09\\
[bn] &= -580104\\
[cn] &= -113.45\\
[dn] &= +268.53\\
[en] &= +94.26\\
[fn] &= -31.81\\
[aa] &= +5.91569\\
[ab] &= +7203.91\\
[ac] &= -0.09344\\
[ad] &= -2.28516\\
[ae] &= -0.34664\\
[af] &= -0.18194\\
[bb] &= +10834225\\
[bc] &= -49.06\\
[bd] &= -3229.77\\
[be] &= -198.64\\
[bf] &= -143.05\\
[cc] &= +0.71917\\
[cd] &= +1.13382\\
[ce] &= +0.06400\\
[cf] &= +0.26341\\
[dd] &= +12.00340\\
[de] &= -0.37137\\
[df] &= -0{,}11762\\
[ee] &= +2.28215\\
[ef] &= -0.36136\\
[ff] &= +5.62456
\end{aligned}
$$

Hinc porro deduxi

$$
\begin{aligned}
[nn,1] &= +125569\\
[bn,1] &= -138534\\
[cn,1] &= -119.31\\
[dn,1] &= -125.18\\
[en,1] &= +72.52\\
[fn,1] &= -43.22\\
[bb,1] &= +2458225\\
[bc,1] &= +62.13\\
[bd,1] &= -510.58\\
[be,1] &= +213.84\\
[bf,1] &= +73.45\\
[cc,1] &= +0.71769\\
[cd,1] &= +1.09773\\
[ce,1] &= -0.05852\\
[cf,1] &= +0.26054\\
[dd,1] &= +11.12064\\
[de,1] &= -0.50528\\
[df,1] &= -0.18790\\
[ee,1] &= +2.26185\\
[ef,1] &= -0.37202\\
[ff,1] &= +5.61905
\end{aligned}
$$

Atque hinc simili modo

$$
\begin{aligned}
[nn,2] &= +117763\\
[cn,2] &= -115.81\\
[dn,2] &= -153.95\\
[en,2] &= +84.57\\
[fn,2] &= -39.08\\
[cc,2] &= +0.71612\\
[cd,2] &= +1.11063\\
[ce,2] &= -0.06392\\
[cf,2] &= +0.25868\\
[dd,2] &= +11.01463\\
[de,2] &= -0.46088\\
[df,2] &= -0.17265\\
[ee,2] &= +2.24325\\
[ef,2] &= -0.37841\\
[ff,2] &= +5.61686
\end{aligned}
$$

Hinc porro

$$
\begin{aligned}
[nn,3] &= +99034\\
[dn,3] &= +25.66\\
[en,3] &= +74.23\\
[fn,3] &= +2.75\\
[dd,3] &= +9.29213\\
[de,3] &= -0.36175\\
[df,3] &= -0.57384\\
[ee,3] &= +2.23754\\
[ef,3] &= -0.35532\\
[ff,3] &= +5.52342
\end{aligned}
$$

Hinc eodem modo

$$
\begin{aligned}
[nn,4] &= +98963\\
[en,4] &= +75.23\\
[fn,4] &= +4.33\\
[ee,4] &= +2.22346\\
[ef,4] &= -0.37766\\
[ff,4] &= +5.48798
\end{aligned}
$$

Hinc

$$
\begin{aligned}
[nn,5] &= +96418\\
[fn,5] &= +17.11\\
[ff,5] &= +5.42383
\end{aligned}
$$

Atque hinc tandem

$$[nn,6] = +96364$$

Habemus itaque ad determinationem incognitarum sex aequationes sequentes:

$$
\begin{aligned}
0 &= +\ 17''11 + 5.42383\,di\\
0 &= +\ 75''23 + 2.22346\,d\Omega - 0.37766\,di\\
0 &= +\ 25''66 + 9.29213\,d\varphi - 0.36175\,d\Omega - 0.57384\,di\\
0 &= -115''81 + 0.71612\,d\Pi + 1.11063\,d\varphi - 0.06392\,d\Omega + 0.25868\,di\\
0 &= -13854'' + 2458225\,d7 + 62.13\,d\Pi - 510.58\,d\varphi + 213{,}84\,d\Omega\\
&\quad + 73.45\,di\\
0 &= -371''09 + 5.91569\,dL + 7203.91\,d7 - 0.09344\,d\Pi - 2.28516\,d\varphi\\
&\quad - 0.34664\,d\Omega - 0.18194\,di
\end{aligned}
$$

**Fig. 12.34** The Latin text on the previous page reads: Formula….. Hence I deduced further on; Formula….. And then in a like manner; Formula….. Hence in turn; Formula….. Hence in the same manner; Formula….. Hence; Formula….. And at last here; The last line in the figure reads: And so we have these six following equations for the determination of the unknowns

$$[nn, 1] - \frac{[bn, 1]^2}{[bb, 1]} = [nn, 2]$$

$$[cn, 1] - \frac{[bn, 1].[bc, 1]}{[bb, 1]} = [cn, 2]$$

$$[cc, 1] - \frac{[bc, 1]^2}{[bb, 1]} = [cc, 2]$$

**Fig. 12.35** Equation

*And so here are the corrected elliptical elements, which satisfy all six oppositions as closely as possible.*

| | |
|---|---|
| *The Epoch of the mean longitude in 1803, at meridian of Goettingen* | *221° 34' 53"64* |
| *The medium tropical daily movement* | *770"5010* |
| *The longitude of the perihelion* | *1803 121° 8' 8"54* |
| *The longitude of the ascending node* | *1803 172 28 12.43* |
| *The inclination of the orbit* | *34 37 28.35* |
| *The eccentricity = sin 14° 9' 59"79* | *0.2447424* |
| *The logarithm of the semimajor axis* | *0.4422071* |

***15.***

*With the values of the corrections substituted* ***dL****, **d7*** etc. *which we have now discovered, in twelve equations in* ***art.*** *12, we obtain the following differences between the observed and calculated values of the heliocentric longitudes and those of the geocentric latitudes:*

| *In the opposition* | *Difference of the year* | *of the longitude of the latitude* |
|---|---|---|
| *1803* | *−111"00* | *−8"31* |
| *1804* | *+ 59. 18* | *−36. 67* |
| *1805* | *+ 19. 92* | *+ 0. 07* |
| *1807* | *+ 85. 77* | *+ 25. 01* |
| *1808* | *+135. 88* | *+ 28. 72* |
| *1809* | *−216. 54* | *+ 83. 01* |

*[end of the treatise by Gauss]*

Gauss's name became associated with elimination through the adoption, by professional computers, of a specialized notation that Gauss devised for his own least-squares calculations. For more on Gaussian elimination, see Grcar (2011). It was Lindenau who procured the 1802 outline of Gauss's first work on the determination of planetary orbits, the *Summarische Uebersicht,* and published it in the Sept. 1809 issue of the *MC.* An English translation of that important text follows. It has been set in different type to ensure fidelity to the original and highly mathematical nature of the work.

See Figs. 12.36, 12.37, 12.38, 12.39, 12.40, 12.41, 12.42, 12.43, 12.44, 12.45, 12.46, 12.47 and 12.48 (sans the first footnote).

# Summary overview of the method which was applied to the determination of the orbits of the two new planets.[1]

Carl Friedrich Gauss

*Monatliche Correspondenz zur Beförderung der Erd- und Himmels-Kunde,*
edited by Baron von Zach. September 1809.

The determination of the orbit of a heavenly body from a short series of observations, independent on the circle and parabola hypotheses, depends on two requirements: **I.** There must be some means to find the orbit which satisfies three given complete observations. **II.** There must be a way to adjust the orbit so found, such that the difference between the calculation and the entire supply of observations becomes as small as possible.

The most convenient means to satisfy the second condition appears to be to reduce it to the first. For the times $t$, $t'$, $t''$, etc., the observed positions would be $m$, $m'$, $m''$, etc. (each of which will have two components); the positions $p$, $p'$, $p''$, etc. calculated according to the known elements $e$ (considered as having six components), and finally, the positions $q$, $q'$, $q''$, etc. calculated according to the (still considered as undetermined) elements $f$. The differences of the [positions calculated according to the] elements $e$ are thus

$$p-m,\ p'-m',\ p''-m'', \text{etc.}$$

On the the other hand, the differences of the [positions calculated according to the] elements $f$ are

$$(p-m)+(q-p),\ (p'-m')+(q'-p'),\ (p''-m'')+(q''-p''), \text{etc.}$$

These latter should thus become as small as possible, and maintain no regularity.

The differences $q-p$, $q'-p'$, etc., are, as long as the elements $f$ are considered as constant, functions of time and, since, properly considered, they will be small by the nature things, it may thus be assumed from the short duration of the observations that they [the observations] are found for the intermediate times with sufficient precision through interpolation, if two outer and one inner [observation] are taken as given. They are denoted for those three times by x, y, and z (each considered as having two components), thus, by the [well] known first principles of interpolation theory, they will have a linear form $\alpha x+\beta y+\gamma z$, where the coefficients $\alpha$, $\beta$, and $\gamma$ are dependant on time. These differences of the elements $f$ will thus have, for the times $t$, $t'$, $t''$, etc. the form:

[1] When I had the pleasure of making the personal acquaintance of Herr Professor Gauss some time back, I saw among his papers the following essay, already outlined many years ago and yet nowhere published, which contained the earlier method of the author for determining the orbit. In my cursory reading of this summary overview I was soon convinced that the method developed here by the author, for making a first approximation of two distances of the planets from the earth, was essentially different from that which the author has now publically expounded upon in his larger work. So I asked him for permission that I might make this treatise known, with the assumption that it would be interesting to all connoisseurs to know the way in which the author succeeded at arriving at a complete solution—which differed from that of which an overview had been communicated to our readers in earlier issues. I originally had the goal to accompany the essay with some remarks for the purpose of making a comparison of the earlier and later methods of the author; but these, had they actually been explained, would be somewhat lengthy, and without reference to the work itself, remain ever unclear. It thus appeared advisable to me to communicate the entire essay without further addenda (which is more intended for connoisseurs who have the work itself at hand) to the astronomical readers of this periodical, just as it was set down by the author in writing six years ago.

-1-

**Fig. 12.36** Equation

$$p - m + \alpha x + \beta y + \gamma z$$
$$p' - m' + \alpha' x + \beta' y + \gamma' z$$
$$p'' - m'' + \alpha'' x + \beta'' y + \gamma'' z$$
$$etc.$$

where everything besides $x$, $y$, $z$ is known. It will then be possible to easily judge which values for $x$, $y$, and $z$ are the most suitable. *This can indeed yield a completely methodical process for finding these values through calculation*: however, a certain instinct will always be just as sure a guide. [insert translators' footnote on *gewisser Tact*]

It is therefore apparent, that as soon as $x$, $y$, and $z$ are determined, the second condition is reduced to the first, and thus are we able to limit ourselves solely to this.

*Determination of the Orbit from Three Complete Observations*

2.

It would indeed not be difficult to represent the relation of the six unknown magnitudes to the given ones in six equations. Only, this would turn out to be far too unhelpful; in order to be in the least bit useful, we must be satisfied with reaching the orbit which exactly represents the three observations stepwise. Obviously all methods at all useful to this [purpose] ultimately yield the same result; the quality of the end results is therefore no measure for the value of the method, but rather only for the acuity of the observations on which it is based. The value of the method can only be appreciated for the number and convenience of the steps, and a method whereby a representation of the three observations could not always be arrived at if so desired would not be a worse one, but rather it would be no method at all.

The investigation is thus split into two parts, a *first approximation*, and a *method of adjustment*. The former will be based upon certain, almost true relations drawn from the nature of the problem, and these are of such a type that the closer the approximations lie to one another, the smaller they err, and speaking mathematically, they are rigorously exact for observations which are infinitely close to one another. In any event, the influence of their deviation from the truth thus diminishes the closer the observations are on which they are based, whereby the method of adjustment is made the easier or wholly superfluous. Only it has to be considered that with close observations small errors in the observations become very serious, and sometimes affect the elements enormously, and hence the subsequent adjustment according to the entire set of observations, which we above called the second requirement, turns out to be all the more difficult. General rules cannot be given for the most appropriate selection of the observations. In order to determine the value, or more or less advantageous positions, of the observations, it is necessary to take the type of orbit into consideration For Ceres, the outermost observations, distant by 41 days, were sufficiently well applied with great success in the first approximation, and the calculations for the adjustment were quite simple. Also, with the calculation of the second Pallas orbit, the preceding approximation was not used, but rather the first method of approximation is well enough applied anew to the 27 day observations. For orbits which were to come nearer to a parabola, and in which the geocentric motion is very fast, the calculation would preferably be begun with somewhat shorter time intervals. Here, a judgement developed by experience is the best guide.

*Principle Points of the First Approximation*

First Point

*Approximate determination of the distances from the Earth for the two outer observations.*

-2-

**Fig. 12.37** Equation

3.

In order to simplify the overview with respect to the great number of symbols necessary for the following investigation, analogous things for the Earth, $P$, and for the observed planet, $p$, should be denoted by the same characters, only the former with capital letters, and the latter with lowercase. If the same letters appear without an apostrophe as well as with two or three, it must therefore be assumed, that the second and third have a similar relation to a second and third observation for the times $\tau'$, $\tau''$, as the first has to the observation for time $\tau$. Moreover, in and of itself it is not necessary that the time $\tau'$ fall between the time $\tau$ and the time $\tau''$; nevertheless, the use of the following prescriptions is most advantageous if $\tau'$ lies approximately halfway between $\tau$ and $\tau''$.

$S$, is the position of the Sun (in space) considered as fixed. $p$ is the position of the planet $p$ at time $\tau$. Similarly $p', p'', P, P', P''$.

X_, Y_, Z_ are three arbitrary fixed planes, which intersect one another perpendicularly at the center of the Sun.

$x, y, z$ are the perpendicular distances of the planet $p$ from these three planes at time $\tau$. Similarly $x', y', z'; x'', y'', z''; X, Y, Z; X', Y', Z'; X'', Y'', Z''$.

| | |
|---|---|
| $\xi = x - X$ | |
| $\eta = y - Y$ | hence $\xi', \eta', \zeta'; \xi'', \eta'', \zeta''$ |
| $\zeta = z - Z$ | |

Thus, $\xi, \eta, \zeta$ are the perpendicular distances of the planet $p$ from three moveable planes laid through $P$ and parallel with X_, Y_, Z_.

| | | | | | |
|---|---|---|---|---|---|
| $r$ | | of $p$ | | $S$ | |
| $R$ | distance | of $P$ | from | $S$ | all positive. Similarly $r'$, etc. |
| $\rho$ | | of $p$ | | $P$ | |

| | | | | |
|---|---|---|---|---|
| $b$ | | $Sp$ | | $Z\_$ |
| $B$ | angle of the line | $SP$ | with the plane | $Z\_$ |
| $\beta$ | | $Pp$ | | which is parallel to Z_ |

| | |
|---|---|
| $d = r\cos b$ | |
| $D = R\cos B$ | i.e. the projected distance on the plane Z_, and that parallel to it. |
| $\delta = \rho\cos\beta$ | |

| | | |
|---|---|---|
| $l$ | | $Y\_$ |
| $L$ | angle of the projection with the plane | $Y\_$ |
| $\lambda$ | | which is parallel to Y_ |

The angles $b$ and $l$ are to be taken as positive on the same side of Z_ and Y_ as $z$ and $y$ are considered to be positive. The angle $b$ can always be taken between the limits -90°, +90° (so that $d$, etc. always remain positive); whereas the angle $l$ can always be allowed to grow from 0° to 360°, and indeed so that it is set $= \left\{ {0^\circ \atop 180^\circ} \right\}$ where $x$ is $\left\{ {\text{positive} \atop \text{negative}} \right\}$. In this way is obtained

-3-

**Fig. 12.38** Equation

$$x = r\cos b\cos l = d\cos l$$
$$y = r\cos b\sin l = d\sin l$$
$$z = r\sin b \quad = d\tan b$$

and similar equations for $x'$ etc., $X$ etc., $\xi$ etc.

We assume the orbits of $p$ and $P$ to be in planes, and in so doing we abstract them from any outside influences which might affect them. We set the longitude of $p$ in the orbit at time $\tau$, $=v$ (similarly $v,'$ $v''$; $V$, $V'$, $V''$) ; and make $\frac{1}{2}r'r''\sin(v''-v') = f$, $\frac{1}{2}r''r\sin(v-v'') = f'$, $\frac{1}{2}rr'\sin(v'-v) = f''$. Thus $f$, $-f'$, $f''$, the areas of the triangles $p'Sp''$, $pSp''$, $pSp'$, are positive (assuming that $p$ runs in its proper direction, and $\tau'$ lies between $\tau$ and $\tau''$; the arrangement of the symbols for other cases presents no difficulty). Similarly $F$, $F'$, $F''$. By $g$, $-g'$, $g''$, $G$, $-G'$, $G''$ we denote the areas of the sectors of the whole orbit, to which these three triangles correspond, whose signs we assume as equal to those of $f$, $-f'$, $f''$, $F$, $-F'$, $F''$. Hence, $g$, $g'$, $g''$ and $G$, $G'$, $G''$ are proportional to the time intervals $\tau''-\tau'$, $\tau-\tau''$, $\tau'-\tau$.

The orbits of $p$ and $P$ are conic sections, whose semimajor axes we denote by $a$, $A$. The eccentricity of the orbit $p$ we set $=e=\sin\varphi$ (for an ellipse); hence $a\cos\varphi^2 = k$ will be the half parameter. Longitude of the aphelion of $p$, in its orbit is $= \pi$. The mean longitude is $=m$ (similarly $m'$, $m''$, $M$, $M'$, $M''$). The other symbols will be indicated in the course of the investigation itself.

4.

Since $p, p', p''$ are in a plane with $S$, it follows from a known theorem that

$$0 = xy'z''+x'y''z+x''yz'-xy''z'-x'yz''-x''y'z$$

and hence, that the upper three of the following nine magnitudes

$$\begin{matrix} y'z''-y''z' & y''z-yz'' & yz'-y'z \\ z'x''-z''x' & z''x-zx'' & zx'-z'x \\ x'y''-x''y' & x''y-xy'' & xy'-x'y \end{matrix}$$

are proportional to the three middle ones and to the three lower, respectively.

It is easily concluded,

I. that these very magnitudes are also proportional to $f, f', f''$, since the three upper, middle and lower simply represent the doubled area of the projection of the triangles whose contents are $f, f', f''$ onto to the fundamental planes X_, Y_, Z_, and thus are proportional to them as the double cosine of the inclination of the orbit of $p$ with respect to these planes is to unity. (In a complete treatment, further remarks regarding the signs would be necessary, which can, however, also be easily circumvented by a mere assessment[2] .)

II. that if the three upper, the three middle, or the three lower are multiplied by $x, x', x''$, $y, y', y''$, or $z, z', z''$, the sum of the products will $= 0$. From this can easily be concluded

$$0 = fx + f'x' + f''x''$$
$$0 = fy + f'y' + f''y''$$
$$0 = fz + f'z' + f''z''$$

[2] *durch blossen Calcul*

-4-

**Fig. 12.39** Equation

Through completely analagous conclusions, is obtained

$$0 = FX + F'X' + F''X'$$
$$0 = FY + F'Y' + F''Y''$$
$$0 = FZ + F'Z' + F''Z''$$

From this, the following three equations can easily be derived

$$(F + F'')(f\xi + f'\xi' + f''\xi'') = (Ff'' - F''f)(X - X'') + [F'(f + f'') - (F + F'')f']X'$$
$$(F + F'')(f\eta + f'\eta' + f''\eta'') = (Ff'' - F''f)(Y - Y'') + [F'(f + f'') - (F + F'')f']Y'$$
$$(F + F'')(f\zeta + f'\zeta' + f''\zeta'') = (Ff'' - F''f)(Z - Z'') + [F'(f + f'') - (F + F'')f']Z'$$

From these three equations we derive four others, by multiplying them

| first with | then with | then with | and finally, with |
|---|---|---|---|
| $\eta\zeta'' - \eta''\zeta$ | $\eta Z' - Y'\zeta$ | $\eta' Z' - Y'\zeta'$ | $\eta'' Z' - Y'\zeta''$ |
| $\zeta\xi'' - \zeta''\xi$ | $\zeta X' - Z'\xi$ | $\zeta' X' - Z'\xi'$ | $\zeta'' X' - Z'\xi''$ |
| $\xi\eta'' - \xi''\eta$ | $\xi Y' - X'\eta$ | $\xi' Y' - X'\eta'$ | $\xi'' Y' - X'\eta''$ |

and adding the products.

For a convenient overview, we denote the sum of the products which arise

| | | | | | after multiplication of these factors with |
|---|---|---|---|---|---|
| | * | * | $\delta'\delta D'[\pi'\pi P']$ | $\delta''\delta D'[\pi''\pi P']$ | $\xi, \eta, \zeta$ |
| $[\pi\pi'\pi'']$ | $\delta\delta'\delta''\times[\pi\pi'\pi'']$ | $\delta\delta' D'[\pi\pi' P']$ | * | $\delta''\delta' D'[\pi''\pi' P']$ | $\xi', \eta', \zeta'$ |
| | * | $\delta\delta'' D'[\pi\pi'' P']$ | $\delta'\delta'' D'[\pi'\pi'' P']$ | * | $\xi'', \eta'', \zeta''$ |
| $[\pi P\pi'']$ | $\delta D\delta''[\pi P\pi'']$ | $\delta DD'[\pi PP']$ | $\delta' DD'[\pi' PP']$ | $\delta'' DD'[\pi'' PP']$ | $X, Y, Z$ |
| $[\pi P'\pi'']$ | $\delta D'\delta''[\pi P'\pi'']$ | * | * | * | $X', Y', Z'$ |
| | $\delta D''\delta''[\pi P''\pi'']$ | $\delta D'' D'[\pi P'' P']$ | $\delta' D'' D'[\pi' P'' P']$ | $\delta'' D'' D'[\pi'' P'' P']$ | $X'', Y'', Z''$ |

It is clear, that in the spots here filled with *, there must be a 0, and that all magnitudes indicated by bracketed symbols are *given*. That is to say

$$[\pi\pi'\pi''] = \tan\beta \cdot \sin(\lambda' - \lambda'') + \tan\beta' \cdot \sin(\lambda'' - \lambda) + \tan\beta'' \cdot \sin(\lambda - \lambda')$$
$$[\pi P\pi''] = \tan\beta \cdot \sin(L - \lambda'') + \tan\beta' \cdot \sin(\lambda'' - \lambda) + \tan\beta'' \cdot \sin(\lambda - L)$$

etc.

It is not necessary to put down all 16 equations here, since they can they can all be derived in an analagous way from the first, by simply exchanging $\beta$ with $\beta'$, $\beta''$, $B$, $B'$, $B''$ and $\lambda$ with $\lambda'$, $\lambda''$, $L$, $L'$, $L''$, if in the place of $\pi$ stands $\pi'$, $\pi''$, $P$, $P'$, $P''$ respectively, and so on. At the same time, it is seen that the 16 magnitudes are reduced to 12, since

-5-

**Fig. 12.40** Equation

$$+[\pi P'\pi''] = -[\pi\pi''P'] = +[\pi''\pi P']$$
$$[\pi\pi'P'] = -[\pi'\pi P']$$
$$[\pi''\pi'P'] = -[\pi'\pi''P']$$

Further, it is easily recognized that the expression $[\pi\pi'\pi'']$, multiplied by the product of the three cosines of the lattitudes which appear in them, is the sixfold volume of a pyramid, whose apex falls in the center, and whose three base angle points fall on the surface of a sphere described with radius 1, such that they correspond to three geocentric positions of $p$, and they will give positive or negative signs to the sixfold value, according as those three geocentric positions lie on the sphere in an opposite or the same sense as the positive poles[3] of the planes X_, Y_, Z_, respectively. And the remaining signs express entirely similar things.

In this way, the following four equations appear:

$$1)\ (F+F'')f'\delta'[\pi\pi'\pi''] = (Ff''-F''f)(D[\pi P\pi''] - D''[\pi P''\pi'']) + (F'(f+f'') - (F+F'')f')D'[\pi P'\pi'']$$

$$2)\ (F+F'')(f'\delta'[\pi\pi'P'] + f''\delta''[\pi\pi''P']) = (Ff''-F''f)(D[\pi PP'] - D''[\pi P''P'])$$

$$3)\ (F+F'')(f\delta[\pi'\pi P'] + f''\delta''[\pi'\pi''P']) = (Ff''-F''f)(D[\pi'PP'] - D''[\pi'P''P'])$$

$$4)\ (F+F'')(f\delta[\pi''\pi P'] + f'\delta'[\pi''\pi'P']) = (Ff''-F''f)(D[\pi''PP'] - D''[\pi''P''P'])$$

5.

We want to examine more closely [*näher betrachten*] these four equations, which are rigorously correct, in order to base our first approximation upon them. If we conceive of the time intervals as infinitely small magnitudes of the first order, then $f, f', f'', G, G', G''$ and all bracketed will also be of the first order, with the exception of $[\pi\pi'\pi'']$ which is of the third order. I omit the proofs, as well as the easily offered [*sich leicht darbietenden*] remarks on special exceptions. Should the inclination of the orbits of $p$ and $P$ relative to one another be considered as magnitudes of the first order, all bracketed magnitudes would stand one order higher. Further

$$Ff''-F''f = \frac{F}{G}\cdot\frac{f''}{g''}Gg'' - \frac{F''}{G''}\cdot\frac{f}{g}G''g$$

or (because $Gg'' = G''g$)

$$= \left(\frac{F}{G}\cdot\frac{f''}{g''} - \frac{F''}{G''}\cdot\frac{f}{g}\right)Gg''$$

Now $G$-$F$ is a magnitude of the third order, hence $1-\frac{F}{G}$ is one of the second, etc., and therefore $\frac{F}{G}\frac{f''}{g''} - \frac{F''}{G''}\frac{f}{g}$ is also of the second, and consequently $Ff''-F''f$ is of the fourth (it would actually be of the fifth, were $\tau'$ to fall halfway between $\tau$ and $\tau''$). Hence, what stands on the right hand side in the second, third, and fourth equations is of the fifth order, and of that which is on the left hand side, both the first and the second part are of the third. Thus, for the first approximation, we can set

from 2) $\qquad f'\delta'[\pi\pi'P'] = -f''\delta''[\pi\pi''P']$

from 4) $\qquad f\delta[\pi\pi''P'] = -f'\delta'[\pi'\pi''P']$

What follows from 3), is identical with these results. To achieve a further shortening [For the sake of

[3] I permit myself this easily understandable expression on account of brevity. The positive pole of X_ lies on the side of this plane where $x$ is considered positive, etc.

-6-

**Fig. 12.41** Equation

brevity? *Zur ferner Abkürzung*] we could set $\frac{f}{g} = \frac{f''}{g''}$, both of which magnitudes differ from unity by only the second order, and differ from each other by just as much (if $\tau'$ falls halfway between $\tau$ and $\tau''$, the difference is only of the third order). Since therefore: $g{:}-g'{:}g'' = \tau''-\tau'{:}\tau''-\tau{:}\tau'-\tau$, we will have

5) $$\delta = \frac{g}{f}\cdot\frac{f'}{g'}\cdot\frac{\tau''-\tau}{\tau''-\tau'}\cdot\frac{[\pi'\pi''P']}{[\pi\pi''P']}\cdot\delta'$$

6) $$\delta'' = \frac{g''}{f''}\cdot\frac{f'}{g'}\cdot\frac{\tau''-\tau}{\tau'-\tau}\cdot\frac{[\pi\pi'P']}{[\pi\pi''P']}\cdot\delta'$$

These formulas give $\delta$ and $\delta''$ from $\delta'$, up to and including the second order, if $\tau'$ lies between $\tau$ and $\tau''$, otherwise, the second order is exluded. In the latter case we can set $\frac{f'}{g'} = 1$, since the difference is only of the second order; on the other hand, in the first case the effort is not unuseful to set

$$f' = g' + \tfrac{4}{3}(f + f' + f'') \quad \text{or} \quad \frac{f'}{g'} = 1 + \frac{4}{3}\,\frac{f+f'+f''}{f'}$$

which will soon become more closely determined and is more exact by one order. (It is easily seen that $f + f' + f''$ is equal to the triangle between the three positions $p, p', p''$; thus, by a known approximation $= \frac{3}{4}$ xsection of the curved surface between the chord $pp''$ and the arc.) Moreover it follows from the above formula

$$\frac{\delta''}{\delta} = \frac{[\pi\pi'P']}{[\pi'\pi''P']}\cdot\frac{\tau''-\tau'}{\tau'-\tau}$$

which, if Z_ is taken for the ecliptic or $B, B', B'' = 0$, transforms itself into

$$\frac{\delta''}{\delta} = \frac{\tan\beta\sin(\lambda'-L')-\tan\beta'\sin(\lambda-L')}{\tan\beta'\sin(\lambda''-L')-\tan\beta''\sin(\lambda'-L')}\cdot\frac{\tau''-\tau'}{\tau'-\tau}$$

that is, the known, Olbersian formula.[4]

6.

According as/after/according to the way that we have derived tractable approximations from formulas 2, 3, and 4, we can undertake the first in a similar way. As is known

$$\frac{1}{r} = \frac{1}{k}(1 - e\cos(v-\pi))$$
$$\frac{1}{r'} = \frac{1}{k}(1 - e\cos(v'-\pi))$$
$$\frac{1}{r''} = \frac{1}{k}(1 - e\cos(v''-\pi))$$

It follows from this, that if these are multiplied with $\sin(v''-v')$, $\sin(v-v'')$, and $\sin(v'-v)$ and added, then

$$\frac{f+f'+f''}{rr'r''} = \frac{1}{k}(\sin(v''-v') + \sin(v-v'') + \sin(v'-v))$$
$$= -\frac{4}{k}\sin\tfrac{1}{2}(v''-v')\sin\tfrac{1}{2}(v-v'')\sin\tfrac{1}{2}(v'-v)$$

[4] Treatise on the Easiest Method etc. p. 45

-7-

**Fig. 12.42** Equation

or

$$\frac{f+f'+f''}{f'} = -\frac{2r'}{k} \cdot \frac{\sin\frac{1}{2}(v''-v')\sin\frac{1}{2}(v'-v)}{\cos\frac{1}{2}(v''-v)}$$

According to a known theorem of theoretical astronomy

$$\frac{\text{traversed space}}{\text{mean motion}} = \frac{a^{\frac{3}{2}} \cdot \sqrt{k}}{2}$$

Consequently

$$k = \frac{4gg'}{a^3(m'-m)(m''-m')} = \frac{4gg''}{A^3(M'-M)(M''-M')}$$

Therefore

$$\frac{f+f'+f''}{f'} = -\frac{A^3(M'-M)(M''-M')}{2\cos^2\frac{1}{2}(v''-v)} \cdot \frac{1}{r'^3} \cdot \frac{r'r'}{rr''} \cdot \frac{r'r''\sin\frac{1}{2}(v''-v')}{g} \cdot \frac{rr'\sin\frac{1}{2}(v'-v)}{g''}$$

$$\frac{f+f'+f''}{g'} = -\frac{r'r''\sin\frac{1}{2}(v''-v')}{g} \cdot \frac{rr''\sin\frac{1}{2}(v''-v)}{g'} \cdot \frac{rr'\sin\frac{1}{2}(v'-v)}{g''} \cdot \frac{A^3(M'-M)(M''-M)}{2r'^3} \cdot \frac{r'r'}{rr''}$$

It is easily concluded from this, since

$$\frac{1}{\cos\frac{1}{2}(v''-v)}, \quad \frac{r'r''\sin\frac{1}{2}(v''-v')}{g}, \quad \frac{rr'\sin\frac{1}{2}(v'-v)}{g''}$$

will only fall short of unity by a magnitude of the second order, as will $\frac{r'r}{rr''}$ if either $\tau'$ falls between $\tau$ and $\tau''$, or the difference between the path of $p$ and a circle can be considered as being of the first order, that we may set approximately

$$\frac{f+f'+f''}{f'} = -\frac{A^3}{2r'^3}(M'-M)(M''-M')$$

In the same manner is approximately

$$\frac{F+F'+F''}{F'} = -\frac{A^3}{2R'^3}(M'-M)(M''-M')$$

If it is so desired, the latter magnitude can also be exactly calculated, since everything relevant to it is given. Both are of the second order, and are determined up to, but excluding the fourth. Thus we have

$$F'(f+f'')-(F+F'')f' = F'(f+f'+f'')-(F+F'+F'')f'$$
$$= Ff'\tfrac{1}{2}A^3(M'-M)(M''-M')(\tfrac{1}{R'^3}-\tfrac{1}{r'^3})$$

which are magnitudes of the fourth order determined up to, and including the sixth. In equation 1 above, the part on the left hand side is of the fifth order; in the part on the right hand side, the first term is of the sixth or seventh order, that is to say $Ff''-F''f$ is of the fourth or fifth, and

-8-

**Fig. 12.43** Equation

$D[\pi P\pi''] - D''[\pi P''\pi'']$ is of the second.[5] The second term is of the fifth order. We omit the former, and thereby obtain

7) $$\frac{[\pi\pi'\pi'']}{[\pi P'\pi'']} \cdot \frac{2}{A^3(M'-M)(M''-M')} = -\left(\frac{1}{R'^3} - \frac{1}{r'^3}\right)\frac{D'}{\delta'}$$

accurate up to and excluding the second [order], if $\tau'$ falls halfway between $\tau$ and $\tau''$; otherwise, it is incorrect only by a magnitude of the first order. This formula, which receives the following form if we take the ecliptic for Z_, is the most important part of the entire method and is its first foundation.

$$\left\{1-\left(\frac{R'}{r'}\right)^3\right\} \cdot \frac{R'}{\delta'} = \frac{-2}{A^3 \cdot (M'-M)(M''-M')} \cdot \frac{\tan\beta'\sin(\lambda''-\lambda) - \tan\beta\sin(\lambda''-\lambda') - \tan\beta''\sin(\lambda'-\lambda)}{\tan\beta\sin(L'-\lambda'') - \tan\beta''\sin(L'-\lambda)}$$

where $L'$ is the longitude of the Sun $+180°$.

Since that which stands here on the right hand side is given, one sees that from the combination of this equation with the following

$$\frac{\frac{R'}{\delta'}}{\frac{R'}{r'}} = \sqrt{\left(1+\tan\beta'^2 + \frac{R'R'}{\delta'\delta'} + 2\frac{R'}{\delta'}\cos(\lambda'-L')\right)}$$

$r'$ can easily be found. The indirect method is here by far the most convenient. After a few trials [*Versuchen*], for which suitable prescriptions can easily be given, the goal is very quickly arrived at. It can also always be seen whether there is more than one value for $r'$, and thus more than one orbit that can represent the observations, which can indeed sometimes be the case.

Otherwise it is further to be remarked, that here the longitudes ought not be computed from the mobile equinoctial point, but rather from a fixed point; in practice, however, this difference is of no meaning. If the time is expressed in days, we obtain

$$\log(M'-M)(M''-M') = \log(\tau'-\tau) + \log(\tau''-\tau') + 6.4711352(-10)$$

(where, $M$, etc. must be expressed not in degrees, but rather in parts of the radius).

If we have $\delta'$ and $r'$, then $\frac{f+f'+f''}{f'}$ can also be determined, and thus also $\delta$ and $\delta''$. Moreover, from the consideration of formula 7), yet more interesting corollaries can be derived, which must be omitted here.

Second Point

*Approximate Determination of the Elements*

7.

We leave out entirely the middle observation for the time $\tau'$, and use instead the distances $\delta$ and $\delta''$, which were approximately determined in the preceding point. It is clear, that from this henceforth the heliocentric longitude, latitude and distance, can be derived; and hence, the longitude of Ω [ascending node symbol] and the inclination of the orbit, and the longitude in the orbit. Thus the problem still remains to determine the remaining elements, namely $a$, $e$, $\pi$, and the epoch,

from the two longitudes in the orbit . . . $v$ $v''$

[5] That is to say, that this is actually subtraction, since $[\pi P\pi'']$, $[\pi P''\pi'']$ have the same sign; this is not the case with the coefficients of $Ff''-F''f$ in equations 2, 3, and 4, but rather the parts will there actually be added. A deeper investigation would be too lengthy here.

-9-

**Fig. 12.44** Equation

the distances from the Sun . . . $r$ $r''$
and the corresponding times . . . $\tau$ $\tau''$

Since the relations of these magnitudes to the given ones are transcendental, we must again depend upon the indirect method. We will consider three here.

*First Method.*

If $\pi$ is taken as given,

$$\frac{r''+r}{r''-r}\tan\tfrac{1}{2}(v''-v)=\tan\zeta$$

so

$$e=\frac{\cos\zeta}{\cos\frac{1}{2}(v''-v)\cos[\pi-\frac{1}{2}(v+v'')-\zeta]}$$

It is therefore most advisable to calculate $k$ in a twofold way:

$$k=r[1-e\cos(v-\pi)]=r''[1-e\cos(v''-\pi)]$$

which will also serve to check the calculation. If $e$ is set $=\sin\varphi$, then $a=\frac{k}{\cos\varphi^2}$. The eccentric and mean anomalies and longitude can then be calculated from the true anomaly either by the usual method or more conveniently by the indirect method; from this, and from the mean motion given by $a$, is obtained a twofold determination of the mean longitude for any epoch whatsoever. If both agree the correct value for $\pi$ has been found; if not, the calculation must be repeated with a somewhat modified value for $\pi$, and the true value found through interpolation. 'It is advisable regarding this to seek the remaining elements *not* through interpolation, but rather from a new calculation from the corrected value of $\pi$, and not stop until both values for the epoch come into complete agreement.'

*Second Method.*

If $e$ is taken as given.

Here the calculation is exactly the same, only the true value must be approximately known, since here $\pi$ must be sought for by the equation

$$\cos[\pi-\tfrac{1}{2}(v+v'')-\zeta]=\frac{\cos\zeta}{e\cos\frac{1}{2}(v''-v)}$$

and two different values belong to the cosine.

Moreover II is prefered to I, and in general both methods are only appropriate if the traversed arc is indeed very large, and the elements are already approximately known. For the first approximation from a short series of observation, we must always depend on the following

*Third Method.*

If $k$ is taken as given.

Here,

-10-

**Fig. 12.45** Equation

$$\frac{\frac{1}{r''}-\frac{1}{r}}{2\sin\frac{1}{2}(v''-v)}=\frac{e}{k}\sin(\tfrac{1}{2}(v+v'')-\pi)$$

$$\frac{\frac{2}{k}-\frac{1}{r''}-\frac{1}{r}}{2\cos\frac{1}{2}(v''-v)}=\frac{e}{k}\cos(\tfrac{1}{2}(v+v'')-\pi)$$

Division thus gives $\tan[\frac{1}{2}(v+v'')-\pi]$, hence $\pi$, and afterwards $e$ from one of the two equations. The rest is entirely the same as in the foregoing methods.

The advantage of this third method consists in the fact that a very approximate value for $k$ can be found directly, if the arc $v''-v$ is not too large. Namely, the sector between the two radii vectors is

$$g'=\frac{a^{\frac{3}{2}}\sqrt{k}}{2}(m''-m)=\tfrac{1}{2}A^{\frac{3}{2}}(M''-M)\sqrt{k}$$

Thus $2g'=\int rrdw$ from $w=v$ to $w=v''$.

Now, however, according to the known approximation-integration formula of Cotes, $\int \varphi w\cdot dw$ from $w=v$ to $w=v''$ is

$$=(\tfrac{1}{2}\varphi v+\tfrac{1}{2}\varphi v'')(v''-v)$$

and still more exact

$$=(\tfrac{1}{6}\varphi v+\tfrac{2}{3}\varphi\tfrac{1}{2}(v+v'')+\tfrac{1}{6}\varphi v'')(v''-v)$$

still more exact

$$=(\tfrac{1}{8}\varphi v+\tfrac{3}{8}\varphi(\tfrac{2}{3}v+\tfrac{1}{3}v'')+\tfrac{3}{8}\varphi(\tfrac{1}{3}v+\tfrac{2}{3}v'')+\tfrac{1}{8}\varphi v'')(v''-v) \text{ etc.}$$

It is sufficient to stick with the first two.

From the first, we thus have

$$2g'=\tfrac{1}{2}(rr+r''r'')(v''-v) \text{ and } \sqrt{k}=\frac{\frac{1}{2}(rr+r''r'')}{A^{\frac{3}{2}}}\cdot\frac{v''-v}{M''-M}$$

$A$ is usually made $=1$; $v''-v$ and $M''-M$ are expressed in seconds, so $\log(M''-M)=\log(\tau''-\tau)+3.5500073$. In order to simplify the calculation, we set $\frac{r'}{r}=\tan(45°\pm\psi)$ whereby $\frac{1}{2}(rr+r''r'')=\frac{1}{\cos 2\psi}$.

From the second integration formula, the radius which belongs to the longitude $\frac{1}{2}(v''+v)$ is set $=r^*$ so

$$\frac{1}{r^*}=\frac{1}{2}(\frac{1}{r}+\frac{1}{r''})+[\frac{1}{2}(\frac{1}{r}+\frac{1}{r''})-\frac{1}{k}]\frac{2\sin^2\frac{1}{4}(v''-v)}{\cos\frac{1}{2}(v''-v)}$$

*-11-*

**Fig. 12.46** Equation

By means of this equation, $r^*$ can be determined by means of the first value of *k*. Then

$$2g' = \tfrac{1}{6}(rr + r''r'') + \tfrac{2}{3}r^*r^*$$

thus the new value of *k* is

$$= k\left\{1 + \frac{\frac{2}{3}\left(r^*r^* - \frac{1}{2}(rr + r''r'')\right)}{rr + r''r''}\right\}^2$$

In practice, it is usually exact enough, and more convenient, to search for the logarithms of the new values of *k* thereby, so that the logarithm of the first will be increased by $\frac{4}{3}\log\frac{r^*r^*}{\frac{1}{2}(rr+r''r'')}$. If it is desired to use the new values of *k* to determine the value of $\frac{1}{r^*}$ still more exactly according to the above equation, and accordingly to correct the value of *k*, thus, the twofold determination of the epoch will almost always agree so well, that absolutely no new assumptions will be necessary. With Ceres and Pallas, since $\tau''-\tau$ was of course 41 and 42 days, they always agreed to within a couple of hundredths of a second.

*Adjustment Methods*

8.

If the position for time $\tau'$ is calculated according to the approximated elements found through the foregoing method, and found to be in agreement with observation, the work is then complete. Usually the agreement will be very close (the difference often amounts to only a few seconds in my calculations) but rarely complete, partly because it is to a degree based on only approximate assumptions, and partly ***because the heliocentric positions which are used in it are not elliptical but rather they include small perturbations.*** The values of the small magnitudes of higher order which were discarded above could indeed be very closely determined from the approximated elements, and so the above formulas and the values of $\delta$ and $\delta''$ could thereby be improved; but I am of the opinion that these calculations would be far more difficult than one of the following methods.

The most simple first method of adjustment, which I first chanced upon at the inducement of Pallas, and which, because the intervals were small enough, I applied with the happiest success, is the following.

According to the approximated elements, ***which were found in the above manner***, let us set the calculated position for the time $\tau$ at longitude $= \lambda' + L\_$ and at latitude $= \beta' + B\_$, since those observed are $\lambda'$ and $\beta'$, so that all of the small inaccuracies in the assumptions conspire such that the longitude turns out to be too large by L_, and the latitude by B_; thus the orbit is calculated entirely anew, and ***in exactly the same way***, by using the observations

$$\lambda,\ \lambda' - L\_,\ \lambda''$$
$$\beta,\ \beta' - B\_,\ \beta''$$

as a basis. The result will be that the positions calculated from the new elements which follow from that will be so little different from $\lambda'$ and $\beta'$ (in my experience, only in parts of seconds) that it will require no further adjustment.

9.

Since the determination of the orbit will be based upon *the same* observations, the just indicated operation applies only in the case that it is applied to the first approximation. If it is desired

*-12-*

**Fig. 12.47** Equation

afterwards to adjust the elements by means of genuine [*lauter*] or partial [*zum Teil*] observations, I have, after many other tests, discovered the following two methods to be the most useful.

I. The heliocentric locations are calculated from two outer geocentric observations according to three hypotheses, by first assuming the approximate distances for these observations, and afterwards slightly adjusting first the one, then the others. The location for the middle observation is calculated according to the elements which were found in all three hypotheses, and is compared with the observed location. The corrected distances are then found through interpolation and, if desired, the corrected elements as well, though it is better not to avoid the effort of calculating these from the new distances through particular calculations, especially if the variations of the elements are yet very large.

$II^a$. Entirely the same procedure is made use of, only with the difference that instead of the approximate distances in the outer observations, the approximate determination of the inclination and of the ascending nodes are used, and each of these is somewhat adjusted.

$II^b$. The heliocentric positions are calculated from the three geocentric positions, partly with approximated and partly with somewhat adjusted determination of the inclination and of the ascending node; from the two outer heliocentric positions the elements are calculated, and from these elements the middle heliocentric observation is calculated, which is then compared with the positions derived from the observed geocentric positions, and then the improved inclination and ascending node are sought through interpolation, etc.

In $II^b$ the ellipse could also be determined from the *three heliocentric* positions according to the known formula, without simultaneously utilizing the times; the two intervals can be calculated from the dimensions of the ellipse and compared with the true, and then the corrected inclination and ascending node could be sought through interpolation, just as before. Only I had to reject this procedure in my experience. In this way an exact representation of the observations would only be attained only after repeated operations with far more trouble. To investigate the causes of this in detail would be too extensive.[6] I will merely remark that in this way the second differential, which we did away with immediately through the manipulations detailed in articles 5 and 6, is again brought about, and that these delicate second differentials can be *massively* distorted through a change in the inclination and ascending node which is itself not very large. It can easily happen here that a change of a few minutes in the ascending node or the inclination can bring about an ellipse which has almost no similarity at all with the previous one, hence, therefore the interpolation can understandably no longer be trusted. This is not the case with our method, which is always based on only two observations. *Sapienti sat.*-- From my repeated experience, I find the first method to be the most suitable and the most general.

Moreover, all of these methods are only applicable as long as the arc remains sufficiently large. If observations are taken of one or many years, other [methods] will again be necessary, upon which I am not able to expound more extensively here. In these cases it is generally not advisable to base the elements on three complete observations, but rather it is far more suitable to use four longitudes and two latitudes.-- If the observations encompass still more years, and if the elements ar determined *to within a small amount of error* [*bis auf kleinigkeiten*], I consider the method of the differential variations [*Differential-Änderungen*] whereby an arbitrary number of observations can be used as a basis, to be the best method.

[6] cf. Theoria motus corporum coelest. art. 93.

-*13*-

**Fig. 12.48** Equation

# References

*Goettingische gelehrte Anzeigen*, no. 198, 13 Dec. issue, pg. 1969-1973 (1810).
Jones, P.: 2001, Ancient and modern. The Spectator 286(9017), June 2 issue, 14.
La Grange, J.-B.: 1675, *Les Principes de la Philosophie*. Georges Josse, Paris.
Plackett, R.: 1972, Studies in the History of Probability and Statistics XXIX. The discovery of the method of least squares. *Biometrika* 59, 239-251.
Ridley, J.: 2001, review of Latin, or, the Empire of a Sign by Waquet. The Spectator 287(9028). Aug. 18 issue, 32.
Szpiro, G.: 2003, *Kepler's Conjecture*. John Wiley & Sons.
Grcar, J.F.: 2011, How ordinary elimination became Gaussian elimination. *Historia Mathematica* 38, 163-218.
Waquet, F.: 2001, Latin, or, The Empire of a Sign. Verso, London.

# Appendix A

## *Papers Relating to the Gauss Anagram*

Gauss published his famous Pallas anagram in the *Goettingische gelehrte Anzeigen* (Learned annnouncements of Goettingen). The *GGA* was founded in 1739 and was published in Gothic script under the auspices of the "Prussian Royal Society of Science. During the eighteenth and nineteenth centuries, it was the foremost review journal in the world." The journal is now published by the "Academy of Science of Goettingen and is the only review journal in existence with such a long history." Many of Gauss's articles about the asteroids were published in the *GGA*. (Szpiro 2003: 105) All papers written about the anagram are presented here, spanning the period 1930–1991. To provide the historical background, given first are extracts from a paper by the French mathematician LeVerrier (1841; Fig. A.1).

> *No precise calculations have been possible up to now on the little planets discovered at the beginning of this century. Their positions given in advance in the ephemerides differ almost always considerably from those obtained later by observation; and the works of the astronomers, far from having solved this difficulty, have barely indicated its cause….*
>
> *Let us consider for example the planet Pallas, whose eccentricity is 0.242 and whose orbit is inclined by 34° 15' on the orbit of Jupiter; and let us see if this large inclination forbids the convergence of the sine and cosine series that represent the perturbing function corresponding to the action of Jupiter. Forgetting for the moment any idea of algebraic development, we see that large changes in the values of the perturbing function for small variations in the longitudes of planets would force to extend the series up to high multiples of the mean longitudes, while conversely a periodic function that would vary very slowly with the longitudes of planets could be represented in general by a small number of terms. Thus, any circumstance such that the variations of the perturbing function are slower would be favourable to the convergence of its development. It appears that the large inclination of the orbit of Pallas on that of Jupiter is precisely such a circumstance.*
>
> *The aphelia of Pallas is at 54° of the intersection [node] of these orbits. Let us assume that when the planet is at its aphelia Jupiter has the same longitude, the case for which the perturbing function varies the most rapidly. Pallas, from the time when it is at the node until that when it is at its aphelia, comes closer to Jupiter, but much less than if its orbit*



C.J. Cunningham, *Studies of Pallas in the Early Nineteenth Century*,
DOI 10.1007/978-3-319-32848-5

**Fig. A.1** Urban LeVerrier

*was in the same plane as that of Jupiter. So, we are led to think that the mutual inclination of orbits can be favourable to the development of the perturbing function and that it can to some extent compensate for the inconvenience of the large eccentricity.*

*The difference between the two angles is only 1631 arc seconds, and the corresponding perturbation is of the eleventh order with respect to the powers of eccentricities and inclinations. But it does not seem that a good classification should be based on the powers of the eccentricities and inclinations if they have large values. Rather, it should be based on the magnitudes of the multipliers of the mean longitudes in the sine and cosine terms...*

*I have shown in this way that the perturbing function coming from the action of Jupiter on Pallas can be effectively developed as a converging series of sines and cosines; that in general the coefficients decrease for higher indices, and that they can be classed using these indices rather than according to their orders with respect to the eccentricities and the inclinations. So, the coefficients of the sine and cosine of the arc (18l'-7l) are much larger than those of the sine and cosine of the arcs (18l'-18l), (18l'-17l), (18l'-16l), although algebraically the first terms are of order eleven with respect to the eccentricities and the inclinations, while the last ones are of low order.*

*I believe that I can conclude that the large inclination of the orbit of Pallas should not bring to its theory as many difficulties as generally believed; that conversely it is possible to determine its perturbations as series of sines and cosines, as astronomers usually do; and, what is of great importance, there is no other equation containing high indices than those with long periods that the smallness of their argument makes sensible. There is only one such perturbation in the mean motion of Pallas: that which depends on the argument 8n' – 7n. But it is very large, as I had foreseen (Comptes rendus des séances de l'Académie des Sciences, t. XI, p701).*

## About Solving the mysterious Gauss Anagram

Wietzke (1930)

*On March 20, 1812, Gauss wrote Olbers among other things: "that the mean motion of Pallas is almost exactly 18/7 times that of Jupiter. It therefore must either be an immense equation, of 7⚴- 18♃ dependent, or both must be equal. I consider the latter more likely (comparable to the fact that there are similar phenomena regarding the rotation of the Moon and Jupiter's satellites); but I can only decide this matter if my work has been carried out with improved elements in twice the dimension".*

*In another letter of March 31, 1812, to Olbers, Gauss said: "I am almost certain that 18 ♃ and 7 ⚴ are equal. This appears to me to be one of the most peculiar discoveries made recently in our Solar System, which I am begging you to keep completely to yourself. My calculations gave me:*
*mot. med. diurn. sid.* ⚴ 769″.20208

| | |
|---|---|
| *7 times* ⚴ | *5384.41455* |
| *18* ♃ | *5384.39227* |

*They could turn out 100 times more different and still the influence of ♃ must lead to a complete mean equity, if my rough calculation is correct. Thus I am impatiently awaiting ⚴ reappearance this year to determine the mean motion better from the observations."*

*Olbers answered on April 5, 1812: "I also consider your discovery that 18♃ exactly = 7 ⚴ one of the most peculiar that have been made in our Solar System for some time."*

*Gauss published in the* Goettingische Gelehrten Anzeigen *of April 25, 1812, his discovery as the following cipher*

$$111000100101001$$

*to which, as he wrote, he would give the key in due time. In order to save the priority of his discovery he enrobed the result in the shape of this mystical anagram. Shortly thereafter, he communicated his discovery to Bessel (on May 5, 1812), and, again, asked to keep it secret.*

*Gauss, however, did not give the key to the above anagram later on. But it is not uninteresting to search for one.*

*M. Brendel says: "Keeping in mind how such ciphers are usually chosen, we dare not hope to find the solution." In a footnote he made the following remark: "In the dyadic system 111 is the number 7, and 10010 the number 18." But apparently, this does not lead far. F. Klein says about Gauss's anagram: "Regarding the cipher it is still today not solved despite Brendel's efforts."*

*I am therefore permitted to give the following solution. The main thing is to connect the numbers 7 and 18 in such a manner that the emerging ratio can be somehow expressed in the above anagram. This indeed is possible. I am not aiming at the true value of some numbers made up of the digits 1 and 0, these only have symbolic meaning. It is*

$$\begin{array}{llll} 1111 & \textit{the} & \textit{number} & 4 \\ 000 & \textit{the} & \textit{number} & 3 \end{array}$$

*and so forth alternately. Then the anagram results in the figure 431211121 from which the following fraction is obtained by breaking off after the fourth digit and using the following numbers as denominator:*

$$\frac{4312}{11121} = 0.3877 \textit{ etc.} \qquad (1)$$

*If one creates the quotient of the mean diurnal motion of ♃ and ⚴, using the values of Gauss mentioned in the above mentioned letter of March 31, 1812, one gets:*

$$\eta'/\eta = 299''.13290/769'.20208 = 0.3888872\ etc. \qquad (2)$$

*and finally:*

$$\frac{7}{18} = 0.3888 \qquad (3)$$

*These three values (1), (2) and (3) are in good correspondence.*

*The following remarks may be added:*

1. *Gauss's value* $(2)<(3)$*, also* $(1)<(3)$*.*
2. *Why is not* $\underline{4312} = \underline{392}$ *?*

11121 1011 – *The zero of the denominator*

*could not be expressed according to the above symbolic terminology, the '9' of the numerator would be impractical, because for this alone 9 digits would be necessary in the anagram. On the other hand in the interest of concealment the fraction should at least contain in the denominator and numerator four respectively five digit numbers since otherwise the value 7/18 is too easily recognized.*

3. *And actually, the fraction 4312/11121 has remarkably small digits. There is hardly any other fraction with such a good approximate value for 7/18 which despite nine digits shows for the quantity of these digits almost minimal values. By this the anagram gets despite its nine digits a minimal dimension, an advantage which certainly is intended.*
4. *And also the digits 0 and 1 were used in the anagram because they are the smallest digits.*

*Bremen, 1930 Nov 10*

## The Anagram of Gauss
MacDonald (1931)

*Let the anagram be written 111, 1000, 10010, 1001. In the dyadic system, this gives 7, 8, 18, 9. At the date (1812), Pallas was the eighth planet from the Sun, Jupiter the ninth. Hence* $7 - 8^{th} = 18 - 9^{th}$*. This simple explanation occurred to me while lecturing on the history of astronomy; I did not think it worth publishing; perhaps it may be of interest now.*
[Dated Glasgow, Scotland Dec. 22, 1930.]

## For a solution of a mysterious anagram by Gauss
Maennchen, Giessen (1934)

*On April 25, 1812, Gauss published in the Goettingischen Gelehrten Anzeigen the cipher:*

1111 0001 0010 1001

*and promised to give the key in due time. But he never did so and even after his death nothing could be found in his papers. But he stated in a letter to Bessel (Briefwechsel Gauss-Bessel, p. 170), that it is about the important discovery that 7* ⚴ = 18 ♃ is, i.e. that the sevenfold mean motion of Pallas equals the eighteenfold of Jupiter. Gauss informed Olbers earlier about this adding, that he considers it one of the most important discoveries made for some time in the Solar System. Olbers agreed to this in his written reply. The attempts to decipher the anagram thus had to tie up this connexion. Of all attempts to find a solution which have become known hitherto, I only want to deal with the one which seems most likely to me. It was published by *MacDonald in 1931 in the Astronomische Nachrichten (vol. 241, p. 31). MacDonald divides the sixteen-digit number into the numbers 111, 1000, 10010, 1001, of which he assumes they were written in the dyadic system. That means 7, 8, 18, 9 in the decimal system. At the time when Gauss published the anagram, Pallas was the* $8^{th}$ *planet in regards of its distance from the Sun and Jupiter the* $9^{th}$*. So if one chooses the first and third number as factors, the second and fourth as numbers, one obtains 7* ⚴ = 8 ♃.

*Certainly, this explanation is extremely simple and by all means effortless, what cannot be said of the other attempts which have to become known. But exactly this extreme simplicity worries me. My studies of the handwritten remains of this ingenious calculator have created my conviction that Gauss never wasted his time with trifles but that he even provided games with difficulties. Therefore, it is my opinion not to be assumed that designed a simple run-of-the-mill anagram but that he also squeezed into it all kinds of other qualities. Thus we now want to try if nothing else can be read out of it. – If one converts the 16-digit number as a whole from the dyadic system into the decimal system, the number 61737 originates. If one assigns to each number of this figure the accordant letter of the alphabet, FAGCG results. If one starts with C and always skips one letter (the last letter is followed by the first again), the sequence CFGGA is obtained; this can be interpreted as: CARL FRIEDRICH GAUSS, Georgia Augusta. If this should be coincidence, it would rather be a very peculiar one. But that is not all. The first four numbers alone result in the number 15, the last four in 9; to the first belongs the letter P (Pallas) and to the latter J (Jupiter). The first three numbers give – as is also assumed by MacDonald– the number 7, the last five in reverse order (from right to left) the number 18 so that even here the important information is included but also the address of the discoverer. – Why I consider this explanation more likely? Not because it is mine; for I already found it in September 1929 and communicated it to Mr. Brendel with sceptical remarks. But I avoided a publication since I was missing the most important until recently: How may the anagram have developed? In order to fathom this, we want to try first to develop a cipher which should include:*

1. *the initial letters of Pallas and Jupiter (at the beginning and the end)*
2. *the number 7 and 18 at the accordant positions and*
3. *the numbers of the letters C, F, G, G, A in any order*

*Let us begin with the first condition: consequently, the anagram has to start with 1111 and end with 1001. Now the second requirement: 111 means 7 and since the number 10010 means 18, but the number has to end with 1001, it remains nothing else than reading it* *backwards. meaning to put a 0 before 1001 2. The third requirement leads to the following task: to build a 5 digit number that begins in the dyadic system with 1111 and ends with 01001 and which, transformed into the decimal system, consists of the number 3, 6, 7, 7, 1 in any order. Since 1111 means the number 15, the fifteenfold of of an appropriate power of two be a five-digit number. Three possibilities occur: 2 16 = 1024, 2 11 = 2048, 2 12 = 4096. In the first case the number written in the dyadic system becomes a 14-digit number, in the second a 15-digit in the third a 16-digit number. Since the number with the most digits offer the most possibilities, we see that the number becomes 16-digit. Thus the number is, inserting for the unknown figures dots (which mean zeros for now): 1111 ……. 01001 = 15 × 4096 + 9 = 61449. The figures 6 and 1 already appear, the figures 7, 7 and 3 are to be obtained by adding an appropriate number, in a way, that either 61773 or 61737 or finally 61377 are obtained. The last case can be ruled out anyway, since 61377 is smaller than 61449. For the usability of the other two cases it is relevant that the number to be added is a multiple of $2^5$ since the last five positions are already taken. Now 61377 – 61449 = 324 cannot be divided by 32, but 61737 – 61449 = 228 = 9 × 32. Consequently, the seven dots have to be replaced by figures that the number 9, dyadic 1001, yields and thus the first three dots have to be zeros. Thus Gauss's anagram develops, or rather, it might have been formed this way. As one can see, three features were forced in this procedure, and the one MacDonald chooses as explanation, arises of its own volition by fortunate chance; for it is coincidence that Pallas was the eighth and Jupiter the ninth planet then. If on the other hand Gauss had proceeded as MacDonald assumed, the three other features would have dropped into his lap as a lucky bonus. But it is more likely, that if one compels the three features a fourth occurs accidentally than rather if one forces one, three others join the one. The latter only happens to a markedly lucky devil which Gauss never was during his entire life.* [Philipp Maennchen, 1869–1945]

## Uber das Anagramm von C. F. Gauss
Goellnitz (1937)

*On April 25, 1812, Gauss published the following anagram*

111 1000 10010 1001

*without leaving a key for deciphering. In a letter to Bessel of May 5, 1812, he mentioned, that it means that the mean motions of the planets Jupiter and Pallas are in the ratio of 7:18. If these mean motions are denominated with the Greek letters ι and π, it is*

$$7 : 18 = \iota : \pi \ or\ 7\pi = 18\iota$$

*The sequence αβγδεζηθι contains nine letters and the sentence ικλμνξοπ eight letters. Therefore symbolically:*

*7.(8) = 18.(9)*
*or in the dyadic system*
*111.(1000) = 10010.(1001).*

*If the arithmetic operators and brackets are omitted, Gauss' anagram emerges.*

**Fig. A.2** Kurt-Reinhard Biermann.

*"Mr. C. Mueller, Hanover, reads the above interpretation of the anagram as follows: i (Jupiter) is the ninth letter of the Latin alphabet, π [παλλας Greek Pallas] is the 17th letter of the Greek numerical alphabet (since for the number 6 the letter ς was inserted). It now is symbolically $\pi \equiv 8(\mathrm{mod}\ i)$, because $17 \equiv 8(\mathrm{mod} 9)$. If it is consequently assumed that i: π = 7:18 it follows symbolically from inserting i = (9) and π = (8): 7.(8) ≈ 18. (9)."*

## On Gauss's Cryptogram of 1812

Biermann (1970; Fig. A2)

*1. In my attempts to interpret Gauss's key words and ciphers*, I already mentioned the cryptogram which Gauss used for expressing that "the mean motions of Jupiter and Pallas are in a ratio of 7:18". In this case the main meaning of the cypher is known, contrary to the other mentioned cipherings, but Gauss did not realize his intention to publish the key later. It is thus understandable that several attempts have been made for an interpretation.*

*Some of these solutions are discussed here.*

*2. The cryptogram is*

1111000100101001

*2.1. In 1919 Brendel pointed out that in the dyadic system 7 is written as 111 and 18 as 10010. By this the first three digits and the $8^{th}$ to $12^{th}$ are interpreted, but Mr. F. Klein could justly announce in his Vorlesungen that the ciphering has not been solved despite Brendel's efforts.*

*2.2. A. Wietzke chose a different approach in 1930. He transformed the number offollowing equal digits into a number of the decadic system: 1111=4, 000 = 3, 1=1, 00=2 etc. and thus found the number 431211121. From this he developed the quotient*

*4312 : 11121 = 0.3877 ≈ 7/18*

*This interpretation however is neither satisfying nor popular.*

*2.3. In 1931 T.L. MacDonald gave an elegant solution in writing the cipher as follows:*

$$111.1000 = 10010.1001,$$

*in the decadic system: 7.8 = 18.9.*

*8 stands for Pallas and 9 for Jupiter since regarding the mean distance from the Sun, Pallas was the 8th and Jupiter the 9th planet.*

*If one accepts this bribingly easy interpretation, the generally accepted name in literature for this ciphering 'anagram' is not correct, for this notion implies displacement. But here every component would be already at the right place. By the way, Gauss himself did not speak of an 'anagram' but 'cipher' [German: chiffre].*

*Especially the lack of intricacy of this interpretation gave reason for doubt.*

*2.4. In 1934 Mr. P. Maennchen explained: "My studies of this numerical genius' unpublished works lead to my conviction that Gauss would never deal with simple matters but even make mathematical games complicated." At the same time Maennchen also published a solution, which he communicated earlier to Brendel, but only in a letter because he wanted to show first how Gauss could have made the cryptogram. After having found the answer to this question, he published it together with his solution that looked like:*

*In the decadic system the cryptogram is 61737. If A = 1, B = 2, C = 3, ..., then*

*61737 = F A G C G*

*II V III I IV*

> *Arranged in the order I, II, ..., V it follows:*
> *[C]arl [F]riedrich G[]auss [G]eogia [A]ugusta.*
> *And the first four digits: 1111 = 15 = P[allas]*
> *the last four digits: 1001 = 9 = I[upiter]*
> *the first three letters: 111 = 7 and*
> *the last five letters read backwards: 10010 = 18.*

*2.5. A 9 was read out by MacDonald and Maennchen and both took it for Jupiter with the difference that Jupiter was for MacDonald the ninth planet according to the knowledge of 1812, and Maennchen came up with the letter I (I and J are not distinguished) as the ninth of the alphabet. Maennchen did not consider MacDonald's solution (1000 = 8 and 1001 = 9) solid because it was according to him too simple and it was "a coincidence that at that time* [1812] *Pallas was the 8th and Jupiter the 9th planet."*

*The second reprimand seems to be highly challengeable; for the word 'coincidence' is totally out of place here. Gauss had to act on the knowledge of 1812; the next discovery of a planetoid (Astraea) was made by Hencke in 1845.*

*And it is hardly necessary for an author to include his name and place of work in the cipher if the place is generally known and the name is already given in Goettingische [!] gelehrten Anzeigen (no. 67 of 4/25/1812). Maennchen's interpretation CFGGA has the little blemish of being unnecessary. Maybe that is why this solution does not convince entirely, even if Maennchen may have reconstructed a method possibly used by Gauss for forming his cipher interpreted as anagram. It did not win recognition.*

*2.6. In 1956 M. F. Subbotin made the same judgement as Brendel in the Soviet commemorative article.* [C. F. Gauss. Moscow, 1956] [Ed[1]]

*3. The key words mentioned above were a fixation of the discoveries and their further consolidation; the cipher is a review of the different stages – both cases are records for private use only in which Gauss let run free his addiction of inserting obstacles. The main topic of the cryptogram in question is different: Gauss's main objective was to secure publicly his claim to a new conclusion, which he considered "one of the most peculiar made in our Solar System" (But he was only "almost certain" and asked Olbers, who was the first to be told, "to keep it secret." Before a publication in plain words he wanted to determine Pallas' mean motion more accurately.)*

*That the result referred to Pallas and Jupiter could be learnt from the accompanying text. We thus consider us justified assuming that Gauss's main objective was, when he started forming his cipher, not the encoding of Pallas and Jupiter but the ratio of 7/18 and the date, which is just as important as the discovery itself, at which he wanted his discovery to be recorded. The almost geometrical law of Titius-Bode (1766 resp. 1772) was a support of Kepler's assumption of a planet between Mars and Jupiter and Herschel*'s *discovery of Uranus in 1781 was further proof of this rule of the distances. The systematic search for a moving star between the minor and major planets was crowned by success – Four small planets had been found in the gap: Piazzi discovered Ceres in 1801, Olbers Pallas in 1802, Harding Juno in 1804 and in 1807 Olbers discovered Vesta. The discovery of a ratio of the period of revolution of a planetoid to that of Jupiter must, if it proved true, be of the utmost interest to any astronomer! Gauss was not yet as relaxed as he was later in his life regarding his claim of priority. This seems to be the only case that he published a result of which he was not entirely certain. Furthermore, he knew from the history of discoveries that in case of such a burning actual question in science the possibility of almost simultaneous discoveries of independently working astronomers could not be dismissed. And keeping in mind the tremendous effort which was associated with the calculations – the repetition of the calculation of the general perturbations of Pallas by Jupiter demanded from Gauss between April 5 and November 25, 1812, to write every day on average 3,500 (!) numbers – it is evident he included a date in his cryptogram. And as a matter of fact, the cipher contains a date: 04/03/1812. Thus there were to include:*

*1812 = 11100010100*
*April = 100*
*3= 11*
*7= 111*
*18 = 10010*

[1]Subbotin, M. F.: 1956, K. F. Gauss's Activity in the Field of Astronomy and Geodesy. *Problems in the History of Natural Sciences and Engineering* 1, 54-69. in Russian.

*Of course a man like Gauss saw at first glance that these numbers already contained*

*Pallas = 8 = 1000 and*
*Jupiter = 9 = 1001*

*a welcome addition, but not a necessity, since as just mentioned, Pallas and Jupiter were mentioned in the context. Now only a displacement of the date has to take place:*

*1812 = 111000 10100*
*April = 100*
*3= 1*
*7= 111*
*8 = 1000*
*18 = 10010*
*9 = 1001*
*therefore:1111000100101001*

*It remains to check whether the date 4/3/1812 is in accord with the known facts. It is as follows: On March 31 Gauss notified Olbers and on April 5 he started checking his discovery! On April 25 the publication followed and on May 5 he informed Bessel about the meaning of the codification. Unfortunately, Gauss's manuscript for the publication in the* Goettingischen gelehrten Anzeigen *is no longer available so that the publication date can no longer be identified. I therefore come to the conclusion thatMacDonald'sinterpretation is probably the correct one, but would like to add the dating contained in the cipher. Since the date appears by displacement it is an anagram indeed.*

*4. This raises the question whether a new discussion of a problem, which eventually is unsolvable – at least as long as no unknown papers by Gauss appear – is useful at all, especially since the meaning of the anagram, as far as of scientific nature, became known by the correspondence between Gauss and Bessel in 1880. I would like to give an affirmative answer to this question, because of the same reasons I gave for my earlier interpretation attempts: Gauss's scientific legacy has not been fully developed; there is still the possibility offinding further actual and important ideas in unpublished papers. Since Gauss loved to cover all tracks of finding solutions and use ciphers, all considerations might be permitted regarding the different possibilities of their decoding. They can help to permeate the train of thoughts and work methods of, as Alexander von Humboldt put it, "the first mathematician of his era, the great astronomer, the ingenious physicist".*

*Kurt-Reinhard Biermann (1919–2000). His two previous papers were published in Mber. Dt. Akad. Wiss. (vol. 5, p. 241, 1963; and vol. 11, p. 526, 1969)

**The Gauss Anagram: An Alternative Solution**
Benham (1974)

[Extracts from Benham's rather fanciful English-language paper follows. This 1974 paper is the first one about the Gauss anagram in English since the single paragraph MacDonald wrote in 1931. After quoting from Gauss's *GGA* paper of 1812 and his letter to Bessel of this same year, Benham continues as follows.]

*Besides conveying the impression that the Gauss anagram must in some way show this ratio, the statement, by its reference to the influence of Jupiter in holding Pallas to an orbital period tied to that of Jupiter, is important in illustrating for the purpose of protecting his priority. This point will be enlarged upon in the final section of this paper. Did Gauss intend a second ratio?*

*The next reference to the meaning of the anagram appears to have been made after the lapse of no less than thirty-eight years from the date of publication of the ratio 7:18. This reference occurs in the seventh volume of a series of eight, published between 1911 and 1929, dealing with and elucidating the considerable achievements of Gauss in both pure and applied mathematics. The series was assembled by F. Klein and M. Brendel, astronomical topics being discussed by the latter. He cited the anagram and noted Gauss promised to give the solution. He then wrote:*

> *Keeping in mind how such ciphers are usually chosen,*
> *we dare not hope to find the solution. Since one*
> *knows from the correspondence what it means,*
> *this also is pointless.*

*In a footnote he made the following remark: "In the dyadic system 111 is the number 7, and 10010 the number 18." This observation, if logically pursued, leads to the numbers 1000 and 1001 being left when the numbers 111 and 10010 are removed, so that an explanation of the former numbers (denary 8 and 9) could give point to the anagram. There are two possible approaches: to try and link the mean motions of the two planets, Pallas and Jupiter, symbolically with the numbers 8 and 9 respectively; and to regard the four separated numbers as all referring to the mean motions (or possibly the time periods) of four different planets, the two not named by Gauss having their orbits intermediate between Jupiter and Pallas. As no planets moving in that zone were known in 1812, this would mean that Gauss was predicting the presence of such planets. This is the preferred approach, since, by his allusion to the influence of Jupiter and to the enormous number of equations involved, the anagram which follows suggests that Gauss was emphasizing the remarkable results flowing from these equations.*

*In respect of the first line of approach, Erich Goellnitz appears to be the first to have had success.* [His paper, included in this appendix, is then printed by Benham.]

*Of the above two solutions, that of Goellnitz is the more straightforward. Although he must have seen the anagram in the undivided state, as shown in the only publication he cites [namely, the letter of 1812 to Bessel] his forthright use of the separated groups of digits indicates his conviction that the number was binary and that Gauss intended this subdivision. His choice of the Greek alphabet gives rise to the query: Would not the Roman alphabet have done just as well? For in Roman inscriptions Jove (=Jupiter) is written IOVE, so, as ι is the ninth letter of the Roman alphabet, one obtains the required number 9. Another point is the use of the difference in position between ι and π rather than the alphabetical position of π. Had Gauss really meant the numbers 8 and 9 to be interpreted by an alphabet, would not he have used 16 instead of 8? The sixteenth letter of the Roman alphabet is p and in the classical Greek alphabet it is π. This would identify Pallas as 16 and overcome a possible objection to Goellnitz's sequences of letters, both of which contain ι. With regard to Mueller's variation, it has the interesting arbitrary feature of using i (not ι), the ninth letter of the Latin alphabet for Jupiter, and π, the seventeenth letter in the Greek number alphabet for Pallas. Using a notation that Gauss had devised in the theory of numbers, namely*

$$D = a(mod\ d),$$

*which signifies that d divides D leaving a remainder a, Mueller writes*

$$17 = 8(mod\ 9),$$

*where 8 and 9 are the mean motions of Pallas and Jupiter respectively. This seems to be a weakness in that Pallas is associated with two numbers, 17 and 8, whereas Jupiter is associated with number 9 only.*

*A third proposal based on the first line of approach will now be offered. It is well known that Gauss used his 'Three Point Method' to evaluate the orbits of minor planets 1 to 4, the order in which they stand today being unaltered from that which prevailed in 1812. Gauss had at first some difficulty in determining the relative mean positions of Ceres and Pallas, which are extremely close together. In his magnum opus (Theoria motus) Gauss is evidently reluctant to mention Pallas by name, though in Art. 56, page 63 of Davis' translation (of 1857), we find mention of an orbital inclination of 34° 38′ 1″.1, which makes one think that Gauss had arrived at this value, or one close to it, for Pallas. In Art. 159, page 321 we find for Ceres a mean daily motion of 771″.5855, or 0°.21432903. Now the earth describes less than one degree of arc when moving in its orbit during one day (or 1/365.25636 sidereal year). During a sidereal year Ceres moves through the angle 0°.21432903 × 365.25636, which on conversion to revolutions per year leads to a mean annual revolution of 0.2174582 of a complete revolution. The reciprocal of this quantity gives the sidereal period of Ceres as 4.598580 years, which is less than the corresponding figure for Pallas. The latter, clearly, on Gauss's own showing is very close to 7/18 of Jupiter's period, any small deviation being caused by perturbations of Pallas by other planets, notably Mars and Saturn. Taking Jupiter's sidereal period as 11.86186 sidereal years (the equivalent of 11.86223 tropical years), one obtains for 7/18 of this value 4.612945 sidereal years (equivalent to 4.613206 tropical years). Comparison with Ceres reveals that Pallas has a slightly longer period, and is therefore further from the Sun. The object of the above discussion is to show what Gauss presumably considered was the order of all planets of the Solar System between Mercury and Jupiter, including the only minor planets known in 1812. The order is thus: (1) Mercury, (2) Venus, (3) Earth, (4) Mars, (5) Vesta, (6) Juno, (7) Ceres, (8) Pallas and (9) Jupiter. Perhaps Gauss wished to associate the planets Pallas and Jupiter with the numbers 8 and 9 in order to draw attention to the fact he had in fact been personally responsible for working out the orbits of the minor planets and placing them in their correct order from the Sun.*

*We now come to the second line of approach. Did Gauss wish to signify an additional ratio in his anagram? The present writer's alternative solution is the following. Taking the anagram as binary, and looking at the separated numbers*

$$7\ 8\ 18\ 9,$$

*it is noticeable that 7 and 18 are not in juxtaposition. Could it be that these four numbers are proportional either to the mean motions, or to the periods of planets Pallas, X, Jupiter and Y, where X and Y refer to two (as then undiscovered) planets in orbit intermediate between Jupiter and Pallas? The sequence of these numbers is not to be regarded in the same manner as was done in the first line of approach. The important point now is to demonstrate that the numbers 8 and 9 are intended to show the ratio of mean motions of two unknown planets.*

*Already in 1802, the year of the discovery of Pallas, Gauss must have realised that his result for these planets gave orbital radii in close agreement, not only with each other, but also with Bode's law. It must have caused considerable embarrassment to supporters of Bode's law when two more minor planets, Juno and Vesta, were found to be much nearer the Sun than the first discovered pair, Ceres and Pallas. This situation may perhaps have caused Gauss to pay even more attention to Ceres and Pallas. The existence of favored orbits, predicted by him in the special case of minor planets perturbed by Jupiter, must have led to a plurality of ratios* pari passu [without partiality] *with the ratio 7:18, and Gauss may have considered this question vital. If we write the numbers in the reverse order:*

$$9 \; 18 \; 8 \; 7,$$

*and divide by 18, we have*

$$\tfrac{1}{2} \; 1 \; \tfrac{4}{9} \, \tfrac{7}{18}$$

*Now the ratio ½ is today recognized as the most unfavorable ratio of periods, and it is hardly likely that Gauss did not know this, despite the fact that the existence of gaps in the zone of minor planets was established later. One may therefore suggest that Gauss elected to show in his anagram that he had arrived mathematically at an expression (given later below) for all ratios of time periods favourable to the maintained presence of minor planets perturbed strongly by Jupiter. This he could do by showing only four terms, which are almost (but not quite) sufficient to cover the accumulations of minor planets known even today. Thus, a fifth term, as we shall see, is one giving the ratio 2/9, and in the zone corresponding to this ratio occurs the Hungaria group. This zone virtually represents planets circulating near the sharp inner edge of the asteroid belt* [this includes 4276 Clifford], *where perturbations by Mars and other interior planets are beginning to be sensible. The zones corresponding to the ratios* $\tfrac{3}{4}$ *and* $\tfrac{2}{3}$ *are represented respectively by Thule (1 planet) and Hilda (23 numbered planets with many unnumbered). Gauss, though not aware of any group other than the important one then consisting only of Pallas and Ceres, could still have arrived at the present writer's sequence F(n) (defined below), though there does not appear to be any record of this having been published.*

*With present-day knowledge, more particularly that of the ratio* $\tfrac{2}{3}$, *it is comparatively easy to obtain the sequence* (Fig. A.3)

*Where in practice a finite number of terms suffices, because as n increases the influence of Jupiter decreases and that of the four interior planets increases, so that the sequence no longer gives the exact periods of the favored orbits. The number of terms for which the observed orbits remain in good agreement with theory is not likely to exceed 6; for when n is 5 we obtain as sixth term the ratio 7/54, which corresponds to an orbit of period 1.5376 years, well inside the orbit of Mars (1.8809). Thus even the sixth term corresponds to no known planet (major or minor), unless one admits the possibility that Mars was in fact in the 'Jupiter controlled' orbit of period 1.5376 years though owing to some other effect has had its orbit changed. The seventh term* (Fig. A.3) *gives a period of just over 7% below that of Earth, for which the orbital radius is thus within 5%. The eighth term gives Venus to 6.2% on time period and 4.1% on radius. The tenth term gives Mercury to within 1%, but as the ninth term gives a period of 0.3684 years, for which no known planet exists, the agreement with Solar System planets other than Mars, though striking, must be regarded as*

$$F(n)=\left\{\frac{n+2^{n-1}}{2(3^{n-1})}\right\}, \quad n=0,\ 1,\ 2,\ 3,\ 4,\ \ldots,$$

**Fig. A3** Equation

*fortuitous, as would be expected in view of the fact that the interior planets all perturb each other. The agreement of the formula for the Hungaria group would certainly not be so good were Mars not a planet of small mass.* (Fig. A.4)

*It remains to justify the discrepancy between the ratios* 1/2 *and* 3/4, *and between* 4/9 *and* 2/3, *the first fractions in each pair being those derived from the anagram. If Gauss were only trying to show the ratio 8/9, then such a discrepancy is not unreasonable, for our terms for n=2 and n=0 will be found to be in that ratio also. Furthermore, this discrepancy (by a factor of 3/2) between our fractions and those of Gauss can be reduced by inverting the Gauss numbers and using some rearrangement of the order, thus making it different from that previously used, but no less legitimate. This procedure must be adopted if the numbers 7, 8, 18, 9 represent mean motions rather than time periods; so that, for example, instead of the number 18 representing the time period of Jupiter, it now represents the mean motion of Pallas. Taking reciprocals and multiplying by 7 throughout, we obtain the following time periods based on Jupiter as 1:*

$$1,\ 7/8,\ 7/18,\ 7/9\ \textit{or}\ 7/8,\ 1,\ 7/9,\ 7/18,$$

*after interchanging terms in pairs. The first and third terms represent Thule and Hilda respectively, subject to a multiplier which is now* $^6/_7$ *and which might even be studied for a possible physical meaning, such as the resolution of the orbital angular momentum of the asteroids unknown in 1812, in the direction perpendicular to the much tilted plane in which Pallas executes its orbit.*

The following paper was written by Eric Forbes (1933–1984) of the University of Edinburgh. His second Ph.D. was on Tobias Mayer's contributions to astronomy.

Table

| $n$ | 0 | 1 | 2 | 3 | 4 | 5 | 6 | 7 | 8 | 9 |
|---|---|---|---|---|---|---|---|---|---|---|
| $F(n)$ | 3/4 | 1 | 2/3 | 7/18 | 2/9 | 7/54 | 19/243 | 71/1458 | 68/2187 | 265/13122 |
| $P$ (years) | | 11·862 | | 4·613 | 2·636 | 1·5376 | 0·9279 | 0·577 | 0·3684 | 0·2393 |
| Group (Chief member in italics) | *Thule* | *Jupiter* and Trojans | *Hilda* and 32 others | *Pallas* Ceres Laetitea and 3 others | *Hungaria* and 22 others | *Mars* (?) | *Earth* (?) | *Venus* (?) | Unknown | *Mercury* (?) |
| Actual period | | | | | | 1·8809 | 1·0000 | 0·6152 | | 0·2409 |

Although the leading member of an asteroid group generally agrees exactly, perturbations from Saturn, Mars and Earth can cause departures for all members.

**Fig. A4** Nine terms of the function and their relationship to the asteroids and planets

## The Astronomical Work of Carl Friedrich Gauss (1777–1855)

Forbes (1978)

*It was while Gauss was engaged on this research (the theory of Pallas' perturbations) that he made a discovery for which he wanted to establish a claim to priority without disclosing it, and followed a precedent set by several other famous scientists before his time in deciding to publish it as an anagram:*

$$1111000100101001$$

*We can infer from letters written about this time to his friends Bessel and Olbers that this anagram refers to his recognition of the period-relation that 7 revolutions of Jupiter are equal to 18 revolutions of Pallas, about the Sun. This suggests that the anagram should be broken up into a sequence of four numbers expressed in binary notation, thus:*

$$111 = 7,\ 1000 = 8,\ 10010 = 18,\ 1001 = 9$$

*The question which this raises is: what is the significance of the numbers 8 and 9, in this context? We may presume that they relate in some way to Pallas and to Jupiter. The answer which I prefer, out of several that various authors have proposed, is that they refer to the positions of those two planets in order of increasing distance from the Sun, as in the case of Bode's law. Five years beforehand Olbers had discovered a fourth minor planet (Vesta), and Gauss's calculations had meanwhile established the following progression in the planetary distances up to Jupiter:*

*Mercury 1, Venus 2, Earth 3, Mars 4, Vesta 5, Juno 6, Ceres 7, Pallas 8, Jupiter 9.*

*Thus the meaning of Gauss's anagram is easily inferred. It should, at the same time, be stated that Bode's Law was not contradicted by the interposition of three minor planets in addition to Ceres, since Olbers' discovery of Vesta had verified his hypothesis that these should all be regarded as fragments of a single major planet which, at some time in the remote past, has become disrupted.*

In the following paper, Schroeder refers to the Golden Section, better known as the Golden Ratio. De Moivre (Fig. A.5), who was mentioned in this book earlier in relation to both probability theory and the Republic of Letters, was the first to discover "the closed-form expression for Fibonacci numbers linking the *n*th power of the golden ratio to the nth Fibonacci number" (source: Wikipedia).

The final paper given here was written by the German physicist Manfred Schroeder (1926–2009; Fig. A.6), who earned his PhD at Goettingen University where Gauss spent his entire career.

## Solution of the Gauss Cryptogram of 1812

Schroeder (1991a, b)

*In my paper "Moderne Anwendung Gauss'sscher Konzepte aus der Zahlentheorie" [Modern applications of Gauss' concepts of number theory; Schroeder* 1991a*] I mention among other things the Gauss Cryptogram: 1111000100101001 which was published by*

**Fig. A5** Abraham de Moivre.

**Fig. A6** Manfred Schroeder.

*this great mathematician in the* Goettingen Gelehrten Anzeigen *of April 25, 1812, regarding Pallas' orbit. Despite his announcement "to give the key to it in his times", this was never done. We know from a letter to F W Bessel of May 5, 1812, that Gauss intended to encrypt the ratio of 7/18 of Pallas' and Jupiter's orbital period. I mentioned this remarkable case of 'mystery-mongering' in my lectures and in a book (Schroeder* 1991b*). One of my graduate students at the* Drittes Physikalisches Institut, *Diplom-Physikerin Inga Holube, came up with a possible explanation and divided the cryptogram – correctly as will be shown – as follows:*

$$111\ 1000\ 10010\ 1001.$$

[Ed: as shown in this book, the division by Holube is not correct] *In binary arithmetic these are the numbers 7, 8, 18, and 9. So both numbers 7 and 18 (which the contemporary reader did not know) had already been 'unmasked'. For the remaining numbers 8 and 9 Ms Holube could not find a plausible explanation. In order to do so she should have known that Pallas and Jupiter were the $8^{th}$ and $9^{th}$ of the known planets. Thus the cryptogram reads like this: the orbital period 7 belongs to the $8^{th}$ planet (Pallas) and the orbital period 18 belongs to the $9^{th}$ planet (Jupiter).*

*Shortly after Ms Holube's attempt Mr Hans-Heinrich Voigt sent me the quotation of a note taken from the* Astronomischen Nachrichten *where Thomas L MacDonald first published the above solution (MacDonald 1931). Other approaches appear rather far fetched and implausible (Wietzke 1930; Behman 1974), also compare Kurt Biermann's (*1970*) detailed analysis.*

*It should also be mentioned that the numbers 7 and 18 are not 'any' numbers but two close Lucas Numbers (Schroeder* 1990b*), which obey the same recursion like the well known Fibonacci Numbers: you get each number by adding together the two previous numbers. Starting with 1 and 3 this results in the Lucas sequence 1, 3, 4, 7, 11, 18, ...like the Fibonacci Numbers (1, 1, 3, 5, 8, 13,...) the Lucas Numbers play a major role in problems of synchronization: ratios of frequencies tend to 'engage' at the ratio of two neighbouring numbers, eg 1/2, 1/3, 2/5, 3/8, 5/13 etc or 1/4, 3/7, 4/11, 7/18 etc. The ratios of neighbouring numbers converge towards the Golden Section that of the neighbours after the next towards its cube . The favourite position of the Golden Section in the dynamic of non-linear systems is caused by the very slowly converging continued fraction representation, since this contains 1's only:* [1] *Thus the Golden Section is sometimes also called the 'most irrational' of all numbers. In general numbers, whose continued fraction results in an infinite series of 1's are called 'noble numbers' (Schroeder* 1990b*).*

*At what ratios of their orbital periods Pallas and Jupiter would preferably 'engage', if the gravitational bond of Pallas with Jupiter were stronger (for instance due to a higher mass of Jupiter)? The real experiment, an increase in Jupiter's mass, is not feasible but it is possible today to study such questions conveniently by computer simulation. And also analytically we can venture some statements. Stronger non-linear couplings in general result in ratios with smaller denominators, especially such that are closer to the root of a family tree, called the 'Farey Tree' . Two such fractions ('parent fractions') are obtained by reducing in the continued fraction development of the preset ratio, 7/18 = [2, 1, 1, 3] the last number (3) by one or omitting it altogether.2) This results in [2,1,1,2] = 5/13 or [2, 1, 1] = 2/5, see figure. Thus the ratio of Pallas' and Jupiter's orbital periods could engage in case of a stronger coupling and dependent on secondary conditions at the value 5/13 ~ ~ 0.3846 or 2/5 = 0.4. Here we find again ratios of Fibonacci Numbers.*

*By the way, the starting fraction (7/18) is the mediant fraction of 5/13 and 2/5. (To form the mediant fraction numerator and denominator are added separately, for this is the 'right' way to form the mean value in regard to ratios of frequencies. In general the mediant fraction of two parent fractions in the Farey Tree yields the value of common descendants.)*

*In case of an even stronger coupling between Pallas and Jupiter we had to look at the 'grand parents' of the ratio 7/8 in the Farey family tree. This yields (according to the mentioned law of 'genealogical research') the ratios 3/8 and 2/5 as well as 1/3 and 1/2 (and again: ratios of Fibonacci Numbers!), see Fig.* A.7. *The fraction 2/5 is both 'mother' and 'grandmother' of 7/18. For humans these are strange family relationships. But among 'gods' everything is possible – especially if Jupiter is involved.*

**Fig. A7** Part of the 'Farey'-family tree.

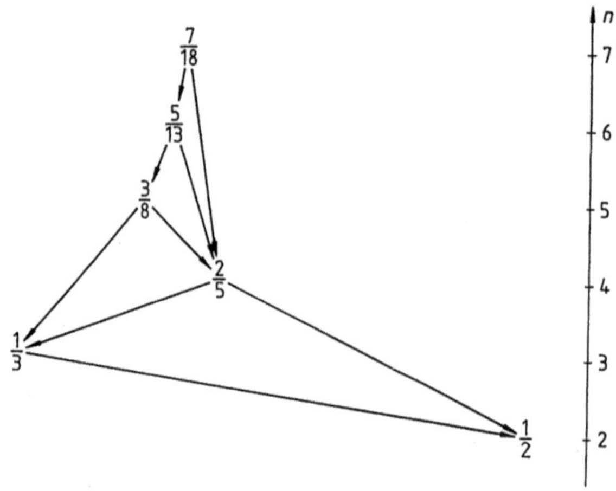


*Footnotes:*

1) *Here we chose the commonly used form to describe continued fractions: 1/(a+1/(b+1/(c+...=[a,b,c...].*
2) *Here the last term of the continued fraction must not equal 1, what can always be prevented because of the identity [a, b, ..., d, 1] = [a, b,..., d+1].*

*Part of the 'Farey'-family tree for 'engaged' orbital periods in coupled, non-linear dynamic systems (Fig. A.6). The arrows show possible transitions in case of an increase of the non-linear coupling. The stronger the coupling the smaller the 'generation number' n, which is obtained from the sum of the digits of the continued fraction. For the ratio of the orbital periods 7/18 = [2, 1, 1, 3] of Pallas and Jupiter the generation number n = 2+1+1 +3 = 7. In case of a stronger coupling the ratio of the orbital periods can engage at the shown 'parents' 5/13 = [2,1, 1, 2] with n = 6 or 2/5 = [2, 2] with n = 4. An even stronger coupling may lead to the 'grand-parents' 3/8 = [2, 1, 2] with n = 5 or 1/3 = [3] with n = 3 or 1/2 = [2] with n = 2.*

# Appendix B

## *The Historical Development of the Orbital Elements of Pallas*

The historical development of the orbital elements of Ceres was given in *Discovery of the First Asteroid, Ceres*. The corresponding history for the elements of Pallas are given here from the same source: *Publ. of the Lick Observatory*, vol. 19 (Leuschner 1935):

> *Discovered 1802 Mar 28 by Olbers at Bremen, 7.0 m, while looking for Ceres, BJ 1805 102, Mon Cor 5.*
>
> *Olbers computed circular and parabolic orbits which did not represent satisfactorily but indicated a large eccentricity and inclination. Burckhardt based an orbit on positions Apr 4-May 20, including perturbations, BJ 1805 181. Gauss derived Elements A from positions Apr 1-Jul 8 and an ephemeris for 1803, BJ 1805 106, 111, 228, Gauss Werke VI 223. Harding reobserved the planet.*

$$1803\ Feb\ 21 + 122'' - 34'' Feb\ 23 + 155'' - 57''$$

> *Gauss derived Elements B by including these positions, average O-C −6" +4", BJ 1806 179. He obtained Elements C by least squares from 3 sets of elements based on 4 oppositions each, 1803-07, 1804-08, 1805-09, Werke VI 3. The large e and I led Gauss to formulate a method of numerically integrating the perturbations in the elements, Werke VII 413. He derived Elements D from 7 oppositions 1803-11, including special perturbations by Jupiter, continued to 1812, O-C heliocentric ΔL 9" ΔB 21", Werke VII 483, VI 329, BJ 1816 234. In 1811 Gauss computed general perturbations with Laplace's elements of Jupiter, and derived mean elements, Werke VII 489. He obtained mean Elements E by repeating the computation and found: a libration in μ within ±0."2153 of 18/7 Jupiter's μ; that 1894 revolutions of Pallas = 737 of Jupiter; and the value 1/1042.86 for Jupiter's mass, Werke VII 529. Assisted by Encke and Nicolai, he tabulated the perturbations by Mars, Jupiter, Saturn, Werke VII 565.*



C.J. Cunningham, *Studies of Pallas in the Early Nineteenth Century*,
DOI 10.1007/978-3-319-32848-5

*Encke derived Elements F, including Jupiter perturbations (1067.09), and an ephemeris, BJ 1828 155. He published elements osculating for 1827 with an ephemeris, BJ 1829 159, 162; and for oppositions 1810, 1831-38, using 1/1053.924 for Jupiter's mass, BJ 1831 250, AN 27 179. The omission of the correction for the secular variation of the obliquity was noted by him, BJ 1838 286, and by Airy, NA 1837.*

| | | |
|---|---|---|
| *1821 +55" − 7"* | *1827 + 21" +44"* | *1836 +32" +33"* |
| *1823 +13 +25* | *1830 + 36 −22* | |
| *1825 +40 −28* | *1834 +294 −16* | |

*Encke suggested that his use of an incorrect value of Jupiter's mass caused the large residual in 1834 (Jupiter conjunction 1832). Accordingly Johann Gottfried Galle derived Elements G from oppositions 1816, 21, 27, 30, 34, 36, including Encke's perturbations multiplied by 1067.09/1048.69; Encke attributed the remaining residuals to perturbations by Mars and Saturn, AN 14 332, BJ 1839 238, AN 27 179. Galle published osculating elements for oppositions 1839-50, using 1/1053.924 for Jupiter's mass, AN 27 179; and continued special perturbations in the elements and ephemerides, BJ 1850-70. O-C heliocentric ΔL ΔB:*

| | | |
|---|---|---|
| *1816 −18" −1"* | *1827 + 4" + 6"* | *1834 + 5" + 1"* |
| *1821 +34 +4* | *1830 −14 −10* | *1836 −13 + 4* |

*About 1806 Galle derived Elements H from 12 oppositions 1816-55, including special perturbations by Jupiter and Guenther's perturbations by Saturn, O-C 1830-59: ΔL −23" to +20", ΔB −6" to +19", AN 55 194.*

*Farley's Elements I include perturbations by Venus, Earth, Mars, Jupiter, Saturn, NA 1860-1914, BJ 1871-1919. O-C from NA ephemerides:*

| | | |
|---|---|---|
| *1883* $-1.^{s}4$ *+3"* | *1906* $-2.^{s}5$ *+ 8"* | *1914* $-2.^{s}1$ *+4"* |
| *1892 −1.2 +1* | *1908 −5.4 −14* | |
| *1895 −1.0 +1* | *1910 −2.0 + 2* | |

*LeVerrier noted that the perturbations ordinarily arising from the large eccentricity were offset by the large inclination which prevented a close approach, the aphelion being 54° from the node with Jupiter's orbit. He used a more convergent form of the series for the reciprocal of the distance, and obtained 895" for the coefficient of the term (7-18), Paris Mem 1 384. To illustrate Augustin-Louis Cauchy's method of finding long period terms of high degree, Victor Puiseux found the value 906", Paris Mem 7 228, Felix Tisserand, Mecanique Celeste IV 294. Houel also computed the coefficient, Paris Mem 8 131.*

*Tisserand found the development of the reciprocal of the distance to be divergent in some parts of the orbit on account of the large e and I, and used his own analytical development, Paris Mem 15:C. Brueck, using Farley's elements for 1878, investigated the secular perturbations by Jupiter with Gauss's method as extended by Hill and Callandreau, BA 12 289.*

*G. Struve based an investigation on 63 normals. "Die Darstellung der Pallasbahn durch die Gauss'sche Theorie fur den Zeitraum 1803 bis 1910", Ebernig, 1910, BA 28 184, Ajb 1910 188. He found* $\mu = 769.''1385$ *and* $18\mu' - 7\mu = 123''$ *per year. The residuals, amounting to 4', attributed to second order perturbations, were somewhat reduced*

| El | Epoch | | M | Eq | ω | Ω | i | ø | μ |
|---|---|---|---|---|---|---|---|---|---|
| A | 1802 Mar 31.0 | S | 41°16'24".8 | .... | 309°12'11" | 172°26'31" | 34°36'59" | 14° 6'58" | 769".5871 |
| B | 1803.0 | S | 100 4 41.0 | 1803 | 308 56 5 | 172 28 8 | 34 38 20 | 14 13 6 | 769.2768 |
| C | 1803.0 | Gö | 100 26 45.1 | 1803 | 308 39 56.1 | 172 28 12.4 | 34 37 28.4 | 14 9 59.8 | 770.3639 |
| D | 1803 Jun 5.0 | Gö | 133 48 44.0 | .... | 308 39 33.3 | 172 28 20.9 | 34 37 44.3 | 14 12 46.3 | 770.7360 |
| E* | 1810.0 | Gö | 286 14 59.8 | 1810 | 308 36 21.0 | 172 31 44.4 | 34 36 9.5 | 14 2 50.0 | 769.15074 |
| F | 1826 Jun 24.5 | P | 133 23 26.0 | Ep | 308 20 58.8 | 172 37 9.7 | 34 35 55.5 | 14 6 13.4 | 769.49107 |
| G | 1810 Jan 0.0 | Gö | 287 51 13.6 | Ep | 308 40 30.3 | 172 33 37.4 | 34 37 42.7 | 14 12 51.0 | 770.73885 |
| H | 1830 Jan 0.0 | B | 48 5 46.8 | .... | 308 21 12.6 | 172 39 5.4 | 34 35 46.5 | 14 3 22.5 | 770.75158 |
| Ia | 1869 Sep 25.0 | B | 267 46 38.3 | Ep | 308 58 38.4 | 172 46 8.3 | 34 42 18.6 | 13 51 4.2 | 769.56817 |
| Ib | 1913 May 5.0 | B | 71 39 31.7 | Ep | 309 0 47.9 | 172 56 47.8 | 34 42 2.5 | 13 46 37.9 | 769.2236 |
| J | 1918 Jun 19.0 | G | 110 11 16.7 | 1918 | 309 38 51.3 | 172 53 45.2 | 34 44 11.3 | 13 33 14.5 | 768.19159 |
| K | 1922 Jun 5.0 | U | 58 42 13.5 | 1925 | 309 48 31.6 | 172 58 10.9 | 34 44 32.6 | 13 34 46.2 | 768.194 |
| L | 1925 Oct 27.0 | U | 323 49 13.7 | 1925e | 323 24 37.5 | 159 56 5.6 | 11 46 33.9 | 13 36 39.6 | 768.53559 |

**Fig. B1** The elements of Pallas from 1802 to 1925.

*empirically. Vilyev computed second order perturbations with the Hansen-Hill method but was unable to reduce Struve's residuals, AN 205 225.*

*Woltjer investigated the libration in μ found by Gauss and concluded that it did not exist, BAN 1 161.*

*Kisseleva based an orbit by Gauss's method on positions in 1918, and then derived Elements J from oppositions 1914, 15, 18, 19, including special perturbations by Venus, Earth, Mars, Jupiter, Saturn, Russ AJ 3 191.*

*Cripps and Merton based an orbit on positions in 1920, 22, revised it with later positions in 1922, 23, and published an ephemeris for 1923, including perturbations in the elements by Jupiter, JBAA 33 326. Cripps again revised the elements with the 'latest observations', but empirically corrected the ephemeris for 1924 from the previous elements to satisfy the 'same recent observations', BAAH 1924.*

*Boyeva computed Elements K from 4 oppositions 1919-23, including the action by Jupiter and Saturn with Numerov's method, VRI 45 102, KP 1927. She published accurate ephemerides, BAAH 1925-26. 1927 $-0.^m4$ +3′ 84°*

*R. T. A. Innes (1925) computed secular perturbations in the elements by Jupiter and confirmed the unusual negative motion of ω found by Gauss, BAN 3 75.*

*Nokonov derived Elements L from 10 condition equations in 5 oppositions 1920-25 (Jupiter conjunction 1922), including the action by Jupiter and Saturn with Numerov's method, O-C 1925 $+1.^s1$ −2″ 334°, KP 1928–, BIA Len 1 191, 2 20. He published accurate ephemerides, BAAH 1927–. The KP ephemeris (and its extension, BON 46) for 1928 was evidently wrong: O-C May 27 $-1.^m3$ +4′, June 22 $-1.^m5$ +4′, Jul 11 $-0.^m3$ −2′, Jul 24 $0.^m0$ 0′, Aug 14 $+0.^m4$ +1′. 1927 $-0.^s5$ +8″ 98° 1928 $0.^m0$ 0′ 176°*

See Fig B.1.

# References


Alder, K.: 2014, *The Measure of All Things*. Simon & Schuster, New York.

Benham, W.: 1974, The Gauss anagram: An alternative solution. *Annals of Science* 31 (5), 449-455.

Bierman, K.: 1970, Zum Gauss'schen Kryptogram von 1812. *Monatberichteder Deutschen Akademie der Wissenschaften* 13 (2), 152-157.

Brendel, M. and Schlesinger, L.: 1870, *Gauss Werke,* 6 volumes. Akademie der Wissenschaften in Goettingen. Goettingen. (see also Klein et al for vol. 7)

Debarbat, S.: 1998, Parisian astronomy at the time of the Gotha meeting. In *The Message of the Angles* (ed. P. Brosche, W. Dick, O. Schwarz and R. Wielen), 31-42. Acta Historica Astronomiae Vol. 3. Verlag Harri Deutsch, Thun.

Forbes, E.: 1978, The astronomical work of Carl Friedrich Gauss (1777–1855). *Historia Mathematica* 5, 167-181.

Franklin, B.: 1787, Letter of May 18 to Mr. Jordain, London. In *The Private Correspondence of Benjamin Franklin,* Vol. 2, 3rd edition (1818), pg. 225-227. Henry Colburn, London.

Garrett, D.: 2007, *Cognition and Commitment in Hume's Philosophy*. Oxford University Press, Oxford.

Goellnitz, E.: 1937, Ueber das Anagramm von Gauss, das die mittleren Beweungen der Planeten Pallas und Jupiter betrifft. *Jahresbericht der Deutschen Mathematiker-Vereinigung*, part 2, pg. 59.

Goldstein, B. and Swerdlow, N.: 1970, Planetary Distances and Sizes in an Anonymous Arabic Treatise Preserved in Bodleian Ms. Marsh 621. *Centaurus* 15(2), 135-170. [This manuscript was later identified as having been written by Urdi. See Saliba (1979)]

Hall, A.R.: 1962a, The History of Science, in *Approaches to History*, pg. 175-196 (H.P. Finberg, ed.). University of Toronto Press, Toronto.

Holton, G.J.: 1988, *Thematic Origins of Scientific Thought: Kepler to Einstein.* Harvard University Press, Cambridge, Massachusetts.

Innes, R.T.A.: 1925, The Secular Perturbations of Pallas by Jupiter. *Bulletin of the Astronomical Institutes of the Netherlands* 3 (90), 75-82.

Leuschner, A.: 1935, Research Surveys of the Orbits and Perturbations of Minor Planets 1 to 1091. *Publ. of the Lick Observatory* 19.

Leverrier, 1841, Sur l'influence des inclinaisons des orbites dans les perturbations des planètes. Determination d'une grande inégalité du moyen mouvement de Pallas. *Compte Rendu des seances de l'Academie des Sciences* 13, 344-348.

MacDonald, T.: 1931, The Anagram of Gauss. *AN* 241, 31-32.

Maennchen, von Philipp: 1934, Zur Loesung eines raetselhaften Gausschen Anagramms. *Unterrichtsblaetter fuer Math U. Naturwiss* 40, 104-106.

Mayer, T.: 1755, *Opera inedita*. Johann Christian Dietrich, Goettingen.

Messier, C.: *Notice de mes comètes* (*Notes on my comets*). Undated manuscript no. C2-19. Paris Observatory.


 

C.J. Cunningham, *Studies of Pallas in the Early Nineteenth Century*,
DOI 10.1007/978-3-319-32848-5

Moivre, A.: 1721, *The Doctrine of Chances, or a Method of calculating probability of Events in Play.* Acta Eruditorium. Leipzig. First English edition in 1736. Third edition, 1756. A. Millar, London.

Saliba, G.: 1979, The First non-Ptolemaic Astronomy at the Maraghah School. *ISIS* 70, 571-576.

Schaefer, C.: 1931, Gauss's Investigations on Electrodynamics. *Nature* 128, 339-341.

Schroeder, M.: 1990a, Moderne Anwendungen Gauss'scher Konzepte aus der Jahlentheorie. *Gauss-Gesellschaft, Mitteilung* Nr. 27, 69-73.

Schroeder, M.: 1990b, *Number Theory in Science and Communication, with Applications in Cryptography, Physics, Digital Information, Computing and Self-Similarity.* Second edition. Springer Berlin, 5th printing.

Schroeder, M.: 1991a, Lösung des Gauss'schen Kryptogramms von 1812. *Gauss-Gesellschaft e. V. Goettingen-Mitteilungen* 28, 87-90.

Schroeder, M.: 1991b, *Fractals, Chaos, Power Laws: Minutes from an Infinite Paradise.* W. H. Freeman, N.Y.

Struve, G.: 1911, *Die Darstellung der Pallasbahn durch die Gauss'sche Theorie fuer den Zeitraum 1803 bis 1910.* Inaugural Dissertation. Ebering, Berlin.

Szpiro, G.: 2003, *Kepler's Conjecture.* John Wiley & Sons.

*The Critical Review*, 38, 15-19 (1803a).

The Monthly Review, *Rivarol on the Universality of the French Language.* Vol. 71, 577-583 (1785).

*The Philosophical Magazine*, French National Institute: Notice of the Labours of the Mathematical and Physical Class of the Sciences, 16, 180-181 (1803).

Ward, A.W., and Waller, A.R. (eds.), 1907–1921, *Cambridge History of English and American Literature. Volume 12.* Cambridge University Press, Cambridge.

Wietzke, A.: 1930, Zur Lösung eines rätselhaften Gauss'sschen Anagramms. *Astron. Nachr.* 240, 403-406.

Wood, P.: 2000, Introduction: Dugald Stewart and the Invention of "The Scottish Enlightenment" in *The Scottish Enlightenment: Essays in Reinterpretation*, pg. 1-35, (P.Wood,ed.). University of Rochester Press, Rochester.

# Index





C.J. Cunningham, *Studies of Pallas in the Early Nineteenth Century*,
DOI 10.1007/978-3-319-32848-5

MIX
Papier aus verantwortungsvollen Quellen
Paper from responsible sources
FSC® C105338

If you have any concerns about our products,
you can contact us on
**ProductSafety@springernature.com**

In case Publisher is established outside the EU,
the EU authorized representative is:
**Springer Nature Customer Service Center GmbH**
**Europaplatz 3, 69115 Heidelberg, Germany**

Printed by Libri Plureos GmbH
in Hamburg, Germany